AF541702

Encyclopaedia

of

GEOLOGICAL SCIENCE AND TECHNOLOGY

Encyclopaedia
of

Geological Science and Technology

Volume 1

Dr. Samayal Varadarajan

ANMOL PUBLICATIONS PVT. LTD.
NEW DELHI-110 002 (INDIA)

ANMOL PUBLICATIONS PVT. LTD.
Regd. Office: 4360/4, Ansari Road, Daryaganj,
New Delhi-110 002 (India)
Ph.: 23278000, 23261597
Branch Office: No. 1015, Ist Main Road, BSK IIIrd Stage
IIIrd Phase, IIIrd Block,
Bangalore-560 085 (India)
Tel.: 080-41723429
Visit us at: www.anmolpublications.com

Encyclopaedia of Geological Science and Technology

First Edition, 2009

ISBN 978-81-261-4125-8 (Set)

PRINTED IN INDIA

Printed at Mehra Offset Press, Delhi.

Contents

Volume 3

Preface

Geology is the science and study of the solid and liquid matter that constitute the Earth. The field of geology encompasses the study of the composition, structure, physical properties, dynamics, and history of Earth materials, and the processes by which they are formed, moved, and changed.

The field is important in academics, industry (due to mineral and hydrocarbon extraction), and for social issues such as geotechnical engineering, the mitigation of natural hazards, and knowledge about past climate and climate change.

The knowledge that geologists have known from the study of the earth a subject so large must be treated very briefly if it is to be presented between the covers of a single book; we have chosen to concentrate on the analysis of processes that are at work upon and within the earth, rather than to present a catalog of descriptive facts and terms.

We have felt, that the student is entitled to know something of the kind of evidence on which geologic conclusions are based, even though its presentation takes valuable pages that might be used to put forth more facts.

Author

Chapter 1

Introduction to Geology

Geology is the science and study of the solid matter that constitutes the Earth. Encompassing such things as rocks, soil and gemstones, geology studies the composition, structure, physical properties, history and the processes that shape Earth's components.

It is one of the Earth sciences. Geologists have established the age of the Earth at about 4.6 billion (4.6×10^9) years and have determined that the Earth's lithosphere, which includes the crust, is fragmented into tectonic plates that move over a rheic upper mantle (asthenosphere) via processes that are collectively referred to as plate tectonics.

Geologists help locate and manage the Earth's natural resources, such as petroleum and coal, as well as metals such as iron, copper and uranium.Additional economic interests include gemstones and many minerals such as asbestos, perlite, mica, phosphates, zeolites, clay, pumice, quartz and silica, as well as elements such as sulfur, chlorine and helium. Geology is also of great importance in the applied fields of civil engineering, soil mechanics, hydrology, environmental engineering and geohazards.

Planetary geology (sometimes known as Astrogeology) refers to the application of geologic principles to other bodies of the solar system. Specialised terms such as *selenology* (studies of the moon), *areology* (of Mars), etc., are also in use. Colloquially, *geology* is most often used with another noun when indicating extra-Earth bodies (e.g. "the geology of Mars"). Geology is the science of the earth in all its aspects, except those which deal with the relationship of the earth to other planets and to the sun of our solar system. That aspect of the earth properly belongs to the science of Astronomy, for our earth is one of the heavenly bodies with which that science is concerned. It is true that this phase of the subject is often treated by the geologist under the name of *astronomical geology*, but in this book we shall consider it as belonging primarily to the field of astronomy.

At the same time we shall recognize the importance to the student of at least a general understanding of these astronomical relations of the earth and consider such an understanding a desirable preliminary preparation.

The term Geology is derived from the Greek words *ge*, science, or logical discourse. The science of geology was preceded by and in a measure grew out of the subject of geography, which literally is the description of the earth and was, of course, chiefly confined to a description of the earth's surface features and their significance in terms of human existence.

But since surface features are generally the expression of underlying structure and of the geological history of the region, it is evident that geography cannot be divorced from geology and that, to no inconsiderable extent, the student of the geography of a region must take account of its geological structure and history.

In its broadest sense, then, geology is the science of the earth and all that pertains to it. It is the study in—detail of one of the planets in one solar system, by the inhabitants of that planet, who are themselves a part of it. If other planets of our solar system or those of other solar systems are inhabited, the study of those planets by their inhabitants would correspond to our geology—it would be the science of a particular planet.

By us little can be ascertained regarding the structure and development of other planets, except in an indirect way and all investigations along such lines fall properly into the domain of the astronomer, though he concerns himself primarily with interrelations of the heavenly bodies. And thus we may consider that the study of the physical universe, *i.e.*, the *cosmos*, which may be called the science of cosmology, has only the two primary divisions, astronomy and geology. This may be summarized as follows:

- *Astronomy*—The science of all the heavenly bodies, including the earth, their character, distribution, interrelations, movements, etc. and the laws which govern them.
- *Geology*—The science which deals with the material, structure, history, etc., of one of those bodies, i.e., the earth.

Such a view of the science of geology gives it an extremely comprehensive scope and we must recognize that in ordinary parlance the term is used in a much more restricted sense Nevertheless it is desirable to take this comprehensive view at the outset and to note the several subdivisions into which such a broad science naturally falls and with all of which the student of any one division should have at least a general acquaintance. We shall best get the proper view-point by first considering the earth in its entirety. Could we view the earth in its entirety from some extraterrestrial vantage point, we would recognize it as an oblate spheroid, that is, a body approaching that of a sphere, but with its polar diameter flattened and its equatorial belt somewhat inflated. By measurement the polar diameter of the earth is found to be 7899.7 miles, while the equatorial diameter is 7926.5 miles.

If the observer in extraterrestrial space is sufficiently removed from the earth's surface, that surface would appear essentially smooth, as does the surface of the moon to our unaided eyes; the irregularities which the dweller on the earth recognizes as mountains and valleys would become of insignificant proportions.

Were we to represent the earth by an accurately scaled model 10 feet in polar diameter, the equatorial diameter would exceed the polar by only a trifle over 4/10 of an inch, while the highest mountains on the earth's surface would form elevations on the model less than 3/32 of an inch in height. Thus viewed, the prominences which appear to us formidable are after all of minor significance and bearing this in mind, we can understand that relatively slight bulgings or crumplings of the earth's surface may produce what to us appear as great elevations.

The wrinklings on the skin of a drying apple constitute far more prominent elevations in proportion to the size of the apple than do the highest mountain chains on the earth's surface, when compared with the size of the earth, while the scratches and notchings formed upon the surface of a glass marble, after a brief play, form vastly greater depressions, in proportion, than do the largest river cañons, such as that of the Colourado, or the deepest ocean depressions.

Thus when the geologist finds evidence that the summit of the high peaks of the Himalaya Mountains or the top of the Apennine chain were once a part of the sea-bottom, he no longer concludes that the sea once covered these mountains, as was formerly supposed, but he infers rightly that these mountain chains were raised by a wrinkling of the earth's crust or by an upward warping which occurred at a time subsequent to that in which this region was beneath the sea. The evidence from the material of the mountain which leads the geologist to such a conclusion and the corroborative evidence from the structure of the mountains will also be given.

The chief lesson which it is intended to impress upon the student at present is that surface features are relatively unimportant when we consider the earth as a whole and that elevations and depressions—except those which we recognize as continental masses and oceanic depressions—are of local significance chiefly and may change from one to the other not once but many times.

INORGANIC SUCCESSIVE

The observer from his extraterrestrial view-point will, however, note clearly that a large part of the earth's surface—to be precise, about 70.8% of it—is covered with water.

This is the sea which is divided into a number of oceans and surrounds all the lands and, moreover, forms a continuous body, the surface of which comprises approximately 137,070,000 square miles or 361.1 million square kilometers.

The land has a surface area of about 59,870,000 square miles or 148.8 million square kilometers, giving a total surface area for the earth of 196,940,000 square miles or 509.9 million square kilometers. Certain portions of the land surfaces are also covered by water bodies, the lakes, which are not in direct connection with the sea. Among these the Great Fresh-water Lakes of North America and the Caspian, a great salt-water lake, are the most prominent examples.

These are to be classed as a part of the land, as continental water bodies, distinct from the sea. Moreover, the upper layers of the land in nearly all parts are water-bearing, this water filling the empty spaces of the soil and occupying the pore-spaces within the solid rock. This is the *ground water*, which is tapped by wells, mines, or borings, or issues on the surface in springs which feed the brooks or rivers, or form swamps, ponds and lakes. It is thus possible to speak of a nearly or quite continuous mantle of water which completely covers some parts of the solid surface of the earth and more or less completely saturates the exposed parts.

This mantle of water may be viewed as an aqueous sphere enveloping the rock sphere of the earth and it is spoken of as the hydrosphere. It in turn is enveloped by the sphere of gas and vapour, the well-known *atmosphere*. These two spheres or shells constitute the outer layers of the earth thus viewed in its entirety and surround the more rigid mass of the earth, with which we are most familiar.

This consists of solid rock and of unconsolidated soil and other loose material, all of which, except the surface film of organic matter, we may recognize, on examination with a magnifier or by other means, as brokendown rock material in a fine state of division. From observation of soundings, dredgings, etc. and by inference, we know that similar material forms the ocean floor and thus we recognize that beneath the hydrosphere is a sphere or a shell of rock materialThis is called the rock sphere, or *lithosphere* and in it are included all of the soil and other unconsolidated rock material of the earth's surface. How thick this shell of rock is, we have no means of knowing from observation.

The deepest borings into the earth's crust have penetrated only a little over a mile in depth, while the deeper mines go less than two miles beneath the surface. But from logical deductions of many observed physical facts, we conclude that this shell has a thickness of at least 75 miles and perhaps much more.

It is indeed generally held that the earth is solid rock to the core, but there are those who have held and some who still hold, that the central portion of the earth consists of fluid or perhaps even of. gaseous matter.

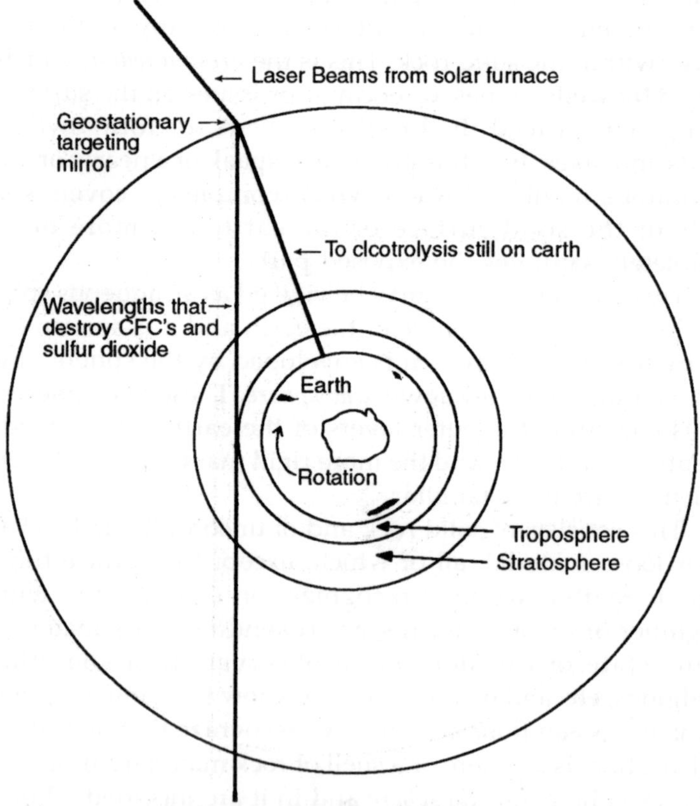

Fig. *Inorganic Spheres of the Earth.*

There are thus three inorganic spheres, the *atmosphere,* the hydrosphere and the lithosphere, open to partial observation.

THE LITHOSPHERE

We now see that the biosphere contributes no inconsiderable portion to the growing lithosphere and this is true to a certain extent also of several of the other spheres. For not only was the carbon of ancient plants, which is now locked up in the earth's crust as coal and oil, contributed by the atmosphere of the past. But the water vapour of the air, freezing and descending as snow or hail, or crystallizing directly upon cold surfaces, forms a not unimportant, though in most regions very transitory, addition to the solid rock of the earth. So, too, the conversion of water to ice forms rock at low temperature, for ice is a rock. More permanent contributions are made by the crystallizing of various salts, chief among them the common salt (sodium chloride) may become buried by other rock material and be preserved for millions of years as an essential part of the earth's crust.

Thus the salt now mined in New York State and in southeastern Michigan was derived from the waters of an ancient sea, the derivation being by a roundabout process. That this occurred in a period of time many millions of years ago is amply shown by the ascertainable facts.

So, too, the great salt deposits of North Germany, which have furnished the world. In the past with most of its potash, are the product of the evaporation of cut-off portions of an ancient sea which covered much of Europe, though the period of its production is not quite so remote as that of the best-known American salts.

That the pyrosphere also contributes its quota to the rock mass of the earth is abundantly shown by the vast extent of ancient lavas and other igneous masses which now form a large proportion of the solid crust of the earth and are visible to us as the result of surface cooling in comparatively modern times, as in the case of the great Columbia lava fields of Northwestern. America and the preserved fragments of similar flows on the British coast, or which were uncovered by the removal, during long geological epochs, of the overlying portions of the earth's crust beneath which the ancient igneous masses assumed their solid form. Throughout the entire history of the earth, the lithosphere has received additions by the cooling of molten rock material injected into it or poured out on its surface and this addition is in progress even to-day.

Inner Spheres of earth

From the observation of many phenomena, geologists and physicists have reasoned that beneath the lithosphere, other spheres characterized by certain peculiarities may be recognized, though their boundaries are variable and probably not very definite. One of these is the sphere or zone in which the temperature of the earth is sufficiently high to permit the fusion of rock if the pressure, which ordinarily keeps it solid by raising the fusing point, were removed or lowered by some structural change in the earth's crust.

There may be regions in which essentially permanent pools of molten rock exist within the crust, forming feeders of volcanoes, but the majority of such feeding areas are more probably temporary. It is convenient to speak of this more or less concentric zone as a distinct sphere and the name *pyrosphere* has been applied to it.

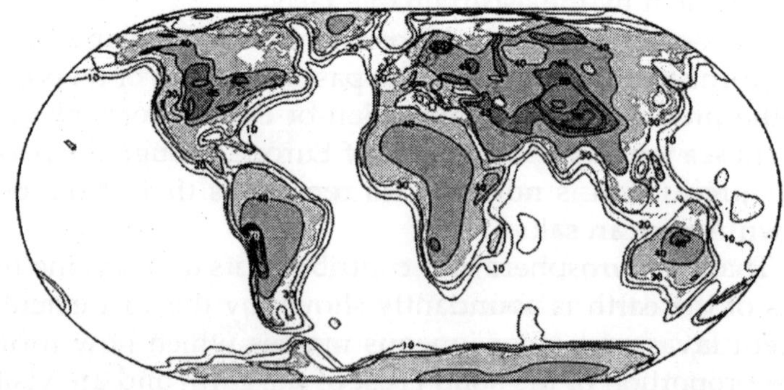

Fig. *The Relative Thickness of the Crust of the Earth*

Finally there remains the central portion of the earth, the *centrosphere*, which is by far the largest part of it and about which we know nothing from observation and but little from inference. From the known rate of temperature increase in borings and mines, it is inferred that the temperature near the centre of the earth may be between 200,000 and 350,000 degrees

Fahrenheit, a temperature so high that were it not for the enormous pressure, all the rocks there would probably be vaporized. Another equally indefinite sphere or zone, variously regarded as lying from 30 to 75 miles beneath the surface (according to our estimate of the thickness of the lithosphere), is one of relative weakness, between a strong external crust and a rigid central mass.

Here the rock yields most readily under longenduring strains of limited magnitude and earth-movements, resulting in the deformation of the rocks, occur. This zone or sphere, which has been called the *tectosphere* or the *asthenosphere,* may in part include the pyrosphere; nevertheless, it is convenient to speak of it as distinct.

From the fact that the average specific gravity of the rocks which constitute the earth's crust is only about 2.6 while that of the earth as a whole is 5.6, it is further inferred that the material of the centrosphere is very heavy and it has been calculated that if there were a steady increase in the specific gravity of the material from the surface toward the centre, this would, at the latter point, become 11.2, which suggests the possibility of a central core of iron, if not of gold and platinum or some other heavy metal. Because of this greater weight, the centrosphere is also spoken of as the *barysphere* or heavy sphere.

ORGANIC SPHERE OR BIOSPHERE OF EARTH

To these inorganic or lifeless spheres there is added the organic sphere, or sphere of life, the *biosphere.* It would surprise the average student could he see the universality of this organic shell of the earth.

In and out among the mass of the water and of the air and upon and through the upper layers of the rocky surface, the thread of living matter is woven, now constituting an almost solid mass of tissue, as in the bodies of peat or in the dense vegetation of a forest, or the almost continuous tangle of seaweed or layer of floating organisms in the sea; again forming a network, the meshes of which vary greatly in size, but are on the whole continuous, except where momentarily broken by the hand of man, or by some abrupt, not to say cataclysmic disturbance, such as an avalanche or outpouring of a mass of lava.

But left for even a short time, the steepest quarry wall, or the most precipitous cliff formed by the sudden dislodgment of a mountain side or by the sinking of a portion of the earth's crust, such as may take place during an earthquake-producing disturbance, will be again taken possession of by some form of plant, if not of animal life. Even the stony lava field will be covered in time by a succession of organic forms of increasing complexity. The apparently barren desert, too, has its wide-meshed net of living beings, except perhaps where the sands are constantly in motion; and the presence of these organisms is often shown by the countless tracks and trails which appear upon the surface of the sand on a dewy morning, or the sudden springing up of vegetation from hidden seeds when moisture furnishes the condition for expansive growth.

And in the sea, life of some form is never wanting long,—here covering the bottom with a continuous living carpet, there forming an ever changing web of floating animal and plant life, through which the swimming world of animals weaves an intricate pattern with the threads of its never ceasing wanderings. Thus it is perfectly in accord with the facts, when we speak of a shell or sphere of living matter, the *biosphere,* which surrounds the lithosphere and penetrates its upper layers as well as the hydrosphere and the atmosphere, but is distinct from all.

That such a biosphere has existed throughout most of the past ages of the earth's history is clearly shown by the countless remains of the hard parts of animals, such as the shells of mollusks, the bones of vertebrates and the like, which fill many of the rocky layers of the lithosphere and indeed sometimes actually compose these layers.

For example, the white chalk of the English and French coastal regions is literally made up of the shells and fragments of shells, of organisms (Foraminifera, coccoliths, etc.) and many of our great limestones are consolidated remains of shells of animals which formerly inhabited the seas, or fresh-water lakes. An enlargement of a group of minute needle-like shells, each one of which was built by a separate marine animal and of which a limestone, traceable over wide areas in the state of New York, is almost entirely composed.

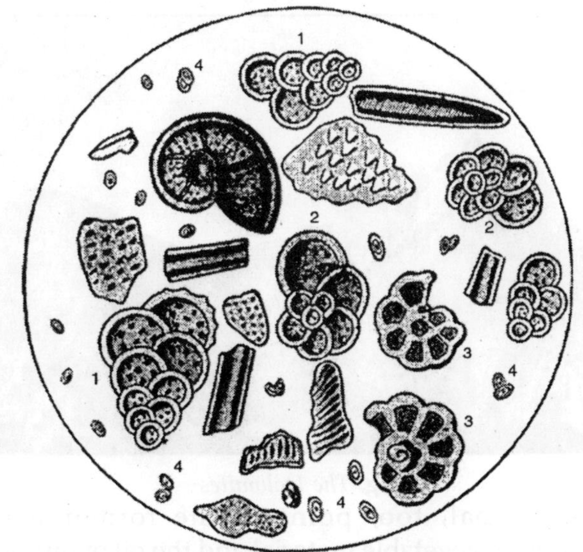

Fig. *Thin Sections of Chalk*

From careful study, it has been determined that a cubic inch of this rock, when pure, contains about 40,000 individual shells.Whole mountains are sometimes made up of rock, which is largely, if not entirely, the product of accumulation of the hard structures of former organisms.

An example of this is found in the famous Dolomites of the Tyrol, the chief rock masses of which are made up of calcareous matter separated from a sea by marine plants and the many centuries of time during which the present forms of these peaks were carved from this rock have witnessed only a partial obliteration of the organic structures, though no inconsiderable portion of the mass has been removed. The remains of animals and plants which compose, or are included in, the rocks of the earth's crust, are known to the geologist as *fossils* (from the Latin *fossilis,* something that is dug up) and their study forms an important and integral part of the science of geology. It is indeed the study of the constantly changing elements of the biosphere, of which the existing animals and plants form only the most modern phase, that has made possible the deciphering of the history of the earth.

Fig. *The Dolomites*

Beds of coal, too, point to the former extensive accumulation of vegetable material and the oil of our oil shales and pools owes its origin largely and perhaps altogether, to the distillation, during many centuries of time, of buried organic matter which constituted a portion of the biosphere of former geological periods.

SUBDIVISIONS OF GEOLOGY

It is but natural that the mind of man, always eager to enter into details, should have developed distinct branches of inquiry into the character and history of the several parts of the earth which is his home. In the beginning of such inquiry, a single master mind may have encompassed the entire field; but with the increasing mass of detail, the individual range became limited to special portions of the field and in the course of time the several lines of inquiry into the nature and history of our earth developed into distinct sciences, each with its host of devoted searchers after truth. We may, then, at this point of our study, endeavor to ascertain the natural subdivisions of the Science of the Earth, or Geology in its broadest sense and after that confine ourselves to those branches which by common consent are reserved as the special field of the geologist, as it is understood at the present time.

Lithology or Geology Proper

As has just been said, the special field of inquiry of the modern geologist is the crust of the earth, or the *lithosphere*. He is thus primarily a student of *Lithology* -though that term has also been used in the past in a narrower sense, the study of rocks *per se*.

It will be well, however, for the sake of unity of terminology—a great desideratum in every science—to adopt this term in its wider sense, namely, that which encompasses all the fields of inquiry which are concerned with the composition, structure and history of the earth's rocky shell, the *lithosphere*, generally assigned as the only legitimate field of the modern geologist. Still, to-day more than ever, the geologist cannot limit his inquiries to the facts disclosed by the rocks alone, but must draw his conclusions with the aid of a wide knowledge of the researches of the atmologist, the hydrologist and even the biologist.

Biology

Biology is the science of living things. It is the study of the *biosphere* and has perhaps been more sedulously developed and has a larger body of votaries, than any other branch of the earth science. Biology, as currently understood, limits itself to the study of modern living beings, the plants and animals of to-day and naturally falls into the two divisions, or separate sciences, of *Botany* (Phytology), the study of plant life and *Zoology*, the study of animal life.

Biologists are, however, aware of the fact that the modern world of plants and animals is only a fragment of the great host of living beings which has inhabited the earth from the remotest ages. Throughout all the past ages, living forms existed upon the earth and their remains are embedded in the rocks of the earth's crust as *fossils*. The study of these was at first left to the geologist, but has now developed into a separate science, that of *Palæontology*.

Atmology or Meteorology

From an inspection of the table of spheres into which the earth may be divided, we form the natural conclusion that the first line of cleavage would be in accord with these natural subdivisions. The study of the atmosphere has now developed into a separate science, which is familiarly known as *Meteorology* because the meteors, or extraterrestrial bodies, the familiar "shooting stars," which on entering our atmosphere become luminous through friction and frequently reach the earth as meteorites, have from the remotest days formed one of the chief attractions for those whose gaze was turned away from the solid earth.

Thus atmospheric phenomena came to be designated as meteoric phenomena and the science of these phenomena became meteorology. But since the sphere with which this science is concerned is the *atmosphere,* a more appropriate name for this science itself is *Atmology*.

Hydrology

That the hydrosphere, that is, the oceans and other water bodies, early attracted the serious-minded student was but natural, since these water bodies form the natural highways of commerce and since they furnish so large a part of the earth's population with the means of a livelihood. Thus the science of the water, or *Hydrology,* was developed, which is variously subdivided again into the sciences of the oceans, or *Oceanography* and into the sciences of the lakes, the ponds and the rivers. That some of these, for example oceanography, have been developed to a remarkable extent, is shown by the great oceanographic institutes of the world, of which that of Berlin under Director Penck and that at Paris, under the direction of the Prince of Monaco, are the best known.

In America, oceanographic studies have chiefly been carried on by the federal government and by private individuals, among these the late Alexander Agassiz; and there is a hydrographic department of our government, the chief business of which is the charting of our sea-coasts, lakes and rivers. In England, too, oceanographic studies have been largely fostered by the government, under whose auspices the famous Challenger Expedition, for the exploration of the oceans, under the leadership of Sir John Murray, was carried to successful completion.

The governments of other countries, too, have fostered exploring expeditions for oceanographic research, these being generally known by the name of the vessel employed in the expedition. The student of geology, even in its narrower sense, must keep abreast of these inquiries, at least to a moderate extent, for without them many of his discoveries lack the clarifying light by the aid of which he must interpret them; and in proportion as the researches of the oceanographers and the other hydrologists become available, the analysis of the structure and composition of the earth's crust, the special field of the geologists in the modern sense and the interpretation of these facts in terms of earth history, become more precise.

Divisions of Geology

When we come to the consideration of the inner spheres of the earth, we must admit that, owing to our inability to make direct observations, no separate sciences worthy of that designation have yet been developed. It is true that a science of earth disturbances or *Seismology* exists and that this properly belongs to the study of the asthenosphere or tectosphere.

But the observations on which this science is based are chiefly possible upon surface manifestations and the effects which these produce, though inferences can also be drawn from the study of disturbances which have occurred in the past, the effects of which are recorded in the rocks. The same is true of the study of the pyrosphere—which can only be approached through observations of modern volcanoes (*Vulcanology*) and of the temperatures in deep borings and mines. Of the interior of the earth we may perhaps never know much through observation and so deductions from physical phenomena alone must guide us. No separate science of the centro- or barysphere has thus been developed. From the foregoing, the student is led to the realization of the truth that all true sciences are based on two great fundamentals, *observation* and *deduction*.

To observe and record facts alone does not constitute the whole of the work of the man of science. Such used to be the sole aim of the older naturalists, so called, who dissented from the attitude of the ancient philosophers, because these based their speculations on mental processes rather than observation of facts. To-day, however, we know that the interpretation of facts, in the light of our growing knowledge of causes, is as legitimate, indeed as important, a function of the student of the earth as the observation of facts and moreover a certain amount of speculation—always held in check by the appeal to facts—is not only desirable but necessary.

The student of nature must be a philosopher in the true sense of the word and so long as he does not lose sight of his fundamental base—the observation of facts -his work will gain in value by allowing his reasoning faculties their fullest play.

THE LITHOSPHERE

Restricting our attention now to the more limited field to which the modern geologist applies himself, namely the study of the material, structure and history of the earth's crust, or the lithosphere, we may note at the outset that there are two phases or aspects of this field, one or the other of which is cultivated more sedulously by the geologist according to his predilections or circumstances, but neither of which can be wholly neglected by the worker in either field.

These two aspects are the purely scientific and the practical or applied. The geologist who works in the pure science field of his domain does so primarily for the intellectual satisfaction derived from the discovery of facts and principles. His aim is chiefly to search for nature's truths irrespective of their bearing on human welfare and his principal endeavor is directed toward widening the boundaries of human knowledge and pushing forward the frontiers of discovery.The chief aim of the practical geologist, on the other hand, is directed toward making the forces and materials of the world available to man, to augment the welfare of the human race and to push forward the boundaries of civilization.

But this work, important and noble as it is, can only be carried on successfully in proportion as the facts and laws of the science are discovered and demonstrated and the practical geologist must forever depend on the worker in pure science, whose reward too often is little more than the satisfaction which is to be derived from a devotion to the search for truth.

We shall for the sake of convenience speak of the pure-science phase of our subject as *the scientific aspect of geology* and of the applied phase of geology as *geology in its relation to man.* Each phase has its special subdivisions, to which attention may now be drawn.

ASPECT OF GEOLOGY

Structural Geology

A logical analysis of this field of investigation reveals a threefold aspect and three corresponding methods of approach in study. In the first place the material of the earth's crust and the composition and the structure of this material, invite attention. This portion of the subject is generally treated under the caption *Structural Geology.* It takes account of the composition, form and architecture of the earth's crust and is primarily an analytic branch of the science. From this point of view the earth's crust may be considered under the following subdivisions, given in the order of their magnitude.

Subdivisions of Structural Geology

These include the following fields:

- Chemical elements and ions.
- The combination of these elements and ions into salts and other compounds which either take on crystalline form or remain in an uncrystalline (amorphous) condition. These compounds as they occur in nature are designated minerals.
- The combination of crystals or fragments of the same

or different minerals into large masses, either bound together into a more or less solid mass, or remaining in an unconsolidated condition. These are the *rocks* of the earth's crust and for their study a recognition of at least the principal component minerals is essential.

- The association of rock masses into original or primary *structural units,* which constitutes the primary *architecture* of the earth's crust; or the impression upon it of *secondary structures,* through deformation or other influences of an outside nature. This is structural geology in the narrower sense—and has also been called *Geotectology*. The original structure corresponds to the initial architecture of a building, the stones or bricks of which correspond to the rock masses of the earth's crust. The secondary structures correspond to the changes subsequently made by additions, removal or modification resulting from sagging with age, etc.
- The surface forms resulting through the activities of destructive as well as reconstructive forces and recognizable in the mountains and valleys, the plains and plateaus and other physical features, generally considered under the subject of *Physical Geography*.

The term *Geomorphology* or the study of the details of the earth's surface forms has been commonly applied to this branch of inquiry, when it is not merely descriptive or analytic, but takes account of the history or genesis of the form as such. It corresponds to the study of the form of a building as a whole, either in its completeness or when changed to a ruin by destructive influences. The study of most of these divisions of structural geology has been carried to such detail that separate sciences have been developed. Thus the study of the natural substances, or minerals, has become the science of *Mineralogy* and that of the rocks the science of *Petrology* (also sometimes designated as lithology in its narrower sense).

While the science of rock structures is *Structural Geology* in the narrower sense and that of the surface forms is *Physiography* and their relation to man is *Geography*. In all such studies, while emphasis is laid on the analytical and descriptive sides, the causal or dynamic side and the historical or developmental side are given their due attention. This is expressed by the preference of terms ending in *ology* (mineralogy, petrology, geotectology, geomorphology) over those ending in *graphy* (petrography, physical geography, etc.), which emphasize mainly the descriptive and analytical side.

Geogenesis

The third method of approach is the historical or evolutionary method, in which emphasis is laid upon development and the causes which underlie this development. It is apparent that this phase of geology is the latest and most specialized aspect and that for its proper prosecution a thorough preparation in the other two phases is needed.

Moreover, since the history of life upon the earth is intimately bound up with the history of the lithosphere, a knowledge of biology and especially of palæontology, is indispensable for the prosecution of any but the most general studies in earth history. Several special branches have been developed within this field.

One of these deals with the origin or genesis of the rocks and their structures, or the origin of the lithosphere. To this branch the name *Lithogenesis* is commonly applied. Another branch considers the character, arrangement, succession, order of formation and age relations of the stratified rocks. This is the science of *Stratigraphy,* primarily a descriptive one.

A third branch deals with the succession and distribution of the organic remains, the petrifactions or fossils in the rocks in so far as they have a bearing on the geological history of the earth, *i.e.,* the *index fossils*. This is the special geological phase of *Palæontology,* which has also been designated by the name *Petrifactology*. Again, there is the science which deals with the development or genesis of the surface forms of the earth not merely in the descriptive manner, but from the view-point of origin. This, as already noted, is *Geomorphology* in its proper sense, though it is more commonly known by the name of *Physiography,* which formerly had a very different meaning.

Finally there is the study of the changes in the earth's surface, or its geography through the successive ages of the earth's history,—together with the changes in climate and the dispersions and migrations of the organisms and the causes which effected these.

This is the latest of the several aspects of Historical Geology and is now termed *Palæogeography* or the geography of the past. In this field the palæogeography of the Pleistocene period has been most extensively studied and a separate branch, that of *glacial geology*, has been developed. *Geography*, in the usual sense of the term, is the geography of the modern or Holocene period of the earth's history

Dynamical Geology

A second mode of approach is that which lays the emphasis upon the forces that work upon and within the earth to produce results, which, in this view, take a secondary rank. This is *Dynamical Geology* and it naturally falls into the two great divisions, the *physical* and *chemical*. The sciences of physics and chemistry, in so far as they lay the stress upon the forces at work, are the specialized development of these aspects. That chemistry, in its analytical as well as its dynamic aspects, is of fundamental importance to the geologist has generally been recognized, but the importance of physics to the student of earth science has only recently received proper recognition by the establishment of geophysical laboratories. Both *chemical* and *physical* forces may be viewed as *constructional*, or those building up rock masses and rock structures and as *destructional*, or those destroying them. A third view of these forces is that which deals with their effects in modifying or deforming the materials and structures and this may be termed the *reconstructional* or deformational aspect.

GEOLOGY IN ITS RELATION TO MAN

So far we have been considering geology in its pure science aspect, that which appeals to the inquiring mind of man in search after truth and knowledge, without ulterior motives of usefulness. There is, however, another aspect of our science and one which in recent times has come strongly to the front. This is *applied geology,* in which geological facts and forces are viewed in their relation to the needs and requirements of man.

The application of any science for any purpose whatsoever is successful in direct proportion to the profundity of knowledge possessed by the applier. No successful exploration of geological products or application of forces is possible without a thorough understanding of the facts and principles of the science and the student who wishes to follow the applied side of his science should not fail to make his preparation in the pure science side as broad and as profound as circumstances will permit.

Among the earliest problems, to the solution of which geological knowledge has been applied, are those of mining. Indeed, the science of geology, in a measure, developed in response to the needs of the miner for accurate knowledge of the conditions of occurrence, distribution and mode of origin of the valuable mineral deposits.

To such an extent has this been carried, that a separate branch of *mining geology* has come into existence. Moreover, as our knowledge increased and the possibility of more detailed application of our science became apparent, special subdivisions of mining geology have been developed and it is found that individuals can profitably devote themselves to the cultivation of a narrow field, to the practical exclusion of the others. Thus there have been developed the branches of *coal geology,* of *petroleum* and *gas geology* and of *salt geology,* including the geology of potash, phosphates, nitrates, borax and other salts and of bauxites and other aluminum ores. In these branches the investigator confines himself to the problems involved in the occurrence of these substances, which are chiefly restricted to the stratified rocks.

It is now well recognized that successful search for such deposits can only be undertaken by one well versed in the science of stratigraphy (including index fossils) and structural geology, while a thorough understanding of the principles of physiography and palæogeography is almost indispensable.

The mining geologist who devotes himself primarily to the problems of the metallic deposits must have not only a thorough knowledge of mineralogy, petrology and structural geology, but of dynamic geology as well and especially of the chemical and physical principles involved in ore deposition. As many ores are also found in stratified deposits, a knowledge of stratigraphy and of index fossils is necessary, while an understanding of the principles of physiography and palæogeography will also be found of value in many cases.

Geology, too, plays an indispensable rôle in the solution of many important engineering problems. In the construction of the Catskill aqueduct for New York City, a force of competent geologists was constantly employed and specialists were frequently called upon for consultation. This same need was felt after the construction of the Panama Canal had been undertaken and a resident geologist was appointed to supervise the later phases of construction.

That many difficulties might have been avoided had such supervision existed from the outset of the undertaking, is now generally conceded. Geological advice has always been employed in the construction of great tunnels such as those piercing mountains or passing under rivers and other water bodies. In many cases, too, the selection of sites for bridges, dams and other great engineering works has been based on geological advice, while in other cases, where such advice has not been sought or has been disregarded, disastrous consequences have resulted.

In consequence of the growing recognition of the need of geological study in the undertaking of engineering problems, the special branch of *engineering geology* has been developed.The geological engineer must be primarily a structural geologist and one who has a thorough grasp of the principles of dynamic geology, including hydrology as well, while physiography too is of great importance to him.

To a lesser degree a knowledge of rock types, of stratigraphic principles and of index fossils will be needed by him and not infrequently a knowledge of palæogeography, especially that phase which deals with the Pleistocene, or the problems of the glacial period, will be found of the greatest value. Finally there has been developed in recent years the special branch of geology which deals with the problems involved in military campaigns and to this the name *war* or *military geology* has been applied. Some of these problems are concerned with the proper location of sites for camps and trenches and with water supply and sanitation and for these a knowledge of structural geology, of rock types, of stratigraphy and of glacial geology has been found necessary.

Other problems are of an engineering type and require the preparation of the geological engineer. Again, the problems involved in military operations need for their solution a well-trained physiographer and a competent meteorologist as well. Problems concerned with naval warfare require the attention of one well versed in hydrology, especially that phase of it which deals with the oceans, or oceanography.

There are other ways in which a knowledge of geology has become useful to man and as the science itself is developed new channels of application into which it may be directed will no doubt be discovered. The student should further realise that the development of his science, the history of geologic thought, cannot be neglected by him. We profit by the mistakes of our predecessors as much as we do by their achievements and the history of the discovery of facts and of the development of geological opinion since the days of the Greek philosophers is fraught with lessons equal in import to those gained from the pursuit of the history of any other department of human thought and endeavor.

Chapter 2

Geological Observation of Earth

The geologist is, above all things, an observer in the great outof-door world. The man whose horizon is bounded by the walls of a city can never be a geologist, though he may gain much scientific knowledge from books and from an inspection of collections in museums and laboratories.

The true geologist, however, goes directly to the earth and there begins his inquiries. Not until observations of natural facts and phenomena were made in extenso was the inquiry of the philosophers regarding the earth and its history placed on a scientific basis.

Scattered observations and more or less accurate deductions were made even in antiquity. Thus Aristotle, in the third century B.C., had a very considerable understanding of the work of rivers and reasoned correctly regarding the changes in the land and sea at former times.

The painter Leonardo da Vinci (1452-1519) correctly reasoned, from an observation of the fossils found in the foothills of the Apennine Mountains, that they were the shells of once living animals, though they were generally regarded either as freaks of nature or as modern shells dropped by the pilgrims in their voyages across these mountains. It is true that the significance of fossil sea shells was recognized by the Greek philosophers but their explanations were generally ridiculed during the Middle Ages.

It was, however, not until the latter part of the eighteenth and the early part of the nineteenth century that systematic investigations of the rocks of the earth and their contained fossils began and from this period we date the birth of geology as a science. Scientific geology arose more or less simultaneously in the different countries of Europe. In France Etienne Guettard and Nicholas Desmarest were among the first to bring observation of facts to the fore, while Buffon indulged in brilliant speculations on the origin of the earth. Later Alexander Brongniart investigated the rocks around Paris and Georges Cuvier described their fossils.

In Germany Abraham Gottlob Werner, who is often called the founder of German geology and who was Proessor at the Mining School at Freiberg i. S., exerted a profound influence on geology especially by his teachings, to which men flocked from all countries.

Being mostly an observer of the details of specimens and rarely venturing beyond his immediate surroundings for field observations, many of his geological deductions have proved erroneous—though his pupils and followers, notably Leopold von Buch, extended their observations over wide areas and added much to the store of geological facts as well as to its philosophy. Switzerland had its enthusiasticstudent of the structure and history of the Alps in the person of H.B. de Saussure (1740-1799) and Russia had in Pierre Simon Pallas (1741-1811) its careful student of the Ural Mountains and the rocks outcropping there and elsewhere in the empire. In Great Britain many men contributed to the discovery of facts and the interpretation of these.

Among them the first rank is given to James Hutton, whose work marks a turning point in the history of geology, for he insisted that "the past history of the globe must be explained by what can be seen happening now, or to have happened only recently," a dictum which has since become the very cornerstone of geology.

In this work are contained many of the fundamental principles with which geologists are concerned to-day and they are illustrated by a wealth of facts gleaned by Hutton from his rambles through Scotland and other countries. Another of the Scottish founders of geology was Sir James Hall, to whom we owe the origination of experimental geology. The best known, however, among the early English geologists was

William Smith, who is generally called the "Father of English Geology." He determined not only the correct successions of the English rock formations and made the first geological map of England, but gave to many of the formations the names which they bear to-day. Finally, the student should remember among English geologists the name of Sir Charles Lyell, as that of a man who has had the most profound influence on geological thought.

His great work, the *Principles of Geology,* has become a classic of geological literature. Among the men who exerted a profound influence on American geology in the early days of its development, the names of William McClure and Amos Eaton stand out prominently. McClure, born in Scotland in 1763, became an American citizen near the close of the century. In 1809 he published the first important work on American Geology, in which appeared the first geological map of the Eastern United States and one of the first geological maps of the country.

Amos Eaton, born in New York state, is known especially for his *Index to the Geology of the Northern States* (1818), which was the first geological textbook published in America and in this and subsequent works he laid the foundation for the New York geological system.

Many of the names of American formations still current were first applied by him. He also made the first geological map of New York. The important influence which McClure and Eaton had on the development of American geology has been recognized by the designation of the first two eras in the history of this science in America as the Maclurean (1785-1819) and the Eatonian (1820-1829).

GEOLOGICAL OBSERVATIONS

"Where, then," the student will ask, "can the facts of geology be observed and how have they become available?" To get at the facts of structural geology the student must go to the rocks.

True, the rocks and the minerals and the fossils are brought to him in museums and laboratories and he will do well to begin his studies of selected examples thus brought together and capable of being examined under the most favorable conditions.

But the knowledge thus gained must be amplified and correlated by repeated visits to the home of the rocks, where alone their larger relations and their true significance in the history of the earth can be ascertained.

THE STUDY OF ROCKS AND ROCK STRUCTURES

Other Exposure of this Type

The type of outcrop just described is found in most regions of flat-lying rocks in our own country as well as abroad. Thus the geologist who starts from Bamboo, Wisconsin and proceeds southeastward to Lake Michigan at Milwaukee, will cross such a series of rocky ridges separated by soilfilled valleys and a similar experience awaits the traveler across Kansas from southeast to northwest, or the one who proceeds southward across Oklahoma, or journeys from central Texas either northwestward or southeastward.

The traveler across England from Liverpool to London also crosses such a series of rocky ridges, which in general extend in a northeast-southwest direction and are formed by a succession of nearly horizontal rock-layers with westward facing cliffs, though these are not generally visible from the train. Similar outcrops of nearly horizontal layers of rock surround Paris in constantly widening circles on three sides. The edges of some of these rocks form cliffs or escarpments facing outward.

If one were to represent the rock formations which underlie Paris and which have a shallow basin-like structure, by a nest of plates, the smallest at the top, the successive rims of these plates would represent the encircling cliffs, while the location of Paris would be in the centre of the uppermost and smallest plate. Many rivers have cut channels through the edges of these rock plates, while others flow in the depressions between successive rims. Thus numerous and excellent exposures of the rocks are found, even though the marginal portions between the edges of successive rock layers are frequently covered by soil and vegetation. These numerous rock exposures made possible the observations which gave the French founders of our science the data on which to build their deductions and the cliffs which they form around Paris were of the utmost importance in the conduct of the Great War so recently closed.

Outcrops in Mountainous Countries

It is, however, in the elevated portions of the earth—the mountains and chains of uplands—that rock outcrops are most frequent on the surface of the land. Here the soil and rock débris lodges only in the depressions, while between these the ledges protrude and give opportunities for observation. Here, too, deep cañons are cut by the rivers and glaciers and thus additional rock-walls are opened for observation.

German, Austrian and Swiss geologists have for the most part been limited to such regions for their observations and the wonderful rock exposures of the Alps and other mountains have made it possible for them to carry their studies along certain lines to great lengths. Werner, the father of German geology, made most of his observations in the subdued mountain district of Saxony, especially the Erzgebirge, which has long been famous for its old and extensive mining operations. Since France and Italy also border on the Alps and have mountain ranges of their own, the geologists of these countries were enabled to avail themselves of the rocks and rock structures thus revealed. British geologists, too, have been able to some extent to resort to this type of exposure, though in these moisture-enveloped islands, as over parts of Scandinavia as well, the dense though low cover of vegetation and the peat accumulations obstruct much of the underlying rock, as all students of Irish geology know only too well.

The Highlands of Scotland, however, furnish many good opportunities for observation, as do also many of the higher English districts and especially the mountain region of Wales. Swedish geologists have frequently had to resort to digging through the surface layers to get at the underlying rock and it is not an uncommon sight to see the Swedish geologist in the southern interior accompanied by a factotum, whose duty it is to wield pick and shovel.

Fig. *Marsten Rock*

In European Russia and adjoining districts there are vast areas of flat-lying rocks covered by soil and drift, so that, except along the coast and in the river channels, outcrops of the bedrock are difficult to find.

But where these rocks are uplifted—in the Ural Mountains on the east, the Caucasus on the south and the Carpathians on the west—outcrops abound and here the true relationships of the rock formations may be ascertained. In America, the New England uplands, the White, Green and Adirondack Mountains and the Appalachian Chain furnish an abundant series of rock outcrops for the eastern geologist.

The old and generally much-subdued mountain system, the rocks of which may be traced by frequent outcrops from New York City to the Highlands of the Hudson and which can be followed southwestward through New Jersey, Pennsylvania and Maryland and indeed all the way down to the Carolinas, where it constitutes the older. Appalachian Chain, is a typical example of an elevated, though for the most part not very mountainous, country and here outcrops abound.

Many of our western mountains are especially well adapted for geological observation, for here the aridity of the climate prevents the growth of much vegetation and the rock structures, frequently developed on a gigantic scale, are visible for many miles. This makes the Rocky Mountains a veritable paradise for the American geologist.

Rock Exposures in Flat Countries

The dwellers in the interior of our country, or the traveler on the broad plains of northern Germany, of Russia, of Hungary, or of China, will find little opportunity to get a view of the rocks which underlie these regions, for an almost continuous mantle of soil and drift covers the solid rock. Only where rivers have cut channels through the surface layer of loose material, the *mantlerock,* or where quarries have been opened, or construction operations have necessitated excavation down to and into the solid bed rock, is there an opportunity for observation of the rocks beneath the mantle-rock. River valleys and gorges are therefore the favourite resorts of the geologist of the plains, while quarries, railroad cuts and other excavations which expose the rock, likewise receive his attention.

Borings and drillings for oil, gas, or water often prove of use to him, though generally, except where the diamond drill is used and the core preserved, the record of such borings is only of doubtful and minor value. Salt shafts, such as those in central New York and that near Detroit, Mich., furnish excellent sections at the time of excavation. Fortunately for the geologist of the plains, however, few regions of any great extent are wholly covered by mantle-rock; more commonly lowlying ridges of rock, formed by resistant layers, are exposed above the general surface of the soil and drift mantle.

Such "ledges" generally mark the edges of beds of hard rock which have a gentle slope away from the edge of the ledge. When the ledge-forming layer is very thick, a cliff of some height may result and this generally furnishes abundant opportunity for observation and deduction. All natural exposures of the rock are called *outcrops* and the outcropping ledges together with the exposures in the stream channels, especially those which cut across the cliffs, furnish to the geologist of the flat countries his most satisfactory data. We may note a few examples.

Illustration from New York State

The state of New York furnishes an illustration of the types of rock outcrop in ledges referred to in the preceding paragraphs. Over the greater part of the state the rocks are so gently inclined that they appear horizontal to the eye and it is only when they are seen in the cliffs of Niagara gorge and along Lake Erie that their gentle southward descent becomes noticeable. Such a section, considerably generalized, is shown in the following diagram.

Where the beds end in the air upon the north, the harder ones, such as limestones and sandstones, form a series of low cliffs, while the ends of the softer shale beds are usually marked by broad flat-bottomed valleys. The largest and deepest of these valleys is occupied by Lake Ontario, while others are filled by soil which conceals the rock. These cliffs or escarpments generally have an abrupt northern face across the edge of the hard rock and in these faces quarries are generally opened.

These cliffs can be traced with more or less interruption across New York state from the Niagara River and Lake Erie to the Hudson, although they become modified because some of the beds seen in the western part of the state die out, or change in character and new ones appear.

This is shown on the geological map of the state of New York, where a series of broad colour bands extends east and west across the state. Each colour band in general represents one rock layer or group of layers and the width of the colour band indicates the amount of exposure of each which would appear were all the covering soil and vegetation removed; or in other words, the amount by which each lower bed projects beyond the next higher one which covers it.

Wherever streams have cut across this series of beds, gorges are formed, in the walls of which the cut edges of the rocks, the soft layers as well as the hard ones, are shown. The most striking examples of such gorges are shown along the Genesee River, which crosses the state from south to north. In the section from Rochester northward it cuts the lower beds, the harder strata producing waterfalls. In the section between Portage and Mt. Morris, it cuts the higher strata and here too several waterfalls are formed by hard layers. Between these two points many smaller tributary streams have cut into the sides of the valley and exposed the rocks. Here, too, are situated several deep shafts which go down vertically to the Salina salt beds and during the cutting of which the succession of the rocky beds was ascertained.

A good understanding of the succession of this series of rocky formations is obtained by the traveler who passes from the Adirondack Mountains southwestward across the state to Elmira, especially if he take advantage of the various sections exposed in the gorges of the streams and the banks of the Finger Lakes. It was by the study of these natural outcrops of the state, supplemented by those made on the sections exposed during the cutting of the Erie Canal in 1817-1825, that the foundation of American geology was laid by such men as Amos Eaton and by James Hall and others associated with him on the geological survey of New York state.

THE COAST OF OUTCROPS

Of all the natural rock exposures, however, those of the coast-line are the most attractive and in many respects the most satisfying.Wherever the sea-coast or the shore of a large lake is formed by rocks which rise above the surface of the water, the cutting work of the waves keeps the exposure fresh. A tramp along a rocky sea-coast is replete with interest to the geologist and many of the choicest bits of geological observation have been made on such sea-cliffs.

Great Britain, with its wonderful rocky sea-coast, probably leads the world in the variety and significance of rocks and rock structures there exposed. No student of geology can afford to neglect the wonderful English and Scottish coast, which has furnished the British geologists so many opportunities for the observation of facts that there, more than elsewhere, geological science has advanced, since the days of William Smith, with phenomenal strides. This is probably the reason why English geology quickly became the standard of comparison for other nations, in whose home countries observation was a more arduous task, because they did not include such marvelous coast exposures.

Northern France, too, has a coast-line of great interest to the geologist and so has Norway. The coast-line of Germany, on the other hand, is mostly sandy and there is little diversity in the types of the facts which it discloses. The Atlantic coast-line of North America is for the most part a sandy one.

Only in New England, in the Canadian coastal Provinces and in Newfoundland can be seen coastal sections comparable to some extent to those of Great Britain. The northern regions, however, are accessible with difficulty and have only recently been investigated. But the New England coast and especially that of Massachusetts, is a Mecca for American geologists and many of the workers in American geology have had their preliminary training through a study of that interesting region.

REGIONS OF DYNAMIC GEOLOGY

Although the principles of dynamic geology, the workings of the chemical and physical forces, may be studied to much advantage in the laboratory—such study, too, is incomplete without recourse to the outdoor field. It is upon the sea-shore that some of the profoundest lessons of erosion by waves, of transportation by currents and of deposition in quieter waters can be learned. Here, too, the method of entombment of fossils and the formation of many original structures, such as ripple-marks and the like, can be observed.

Rocky as well as sandy and muddy shores should be visited. Shores of large lakes may serve as a substitute in inland regions, but lakes have in addition many characters of their own. Ponds and temporary pools also teach their lessons. River valleys and gorges, rapids and waterfalls, brooks and even the roadside gutter, furnish lessons in dynamic geology, as do also the hillside, the mountain slopes and the elevated peaks, where rocks are shattered by frost and decay under atmospheric influence. Glaciers present many illustrations of dynamic geology, while caverns and underground channels have special lessons to teach. The deserts and all regions where wind is at work furnish illustrations of the mechanical activities of the wind, while pools and salt pans in and regions furnish illustrations of chemical activities and of precipitation of salts through condensation of the water under evaporation. Springs, too, illustrate dynamical activities, both physical and chemical and artesian wells, oil wells, geysers and similar phenomena are replete with them.

Finally, volcanoes and other such phenomena furnish the means for the study of igneous activities. The great English geologist, Sir Charles Lyell, whom we sometimes call the " Father of Modern Geology," has said that the geologist must be primarily a traveler—he must go to other lands than his own and so widen the scope of his experience.

Werner, the founder of German geology, confined his observations mainly to his limited Saxon district and attempted to formulate from these observations laws which should govern the rest of the world. Naturally he fell into many and profound errors, so that to-day scarcely one of his theories is held. Since his day German geologists have, however, become great travelers, not only in their own but in most other lands. As a result, their observations have become of wide scope and they have added much to geological knowledge. British and American geologists have only recently begun to follow the advice of Lyell, but already their efforts have been crowned with considerable success.

Let the student of geology, then, come to realise that the value of his deductions increases in proportion to the range of his observations and that no single country or region of the world will give him all he needs.

The American student, owing to the wide extent and diversity of his country, is perhaps more favored in this respect than is the geologist of any other nationality, but at present only a limited portion of our country has become sufficiently accessible to make extended observations possible.

Finally, it must not be overlooked that the observations of our predecessors are recorded in the literature of the science and that here we find a mine of information, the value of which cannot be overestimated. No one can repeat all of the observations which have been made in the past, even were such repetition desirable. In addition to the laboratory and field, then, the student of geology must go to the library and a thorough understanding of the literature on his special field is of fundamental importance to the worker.

Besides special books on different aspects of the science, the student should gain familiarity in the use of the official publications issued by the governments of the various countries, the proceedings of scientific societies and the special journals devoted to geology and kindred sciences.

Chapter 3

Structure of the Earth

THE EARTH

The word " earth " has been used in several connexions - from that of soil or ground to that of the planet which we inhabit, but it is difficult to trace the exact historic sequence of the diverse usages.

Fig. *The Earth*

In the cosmogony of the Pythagoreans, Platonists and other philosophers, the term or its equivalent denoted an element or fundamental quality which conferred upon matter the character of earthiness; and in the subsequent development of theories as to the ultimate composition of matter by the alchemists, iatrochemists and early phlogistonists an element of the same name was retained. In modern chemistry, the common term " earth " is applied to certain oxides: - the alkaline earths " are the oxides of calcium (lime), barium. (baryta) and strontium (strontia); the " rare earths " are the oxides of a certain class of rare metals.

The terrestrial globe is a member of the Solar system, the third in distance from the Sun and the largest within the orbit of Jupiter.

In the wider sense it may be regarded as composed of a gaseous atmosphere, which encircles the crust or lithosphere and surface waters or hydrosphere.

The description of the surface features is a branch of Geography and the discussions as to their origin and permanence belongs to Physiography, physiographical geology, or physical geography.

The investigation of the crust belongs to geology and of rocks in particular to petrology.

In the present article we shall treat the subject matter of the Earth as a planet under the following headings:

- Figure and Size,
- Mass and Density,
- Astronomical Relations,
- Evolution and Age.

FIGURE AND SIZE OF EARTH

To primitive man the Earth was a flat disk with its surface diversified by mountains, rivers and seas. In many cosmogonies this disk was encircled by waters, unmeasurable by man and extending to a junction with the sky; and the disk stood as an island rising up through the waters from the floor of the universe, or was borne as an immovable ship on the surface.

Of such a nature was the cosmogony of the Babylonians and Hebrews; Homer states the same idea, naming the encircling waters and Hesiod regarded it as a disk midway between the sky and the infernal regions.

The theory that the Earth extended downwards to the limit of the universe was subjected to modification when it was seen that the same sun and stars reappeared in the east after their setting in the west. But man slowly realized that the earth was isolated in space, floating freely as a balloon and much speculation was associated about that which supported the Earth. Tunnels. in the foundations to permit the passage of the sun and stars were suggested; the Greeks considered twelve columns to support the heavens and in their mythology the god Atlas appears condemned to support the columns; while the Egyptians had the Earth supported by four elephants, which themselves stood on a tortoise swimming on a sea. Earthquakes were regarded as due to a movement of these foundations; in Japan this was considered to be due to the motion of a great spider, an animal subsequently replaced by a cat-fish; in Mongolia it is a hog; in India, a mole; in some parts of South America, a whale; and among some of the North American Indians,. a giant tortoise.

The doctrine of the spherical form has been erroneously assigned to Thales; but he accepted the Semitic conception of the disk and regarded the production of springs after earthquakes as due to the inrushing of the waters under the Earth into fissures in the surface.

His pupil, Anaximander, according to Diogenes Laertius, believed it to be spherical and Anaximenes probably held a similar view. The spherical form is undoubtedly a discovery of Pythagoras and was taught by the Pythagoreans and by the Eleatic Parmenides. The expositor of greatest moment was Aristotle; his arguments are those which we employ to-day:- the ship gradually disappearing from hull to mast as it recedes from the harbour to the horizon; the circular shadow cast by the Earth on the Moon during an eclipse and the alteration in the appearance of the heavens as one passes from point to point on the Earth's surface.

He records attempts made to determine the circumference; but the first scientific investigation in this 1 Aristotle regarded the Earth as having an upper inhabited half and a lower uninhabited one and the air on the lower half as tending to flow upwards through the Earth. The obstruction of this passage brought about an accumulation of air within the Earth and the increased pressure may occasion oscillations of the surface, which may be so intense as to cause earthquakes. The historical development of the methods for determining the figure of the Earth (by which we mean a theoretical surface in part indicated by the ocean at rest and in other parts by the level to which water freely communicating with the oceans by canals traversing the land masses would rise) and the mathematical investigation of this problem are treated in the articles Figure of the Earth and Geodesy; here the results are summarized. Sir Isaac Newton deduced from the mechanical consideration of the figure of equilibrium of a mass of rotating fluid, the form of an oblate spheroid, the ellipticity of a meridian section being 1/231 and the axes in the ratio 230:231.

Geodetic measurements by the Cassinis and other French astronomers pointed to a prolate form, but the Newtonian figure was proved to be correct by the measurement of meridional arcs in Peru and Lapland by the expeditions organized by the French Academy of Sciences. The position of the longer axis is somewhat uncertain; it is certainly in Africa, Clarke placing it in longitude 8° 15' W. and Schubert in longitude 41° 4' E.; arguing from terrestrial symmetry, has chosen the position lat. 6° N., long. 28° E., *i.e.* between Clarke's and Schubert's positions. For the lengths of the axes and the ellipticity of the Earth.

Astronomical Relations

The grandest achievements of astronomical science are undoubtedly to be associated with the elucidation of the complex motion of our planet. The notion that the Earth was fixed and immovable at the centre of an immeasurable universe long possessed the minds of men; and we find the illustrious Ptolemy accepting this view in the 2nd century A.D.. and rejecting the notion of a rotating Earth - a theory which had been proposed as early as the 5th century B.C. by Philolaus on philosophical grounds and in the 3rd century B.C. by the astronomer Aristarchus of Samos.

He argued that if the Earth rotated then points at the equator had the enormous velocity of about 1000 m. per hour and as a consequence there should be terrific gales from the east; the fact that there were no such gales invalidated, in his opinion, the theory.

The Ptolemaic theory was unchallenged until 1543, in which year the *De Revolutionibus orbium Celestium* of Copernicus was published. In this work it was shown that the common astronomical phenomena could be more simply explained by regarding the Earth as annually revolving about a fixed Sun and daily rotating about itself.

A clean sweep was made of the geocentric epicyclic motions of the planets which Ptolemy's theory demanded and in place there was substituted a procession of planets about the Sun at different distances.The Earth has two principal motions - revolution about the Sun, rotation about its axis; there are in addition a number of secular motions.

Mass and Density

The earliest scientific investigation on the density and mass of the Earth (the problem is really single if the volume of the Earth be known) was made by Newton, who, mainly from astronomical considerations, suggested the limiting densities 5 and 6; it is remarkable that this prophetic guess should be realized, the mean value from subsequent researches being about 5.2. The density of the Earth has been determined by several experimenters within recent years by methods.

ROTATION

The Earth rotates about an axis terminating at the north and south geographical poles and perpendicular to the equator; the period of rotation is termed the day, of which several kinds are distinguished according to the body or point of reference. The rotation is performed from west to east; this daily rotation occasions the *diurnal* motion of the celestial sphere, the rising of the Sun and stars in the east and their setting in the west and also the phenomena of day and night. The inclination of the axis to the ecliptic brings about the presentation of places in different latitudes to the more direct rays of the sun; this is revealed in the variation in the length of daylight with the time of the year and the phenomena of seasons. Although the rotation of the Earth was an accepted fact soon after its suggestion by Copernicus, an experimental proof was wanting until 1851, when Foucault performed his celebrated pendulum experiment at the Pantheon, Paris.

A pendulum about 200 ft. long, composed of a flexible wire carrying a heavy iron bob, was suspended so as to be free to oscillate in any direction. The bob was provided with a style which passed over a table strewn with fine sand, so that the style traced the direction in which the bob was swinging.

It was found that the oscillating pendulum never retraced its path, but at each swing it was apparently deviated to the right and moreover the deviations in equal times were themselves equal. This means that the floor of the Pantheon was moving and therefore the Earth was rotating.

If the pendulum were swung in the southern hemisphere, the deviation would be to the left; if at the equator it would not deviate, while at the poles the plane of oscillation would traverse a complete circle in 24 hours. The rotation of the Earth appears to be perfectly uniform, comparisons of the times of transits, eclipses, point to a variation of less than yhth of a second since the time of Ptolemy. Theoretical investigations on the phenomena of tidal friction point, however, to a retardation, which may to some extent be diminished by the accelerations occasioned by the shrinkage of the globe and some other factors difficult to evaluate.

Precession

The axis of the earth does not preserve an invariable direction in space, but in a certain time it describes a cone, in much the same manner as the axis of a top spinning out of the vertical. The equator, which preserves approximately the same inclination to the ecliptic must move so that its intersections with the ecliptic, or equinoctial points, pass in a retrograde direction, *i.e.* opposite to that of the Earth.

This motion is termed the precession of the equinoxes and was observed by Hipparchus in the 2nd century B.C.; Ptolemy corrected the catalogue of Hipparchus for precession by adding 2° 40' to the longitudes, the latitudes being unaltered by this motion, which at the present time is 50.26" annually, the complete circuit being made in about 26,000 years. Owing to precession the signs of the zodiac are traversing paths through the constellations, or, in other words, the constellations are continually shifting with regard to the equinoctial points; at one time the vernal equinox Aries was in the constellations of that name; it is now in Pisces and will then pass into Aquarius.

The pole star, *i.e.* the star towards which the Earth's axis points, is also shifting owing to precession; in about 2700 B.C. the Chinese observed a Draconis as the pole star (at present a Ursae minoris occupies this position and will do so until 3500); in 13600 Vega (a Lyrae) the brightest star in the Northern hemisphere, will be nearest. Precession is the result of the Sun and the Moon's attraction on the Earth not being a single force through its centre of gravity.

If the Earth were a homogeneous sphere the attractions would act through the centre and such forces would have no effect upon the rotation about the centre of gravity, but the Earth being spheroidal the equatorial band which stands up as it were beyond the surface of a sphere is more strongly attracted, with the result that the axis undergoes a tilting. The precession due to the Sun is termed the *solar precession* and that due to the Moon the *lunar precession;* the joint effect (two-thirds of which is due to the Moon) is the *luni-solar* precession. Solar precession is greatest at the solstices and zero at the equinoxes; the part of luni-solar precession due to the Moon varies with the position of the Moon in its orbit. The obliquity is unchanged by precession.

Nutation

In treating precession we have stated that the axis of the Earth traces a cone and it follows that the pole describes a circle on the celestial sphere, about the pole of the ecliptic. This is not quite true. Irregularities in the attracting forces which occasion precession also cause a slight oscillation backwards and forwards over the mean precessional path of the pole, the pole tracing a wavy line or nodding. Both the Sun and Moon contribute to this effect. Solar nutation depends upon the position of the Sun on the ecliptic; its period is therefore I year and in extent it is only I. 2"; lunar nutation depends upon the position of the Moon's nodes; its period is therefore about 18.6 years, the time of revolution of the nodes and its extent is 9. 2". There is also given to the obliquity a small oscillation to and fro. Nutation is one of the great discoveries of James Bradley.

Planetary Precession

So far we have regarded the ecliptic as absolutely fixed and treated precession as a real motion of the equator. The ecliptic, however, is itself subject to a motion, due to the attractions of the planets on the Earth. This effect also displaces the equinoctial points. Its annual value is 0.13".

The term General Precession in longitude is given to the displacement of the intersection of the equator with the apparent ecliptic on the latter. The standard value is 50.2453", which prevailed in 1850 and the value at 1850+t, *i.e.* the constant of precession, is 5 0.2 453" + 0.0002225" *t*. This value is also liable to a very small change. The nutation of the obliquity at time 1850 + t is given by the formula 23° 27' 32. o" -0.47" I. Complete expressions for these functions are given in Newcomb's *Spherical Astronomy* and in the *Nautical Almanac*. The variation of the *line of apsides* is the name given to the motion of the major axis of the Earth's orbit along the ecliptic. It is due to the general influence of the planets and the revolution is effected in 21,000 years.

The variation of the eccentricity denotes an oscillation of the form of the Earth's orbit between a circle and ellipse. This followed the mathematical researches of Lagrange and Leverrier. It was suggested by Sir John Herschel in 1830 that this variation might occasion great climatic changes and James Croll developed the theory as affording a solution of the glacial periods in geology.

Variation of Latitude

Another secular motion of the Earth is due to the fact that

the axis of rotation is not rigidly fixed within it, but its polar extremities wander in a circle of about 50 ft. diameter. This oscillation brings about a variability in terrestrial latitudes, hence the name. Euler showed mathematically that such an oscillation existed and, making certain assumptions as to the rigidity of the Earth, deduced that its period was 305 days; S. C. Chandler, from 1890 onwards, deduced from observations of the stars a period of 428 days;. and Simon Newcomb explained the deviation of these periods by pointing out that Euler's assumption of a perfectly rigid Earth is not in accordance with fact.

REVOLUTION

The Earth revolves about the Sun in an elliptical orbit having the Sun at one focus. The plane of the orbit is termed the ecliptic; it is inclined to the Earth's equator at an angle termed the obliquity and the points of intersection of the equator and ecliptic are termed the equinoctial points. The major axis of the ellipse is the line of apsides; when the Earth is nearest the Sun it is said to be in perihelion, when farthest it is in aphelion.

The mean distance of the Earth from the Sun is a most important astronomical constant, since it is the unit of linear measurement; its value is about 93,000,000 m. and the difference between the perihelion and aphelion distances is about 3,000,000 m. The eccentricity of the orbit is o 016751. A tabular comparison of the orbital constants of the Earth and the other planets is given in the article Planet.

The period of revolution with regard to the Sun, or, in other words, the time taken by the Sun apparently to pass from one equinox to the same equinox, is the tropical or equinoctial year; its length is 365 d. 5 hrs. 48 m. 46 secs. It is about 20 minutes shorter than the true or sidereal year, which is the time taken for the Sun apparently to travel from one star to it again. The difference in these two years is due to the secular variation termed precession. A third year is named the *anomalistic year,* which is the time occupied in the passage from perihelion to perihelion; it is a little longer than the sidereal.

EVOLUTION AND AGE

In its earliest history the mass now consolidated as the Earth and Moon was part of a vast nebulous aggregate, which in the course of time formed a central nucleus - our Sun - which shed its outer layers in such a manner as to form the solar system. The moon may have been formed from the Earth in a similar manner, but the theory of tidal friction suggests the elongation of the Earth along an equatorial axis to form a pear-shaped figure and that in the course of time the protuberance shot off to form the Moon. The age of the Earth has been investigated from several directions, as have also associated questions related to climatic changes, internal temperature, orientation of the land and water (permanence of oceans and continents), &c. These problems are treated in the articles Geology and Geography.

Figure of the Earth

The determination of the figure of the earth is a problem of the highest importance in astronomy, inasmuch as the diameter of the earth is the unit to which all celestial distances must be referred. Scientists have been able to reconstruct detailed information about the planet's past. Earth and the other planets in the Solar System formed 4.54 billion years ago out of the solar nebula, a disk-shaped mass of dust and gas left over from the formation of the Sun. Initially molten, the outer layer of the planet Earth cooled to form a solid crust when water began accumulating in the atmosphere.

The Moon formed soon afterwards, possibly as the result of a Mars-sized object (sometimes called Theia) with about 10% of the Earth's mass impacting the Earth in a glancing blow. Some of this object's mass would have merged with the Earth and a portion would have been ejected into space, but enough material would have been sent into orbit to form the Moon.Outgassing and volcanic activity produced the primordial atmosphere.

Condensing water vapour, augmented by ice and liquid water delivered by asteroids and the larger proto-planets, comets and trans-Neptunian objects produced the oceans. The highly energetic chemistry is believed to have produced a self-replicating molecule around 4 billion years ago and half a billion years later, the last common ancestor of all life existed.

The development of photosynthesis allowed the Sun's energy to be harvested directly by life forms; the resultant oxygen accumulated in the atmosphere and resulted in a layer of ozone (a form of molecular oxygen [O_3]) in the upper atmosphere. The incorporation of smaller cells within larger ones resulted in the development of complex cells called eukaryotes. True multicellular organisms formed as cells within colonies became increasingly specialized. Aided by the absorption of harmful ultraviolet radiation by the ozone layer, life colonized the surface of Earth.

Beginning with almost no dry land, the total amount of surface lying above the oceans has steadily increased. During the past two billion years, for example, the total size of the continents has doubled. As the surface continually reshaped itself, over hundreds of millions of years, continents formed and broke up. The continents migrated across the surface, occasionally combining to form a supercontinent. Roughly 750 million years ago, the earliest known supercontinent, Rodinia, began to break apart. The continents later recombined to form Pannotia, 600–540 mya, then finally Pangaea, which broke apart 180 mya.

Since the 1960s, it has been hypothesized that severe glacial action between 750 and 580 mya, during the Neoproterozoic, covered much of the planet in a sheet of ice. This hypothesis has been termed "Snowball Earth" and is of particular interest because it preceded the Cambrian explosion, when multicellular life forms began to proliferate.

Following the Cambrian explosion, about 535 mya, there have been five mass extinctions. The last extinction event occurred 65 mya, when a meteorite collision probably triggered the extinction of the (non-avian) dinosaurs and other large reptiles, but spared small animals such as mammals, which then resembled shrews.

Over the past 65 million years, mammalian life has diversified and several mya, an African ape-like animal gained the ability to stand upright. This enabled tool use and encouraged communication that provided the nutrition and stimulation needed for a larger brain. The development of agriculture and then civilization, allowed humans to influence the Earth in a short time span as no other life form had, affecting both the nature and quantity of other life forms.

The present pattern of ice ages began about 40 mya, then intensified during the Pleistocene about 3 mya. The polar regions have since undergone repeated cycles of glaciation and thaw, repeating every 40–100,000 years. The last ice age ended 10,000 years ago.

COMPOSITION AND STRUCTURE

Earth is a terrestrial planet, meaning that it is a rocky body, rather than a gas giant like Jupiter. It is the largest of the four solar terrestrial planets, both in terms of size and mass. Of these four planets, Earth also has the highest density, the highest surface gravity and the strongest magnetic field.

SHAPE

Chemical Composition

The mass of the Earth is approximately 5.98×10 kg. It is composed mostly of iron (32.1%), oxygen (30.1%), silicon (15.1%), magnesium (13.9%), sulfur (2.9%), nickel (1.8%), calcium (1.5%) and aluminium (1.4%); with the remaining 1.2% consisting of trace amounts of other elements. Due to mass segregation, the core region is believed to be primarily composed of iron (88.8%), with smaller amounts of nickel (5.8%), sulfur (4.5%) and less than 1% trace elements.

The geochemist F. W. Clarke calculated that a little more than 47% of the Earth's crust consists of oxygen. The more common rock constituents of the Earth's crust are nearly all oxides; chlorine, sulfur and fluorine are the only important exceptions to this and their total amount in any rock is usually much less than 1%.

The principal oxides are silica, alumina, iron oxides, lime, magnesia, potash and soda. The silica functions principally as an acid, forming silicates and all the commonest minerals of igneous rocks are of this nature. From a computation based on 1,672 analyses of all kinds of rocks, Clarke deduced that 99.22% were composed of 11 oxides. All the other constituents occur only in very small quantities.

Internal Structure

The interior of the Earth, like that of the other terrestrial planets, is chemically divided into layers. The Earth has an outer silicate solid crust, a highly viscous mantle, a liquid outer core that is much less viscous than the mantle and a solid inner core.

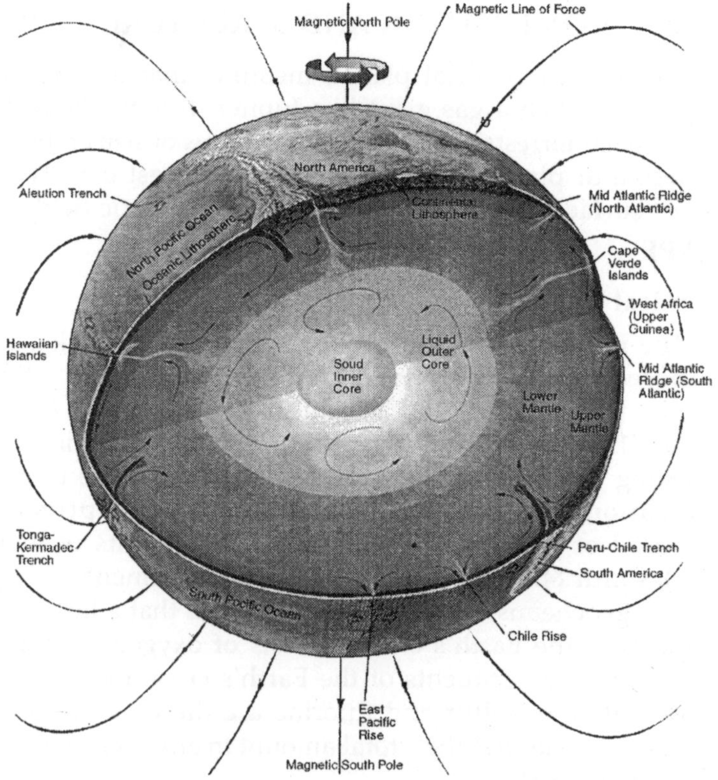

Fig. *Internal Structure of the Earth*

The crust is separated from the mantle by the Mohorovièiæ discontinuity and the thickness of the crust varies: averaging 6 km under the oceans and 30–50 km on the continents.The internal heat of the planet is probably produced by the radioactive decay of potassium-40, uranium-238 and thorium-232 isotopes.

All three have half-life decay periods of more than a billion years. At the centre of the planet, the temperature may be up to 7,000 K and the pressure could reach 360 GPa. A portion of the core's thermal energy is transported toward the crust by Mantle plumes; a form of convection consisting of upwellings of higher-temperature rock. These plumes can produce hotspots and flood basalts.

Tectonic Plates

According to plate tectonics theory, the outermost part of the Earth's interior is made up of two layers: the lithosphere, comprising the crust and the solidified uppermost part of the mantle. Below the lithosphere lies the asthenosphere, which forms the inner part of the upper mantle. The asthenosphere behaves like a superheated material that is in a semi-fluidic, plastic-like state.The lithosphere essentially *floats* on the asthenosphere and is broken up into what are called tectonic plates. These plates are rigid segments that move in relation to one another at one of three types of plate boundaries: convergent, divergent and transform. The last occurs where two plates move laterally relative to each other, creating a strike-slip fault. Earthquakes, volcanic activity, mountain-building and oceanic trench formation can occur along these plate boundaries.

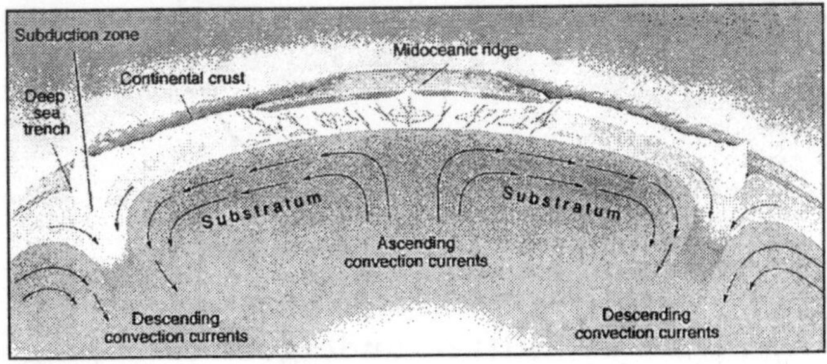

Fig. *Plate Tectonics*

Notable minor plates include the Indian Plate, the Arabian Plate, the Caribbean Plate, the Nazca Plate off the west coast of South America and the Scotia Plate in the southern Atlantic Ocean. The Australian Plate actually fused with Indian Plate between 50 and 55 million years ago. The fastest-moving plates are the oceanic plates, with the Cocos Plate advancing at a rate of 75 mm/yr and the Pacific Plate moving 52–69 mm/yr. At the other extreme, the slowest-moving plate is the Eurasian Plate, progressing at a typical rate of about 21 mm/yr.

Surface

The Earth's terrain varies greatly from place to place. About 70.8% of the surface is covered by water, with much of the continental shelf below sea level. The submerged surface has mountainous features, including a globe-spanning mid-ocean ridge system, as well as undersea volcanoes, oceanic trenches, submarine canyons, oceanic plateaus and abyssal plains. The remaining 29.2% not covered by water consists of mountains, deserts, plains, plateaus and other geomorphologies.

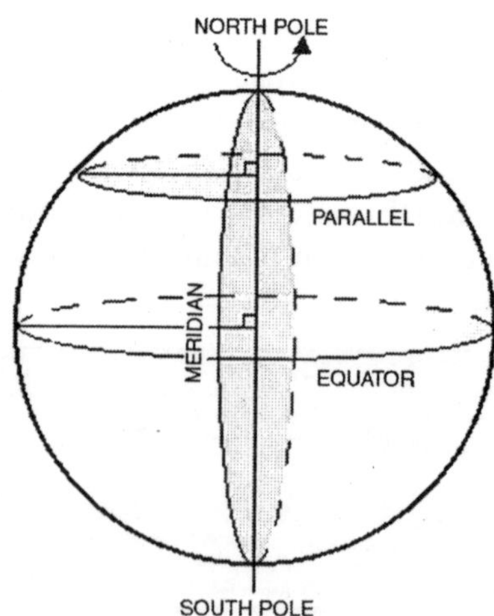

Fig. *Surface of the Earth*

The planetary surface undergoes reshaping over geological time periods due to the effects of tectonics and erosion. The surface features built up or deformed through plate tectonics are subject to steady weathering from precipitation, thermal cycles and chemical effects. Glaciation, coastal erosion, the build-up of coral reefs and large meteorite impacts also act to reshape the landscape.

As the tectonic plates migrate across the planet, the ocean floor is subducted under the leading edges. At the same time, upwellings of mantle material create a divergent boundary along mid-ocean ridges. The combination of these processes continually recycles the oceanic crustal material. Most of the ocean floor is less than 100 million years in age. The oldest oceanic crust is located in the Western Pacific and has an estimated age of about 200 million years.

By comparison, the oldest fossils found on land have an age of about 3 billion years.The continental crust consists of lower density material such as the igneous rocks granite and andesite. Less common is basalt, a denser volcanic rock that is the primary constituent of the ocean floors. Sedimentary rock is formed from the accumulation of sediment that becomes compacted together. Nearly 75% of the continental surfaces are covered by sedimentary rocks, although they form only about 5% of the crust.

The third form of rock material found on Earth is metamorphic rock, which is created from the transformation of pre-existing rock types through high pressures, high temperatures, or both. The most abundant silicate minerals on the Earth's surface include quartz, the feldspars, amphibole, mica, pyroxene and olivine.

Common carbonate minerals include calcite (found in limestone), aragonite and dolomite. The pedosphere is the outermost layer of the Earth that is composed of soil and subject to soil formation processes. It exists at the interface of the lithosphere, atmosphere, hydrosphere and biosphere. Currently the total arable land is 13.31% of the land surface, with only 4.71% supporting permanent crops. Close to 40% of the Earth's land surface is presently used for cropland and pasture, or an estimated 1.3×10 km^2 of cropland and 3.4×10 km^2 of pastureland.

The elevation of the land surface of the Earth varies from the low point of "418 m at the Dead Sea, to a 2005-estimated maximum altitude of 8,848 m at the top of Mount Everest. The mean height of land above sea level is 840 m.

Hydrosphere

The abundance of water on Earth's surface is a unique feature that distinguishes the "Blue Planet" from others in the solar system. The Earth's hydrosphere consists chiefly of the oceans, but technically includes all water surfaces in the world, including inland seas, lakes, rivers and underground waters down to a depth of 2,000 m. The deepest underwater location is Challenger Deep of the Mariana Trench in the Pacific Ocean with a depth of "10,911.4 m.

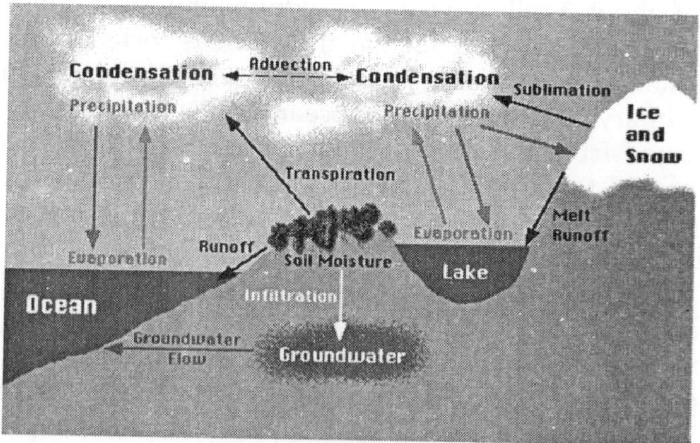

Fig. *Hydrosphere*

The average depth of the oceans is 3,800 m, more than four times the average height of the continents. The mass of the oceans is approximately 1.35×10 metric tons, or about 1/4400 of the total mass of the Earth and occupies a volume of 1.386×10 km³. If all of the land on Earth were spread evenly, water would rise to an altitude of more than 2.7 km. About 97.5% of the water is saline, while the remaining 2.5% is fresh water.

The majority of the fresh water, about 68.7%, is currently in the form of ice.About 3.5% of the total mass of the oceans consists of salt. Most of this salt was released from volcanic activity or extracted from cool, igneous rocks. The oceans are also a reservoir of dissolved atmospheric gases, which are essential for the survival of many aquatic life forms. Sea water has an important influence on the world's climate, with the oceans acting as a large heat reservoir. Shifts in the oceanic temperature distribution can cause significant weather shifts, such as the El Niño-Southern Oscillation.

Atmosphere

The atmospheric pressure on the surface of the Earth averages 101.325 kPa, with a scale height of about 8.5 km. It is 78% nitrogen and 21% oxygen, with trace amounts of water vapour, carbon dioxide and other gaseous molecules. The height of the troposphere varies with latitude, ranging between 8 km at the poles to 17 km at the equator, with some variation due to weather and seasonal factors.

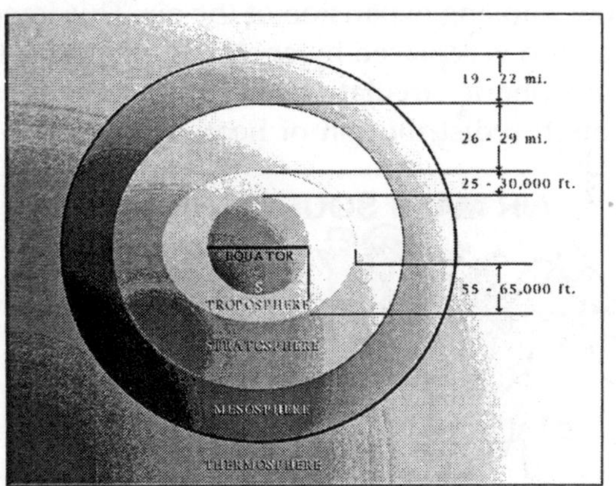

Fig. *Atmosphere of the Earth*

Earth's biosphere has significantly altered its atmosphere. Oxygenic photosynthesis evolved 2.7 billion years ago, forming the primarily nitrogen-oxygen atmosphere that exists today. This change enabled the proliferation of aerobic organisms as well as the formation of the ozone layer which, together with Earth's magnetic field, blocks ultraviolet solar radiation, permitting life on land. Other atmospheric functions important to life on Earth's include transporting water vapour, providing useful gases, causing small meteors to burn up before they strike the surface and moderating temperature.

This last phenomenon is known as the greenhouse effect: trace molecules within the atmosphere serve to capture thermal energy emitted from the ground, thereby raising the average temperature. Carbon dioxide, water vapour, methane and ozone are the primary greenhouse gases in the Earth's atmosphere. Without this heat-retention effect, the average surface temperature would be 18°C and life would likely not exist.

WEATHER AND CLIMATE

The Earth's atmosphere has no definite boundary, slowly becoming thinner and fading into outer space. Three-quarters of the atmosphere's mass is contained within the first 11 km of the planet's surface. This lowest layer is called the troposphere. Energy from the Sun heats this layer and the surface below, causing expansion of the air. This lower density air then rises and is replaced by cooler, higher density air. The result is atmospheric circulation that drives the weather and climate through redistribution of heat energy.

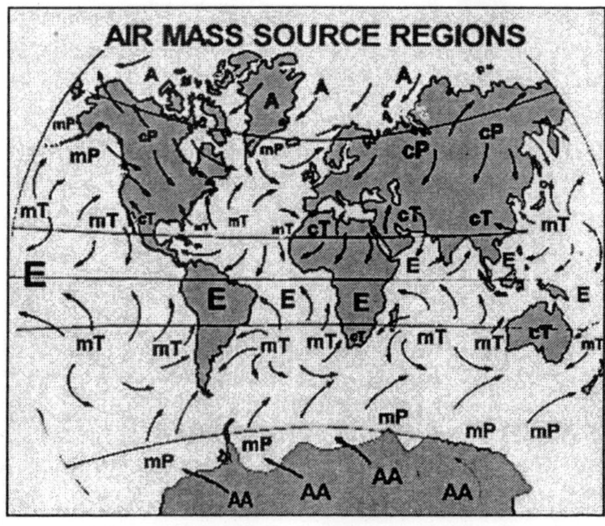

Fig. *Source Regions of Global air Masses*

The primary atmospheric circulation bands consist of the trade winds in the equatorial region below 30° latitude and the westerlies in the mid-latitudes between 30° and 60°. Ocean currents are also important factors in determining climate, particularly the thermohaline circulation that distributes heat energy from the equatorial oceans to the polar regions

Water vapour generated through surface evaporation is transported by circulatory patterns in the atmosphere. When atmospheric conditions permit an uplift of warm, humid air, this water condenses and settles to the surface as precipitation. Most of the water is then transported back to lower elevations by river systems, usually returning to the oceans or being deposited into lakes.

This water cycle is a vital mechanism for supporting life on land and is a primary factor in the erosion of surface features over geological periods. Precipitation patterns vary widely, ranging from several meters of water per year to less than a millimeter. Atmospheric circulation, topological features and temperature differences determine the average precipitation that falls in each region.The Earth can be sub-divided into specific latitudinal belts of approximately homogeneous climate. Ranging from the equator to the polar regions, these are the tropical (or equatorial), subtropical, temperate and polar climates. Climate can also be classified based on the temperature and precipitation, with the climate regions characterized by fairly uniform air masses. The commonly-used Köppen climate classification system has five broad groups (humid tropics, arid, humid middle latitudes, continental and cold polar), which are further divided into more specific subtypes.

Upper Atmosphere

Above the troposphere, the atmosphere is usually divided into the stratosphere, mesosphere and thermosphere. Each of these layers has a different lapse rate, defining the rate of change in temperature with height. Beyond these, the exosphere thins out into the magnetosphere (where the Earth's magnetic fields interact with the solar wind). An important part of the atmosphere for life on Earth is the ozone layer, a component of the stratosphere that partially shields the surface from ultraviolet light.

The Kármán line, defined as 100 km above the Earth's surface, is a working definition for the boundary between atmosphere and space.Due to thermal energy, some of the molecules at the outer edge of the Earth's atmosphere have their velocity increased to the point where they can escape from the planet's gravity.

This results in a slow but steady leakage of the atmosphere into space. Because unfixed hydrogen has a low molecular weight, it can achieve escape velocity more readily and it leaks into outer space at a greater rate.

For this reason, the Earth's current environment is oxidizing, rather than reducing, with consequences for the chemical nature of life which developed on the planet. The oxygen-rich atmosphere also preserves much of the surviving hydrogen by locking it up in water molecules.

MAGNETIC FIELD OF EARTH

The Earth's magnetic field is shaped roughly as a magnetic dipole, with the poles currently located proximate to the planet's geographic poles.

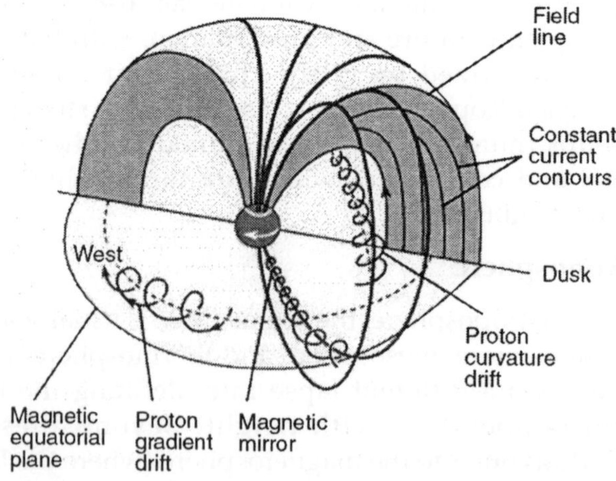

Fig. *The Earth's Magnetic Field*

According to dynamo theory, the field is generated within the molten outer core region where heat creates convection motions of conducting materials, generating electric currents. These in turn produce the Earth's magnetic field. The convection movements in the core are chaotic in nature and periodically change alignment. This results in field reversals at irregular intervals averaging a few times every million years.

The most recent reversal occurred approximately 700,000 years ago.The field forms the magnetosphere, which deflects particles in the solar wind. The sunward edge of the bow shock is located at about 13 times the radius of the Earth. The collision between the magnetic field and the solar wind forms the Van Allen radiation belts, a pair of concentric, torus-shaped regions of energetic charged particles. When the plasma enters the Earth's atmosphere at the magnetic poles, it forms the aurora.

ORBIT AND ROTATION OF EARTH

Relative to the background stars, it takes the Earth, on average, 23 hours, 56 minutes and 4.091 seconds (one sidereal day) to rotate around the axis that connects the north and the south poles from west to east. From Earth, the main apparent motion of celestial bodies in the sky (except that of meteors within the atmosphere and low-orbiting satellites) is to the west at a rate of 15°/h = 15'/min. This is equivalent to an apparent diameter of the Sun or Moon every two minutes. Earth orbits the Sun at an average distance of about 150 million kilometers every 365.2564 mean solar days (1 sidereal year).

From Earth, this gives an apparent movement of the Sun with respect to the stars at a rate of about 1°/day (or a Sun or Moon diameter every 12 hours) eastward. Because of this motion, on average it takes 24 hours—a solar day—for Earth to complete a full rotation about its axis so that the Sun returns to the meridian. The orbital speed of the Earth averages about 30 km/s (108,000 km/h), which is fast enough to cover the planet's diameter (about 12,600 km) in seven minutes and the distance to the Moon (384,000 km) in four hours.

The Moon revolves with the Earth around a common barycenter every 27.32 days relative to the background stars.

When combined with the Earth–Moon system's common revolution around the Sun, the period of the synodic month, from new moon to new moon, is 29.53 days. Viewed from the celestial north pole, the motion of Earth, the Moon and their axial rotations are all counter-clockwise. The orbital and axial planes are not precisely aligned: Earth's axis is tilted some 23.5 degrees from the perpendicular to the Earth–Sun plane (which causes the seasons); and the Earth–Moon plane is tilted about 5 degrees against the Earth-Sun plane (without this tilt, there would be an eclipse every two weeks, alternating between lunar eclipses and solar eclipses).

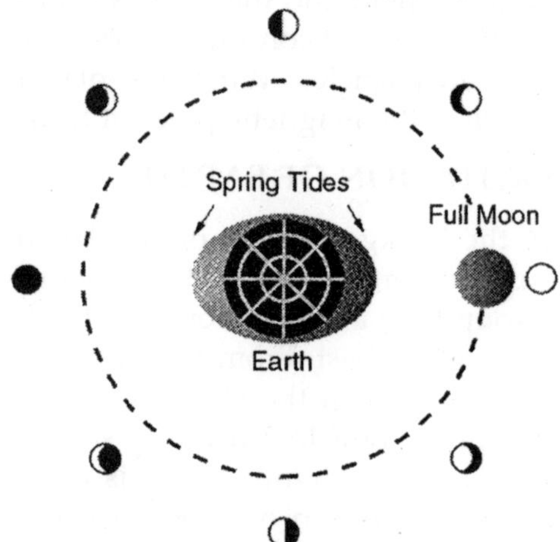

Fig. *Moon's Orbit Around Earth*

Because of the axial tilt of the Earth, the position of the Sun in the sky varies over the course of the year. For an observer at a northern latitude, when the northern pole is tilted toward the Sun the day lasts longer and the Sun climbs higher in the sky. This results in warmer average temperatures from the increase in solar radiation reaching the surface. When the northern pole is tilted away from the Sun, the reverse is true and the climate is generally cooler. Above the arctic circle, an extreme case is reached where there is no daylight at all for part of the year.

This variation in the climate (because of the direction of the Earth's axial tilt) results in the seasons. By astronomical convention, the four seasons are determined by the solstices—the point in the orbit of maximum axial tilt toward or away from the Sun—and the equinoxes, when the direction of the tilt and the direction to the Sun are perpendicular. Winter solstice occurs on about December 21, summer solstice is near June 21, spring equinox is around March 20 and autumnal equinox is about September 23. The axial tilt in the southern hemisphere is exactly the opposite of the direction in the northern hemisphere. Thus the seasonal effects in the south are reversed.The angle of the Earth's tilt is relatively stable over long periods of time.

However, the tilt does undergo a slight, irregular motion (known as nutation) with a main period of 18.6 years. The orientation (rather than the angle) of the Earth's axis also changes over time, precessing around in a complete circle over each 25,800 year cycle; this precession is the reason for the difference between a sidereal year and a tropical year. Both of these motions are caused by the varying attraction of the Sun and Moon on the Earth's equatorial bulge. From the perspective of the Earth, the poles also migrate a few meters across the surface. This polar motion has multiple, cyclical components, which collectively are termed quasiperiodic motion. In addition to an annual component to this motion, there is a 14-month cycle called the Chandler wobble. The rotational velocity of the Earth also varies in a phenomenon known as length of day variation.

The changing Earth-Sun distance results in an increase of about 6.9% in solar energy reaching the Earth at perihelion relative to aphelion. Since the southern hemisphere is tilted toward the Sun at about the same time that the Earth reaches the closest approach to the Sun, the southern hemisphere receives slightly more energy from the Sun than does the northern over the course of a year. However, this effect is much less significant than the total energy change due to the axial tilt and most of the excess energy is absorbed by the higher proportion of water in the southern hemisphere.

The Hill sphere (gravitational sphere of influence) of the Earth is about 1.5 Gm (or 1,500,000 kilometers) in radius. This is maximum distance at which the Earth's gravitational influence is stronger than the more distant Sun and planets. Objects must orbit the Earth within this radius, or they can become unbound by the gravitational perturbation of the Sun.Earth, along with the Solar System, is situated in the Milky Way galaxy, orbiting about 28,000 light years from the centre of the galaxy and about 20 light years above the galaxy's equatorial plane in the Orion spiral arm.

MOON

The Moon is a relatively large, terrestrial, planet-like satellite, with a diameter about one-quarter of the Earth's. It is the largest moon in the solar system relative to the size of its planet. (Charon is larger relative to the dwarf planet Pluto.) The natural satellites orbiting other planets are called "moons" after Earth's Moon.The gravitational attraction between the Earth and Moon causes tides on Earth.

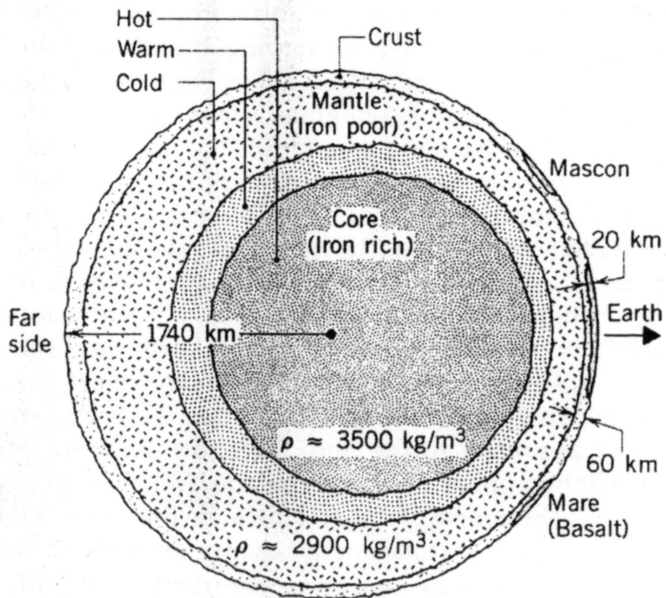

Fig. *Interior Structure of the Moon*

The same effect on the Moon has led to its tidal locking: its rotation period is the same as the time it takes to orbit the

Earth. As a result, it always presents the same face to the planet. As the Moon orbits Earth, different parts of its face are illuminated by the Sun, leading to the lunar phases.

The dark part of the face is separated from the light part by the solar terminator.Because of their tidal interaction, the Moon recedes from Earth at the rate of approximately 38 mm a year. Over millions of years, these tiny modifications—and the lengthening of Earth's day by about 23 μs a year—add up to significant changes. During the Devonian period, for example, (approximately 410 million years ago) there were 400 days in a year, with each day lasting 21.8 hours.

The Moon may have dramatically affected the development of life by moderating the planet's climate. Paleontological evidence and computer simulations show that Earth's axial tilt is stabilized by tidal interactions with the Moon. Some theorists believe that without this stabilization against the torques applied by the Sun and planets to the Earth's equatorial bulge, the rotational axis might be chaotically unstable, as it appears to be for Mars.

If Earth's axis of rotation were to approach the plane of the ecliptic, extremely severe weather could result from the resulting extreme seasonal differences. One pole would be pointed directly toward the Sun during *summer* and directly away during *winter*.

Planetary scientists who have studied the effect claim that this might kill all large animal and higher plant life. However, this is a controversial subject and further studies of Mars—which has a similar rotation period and axial tilt as Earth, but not its large Moon or liquid core—may settle the matter. Viewed from Earth, the Moon is just far enough away to have very nearly the same apparent-sized disk as the Sun.

The angular size (or solid angle) of these two bodies match because, although the Sun's diameter is about 400 times as large as the Moon's, it is also 400 times more distant. This allows total and annular eclipses to occur on Earth.The most widely accepted theory of the Moon's origin, the giant impact theory, states that it formed from the collision of a Mars-size protoplanet called Theia with the early Earth. This hypothesis explains (among other things) the Moon's relative lack of iron and volatile elements and the fact that its composition is nearly identical to that of the Earth's crust. Earth has at least two co-orbital asteroids, 3753 Cruithne and 2002 AA.

HABITABILITY

A planet that can sustain life is termed habitable, even if life did not originate there. The Earth provides the (currently understood) requisite conditions of liquid water, an environment where complex organic molecules can assemble and sufficient energy to sustain metabolism. The distance of the Earth from the Sun, as well as its orbital eccentricity, rate of rotation, axial tilt, geological history, sustaining atmosphere and protective magnetic field all contribute to the conditions necessary to originate and sustain life on this planet.

Biosphere

The planet's life forms are sometimes said to form a "Biosphere". This biosphere is generally believed to have begun evolving about 3.5 billion years ago. Earth is the only place in the universe where life is known to exist. Some scientists believe that Earth-like biospheres might be rare.The biosphere is divided into a number of biomes, inhabited by broadly similar plants and animals.

On land primarily latitude and height above the sea level separates biomes. Terrestrial biomes lying within the Arctic, Antarctic Circle or in high altitudes are relatively barren of plant and animal life, while the greatest latitudinal diversity of species is found at the Equator.

Natural Resources and Land Use

The Earth provides resources that are exploitable by humans for useful purposes. Some of these are non-renewable resources, such as mineral fuels, that are difficult to replenish on a short time scale.Large deposits of fossil fuels are obtained from the Earth's crust, consisting of coal, petroleum, natural gas and methane clathrate. These deposits are used by humans both for energy production and as feedstock for chemical production. Mineral ore bodies have also been formed in Earth's crust through a process of Ore genesis, resulting from actions of erosion and plate tectonics.

These bodies form concentrated sources for many metals and other useful elements. The Earth's biosphere produces many useful biological products for humans, including (but far from limited to) food, wood, pharmaceuticals, oxygen and the recycling of many organic wastes. The land-based ecosystem depends upon topsoil and fresh water and the oceanic ecosystem depends upon dissolved nutrients washed down from the land. Humans also live on the land by using building materials to construct shelters. In 1993, human use of land is approximately:

Natural and Environmental Hazards

Large areas are subject to extreme weather such as tropical cyclones, hurricanes, or typhoons that dominate life in those areas. Many places are subject to earthquakes, landslides, tsunamis, volcanic eruptions, tornadoes, sinkholes, blizzards, floods, droughts and other calamities and disasters. Many localized areas are subject to human-made pollution of the air and water, acid rain and toxic substances, loss of vegetation (overgrazing, deforestation, desertification), loss of wildlife, species extinction, soil degradation, soil depletion, erosion and introduction of invasive species.

A scientific consensus exists linking human activities to global warming due to industrial carbon dioxide emissions. This is predicted to produce changes such as the melting of glaciers and ice sheets, more extreme temperature ranges, significant changes in weather conditions and a global rise in average sea levels.

Human Geography

Earth has approximately 6,707,000,000 human inhabitants as of July 2008. Projections indicate that the world's human population will reach seven billion in 2013 and 9.2 billion in 2050. Most of the growth is expected to take place in developing nations. Human population density varies widely around the world, but a majority live in Asia. By 2020, 60% of the world's population is expected to be living in urban, rather than rural, areas. It is estimated that only one eighth of the surface of the Earth is suitable for humans to live on-three-quarters is covered by oceans and half of the land area is either desert (14%), high mountains (27%), or other less suitable terrain. The northernmost permanent settlement in the world is Alert, on Ellesmere Island in Nunavut, Canada. (82°282 N) The southernmost is the Amundsen-Scott South Pole Station, in Antarctica, almost exactly at the South Pole. (90°S)

Independent sovereign nations claim all of the planet's land surface, with the exception of some parts of Antarctica. As of 2007 there are 201 sovereign states, including the 192 United Nations member states. In addition, there are 59 dependent territories and a number of autonomous areas, territories under dispute and other entities. Historically, Earth has never had a sovereign government with authority over the entire globe, although a number of nation-states have striven for world domination and failed.

The United Nations is a worldwide intergovernmental organization that was created with the goal of intervening in the disputes between nations, thereby avoiding armed conflict. It is not, however, a world government. While the U.N. provides a mechanism for international law and, when the consensus of the membership permits, armed intervention, it serves primarily as a forum for international diplomacy. In total, about 400 people have been outside the Earth's atmosphere as of 2004 and, of these, twelve have walked on the Moon. Normally the only humans in space are those on the International Space Station. The station's crew of three people is usually replaced every six months.

CULTURAL VIEWPOINT

Etymology

The name *Earth* originates from the 8th century Anglo-Saxon word *erda*, which means ground or soil. In Old English the word became *eorthe*, then *erthe* in Middle English. Earth was first used as the name of the sphere of the Earth around 1400. It is the only planet whose name in English is not derived from Greco-Roman mythology. The standard astronomical symbol of the Earth consists of a cross circumscribed by a circle. This symbol is known as the wheel cross, sun cross, Odin's cross or Woden's cross.

Although it has been used in various cultures for different purposes, it came to represent the compass points, earth and the land. Another version of the symbol is a cross on top of a circle; a stylized globus cruciger that was also used as an early astronomical symbol for the planet Earth.

Religious Beliefs

Earth has often been personified as a deity, in particular a goddess. In many cultures the mother goddess, also called the Mother Earth, is also portrayed as a fertility deity. To the Aztec, Earth was called Tonantzin—"our mother"; to the Incas, Earth was called Pachamama—"mother earth". The Chinese Earth goddess Hou-T'u is similar to Gaia, the Greek goddess personifying the Earth. To Hindus it is called Bhuma Devi, the Goddess of Earth. In Norse mythology, the Earth goddess Jord was the mother of Thor and the daughter of Annar.

Ancient Egyptian mythology is different from that of other cultures because Earth is male, Geb and sky is female, Nut. Creation myths in many religions recall a story involving the creation of the Earth by a supernatural deity or deities. A variety of religious groups, often associated with fundamentalist branches of Protestantism or Islam, assert that their interpretations of the accounts of creation in sacred texts are literal truth and should be considered alongside or replace conventional scientific accounts of the formation of the Earth and the origin and development of life.

Such assertions are opposed by the scientific community and other religious groups. A prominent example is the creation-evolution controversy.

Exploration and Mapping

In the ancient past there were varying levels of belief in a flat Earth, with the Mesopotamian culture portraying the world as a flat disk afloat in an ocean. The spherical form of the Earth was suggested by early Greek philosophers; a belief espoused by Pythagoras. By the Middle Ages—as evidenced by thinkers such as Thomas Aquinas—European belief in a spherical Earth was widespread. Prior to circumnavigation of the planet and the introduction of space flight, belief in a spherical Earth was based on observations of the secondary effects of the Earth's shape and parallels drawn with the shape of other planets.

Cartography, the study and practice of map making and vicariously geography, have historically been the disciplines devoted to depicting the Earth. Surveying, the determination of locations and distances, to a lesser extent navigation, the determination of position and direction, have developed alongside cartography and geography, providing and suitably quantifying the requisite information.

Modern Perspective

The technological developments of the latter half of the 20th century are widely considered to have altered the public's perception of the Earth. Before space flight, the popular image of Earth was of a green world. Science fiction artist Frank R. Paul provided perhaps the first image of a cloudless *blue* planet (with sharply defined land masses) on the back cover of the July 1940 issue of *Amazing Stories,* a common depiction for several decades tnereafter.Earth was first photographed from space by Explorer 6 in 1959. Yuri Gagarin became the first human to view Earth from space in 1961.

The crew of the Apollo 8 was the first to view an Earth-rise from lunar orbit in 1968. In 1972 the crew of the Apollo 17 produced the famous "Blue Marble" photograph of the planet Earth from cislunar space. This became an iconic image of the planet as a marble of cloud-swirled blue ocean broken by green-brown continents. NASA archivist Mike Gentry has speculated that "The Blue Marble" is the most widely distributed image in human history. A photo taken of a distant Earth by *Voyager 1* in 1990 inspired Carl Sagan to describe the planet as a "Pale Blue Dot." Since the 1960s, Earth has also been described as a massive "Spaceship Earth," with a life support system that requires maintenance, or, in the Gaia hypothesis, as having a biosphere that forms one large organism. Over the past two centuries a growing environmental movement has emerged that is concerned about humankind's effects on the Earth.

The key issues of this socio-political movement are the conservation of natural resources, elimination of pollution and the usage of land. Environmentalists advocate sustainable management of resources and stewardship of the environment through changes in public policy and individual behaviour. Of particular concern is the large-scale exploitation of non-renewable resources. Changes sought by the environmental movements are sometimes in conflict with commercial interests due to the additional costs associated with managing the environmental impact of those interests.

Future

The future of the planet is closely tied to that of the Sun. As a result of the steady accumulation of helium ash at the Sun's core, the star's total luminosity will slowly increase. The luminosity of the Sun will increase by 10 percent over the next 1.1 Gyr (1.1 billion years) and by 40% over the next 3.5 Gyr. Climate models indicate that the rise in radiation reaching the Earth is likely to have dire consequences, including the possible loss of the planet's oceans.

The Earth's increasing surface temperature will accelerate the inorganic CO_2 cycle, reducing its concentration to the lethal levels for plants (10 ppm for C4 photosynthesis) in 900 million years. The lack of vegetation will result in the loss of oxygen in the atmosphere, so animal life will become extinct within several million more years.

But even if the Sun were eternal and stable, the continued internal cooling of the Earth would have resulted in a loss of much of its atmosphere and oceans (due to lower volcanism). After another billion years the surface water will have completely disappeared and the mean global temperature will reach 70°C. The Earth is expected to be effectively habitable for another 500 million years or so. The Sun, as part of its evolution, will expand to a red giant in about 5 Gyr. Models predict that the Sun will expand out to about 250 times its present size, roughly 1 AU (150,000,000 km). Earth's fate is less clear. As a red giant, the Sun will lose roughly 30% of its mass, so, without tidal effects, the Earth will be in an orbit 1.7 AU (250,000,000 km) from the Sun when the star reaches it maximum radius.

Therefore, the planet is expected to escape envelopment by the expanded Sun's sparse outer atmosphere, though most, if not all, existing life will be destroyed because of the Sun's increased luminosity. However, a more recent simulation indicates that Earth's orbit will decay due to tidal effects and drag, causing it to enter the red giant Sun's atmosphere and be destroyed. The Earth's shape is very close to an oblate spheroid—a rounded shape with a bulge around the equator—although the precise shape (the geoid) varies from this by up to 100 meters. The average diameter of the reference spheroid is about 12,742 km. More approximately the distance is 40,000 km/ð because the meter was originally defined as 1/10,000,000 of the distance from the equator to the north pole through Paris, France.

The rotation of the Earth creates the equatorial bulge so that the equatorial diameter is 43 km larger than the pole to pole diameter. The largest local deviations in the rocky surface of the Earth are Mount Everest (8,848 m above local sea level) and the Mariana Trench (10,911 m below local sea level). Hence compared to a perfect ellipsoid, the Earth has a tolerance of about one part in about 584, or 0.17%, which is less than the 0.22% tolerance allowed in billiard balls. Because of the bulge, the feature farthest from the centre of the Earth is actually Mount Chimborazo in Ecuador.

MECHANICAL THEORY

Newton, by applying his theory of gravitation, combined with the so-called centrifugal force, to the earth and assuming that an oblate ellipsoid of rotation is a form of equilibrium for a homogeneous fluid rotating with uniform angular velocity, obtained the ratio of the axes 229:230 and the law of variation of gravity on the surface.

A few years later Huygens published an investigation of the figure of the earth, supposing the attraction of every particle to be towards the centre of the earth, obtaining as a result that the proportion of the axes should be 578:579. In 1740 Colin Maclaurin, in his *De causa physica fluxus et refluxus maris,* demonstrated that the oblate ellipsoid of revolution is a figure which satisfies the conditions of equilibrium in the case of a revolving homogeneous fluid mass, whose particles attract one another according to the law of the inverse square of the distance; he gave the equation connecting the ellipticity with the proportion of the centrifugal force at the equator to gravity and determined the attraction on a particle situated anywhere on the surface of such a body.

Assuming that the earth is composed of concentric ellipsoidal strata having a common axis of rotation, each stratum homogeneous in itself, but the ellipticities and densities of the successive strata varying according to any law and that the superficial stratum has the same form as if it were fluid, he proved that g where *g*, *g'* are the amounts of gravity at the equator and at the pole respectively, *e* the ellipticity of the meridian (or " flattening ") and m the ratio of the centrifugal force at the equator to *g*.

He also proved that the increase of gravity in proceeding from the equator to the poles is as the square of the sine of the latitude. This, taken with the former theorem, gives the means of determining the earth's ellipticity from observation of the relative force of gravity at any two places. P. S. Laplace, who devoted much attention to the subject, remarks on Clairault's work that " the importance of all his results and the elegance with which they are presented place this work amongst the most beautiful of mathematical productions ".

The problem of the figure of the earth treated as a question of mechanics or hydrostatics is one of great difficulty and it would be quite impracticable but for the circumstance that the surface differs but little from a sphere. In order to express the forces at any point of the body arising from the attraction of its particles, the form of the surface is required, but this form is the very one which it is the object of the investigation to discover; hence the complexity of the subject and even with all the present resources of mathematicians only a partial and imperfect solution can be obtained.

We may here briefly indicate the line of reasoning by which some of the most important results may be obtained. If X, Y, Z be the components parallel to three rectangular axes of the forces acting on a particle of a fluid mass at the point *x, y,* z, then, *p* being the pressure there and *p* the density, $dp = p(Xdx+Ydy+Zdz)$ and for equilibrium the necessary conditions are, that $p(Xdx+ Ydy+Zdz)$ be a complete differential and at the free surface $Xdx+ Ydy+Zdz=o$.

This equation implies that the resultant of the forces is normal to the surface at every point and in a homogeneous fluid it is obviously the differential equation of all surfaces of equal pressure.

If the fluid be heterogeneous then it is to be remarked that for forces of attraction according to the ordinary law of gravitation, if X, Y, Z be the components of the attraction of a mass whose potential is V, then Xdx+Ydy+Zdz = d_V d_V *dx* dx *d–Ty – dy* + dz dz, which is a complete differential. And in the case of a fluid rotating with uniform velocity, in which the so–called centrifugal force enters as a force acting on each particle proportional to its distance from the axis of rotation, the corresponding part of $Xdx+Ydy+Zdz$ is obviously a complete differential.

Therefore for the forces with which we are now concerned $Xdx+Ydy+Zdz=dU$, where U is some function of x, *y,* z and it is necessary for equilibrium that $dp = pdU$ be a complete differential; that is, *p* must be a function of U or a function of *p* and so also *p* a function of U. So that dU=0 is the differential equation of surfaces of equal pressure and density.

We may now show that a homogeneous fluid mass in the form of an oblate ellipsoid of revolution having a uniform velocity of rotation can be in equilibrium. It may be proved that the attraction of the ellipsoid $x_2 + y_2 + z_2 (1 + E_2) = c_2$

$(I+E_2)$ upon a particle P of its mass at x, *y*, z has for components X = – Ax, Y = – Ay, Z = – Cz. Besides the attraction of the mass of the ellipsoid, the centrifugal force at P has for components d–xw 2, +*yw* 2, o; then the condition of fluid equilibrium is (A – w 2) xdx+ (A – *w* 2) *ydy+Czdz* =0, which by integration gives (A – w 2) (x 2 +y 2) +Cz 2 = constant.

This is the equation of an ellipsoid of rotation and therefore the equilibrium is possible. The equation coincides with that of the surface of the fluid mass if we make A – w2= C/(1 +€2), which gives *k2p '3'* E_2 In the case of the earth, which is nearly spherical, we obtain by expanding the expression for 3) 2 in powers of E_2, rejecting the higher powers and remarking that the ellipticity *e*= w 2/27rk_2 p =4E_2/15 = 8e/15. Now if *m* be the ratio of the centrifugal force to the intensity of gravity at the equator and *a* = *c (1 +e)*, then *m* =aw 2/31rk 2 pa,.. 0/27rk_2 p = m.

In the case of the earth it is a matter of observation that m =1/289, hence the ellipticity e=5m/4=1/231, so that the ratio of the axes on the supposition of a homogeneous fluid earth is 230:231, as stated by Newton.

Now, to come to the case of a heterogeneous fluid, we shall assume that its surfaces of equal density are spheroids, concentric and having a common axis of rotation and that the ellipticity of these surfaces varies from the centre to the outer surface, the density also varying.

In other words, the body is composed of homogeneous spheroidal shells of variable density and ellipticity. On this supposition we shall express the attraction of the mass upon a particle in its interior and then, taking into account the centrifugal force, form the equation expressing the condition of fluid equilibrium.

The attraction of the homogeneous spheroid $x_2 + y_2 + z_2$ (1 +2*e)* = c2 (1 + 2*e)*, where *e* is the ellipticity (of which the square is neglected), on an internal particle, whose co–ordinates are *x = f, y =o,* z = *h,* has for its x and z components X'= *pf(1 – fe),* Z'= – *431rk2ph(1+5e),* the Y component being of course zero. Hence we infer that the attraction of a shell whose inner surface has an ellipticity *e* and its outer surface an ellipticity *e+de,* the density being *p*, is expressed by dX′ = i rk_2 pfde, *dZ′* = –*S*– irk_2 phde. To apply this to our heterogeneous spheroid; if we put *c 1* for the semiaxis of that surface of equal density on which is situated the attracted point P and co for the semiaxis of the outer surface, the attraction of that portion of the body which is exterior to P, namely, of all the shells which enclose P, has for components.

Again the attraction of a homogeneous spheroid of density *p* on an *external* point *f, h* has the components X″ = – 3,rk_2 pfr_3 {c_3 (I +2*e*) – *Xec 5 l*, Z″ = – 3,rk_2 phr_3 {c_3 (I +2*e*) – *A′ec* 5}, where A= t (4h_2 – f_2)/r_4, v= 5(Zh_2 – 3f_2)/ r_4 and r_2=f +h_2. Now *e* being considered a function of *c*, we can at once express the attraction of a shell (density *p*) contained between the surface defined by *c+dc, e+de* and that defined by *c, e* upon an external point; the differentials with respect to *c*, viz. dX″ dZ″, must then be integrated with *p* under the integral sign as being a function of *c*. The integration will extend from *c* = *o* to *c* = *c l*.

Thus the components of the attraction of the heterogeneous spheroid upon a particle within its mass, whose co–ordinates are *f*, o, *h*, fcx=–3rk_2fy_3 d{c_3 (1+2e)} – $y_1$$O_1$ *pd(eG5)* scl o *pde* Z= – 3rkhd{c_3(1+2*e*)) –3*pd*(*e*G_5) +*fpde*]. We take into account the rotation of the earth by adding the centrifugal force fw 2 = F to X.

Now, the surface of constant density upon which the point *f*, o, *h* is situated gives (1 – 2e) *fdf* + *hdh* = o; and the condition of equilibrium is that (x+*F*)*df*+*Zdh* = o. Therefore, (X+F)h = Zf (1 – 2e), which, neglecting small quantities of the order e 2 and putting *w′*t_2 = 4,$r_2$$k_2$, gives 2ec1 36*Cc′* 56% 3ry_3 o pd{c (1–}–2e)) – 5r_5 *pd*(*e*c_5) *SJci* pde = 2 Here we must now put c for c1, c for r; and 1 +2e under the first integral sign may be replaced by unity, since small quantities of the second order are neglected. Two differentiations lead us to the following very important differential equation (Clairault): d_2e 2*p*c_2 *de*(2 *p*c_6) *d*c_2+ *fp*c_2 *dcdc* + fpcdc c_2e = o.

When *p* is expressed in terms of *c*, this equation can be integrated. We infer then that a rotating spheroid of very small ellipticity, composed of fluid homogeneous strata such as we have specified, will be in equilibrium; and when the law of the density is expressed, the law of the corresponding ellipticities will follow. If we put M for the mass of the spheroid, then M;andm= e 3 *t* 472 and putting *c=co* in the equation expressing the condition of equilibrium, we find M(2e – *m)* =3,*r.5 c* 2 o *pd(ec5).* Making these substitutions in the expressions for the forces at the surface and putting *r/c = I +e – e(h/c)2,* we get G sin = ak2 I (52–m–2e) *h2 S h.* Here G is gravity in the latitude *4 a*nd a the radius of the equator. Since G= Mksec *4,* = *(c/f) {'* + e + *(eh '/ c2)),* t) 13 –m+ jm – *e) sin e 4,* an expression which contains the theorems we have referred to as discovered by Clairault.

The theory of the figure of the earth as a rotating ellipsoid has been especially investigated by Laplace in his *Mecanique celeste.* The principal English works are: – Sir George Airy, *Mathematical Tracts,* a lucid treatment without the use of Laplace's coefficients; Archdeacon Pratt's *Attractions and Figure of the Earth;* and O'Brien's *Mathematical Tracts;* in the last two Laplace's coefficients are used.

In 1845 Sir G. G. Stokes proved that if the external form of the sea-imagined to percolate the land by canals - be a spheroid with small ellipticity, then the law of gravity is that which we have shown above; his proof required no assumption as to the ellipticity of the internal strata, or as to the past or present fluidity of the earth.

This investigation admits of being regarded conversely, viz. as determining the elliptical form of the earth from measurements of gravity; if G, the observed value of gravity in latitude 4), be expressed in the form G = g(sine 0), where g is the value at the equator and a coefficient. In this investigation, the square and higher powers of the ellipticity are neglected; the solution was completed by F. R. Helmert with regard to the square of the ellipticity, who showed that a term with sin 2 24) appeared.

For the coefficient of this term, the gravity measurements give a small but not sufficiently certain value; we therefore assume a value which agrees best with the hypothesis of the fluid state of the entire earth; this assumption is well supported, since even at a depth of only 50 km. the pressure of the superincumbent crust is so great that rocks become plastic and behave approximately as fluids and consequently the crust of the earth floats, to some extent, on the interior (even though this may not be fluid in the usual sense of the word). This is the geological theory of " Isostasis " it agrees with the results of measurements of gravity and was brought forward in the middle of the 19th century by J. H. Pratt, who deduced it from observations made in India.

The sin 2 24) term in the expression for G and the corresponding deviation of the meridian from an ellipse, have been analytically established by Sir G. H. Darwin and E. Wiechert; earlier and less complete investigations were made by Sir G. B. Airy and O. Callandreau.

In consequence of the sin 2 24) term, two parameters of the level surfaces in the interior of the earth are to be determined; for this purpose, Darwin develops two differential equations in the place of the one by Clairault. By assuming Roche's law for the variation of the density in the interior of the Earth, viz. *p = p i - k (c/c,) 2, k* being a coefficient, it is shown that in latitude 45°, the meridian is depressed about 34 metres from the ellipse and the coefficient of the term sin'4) cos' 0(=1 sin224)) is - (3 0000295.

According to Wiechert the earth is composed of a kernel and a shell, the kernel being composed of material, chiefly metallic iron, of density near 8.2 and the shell, about 900 miles thick, of silicates, &c., of density about 3.2. On this assumption the depression in latitude 45° is 24 metres and the coefficient of sin e 4) cos 2 4) is, in round numbers, - 0000280.1 To this additional term in the formula for G, there corresponds an extension of Clairault's formula for the calculation of the flattening from with terms of the higher orders; this was first accomplished by Helmert.

For a long time the assumption of an ellipsoid with three unequal axes has been held possible for the figure of the earth, in consequence of an important theorem due to K. G. Jacobi, who proved that for a homogeneous fluid in rotation a spheroid is not the only form of equilibrium; an ellipsoid rotating round its least axis may with certain proportions of the axes and a certain time of revolution be a form of equilibrium. It has been objected to the figure of three unequal axes that it does not satisfy, in the proportions of the axes, the conditions brought out in Jacobi's theorem.

Admitting this, it has to be noted, on the other hand, that Jacobi's theorem contemplates a homogeneous fluid and this is certainly far from the actual condition of our globe; indeed the irregular distribution of continents and oceans suggests the possibility of a sensible divergence from a perfect surface of revolution. We may, however, assume the ellipsoid with three unequal axes to be an interpolation form. More plausible forms are little adapted for computation.' Consequently we now generally take the ellipsoid of rotation as a basis, especially so because measurements of gravity have shown that the deviation from it is but trifling.

GRAVITATION-MEASUREMENTS

According to Clairault's theorem the ellipticity *e* of the mathematical surface of the earth is equal to the difference 2m–13, where *m* is the ratio of the centrifugal force at the equator to gravity at the equator and s is derived from the formula G= g(1 +0 sin 2 0). Since the beginning of the 19th century many efforts have been made to determine the constants of this formula and numerous expeditions undertaken to investigate the intensity of gravity in different latitudes.

If *m* be known, it is only necessary to determine/3 for the evaluation of *e;* consequently it is unnecessary to determine G absolutely, for the relative values of G at two known latitudes suffice. Such relative measurements are easier and more exact than absolute ones. In some cases the ordinary thread pendulum, *i.e.* a spherical bob suspended by a wire, has been employed; but more often a rigid metal rod, bearing a weight and a knifeedge on which it may oscillate, has been adopted. The main point is the constancy of the pendulum. From the formula for the time of oscillation of the mathematically ideal pendulum, t= 27r1/ 11G, *1* being the length, it follows that for two points GI/G2=4/ti.

Arc of Parallel in 52° Lat

Many measurements of degrees of longitudes along central parallels in Europe were projected and partly carried out as early as the first half of the 19th century; these, however, only became of importance after the introduction of the electric telegraph, through which calculations of astronomical longitudes obtained a much higher degree of accuracy.

Of the greatest moment is the measurement near the parallel of 52° lat., which extended from Valentia in Ireland to Orsk in the southern Ural mountains over 69° long. (about 6750 km.). F. G. W. Struve, who is to be regarded as the father of the Russo-Scandinavian latitude-degree measurements, was the originator of this investigation. Having made the requisite arrangements with a governments in 1857, he transferred them to his son Otto, who, in 1860, secured the co-operation of England.

A new connexion of England with the continent, via the English Channel, was accomplished in the next two years; whereas the requisite triangulations in Prussia and Russia extended over several decennaries.

The number of longitude stations originally arranged for was 15; and the determinations of the differences in longitude were uniformly commenced by the Russian observers E. I. von Forsch, J. I. Zylinski, B. Tiele and others; Feaghmain (Valentia) being reserved for English observers. With the concluding calculation of these operations, newer determinations of differences of longitudes were also applicable, by which the number of stations was brought up to 29. Since local deflections of the plumb-line were suspected at Feaghmain, the most westerly station, the longitude (with respect to Greenwich) of the trigonometrical station Killorglin at the head of Dingle Bay was shortly afterwards determined.

Deviation of the Astronomical Zenith

These deviations of the plumb-line correspond to an ellipsoid having an equatorial radius of nearly 6,378,000 metres (prob. error ° 70 metres) and an ellipticity 1/299.15. The latter was taken for granted; it is nearly equal to the result from the gravity-measurements; the value for a then gives/77 2 a minimum (nearly). The astronomical values of the geographical longitudes are assumed, according to the compensation of longitude differences carried out by van de Sande Bakhuyzen.

Recent determinations have introduced only small alterations in the deviations, a being slightly increased. In 1875 a number of European states concluded the metre convention and in 1877 an international weights-and-measures bureau was established at Breteuil. Until this time the metre was determined by the end-surfaces of a platinum rod subsequently, rods of platinum-iridium, of cross-section H, were constructed, having engraved lines at both ends of the bridge, which determine the distance of a metre. There were thirty of the rods which gave as accurately as possible the length of the metre; and these were distributed among the different states.

Careful comparisons with several standard toises showed that the metre was not exactly equal to 443,296 lines of the toise, but, in round numbers, 1/75000 of the length smaller. The metre according to the older relation is called the " legal metre," according to the new relation the "international metre." The values are: Legal metre =3.2808-6933 ft., International metre =3.2808257 ft. The values of a given above are in terms of the international metre; the earlier ones in legal metres, while the gravity formulae are in international metres.

Chapter 4

Tectonic Plate of the Earth

A tectonic plate (also called lithospheric plate) is a massive, irregularly shaped slab of solid rock, generally composed of both continental and oceanic lithosphere. Plate size can vary greatly, from a few hundred to thousands of kilometers across; the Pacific and Antarctic Plates are among the largest. Plate thickness also varies greatly, ranging from less than 15 km for young oceanic lithosphere to about 200 km or more for ancient continental lithosphere.How do these massive slabs of solid rock float despite their tremendous weight? The answer lies in the composition of the rocks.

Continental crust is composed of granitic rocks which are made up of relatively lightweight minerals such as quartz and feldspar. By contrast, oceanic crust is composed of basaltic rocks, which are much denser and heavier. The variations in plate thickness are nature's way of partly compensating for the imbalance in the weight and density of the two types of crust. Because continental rocks are much lighter, the crust under the continents is much thicker (as much as 100 km) whereas the crust under the oceans is generally only about 5 km thick. Like icebergs, only the tips of which are visible above water, continents have deep "Roots" to support their elevations.Most of the boundaries between individual plates cannot be seen, because they are hidden beneath the oceans.

Yet oceanic plate boundaries can be mapped accurately from outer space by measurements from GEOSAT satellites. Earthquake and volcanic activity is concentrated near these boundaries. Tectonic plates probably developed very early in the Earth's 4.6-billion-year history and they have been drifting about on the surface ever since-like slow-moving bumper cars repeatedly clustering together and then separating. Like many features on the Earth's surface, plates change over time. Those composed partly or entirely of oceanic lithosphere can sink under another plate, usually a lighter, mostly continental plate and eventually disappear completely. This process is happening now off the coast of Oregon and Washington.

The small Juan de Fuca Plate, a remnant of the formerly much larger oceanic Farallon Plate, will someday be entirely consumed as it continues to sink beneath the North American Plate.One of the first steps in performing a scientific investigation is to understand what is already known. The investigation could be posed in a number of ways, but I have settled on asking students to find evidence for the theory of plate tectonics.

This approach has worked well in a class of general education majors. With more class time it might be possible to lead students to the discovery that the earth's surface must actually be moving around. The seafloor age data requires it, as do the existence of earthquakes in well defined patterns. Nevertheless, the search for this evidence provides rich investigative opportunities. Furthermore, the real earth is uncooperative enough to provide many complications for deeper investigation.

The first step is to familiarize students with the theory as described in the simplistic cartoons that depict spreading centers, transform faults and subduction zones. A clear understanding of the basic theory is essential if students are to successfully find data in support of it.

The "Plate Tectonics Lecture" does an excellent job of animating these cartoons to aid in visualization. In some instances, clay models might also prove useful. The theory of plate tectonics states that the earth's surface has a number of rigid, brittle segments that move around. The outlines of these plates are easy to see on a world map of earthquakes. Earthquakes occur where brittle pieces of the earth are slipping past one another. When you break a dish, a "crack" occurs and there is a separation of two or more pieces. The noise is analogous to the earthquake and the crack is where relative motion between two pieces takes place. The plate is called of "Lithosphere". Within the lithosphere is the crust and upper portion of the mantle. There are three kinds of boundaries between plates.

TRANSFORM BOUNDARIES

Transform boundaries are where the plates slide past one another. The San Andreas Fault is a transform boundary between the Pacific and North American Plates. Transform boundaries are characterized by many earthquakes at relatively shallow depths (less that 50 km deep). The location between the two spreading centers which is the transform fault. This is the only region where there is relative sliding motion. Each of the offset spreading centers is at a higher elevation, so when the offset spreading centers are joined by a transform fault, there are elevation differences at the boundary..

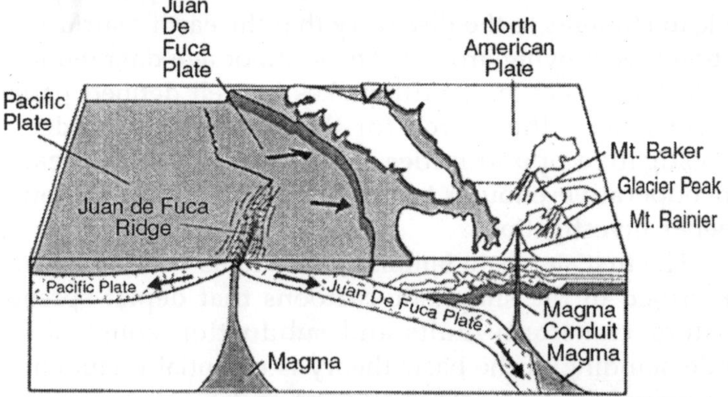

Fig. *Offset Spreading Centers, Joined by a Transform Fault*

Boundary segments marked by A and C do not have relative motion across the "Fault" (dotted line segment). Yet the original elevation differences caused by the offset are still present. These segments are called "Fracture Zones" and can be observed in the Map elevation data in the eastern Pacific Ocean, where they may persist for thousands of kilometers. In map view, the representation of offset spreading centers, joined by a transform fault. The transform fault is indicated by the B and A and C are the fracture zones, which carry the scar of the old transform fault along as the plate spreads away from the spreading centre.

Convergent Boundaries

A boundary where plates are coming together, or "Converging". There are a number of important variations to converging boundaries. Collision between an oceanic plate and a continent. The collision between India and Eurasia is an example of a plate convergent boundary where two continents collide.

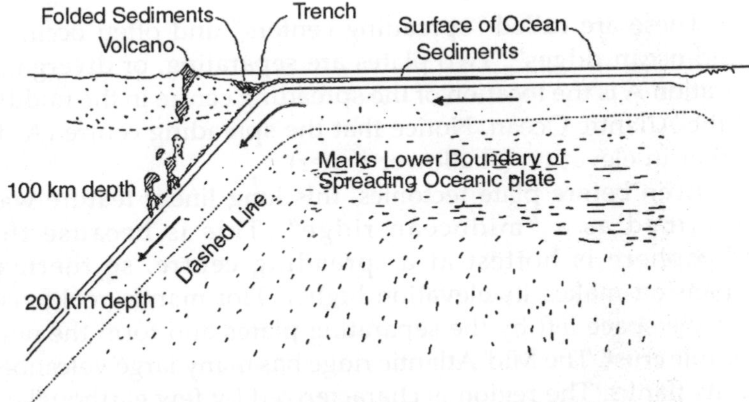

Fig. *Cross Section of a Convergent Margin between two Oceanic Plates*

The continents are both too buoyant to sink, so their collision causes a thickening of the lithosphere, which is manifested in the Himalayan Mountains and Tibetan Plateau. When an oceanic plate collides and subducts beneath another oceanic plate, an island arc forms. The islands are caused by the volcanoes created by the subduction process. Behind the island arc, a basin is formed. Often, a small spreading centre occurs, caused by the pull of the subducting slab.

This is called "back-arc spreading".There are many "real world" examples of the simple structures described above. You will be able to find these using the map software. However, there are also many complications, which are beyond the scope of this treatment. One interesting complication occurs at the northern end of the San Andreas Fault. The western margin of North America is a combination of subduction zone and transform fault. North of Mendocino, beneath Northern California, Oregon, Washington and southern Canada, there is a very slow subduction zone. This subduction zone is responsible for the active volcanoes: Mt. St. Helens, Mt. Ranier, Mt. Hood, Mt. Adams and others. South of Mendocino is the San Andreas Fault, which veers offshore to the Gorda Rise, which is a spreading centre.

DIVERGENT BOUNDARIES

These are called "spreading centers" and often occur at "mid-ocean ridges". Two plates are separating, or diverging. Location A is the location of the spreading centre in the middle of the Atlantic Ocean. Notice that the spreading centre (A) is higher in elevation (shallower water).

Even before plate tectonics, this long linear feature was described as a "midocean ridge". This is because the lithosphere is hottest at a spreading centre, so thermal expansion makes its elevation higher. Hot mantle rocks rise into the space left by the separating plates and form the new oceanic crust. The Mid-Atlantic ridge has many large volcanoes on its flanks. The region is characterized by few earthquakes, most of which are on short "transform" segments.

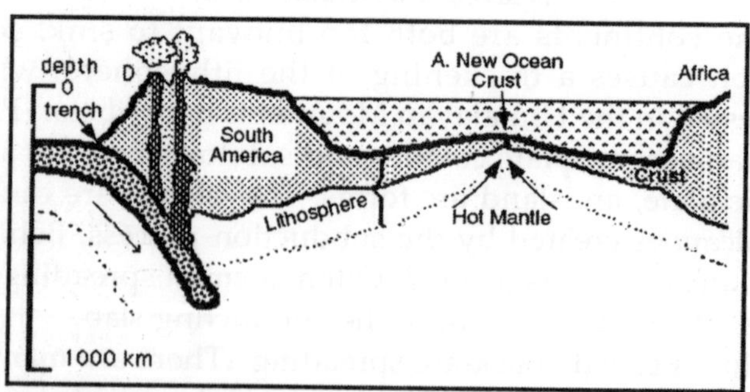

Fig. *Cross section diagram of the Eastern Pacific and Atlantic Ocean basins.*

In geologic terms, a *plate* is a large, rigid slab of solid rock. The word *tectonics* comes from the Greek root "to build."

Putting these two words together, we get the term *plate tectonics,* which refers to how the Earth's surface is built of plates. The *theory of plate tectonics* states that the Earth's outermost layer is fragmented into a dozen or more large and small plates that are moving relative to one another as they ride atop hotter, more mobile material.

Before the advent of plate tectonics, however, some people already believed that the present-day continents were the fragmented pieces of preexisting larger landmasses ("supercontinents"). The diagrams below show the break-up of the supercontinent *Pangaea* (meaning "all lands" in Greek), which figured prominently in the *theory of continental drift* the forerunner to the theory of plate tectonics.

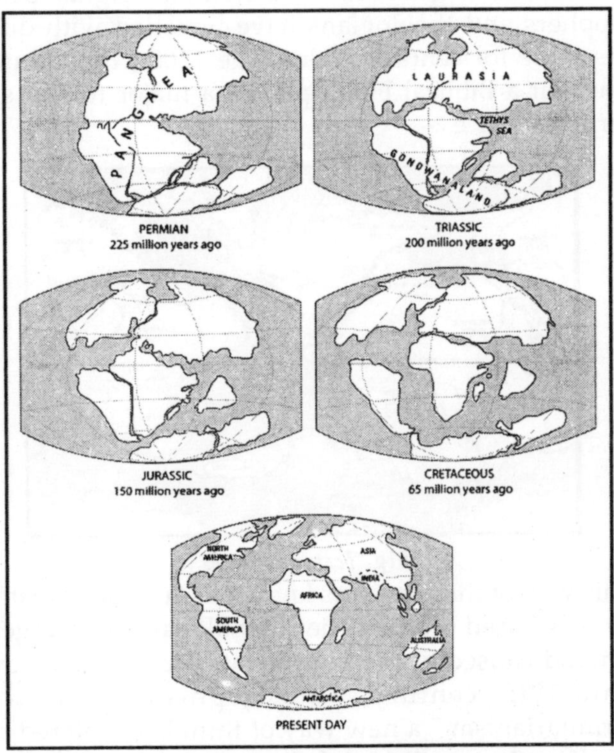

Fig. *The Supercontinent Pangaea Began to Break up about 225-200 Million Years Ago.*

Plate tectonics is a relatively new scientific concept, introduced some 30 years ago, but it has revolutionized our understanding of the dynamic planet upon which we live. The theory has unified the study of the Earth by drawing together many branches of the earth sciences, from *paleontology* (the study of fossils) to *seismology*.

It has provided explanations to questions that scientists had speculated upon for centuries — such as why earthquakes and volcanic eruptions occur in very specific areas around the world and how and why great mountain ranges like the Alps and Himalayas formed.

Why is the Earth so restless? What causes the ground to shake violently, volcanoes to erupt with explosive force and great mountain ranges to rise to incredible heights? Scientists, philosophers and theologians have wrestled with questions such as these for centuries. Until the 1700s, most Europeans thought that a Biblical Flood played a major role in shaping the Earth's surface.

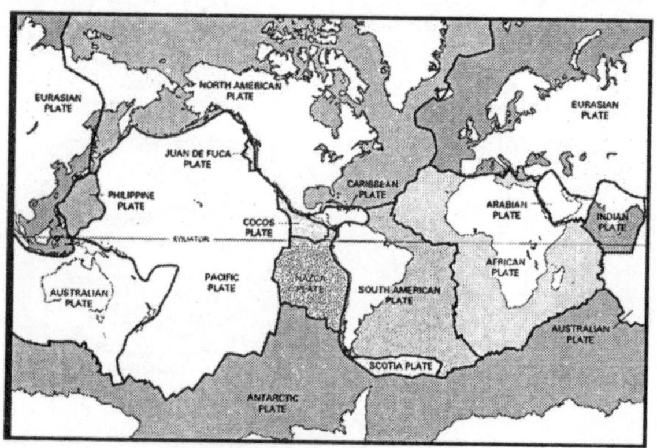

Fig. *Tectonic Plates*

This way of thinking was known as "catastrophism," and *geology* was based on the belief that all earthly changes were sudden and caused by a series of catastrophes. However, by the mid-19th century, catastrophism gave way to "uniformitarianism," a new way of thinking centered around the "Uniformitarian Principle" proposed in 1785 by James Hutton, a Scottish geologist. This principle is commonly stated as follows: *The present is the key to the past.* Those holding this viewpoint assume that the geologic forces and processes—gradual as well as catastrophic — acting on the Earth today are the same as those that have acted in the geologic past.

The belief that continents have not always been fixed in their present positions was suspected long before the 20th century; this notion was first suggested as early as 1596 by the Dutch map maker Abraham Ortelius in his work *Thesaurus Geographicus*. Ortelius suggested that the Americas were "torn away from Europe and Africa... by earthquakes and floods" and went on to say: "The vestiges of the rupture reveal themselves, if someone brings forward a map of the world and considers carefully the coasts of the three [continents]." Ortelius' idea surfaced again in the 19th century.

However, it was not until 1912 that the idea of moving continents was seriously considered as a full-blown scientific theory — called *Continental Drift* — introduced in two articles published by a 32-year-old German meteorologist named Alfred Lothar Wegener. He contended that, around 200 million years ago, the supercontinent Pangaea began to split apart. Alexander Du Toit, Professor of Geology at Johannesburg University and one of Wegener's staunchest supporters, proposed that Pangaea first broke into two large continental landmasses, *Laurasia* in the northern hemisphere and *Gondwanaland* in the southern hemisphere. Laurasia and Gondwanaland then continued to break apart into the various smaller continents that exist today.

Wegener's theory was based in part on what appeared to him to be the remarkable fit of the South American and African continents, first noted by Abraham Ortelius three centuries earlier. Wegener was also intrigued by the occurrences of unusual geologic structures and of plant and animal fossils found on the matching coastlines of South America and Africa, which are now widely separated by the Atlantic Ocean.

He reasoned that it was physically impossible for most of these organisms to have swum or have been transported across the vast oceans. To him, the presence of identical fossil species along the coastal parts of Africa and South America was the most compelling evidence that the two continents were once joined. In Wegener's mind, the drifting of continents after the break-up of Pangaea explained not only the matching fossil occurrences but also the evidence of dramatic climate changes on some continents. For example, the discovery of fossils of tropical plants (in the form of coal deposits) in Antarctica led to the conclusion that this frozen land previously must have been situated closer to the equator, in a more temperate climate where lush, swampy vegetation could grow. Other mismatches of geology and climate included distinctive fossil ferns discovered in now-polar regions and the occurrence of glacial deposits in present-day arid Africa, such as the Vaal River valley of South Africa.

The *theory of continental drift* would become the spark that ignited a new way of viewing the Earth. But at the time Wegener introduced his theory, the scientific community firmly believed the continents and oceans to be permanent features on the Earth's surface. Not surprisingly, his proposal was not well received, even though it seemed to agree with the scientific information available at the time.

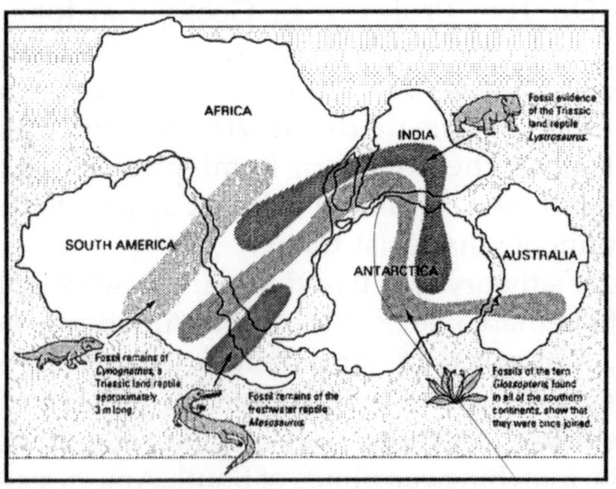

Fig. *Rejoined Continents*

A fatal weakness in Wegener's theory was that it could not satisfactorily answer the most fundamental question raised by his critics: What kind of forces could be strong enough to move such large masses of solid rock over such great distances? Wegener suggested that the continents simply plowed through the ocean floor, but Harold Jeffreys, a noted English geophysicist, argued correctly that it was physically impossible for a large mass of solid rock to plow through the ocean floor without breaking up. Undaunted by rejection, Wegener devoted the rest of his life to doggedly pursuing additional evidence to defend his theory. He froze to death in 1930 during an expedition crossing the Greenland ice cap, but the controversy he spawned raged on. However, after his death, new evidence from ocean floor exploration and other studies rekindled interest in Wegener's theory, ultimately leading to the development of the *theory of plate tectonics.* Plate tectonics has proven to be as important to the earth sciences as the discovery of the structure of the atom was to physics and chemistry and the theory of evolution was to the life sciences.

Even though the theory of plate tectonics is now widely accepted by the scientific community, aspects of the theory are still being debated today. Ironically, one of the chief outstanding questions is the one Wegener failed to resolve: What is the nature of the forces propelling the plates? Scientists also debate how plate tectonics may have operated earlier in the Earth's history and whether similar processes operate, or have ever operated, on other planets in our solar system.

INSIDE SURFACE OF THE EARTH

The size of the Earth — about 12,750 kilometers (km) in diameter-was known by the ancient Greeks, but it was not until the turn of the 20th century that scientists determined that our planet is made up of three main layers: *crust, mantle* and *core.* This layered structure can be compared to that of a boiled egg. The *crust,* the outermost layer, is rigid and very thin compared with the other two. Beneath the oceans, the crust varies little in thickness, generally extending only to about 5 km.

The thickness of the crust beneath continents is much more variable but averages about 30 km; under large mountain ranges, such as the Alps or the Sierra Nevada, however, the base of the crust can be as deep as 100 km. Like the shell of an egg, the Earth's crust is brittle and can break.

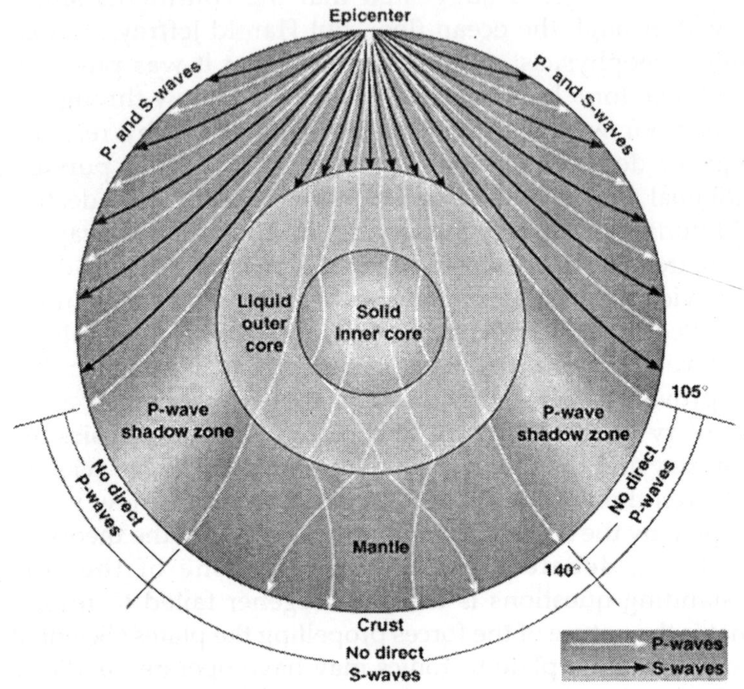

Fig. *Internal Structure of the Earth*

Below the crust is the *mantle,* a dense, hot layer of semi-solid rock approximately 2,900 km thick. The mantle, which contains more iron, magnesium and calcium than the crust, is hotter and denser because temperature and pressure inside the Earth increase with depth.

As a comparison, the mantle might be thought of as the white of a boiled egg. At the centre of the Earth lies the *core,* which is nearly twice as dense as the mantle because its composition is metallic (iron-nickel alloy) rather than stony.

Unlike the yolk of an egg, however, the Earth's core is actually made up of two distinct parts: a 2,200 km-thick *liquid* outer core and a 1,250 km-thick *solid* inner core. As the Earth rotates, the liquid outer core spins, creating the Earth's magnetic field.

Not surprisingly, the Earth's internal structure influences plate tectonics. The upper part of the mantle is cooler and more rigid than the deep mantle; in many ways, it behaves like the overlying crust. Together they form a rigid layer of rock called the *lithosphere* (from *lithos,* Greek for stone).

The lithosphere tends to be thinnest under the oceans and in volcanically active continental areas, such as the Western United States. Averaging at least 80 km in thickness over much of the Earth, the lithosphere has been broken up into the moving plates that contain the world's continents and oceans.

Scientists believe that below the lithosphere is a relatively narrow, mobile zone in the mantle called the *asthenosphere* (from *asthenes,* Greek for weak). This zone is composed of hot, semi-solid material, which can soften and flow after being subjected to high temperature and pressure over geologic time. The rigid lithosphere is thought to "float" or move about on the slowly flowing asthenosphere.

DEVELOPING THE THEORY

Continental drift was hotly debated off and on for decades following Wegener's death before it was largely dismissed as being eccentric, preposterous and improbable. However, beginning in the 1950s, a wealth of new evidence emerged to revive the debate about Wegener's provocative ideas and their implications. In particular, four major scientific developments spurred the formulation of the plate-tectonics theory:

- Demonstration of the ruggedness and youth of the ocean floor;
- Confirmation of repeated reversals of the Earth magnetic field in the geologic past;
- Emergence of the seafloor-spreading hypothesis and associated recycling of oceanic crust; and
- Precise documentation that the world's earthquake and volcanic activity is concentrated along oceanic trenches and submarine mountain ranges.

Ocean Floor Mapping

About two thirds of the Earth's surface lies beneath the oceans. Before the 19th century, the depths of the open ocean were largely a matter of speculation and most people thought that the ocean floor was relatively flat and featureless. However, as early as the 16th century, a few intrepid navigators, by taking soundings with hand lines, found that the open ocean can differ considerably in depth, showing that the ocean floor was not as flat as generally believed. Oceanic exploration during the next centuries dramatically improved our knowledge of the ocean floor. We now know that most of the geologic processes occurring on land are linked, directly or indirectly, to the dynamics of the ocean floor.

"Modern" measurements of ocean depths greatly increased in the 19th century, when deep-sea line soundings (*bathymetric* surveys) were routinely made in the Atlantic and Caribbean. In 1855, a bathymetric chart published by U.S. Navy Lieutenant Matthew Maury revealed the first evidence of underwater mountains in the central Atlantic (which he called "Middle Ground"). This was later confirmed by survey ships laying the trans-Atlantic telegraph cable. Our picture of the ocean floor greatly sharpened after World War I, when echo-sounding devices—primitive sonar systems—began to measure ocean depth by recording the time it took for a sound signal (commonly an electrically generated "ping") from the ship to bounce off the ocean floor and return.

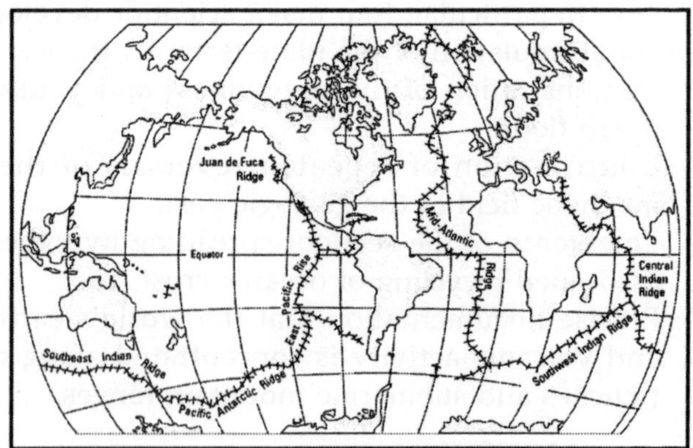

Fig. *The Mid-ocean Ridge (shown in Red) Winds its Way between the Continents much like the Seam on a Baseball*

Time graphs of the returned signals revealed that the ocean floor was much more rugged than previously thought.

Such echo-sounding measurements clearly demonstrated the continuity and roughness of the submarine mountain chain in the central Atlantic (later called the *Mid-Atlantic Ridge*) suggested by the earlier bathymetric measurements.

In 1947, seismologists on the U.S. research ship *Atlantis* found that the sediment layer on the floor of the Atlantic was much thinner than originally thought. Scientists had previously believed that the oceans have existed for at least 4 billion years, so therefore the sediment layer should have been very thick. Why then was there so little accumulation of sedimentary rock and debris on the ocean floor?

The answer to this question, which came after further exploration, would prove to be vital to advancing the concept of plate tectonics. In the 1950s, oceanic exploration greatly expanded. Data gathered by oceanographic surveys conducted by many nations led to the discovery that a great mountain range on the ocean floor virtually encircled the Earth.

Called the *global mid-ocean ridge,* this immense submarine mountain chain more than 50,000 kilometers (km) long and, in places, more than 800 km across—zig-zags between the continents, winding its way around the globe like the seam on a baseball. Rising an average of about 4,500 meters(m) above the sea floor, the mid-ocean ridge overshadows all the mountains in the United States except for Mount McKinley (Denali) in Alaska (6,194 m). Though hidden beneath the ocean surface, the global mid-ocean ridge system is the most prominent topographic feature on the surface of our planet.

Magnetic Striping and Polar Reversals

Beginning in the 1950s, scientists, using magnetic instruments *(magnetometers)* adapted from airborne devices developed during World War II to detect submarines, began recognizing odd magnetic variations across the ocean floor. This finding, though unexpected, was not entirely surprising because it was known that *basalt* the iron-rich, volcanic rock making up the ocean floor contains a strongly magnetic mineral *(magnetite)* and can locally distort compass readings. This distortion was recognized by Icelandic mariners as early as the late 18th century. More important, because the presence of magnetite gives the basalt measurable magnetic properties, these newly discovered magnetic variations provided another means to study the deep ocean floor.

Early in the 20th century, *paleomagnetists* (those who study the Earth's ancient magnetic field) such as Bernard Brunhes in France (in 1906) and Motonari Matuyama in Japan (in the 1920s) — recognized that rocks generally belong to two groups according to their magnetic properties. One group has so-called *normal polarity,* characterized by the magnetic minerals in the rock having the same polarity as that of the Earth's present magnetic field. This would result in the north end of the rock's "compass needle" pointing toward magnetic north.

The other group, however, has *reversed polarity,* indicated by a polarity alignment opposite to that of the Earth's present magnetic field. In this case, the north end of the rock's compass needle would point south. How could this be? This answer lies in the magnetite in volcanic rock. Grains of magnetite — behaving like little magnets—can align themselves with the orientation of the Earth's magnetic field. When *magma* (molten rock containing minerals and gases) cools to form solid volcanic rock, the alignment of the magnetite grains is "locked in," recording the Earth's magnetic orientation or polarity (normal or reversed) at the time of cooling.

As more and more of the seafloor was mapped during the 1950s, the magnetic variations turned out not to be random or isolated occurrences, but instead revealed recognizable patterns. When these magnetic patterns were mapped over a wide region, the ocean floor showed a zebra-like pattern. Alternating stripes of magnetically different rock were laid out in rows on either side of the mid-ocean ridge: one stripe with normal polarity and the adjoining stripe with reversed polarity. The overall pattern, defined by these alternating bands of normally and reversely polarized rock, became known as magnetic striping.

RECYCLING OF OCEANIC CRUST

The discovery of magnetic striping naturally prompted

more questions: How does the magnetic striping pattern form? And why are the stripes symmetrical around the crests of the mid-ocean ridges? These questions could not be answered without also knowing the significance of these ridges. In 1961, scientists began to theorize that mid-ocean ridges mark structurally weak zones where the ocean floor was being ripped in two lengthwise along the ridge crest.

New magma from deep within the Earth rises easily through these weak zones and eventually erupts along the crest of the ridges to create new oceanic crust. This process, later called *seafloor spreading,* operating over many millions of years has built the 50,000 km-long system of mid-ocean ridges. This hypothesis was supported by several lines of evidence:

- At or near the crest of the ridge, the rocks are very young and they become progressively older away from the ridge crest
- The youngest rocks at the ridge crest always have present-day (normal) polarity; and
- Stripes of rock parallel to the ridge crest alternated in magnetic polarity (normal-reversed-normal, etc.), suggesting that the Earth's magnetic field has flip-flopped many times. By explaining both the zebralike magnetic striping and the construction of the mid-ocean ridge system, the seafloor spreading hypothesis quickly gained converts and represented another major advance in the development of the plate-tectonics theory. Furthermore, the oceanic crust now came to be appreciated as a natural "tape recording" of the history of the reversals in the Earth's magnetic field.

Additional evidence of seafloor spreading came from an unexpected source: petroleum exploration. In the years following World War II, continental oil reserves were being depleted rapidly and the search for offshore oil was on. To conduct offshore exploration, oil companies built ships equipped with a special drilling rig and the capacity to carry many kilometers of drill pipe.

This basic idea later was adapted in constructing a research vessel, named the *Glomar Challenger,* designed specifically for marine geology studies, including the collection of drill-core samples from the deep ocean floor. In 1968, the vessel embarked on a year-long scientific expedition, criss-crossing the Mid-Atlantic Ridge between South America and Africa and drilling core samples at specific locations. When the ages of the samples were determined by paleontologic and isotopic dating studies, they provided the clinching evidence that proved the seafloor spreading hypothesis.

A profound consequence of seafloor spreading is that new crust was and is now, being continually created along the oceanic ridges. This idea found great favour with some scientists who claimed that the shifting of the continents can be simply explained by a large increase in size of the Earth since its formation. However, this so-called "expanding Earth" hypothesis was unsatisfactory because its supporters could offer no convincing geologic mechanism to produce such a huge, sudden expansion. Most geologists believe that the Earth has changed little, if at all, in size since its formation 4.6 billion years ago, raising a key question: how can new crust be continuously added along the oceanic ridges without increasing the size of the Earth?

This question particularly intrigued Harry H. Hess, a Princeton University geologist and a Naval Reserve Rear Admiral and Robert S. Dietz, a scientist with the U.S. Coast and Geodetic Survey who first coined the term *seafloor spreading*. Dietz and Hess were among the small handful who really understood the broad implications of sea floor spreading. If the Earth's crust was expanding along the oceanic ridges, Hess reasoned, it must be shrinking elsewhere. He suggested that new oceanic crust continuously spread away from the ridges in a conveyor belt-like motion.

Many millions of years later, the oceanic crust eventually descends into the oceanic *trenches* — very deep, narrow canyons along the rim of the Pacific Ocean basin. According to Hess, the Atlantic Ocean was expanding while the Pacific Ocean was shrinking. As old oceanic crust was consumed in the trenches, new magma rose and erupted along the spreading ridges to form new crust. In effect, the ocean basins were perpetually being "recycled," with the creation of new crust and the destruction of old oceanic lithosphere occurring simultaneously. Thus, Hess' ideas neatly explained why the Earth does not get bigger with sea floor spreading, why there is so little sediment accumulation on the ocean floor and why oceanic rocks are much younger than continental rocks.

CONCENTRATION OF EARTHQUAKES

During the 20th century, improvements in seismic instrumentation and greater use of earthquake-recording instruments *(seismographs)* worldwide enabled scientists to learn that earthquakes tend to be concentrated in certain areas, most notably along the oceanic trenches and spreading ridges. By the late 1920s, seismologists were beginning to identify several prominent earthquake zones parallel to the trenches that typically were inclined 40-60° from the horizontal and extended several hundred kilometers into the Earth.

These zones later became known as *Wadati-Benioff zones,* or simply *Benioff zones,* in honour of the seismologists who first recognized them, Kiyoo Wadati of Japan and Hugo Benioff of the United States. The study of global seismicity greatly advanced in the 1960s with the establishment of the Worldwide Standardized Seismograph Network monitor the compliance of the 1963 treaty banning above-ground testing of nuclear weapons. The much-improved data from the WWSSN instruments allowed seismologists to map precisely the zones of earthquake concentration worldwide.

But what was the significance of the connection between earthquakes and oceanic trenches and ridges? The recognition of such a connection helped confirm the seafloor-spreading hypothesis by pin-pointing the zones where Hess had predicted oceanic crust is being generated (along the ridges) and the zones where oceanic lithosphere sinks back into the mantle (beneath the trenches).

CONVERGENT BOUNDARIES

The size of the Earth has not changed significantly during the past 600 million years and very likely not since shortly after its formation 4.6 billion years ago. The Earth's unchanging size implies that the crust must be destroyed at about the same rate as it is being created, as Harry Hess surmised. Such destruction (recycling) of crust takes place along convergent boundaries where plates are moving toward each other and sometimes one plate sinks (is *subducted*) under another.

The location where sinking of a plate occurs is called a *subduction zone.* The type of convergence called by some a very slow "collision" that takes place between plates depends on the kind of lithosphere involved. Convergence can occur between an oceanic and a largely continental plate, or between two largely oceanic plates, or between two largely continental plates.

Oceanic-continental Convergence

If by magic we could pull a plug and drain the Pacific Ocean, we would see a most amazing sight — a number of long narrow, curving *trenches* thousands of kilometers long and 8 to 10 km deep cutting into the ocean floor. Trenches are the deepest parts of the ocean floor and are created by subduction.

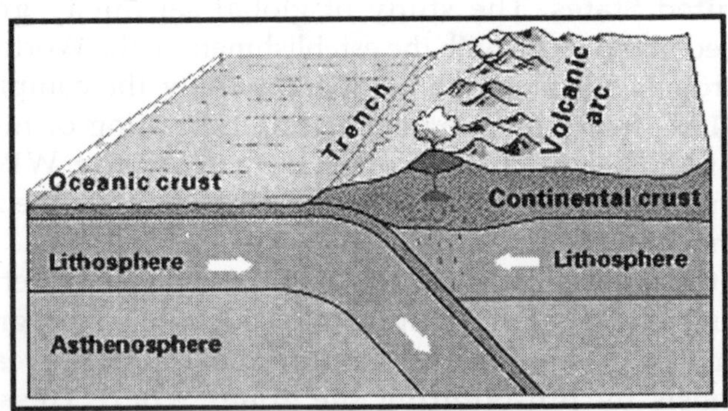

Fig. *Oceanic-continental Covergence*

Off the coast of South America along the Peru-Chile trench, the oceanic Nazca Plate is pushing into and being subducted under the continental part of the South American Plate. In turn, the overriding South American Plate is being lifted up, creating the towering Andes mountains, the backbone of the continent. Strong, destructive earthquakes and the rapid uplift of mountain ranges are common in this region. Even though the Nazca Plate as a whole is sinking smoothly and continuously into the trench, the deepest part of the subducting plate breaks into smaller pieces that become locked in place for long periods of time before suddenly moving to generate large earthquakes. Such earthquakes are often accompanied by uplift of the land by as much as a few meters.

On 9 June 1994, a magnitude-8.3 earthquake struck about 320 km northeast of La Paz, Bolivia, at a depth of 636 km. This earthquake, within the subduction zone between the Nazca Plate and the South American Plate, was one of deepest and largest subduction earthquakes recorded in South America. Fortunately, even though this powerful earthquake was felt as far away as Minnesota and Toronto, Canada, it caused no major damage because of its great depth.

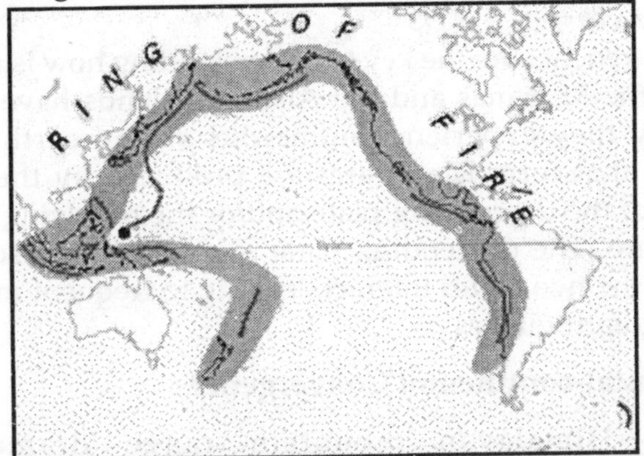

Fig. *Ring of Fire*

Oceanic-continental convergence also sustains many of the Earth's active volcanoes, such as those in the Andes and the Cascade Range in the Pacific Northwest. The eruptive activity is clearly associated with subduction, but scientists vigorously debate the possible sources of magma: Is magma generated by the partial melting of the subducted oceanic slab, or the overlying continental lithosphere, or both?

Oceanic-oceanic Convergence

As with oceanic-continental convergence, when two

oceanic plates converge, one is usually subducted under the other and in the process a trench is formed. The Marianas Trench for example, marks where the fast-moving Pacific Plate converges against the slower moving Philippine Plate. The Challenger Deep, at the southern end of the Marianas Trench, plunges deeper into the Earth's interior (nearly 11,000 m) than Mount Everest, the world's tallest mountain, rises above sea level (about 8,854 m).

Subduction processes in oceanic-oceanic plate convergence also result in the formation of volcanoes. Over millions of years, the erupted lava and volcanic debris pile up on the ocean floor until a submarine volcano rises above sea level to form an island volcano. Such volcanoes are typically strung out in chains called *island arcs*. As the name implies, volcanic island arcs, which closely parallel the trenches, are generally curved.

The trenches are the key to understanding how island arcs such as the Marianas and the Aleutian Islands have formed and why they experience numerous strong earthquakes. Magmas that form island arcs are produced by the partial melting of the descending plate and/or the overlying oceanic lithosphere. The descending plate also provides a source of stress as the two plates interact, leading to frequent moderate to strong earthquakes.

Continental-continental Convergence

The Himalayan mountain range dramatically demonstrates one of the most visible and spectacular consequences of plate tectonics. When two continents meet head-on, neither is subducted because the continental rocks are relatively light and, like two colliding icebergs, resist downward motion. Instead, the crust tends to buckle and be pushed upward or sideways. The collision of India into Asia 50 million years ago caused the Eurasian Plate to crumple up and override the Indian Plate.

After the collision, the slow continuous convergence of the two plates over millions of years pushed up the Himalayas and the Tibetan Plateau to their present heights. Most of this growth occurred during the past 10 million years. The Himalayas, towering as high as 8,854 m above sea level, form the highest continental mountains in the world. Moreover, the neighboring Tibetan Plateau, at an average elevation of about 4,600 m, is higher than all the peaks in the Alps except for Mont Blanc and Monte Rosa and is well above the summits of most mountains in the United States.

UNDERSTANDING PLATE MOTION

Scientists now have a fairly good understanding of how the plates move and how such movements relate to earthquake activity. Most movement occurs along narrow zones between plates where the results of plate-tectonic forces are most evident.

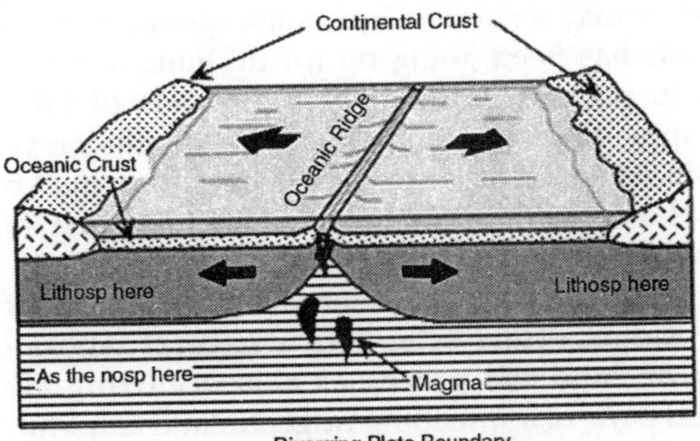

Fig. *The Main Types of Plate Boundaries*

There are four types of plate boundaries:

- Divergent boundaries where new crust is generated as the plates pull away from each other.
- Convergent boundaries where crust is destroyed as one plate dives under another.
- Transform boundaries where crust is neither produced nor destroyed as the plates slide horizontally past each other.
- Plate boundary zones broad belts in which boundaries are not well defined and the effects of plate interaction are unclear.

Divergent Boundaries

Divergent boundaries occur along spreading centers where plates are moving apart and new crust is created by magma pushing up from the mantle. Picture two giant conveyor belts, facing each other but slowly moving in opposite directions as they transport newly formed oceanic crust away from the ridge crest.

Perhaps the best known of the divergent boundaries is the Mid-Atlantic Ridge. This submerged mountain range, which extends from the Arctic Ocean to beyond the southern tip of Africa, is but one segment of the global mid-ocean ridge system that encircles the Earth. The rate of spreading along the Mid-Atlantic Ridge averages about 2.5 centimeters per year (cm/yr), or 25 km in a million years.

This rate may seem slow by human standards, but because this process has been going on for millions of years, it has resulted in plate movement of thousands of kilometers. Seafloor spreading over the past 100 to 200 million years has caused the Atlantic Ocean to grow from a tiny inlet of water between the continents of Europe, Africa and the Americas into the vast ocean that exists today.

The volcanic country of Iceland, which straddles the Mid-Atlantic Ridge, offers scientists a natural laboratory for studying on land the processes also occurring along the submerged parts of a spreading ridge. Iceland is splitting along the spreading centre between the North American and Eurasian Plates, as North America moves westward relative to Eurasia. The consequences of plate movement are easy to see around Krafla Volcano, in the northeastern part of Iceland. Here, existing ground cracks have widened and new ones appear every few months. From 1975 to 1984, numerous episodes of *rifting* (surface cracking) took place along the Krafla fissure zone. Some of these rifting events were accompanied by volcanic activity; the ground would gradually rise 1-2 m before abruptly dropping, signalling an impending eruption.

Between 1975 and 1984, the displacements caused by rifting totalled about 7 m. In East Africa, spreading processes have already torn Saudi Arabia away from the rest of the

African continent, forming the Red Sea. The actively splitting African Plate and the Arabian Plate meet in what geologists call a *triple junction,* where the Red Sea meets the Gulf of Aden.

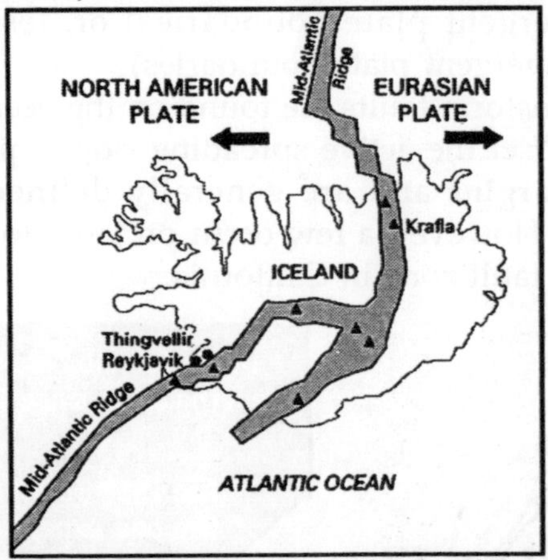

Fig. *Mid-Atlantic Ridge splitting Iceland and Separating the North American and Eurasian Plates.*

A new spreading centre may be developing under Africa along the East African Rift Zone. When the continental crust stretches beyond its limits, tension cracks begin to appear on the Earth's surface. Magma rises and squeezes through the widening cracks, sometimes to erupt and form volcanoes. The rising magma, whether or not it erupts, puts more pressure on the crust to produce additional fractures and, ultimately, the rift zone.

East Africa may be the site of the Earth's next major ocean. Plate interactions in the region provide scientists an opportunity to study first hand how the Atlantic may have begun to form about 200 million years ago. Geologists believe that, if spreading continues, the three plates that meet at the edge of the present-day African continent will separate completely, allowing the Indian Ocean to flood the area and making the easternmost corner of Africa a large island.

TRANSFORM BOUNDARIES

The zone between two plates sliding horizontally past one another is called a *transform-fault boundary,* or simply a *transform boundary*. The concept of transform faults originated with Canadian geophysicist J. Tuzo Wilson, who proposed that these large faults or *fracture zones* connect two spreading centers (divergent plate boundaries) or, less commonly, trenches (convergent plate boundaries).

Most transform faults are found on the ocean floor. They commonly offset the active spreading ridges, producing zig-zag plate margins and are generally defined by shallow earthquakes. However, a few occur on land, for example the San Andreas fault zone in California.

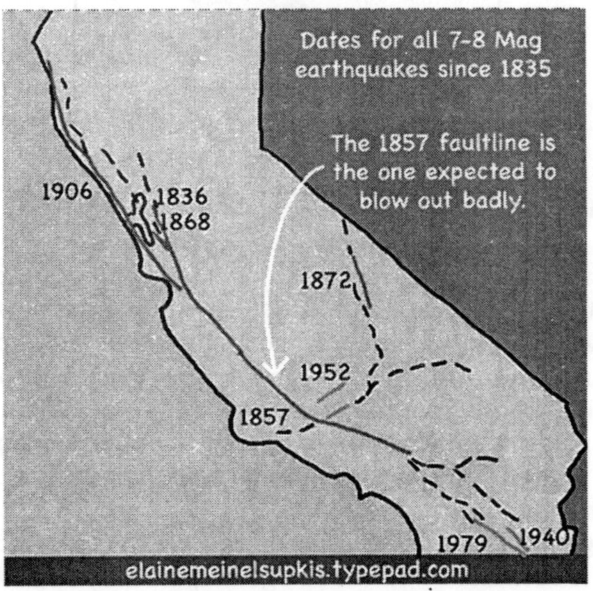

Fig. *San Andreas Fault*

This transform fault connects the East Pacific Rise, a divergent boundary to the south, with the South Gorda — Juan de Fuca Explorer Ridge, another divergent boundary to the north. The San Andreas fault zone, which is about 1,300 km long and in places tens of kilometers wide, slices through two thirds of the length of California.

Along it, the Pacific Plate has been grinding horizontally past the North American Plate for 10 million years, at an average rate of about 5 cm/yr. Land on the west side of the fault zone (on the Pacific Plate) is moving in a northwesterly direction relative to the land on the east side of the fault zone. Oceanic fracture zones are ocean-floor valleys that horizontally offset spreading ridges; some of these zones are hundreds to thousands of kilometers long and as much as 8 km deep. Examples of these large scars include the Clarion, Molokai and Pioneer fracture zones in the Northeast Pacific off the coast of California and Mexico. These zones are presently inactive, but the offsets of the patterns of magnetic striping provide evidence of their previous transform-fault activity.

PLATE-BOUNDARY ZONES

Not all plate boundaries are as simple as the main types. In some regions, the boundaries are not well defined because the plate-movement deformation occurring there extends over a broad belt (called a *plate-boundary zone*).

One of these zones marks the Mediterranean-Alpine region between the Eurasian and African Plates, within which several smaller fragments of plates *(microplates)* have been recognized. Because plate-boundary zones involve at least two large plates and one or more microplates caught up between them, they tend to have complicated geological structures and earthquake patterns.

Rates of Motion

We can measure how fast tectonic plates are moving today, but how do scientists know what the rates of plate movement have been over geologic time? The oceans hold one of the key pieces to the puzzle. Because the ocean-floor magnetic striping records the flip-flops in the Earth's magnetic field, scientists, knowing the approximate duration of the reversal, can calculate the average rate of plate movement during a given time span.

These average rates of plate separations can range widely. The Arctic Ridge has the slowest rate (less than 2.5 cm/yr) and the East Pacific Rise near Easter Island, in the South Pacific about 3,400 km west of Chile, has the fastest rate (more than 15 cm/yr). Evidence of past rates of plate movement also can be obtained from geologic mapping studies. If a rock formation of known age with distinctive composition, structure, or fossils mapped on one side of a plate boundary can be matched with the same formation on the other side of the boundary, then measuring the distance that the formation has been offset can give an estimate of the average rate of plate motion. This simple but effective technique has been used to determine the rates of plate motion at divergent boundaries, for example the Mid-Atlantic Ridge and transform boundaries, such as the San Andreas Fault.

Current plate movement can be tracked directly by means of ground-based or space-based *geodetic* measurements; *geodesy* is the science of the size and shape of the Earth. Ground-based measurements are taken with conventional but very precise ground-surveying techniques, using laser-electronic instruments. However, because plate motions are global in scale, they are best measured by satellite-based methods.

The late 1970s witnessed the rapid growth of *space geodesy,* a term applied to space-based techniques for taking precise, repeated measurements of carefully chosen points on the Earth's surface separated by hundreds to thousands of kilometers. The three most commonly used space-geodetic techniques very long baseline interferometry (VLBI), satellite laser ranging (SLR) and the Global Positioning System (GPS) are based on technologies developed for military and aerospace research, notably radio astronomy and satellite tracking. Among the three techniques, to date the GPS has been the most useful for studying the Earth's crustal movements. Twenty-one satellites are currently in orbit 20,000 km above the Earth as part of the NavStar system of the U.S. Department of Defence. These satellites continuously transmit radio signals back to Earth. To determine its precise position on Earth (longitude, latitude, elevation), each GPS ground site must simultaneously receive signals from at least four satellites, recording the exact time and location of each satellite when its signal was received.

By repeatedly measuring distances between specific points, geologists can determine if there has been active movement along faults or between plates. The separations between GPS sites are already being measured regularly around the Pacific basin. By monitoring the interaction between the Pacific Plate and the surrounding, largely continental plates, scientists hope to learn more about the events building up to earthquakes and volcanic eruptions in the circum-Pacific Ring of Fire.

Space-geodetic data have already confirmed that the rates and direction of plate movement, averaged over several years, compare well with rates and direction of plate movement averaged over millions of years.

MANTLE THERMAL PLUMES

The vast majority of earthquakes and volcanic eruptions occur near plate boundaries, but there are some exceptions. For example, the Hawaiian Islands, which are entirely of volcanic origin, have formed in the middle of the Pacific Ocean more than 3,200 km from the nearest plate boundary. How do the Hawaiian Islands and other volcanoes that form in the interior of plates fit into the plate-tectonics picture?

In 1963, J. Tuzo Wilson, the Canadian geophysicist who discovered transform faults, came up with an ingenious idea that became known as the "hotspot" theory. Wilson noted that in certain locations around the world, such as Hawaii, volcanism has been active for very long periods of time. This could only happen, he reasoned, if relatively small, long-lasting and exceptionally hot regions called *hotspots* existed below the plates that would provide localized sources of high heat energy *(thermal plumes)* to sustain volcanism.

Specifically, Wilson hypothesized that the distinctive linear shape of the Hawaiian Island-Emperor Seamounts chain resulted from the Pacific Plate moving over a deep, stationary hotspot in the mantle, located beneath the present-day position of the Island of Hawaii. Heat from this hotspot produced a persistent source of magma by partly melting the overriding Pacific Plate. The magma, which is lighter than the surrounding solid rock, then rises through the mantle and crust to erupt onto the seafloor, forming an active seamount. Over time, countless eruptions cause the seamount to grow until it finally emerges above sea level to form an island volcano. Wilson suggested that continuing plate movement eventually carries the island beyond the hotspot, cutting it off from the magma source and volcanism ceases.

As one island volcano becomes extinct, another develops over the hotspot and the cycle is repeated. This process of volcano growth and death, over many millions of years, has left a long trail of volcanic islands and seamounts across the Pacific Ocean floor.

According to Wilson's hotspot theory, the volcanoes of the Hawaiian chain should get progressively older and become more eroded the farther they travel beyond the hotspot. The oldest volcanic rocks on Kauai, the northwesternmost inhabited Hawaiian island, are about 5.5 million years old and are deeply eroded. By comparison, on the "Big Island" of Hawaii — southeasternmost in the chain and presumably still positioned over the hotspot — the oldest exposed rocks are less than 0.7 million years old and new volcanic rock is continually being formed.

The possibility that the Hawaiian Islands become younger to the southeast was suspected by the ancient Hawaiians, long before any scientific studies were done. During their voyages, sea-faring Hawaiians noticed the differences in erosion, soil formation and vegetation and recognized that the islands to the northwest were older than those to the southeast.

This idea was handed down from generation to generation in the legends of Pele, the fiery Goddess of Volcanoes. Pele originally lived on Kauai. When her older sister Namakaokahai, the Goddess of the Sea, attacked her, Pele fled to the Island of Oahu.

When she was forced by Namakaokahai to flee again, Pele moved southeast to Maui and finally to Hawaii, where she now lives in the Halemaumau Crater at the summit of Kilauea Volcano. The mythical flight of Pele from Kauai to Hawaii, which alludes to the eternal struggle between the growth of volcanic islands from eruptions and their later erosion by ocean waves, is consistent with geologic evidence obtained centuries later that clearly shows the islands becoming younger from northwest to southeast. Although Hawaii is perhaps the best known hotspot, others are thought to exist beneath the oceans and continents. More than a hundred hotspots beneath the Earth's crust have been active during the past 10 million years. Most of these are located under plate interiors but some occur near diverging plate boundaries. Some are concentrated near the mid-oceanic ridge system, such as beneath Iceland, the Azores and the Galapagos Islands.

A few hotspots are thought to exist below the North American Plate. Perhaps the best known is the hotspot presumed to exist under the continental crust in the region of Yellowstone National Park in northwestern Wyoming. Here are several *calderas* (large craters formed by the ground collapse accompanying explosive volcanism) that were produced by three gigantic eruptions during the past two million years, the most recent of which occurred about 600,000 years ago. Ash deposits from these powerful eruptions have been mapped as far away as Iowa, Missouri, Texas and even northern Mexico.

The thermal energy of the presumed Yellowstone hotspot fuels more than 10,000 hot pools and springs, geysers (like Old Faithful) and bubbling *mudpots* (pools of boiling mud).

A large body of magma, capped by a *hydrothermal system* (a zone of pressurized steam and hot water), still exists beneath the caldera. Recent surveys demonstrate that parts of the Yellowstone region rise and fall by as much as 1 cm each year, indicating the area is still geologically restless. However, these measurable ground movements, which most likely reflect hydrothermal pressure changes, do not necessarily signal renewed volcanic activity in the area.

Chapter 5

Effect of Earth's Rotation

CLIMATE ON EARTH

THE WORD "climate," from the Greek for "incline," preserves one of the oldest scientific discoveries, that year in and year out weather is largely controlled by the tilting of the earth's axis of rotation to the plane of its path around the sun. The Greeks noticed that, in general, the climates are colder at higher latitudes; they correlated this with the obvious fact that the air is generally warmer when the sun is high above the horizon.

As the sun rises higher in summer than in winter (a fact that indicates tilting of the earth's axis), it was natural to attribute seasonal weather differences to this tilting. Modern studies of *weather* (the day-to-day variation) and *climate* (the long-period weather, including seasonal changes) show that many other factors besides tilting of the earth's axis are involved. Weather and climate involve not only heat and cold, but also such features as wind and calm, the direction and intensity of the winds, cloudiness and precipitation either as rain or snow.

All of these aspects are closely interrelated and depend on such factors as heat supplied to the atmosphere by the sun's radiation, the movements of the atmosphere and the distribution of land, sea and mountains. A full treatment of these interrelations is the subject of a separate earth science-meteorology, the study of the atmosphere—but any geologic study requires some acquaintance with them. Not only do geologic processes differ greatly in different climates, but the clear evidences of drastic climatic changes in the geologic past cannot be appreciated without some understanding of the factors involved. We begin our study with an analysis of the nature and movements of the atmosphere.

The Atmosphere

The atmosphere is an envelope of gas around the earth. At the earth's surface the air contains about 79 per cent nitrogen, 21 per cent oxygen, 0.03 per cent carbon dioxide, small but variable amounts of water vapour and small amounts of chemically inert gases.

Air Pressure

As in all gases, every portion of the atmosphere exerts pressure against every adjacent portion. Such pressure is one of the evidences that gases are made up of particles in constant motion, colliding with each other and rebounding. The atmospheric pressure can be measured by the height of the column of mercury in an inverted evacuated tube.

One of the first discoveries made with the aid of Torricelli's device was that the pressure of the air is less at high elevations than at low. The pressure at any elevation results from the interplay of two tendencies:

- That of the gas to expand equally in all directions,
- That of each particle to fall toward the earth's surface because of gravity.

The barometric pressure, actually a measure of the weight of the air particles above the barometer, decreases rapidly with height and has only about half its sea-level value at a height of 17,500 feet, thus indicating that only half the mass of the atmosphere is above that elevation, even though the atmosphere extends upward for several hundred miles.

Movements in the Atmosphere

The energy to drive all atmospheric motion is ultimately derived from the sun's radiation. The actual motion is in response to differences in the gravitational attraction of the earth upon air masses of different densities. but the density differences are due to changes in temperature set up by the sun's radiation. Over a long time the amount of heat received from the sun must be balanced by the heat lost to outer space, or the earth would be either heating up or cooling down.

True, climatic fluctuations are not negligible but the fossil record strongly indicates that these are minor; roughly, the heat budget of the earth (amount received minus the amount lost) must be nearly balanced.

Much of the sun's radiation striking the outer atmosphere is reflected back into space; it does not affect the climate or temperature of the earth's surface. Of the radiation entering the atmosphere, part is radiated back into space as heat waves, part is reflected and a part is absorbed, raising the temperature of the land, air and water. Measurements and calculations of heat received at the earth's surface indicate that the amount varies with latitude, much as the ancient Greeks thought.

Losses by reflection and back-radiation into space also vary with latitude because a warm body radiates more than a cold one, but these differences are not so great as those in the amounts received.

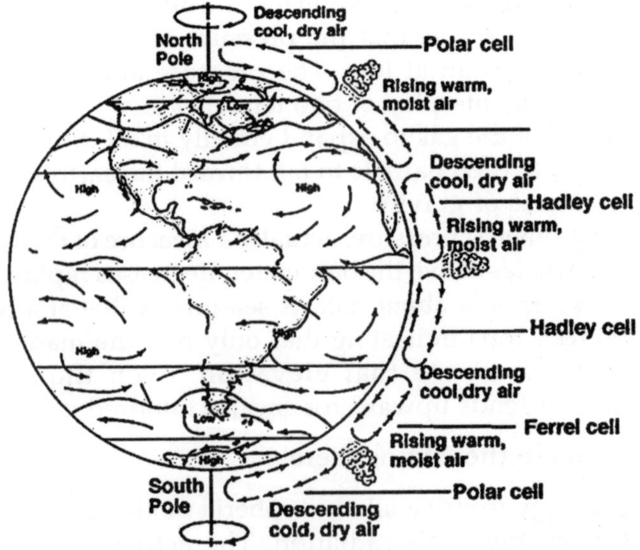

Fig. *Atmospheric Circulation in the Northern Hemisphere*

Thus, in low latitudes, more heat is received than lost and in high latitudes.Hotter and the polar cooler, the net heat transported from the tropics to higher latitudes must compensate nearly exactly for these differences the last column of the table. Most of this transport is by movements of warmed air (winds) but some is by ocean currents.

Consider first the condition in the tropics: Air heated near the equator must expand (become less dense). A column of air in the tropics thus weighs less than a column of cooler air of equal height in high latitudes.

Such a system is unstable and, since no gas has rigidity, these pressure differences cannot be sustained. The warm equatorial air column rises and flows poleward, while the denser cool air from higher latitudes flows equatorward beneath it. This establishes a convective circulation like that seen in a pot of soup slowly simmering on the stove. But flow cannot be directly toward the equator because, due to its inertia, the air is deflected by the rotation of the earth.

EFFECT OF THE EARTH'S ROTATION

The equatorial circumference of the earth is about 25,000 miles; a point on the equator thus moves eastward at the rate of more than 1,000 miles per hour. At latitude 40° the eastward velocity is about 750 miles per hour, but at latitude 45° only about 700. An air mass moving with the earth (on a windless day) has the velocity appropriate to its latitude. If it later moves southward from a point in north latitude toward the equator, its speed is less than that of the ground at the lower latitude and to an observer there results a northeast wind. In general, if an observer stands with his back to the wind, he finds that in the northern hemisphere all winds are deflected to his right and those in the southern hemisphere to his left, because of the earth's rotation.

One result of these two factors—the equatorial heating and the earth's rotation—is a prevailing wind called the *trade wind,*, blowing obliquely toward the equator in each hemisphere. In the northern hemisphere this is the *northeast trade;* in the southern, the *southeast trade* wind. Between lies the equatorial belt of calms. If the circulation were a single pair of simple circuits, one in each hemisphere, trade winds would blow all the way from the poles. Actually two (or three) circuits appear in each hemisphere, the upper air coming down in middle latitudes and moving along the earth's surface both toward the equator and toward a pole.

Air Masses

On a smaller scale, air masses, commonly 500 to 1,000 miles across, move slowly in the lower levels of the atmosphere in response to regional or local differences of pressure. Thus, a meteorologist speaks of a cold, dry, polar air mass penetrating beneath a warm, moist, tropical air mass.

The advancing margin of such a polar mass is called a "cold front." Rain is likely to occur behind a cold front. A large polar mass, twisting southeast (dashed arrows) across North America, insinuated a thin advancing wedge beneath a mass of warm moist air moving north and northeast (solid arrows) from the Gulf of Mexico.

The lower part of the water-saturated tropic air mass was lifted above the cold heavy mass, just as air rises when a liquid flows into a pan. In rising into a region of lower pressure, the air expanded and cooled. As cold air can hold less water vapour than warm, the moisture condensed as a torrential rain, bringing devastating floods to the Ohio River Valley and snow to the northern Midwest.

EFFECT OF MOUNTAIN RANGES

Mountain ranges profoundly modify local climates. Moist oceanic winds lose much of their water content as rain or snow when they rise over a mountain range. Such an ascent has the same effect as rise over a cold front. For example, the windward sides of the Hawaiian Islands may have 50 to 200 inches of rainfall a year; the sheltered sides 5 to 10. Similarly, the deserts of Turkestan are behind the Himalayas, whose seaward slopes receive more rain. The air that pours down the inland slope of a mountain range is compressed and heated, so that precipitation is usually prevented.

CLIMATE TYPES

If one considers rainfall alone, climates may be called *humid* if the mean annual precipitation exceeds 20 inches and if less than 10 inches and *semiarid* if between 10 and 20 inches. From another point of view, *maritime climates* are characterized by relatively uniform temperatures, moist air, fog and cloudiness, rainy winter, high wind velocities and freedom from dust.

In *continental climates*, both daily and seasonal temperature ranges are greater, windless days more frequent, the winter rainfall less and the air drier and often dust-filled. The extreme continental climate is the and or desert climate, which is much windier than other continental types, partly because of the great daily ranges in temperature.

Many deserts lie in "rainshadows" behind high mountains, but some of the greatest—the Sahara, Australian and southwest African deserts—extend to the ocean shores in the subtropical high-pressure belts.

WEATHERING AND SOILS

The principal significance of the different climatic types to the student of geology lies in their influence on soil formation and on erosion.

Each climatic environment places its own stamp upon the soils developed there and each influences, through its control over vegetation, amount of rainfall and evaporation losses. The geological processes involved in molding the details of the earth's surface.

Erosion, the process of removal of rock waste; here, we will examine the influence of several different climatic environments upon the weathering of different rock types. The most familiar example of *weathering* is the etching and discolouration of the surface of an unpainted board left out-of-doors. Rock, exposed on the earth's surface, also decays and leaches, but much more slowly.

If the product of rock decay is merely broken and discoloured, it is called *mantle* rock; but if it is loose and porous enough for plants to find a foothold, it is called *soil*.

Soil is more common than rock at the earth's surface. Almost all outcrops of rock are less firm—more easily crumbled and broken—than is the same rock at a depth of 20 or 100 feet. Many rocks that are black or steel gray where penetrated in mines, wells, or deep quarries are yellow or brown in outcrops.

In some, the yellow colour is a mere stain on or near cracks, but in others it is more pervasive and is accompanied by drastic changes in mineral composition of the rock. That the changes result from weathering is shown by observations on building stones.

For example, the exposed faces of the sandstone used in the older buildings at Stanford University turned yellow in 5 to 10 years and, where exposed to repeated wettings from garden sprinklers, began to crumble in 20 to 30 years.

Analysis of Weathering

Weathering results chiefly from exposure to the air or to the action of substances derived from the air, the most important of which are water, oxygen and carbon dioxide. Most studies of weathering have been made on one of two very different materials:

- Soils clearly derived from immediately underlying rocks.
- Building or monumental stones weathered superficially after years or centuries of exposure under various conditions. Such studies have shown that weathering usually includes both *mechanical disintegration* and *chemical decomposition. Disintegration* includes only loss of coherence, with little or no changes in composition. It does not include abrasion and removal of the constituents; such movements are part of erosion. *Decomposition* or *chemical weathering* is the term for change in chemical (and mineralogical) composition. The complex silicates that make up the bulk of most crystalline rocks alter into other substances, such as hydrous silicates, hydrous oxides and carbonates.

EXAMPLES OF MECHANICAL WEATHERING

Frost Action

Wherever water freezes, the 9 per cent expansion is an effective disruptive force. The expanding of ice to break rock is called *frost action* or *wedging*. Water freezes in the soil and in cracks or pores of rocks. The colder the climate, the deeper the freeze. Thawing and repeated freezing add to the effect, however, so that frost action is more effective in cold temperate zones than under extreme arctic conditions.

In soil, especially, because of the high water content and the complexity of the pore systems, freezing increases the volume considerably. After thawing, the soil is open and spongy. To break rocks, freezing water must be confined. As the water in a crack begins to freeze at or near the ground surface, it may be completely confined in the lower part of the crevice in the rock. Over a long time, rocks become extensively shattered by frost action. Many high mountain peaks are covered with rubble formed in this way.

Plants and Animals as Aids to Weathering

Organisms assist in the breakdown of rock to soil. Their action is in part disintegration and in part chemical effects. The roots of growing plants powerfully wedge the soil aside, raising and breaking slabs of rock (or concrete) and widening cracks. Burrowing animals move and mix the soil effectively. Charles Darwin, in his last published work (1881), showed that English earthworms may spread their casts over the ground to the average thickness of 0.1-0.2 inch per year. The soil is thus loosened, aerated and subjected to the chemical action of the worms' digestive processes. This mixing probably accounts for the uniformity of the humid and especially the forest or meadow, soils. There are practically no earthworms in arid regions.

Most plant tissues are carbohydrates (compounds of carbon, hydrogen and oxygen). Microscopic organisms, such as molds and bacteria, in the absence of oxygen, change leaves, fruit and wood on or in the soil into the dark organic substance known as *humus*. In the presence of oxygen, the organic substances are destroyed and CO_2 and H_2O are formed from them. Some H_2CO_3 is formed by both processes and contributes to chemical weathering. Living or dead, land plants and animals are aids to weathering.

Other Agents of Disintegration

Frost action and wedging by plants are the most effective agents of disintegration, but many other processes play a role.

Chemical decomposition is one cause of disintegration and possibly a major one. Feldspars swell as they weather into clay and many other minerals decompose to substances that occupy more volume. In a granite composed of feldspar, quartz and mica, this swelling of the rotting feldspar may disintegrate the massive rock to a pile of granitic sand. Alternate swelling and shrinking of the clay as it is wet by the rains and heated by the sun aid the process.

Rain may also wash soluble salts into cracks in rocks and soils where they crystallize. The growing crystals may exert enough force to break and disintegrate the rock still further. Telephone poles set in the Bonneville salt flats west of Salt Lake City show this effect to a striking degree. During the wet season the pole acts as a wick, drawing the salt brine up into the wood by capillarity. Here, the brine evaporates, precipitating salt crystals so abundantly that, after a few years, the pole up to about 18 inches above ground is swelled to twice its normal size and the wood fibers are completely shattered.

A forest fire often shatters and spalls the edges from exposed rock ledges and boulders. Rock is a poor conductor of heat, so the interior is not appreciably warmed by the fire. Consequently, the highly heated outside expands and pulls away from it. Lightning also breaks and shatters rocks.

It is possible that the drastic day-to-night changes of temperature that occur in deserts may cause enough expansion and contraction to weaken or even break some minerals and rocks. However, specimens of granite have been placed in electric ovens and run through the range of temperatures characteristic of deserts many thousands of times without showing any appreciable loss in strength. Doubtless, the swelling effects of decomposing minerals and the expansion of growing salt crystals are far more important in producing disintegration in deserts than are the diurnal temperature changes.

EXAMPLES OF CHEMICAL WEATHERING

We will now consider a few details of the chemical processes whereby rockforming minerals are decomposed into hydrous silicates, oxides and carbonates. Our first example is the chemically simple rock, limestone. Weathering of Limestone. Limestone is almost wholly calcite, but generally contains a few particles of clay. Calcite dissolves very slightly in pure water, sending some Ca^{2+} and CO_3^{2-} ions into solution. As indicated in chemical notation:

- $CaCO_3 \rightleftharpoons CO^{2+} + CO_3^{2-}$
  calcite calcium ion carbonate
  ion

In water containing dissolved carbon dioxide (as all rain water does), the calcite becomes much more soluble. The water and carbon dioxide react to yield carbonic acid and this in turn furnishes ions of hydrogen and bicarbonate. Thus:

- $H_2O + CO_2 \rightleftharpoons H_2CO_3$ $H + HCO_3^-$
  water carbon carbonic hydrogen bicarbonate dioxide
  acid ion ion

But the hydrogen ion thus formed reacts with carbonate ion from the calcite to form more bicarbonate ion:

- $H^+ + CO_3^{2-} \rightleftharpoons HCO_3^-$
  hydrogen ion carbonate ion bicarbonate ion

The arrows in these equations are the chemist's shorthand for a reversible process. In solutions the combination and dissociation of ions are constantly proceeding at rates governed by the abundance of the several ionic species present. For example, at equilibrium in reaction as much calcite is being precipitated from solution as is being dissociated into ions in that solution. But if the ions on one side of the reaction are selectively removed, they obviously cannot furnish as many products on the other side as they formerly did—the reaction will take place in the direction of yielding more of the ions being removed. When calcite is dissolved in water containing carbon dioxide, the carbon dioxide enters into new combinations with the water, as in equations, so that more of the products on the right side of equation are formed than would be in pure water; that is, more calcite is dissolved.

Thus, in the weathering of limestone the most important reaction is the slow solution of the calcite in percolating water that has absorbed carbon dioxide from the air. Calcite is dissolved not only at the surface, but along every crack and fissure into which the solutions penetrate.

The clay in the limestone accumulates on the surface as the underlying rock slowly dissolves. A few inches of clay soil may thus represent the residue of tens or scores of feet of dissolved limestone. Where underground circulation is vigorous the limestone may be extensively dissolved, resulting in the formation of great subterranean caves.

It is noteworthy that, despite the differences in composition of the underlying rock, soils derived from limestone and granodiorite are both rich in clay. The clay minerals are among the stablest under surficial conditions, both in temperate and in arctic regions.

From this analysis of the processes of weathering we can see that disintegration and decomposition take place together, each aiding the other.

In moist temperate regions, which are the areas that have been most studied by soil scientists, decomposition appears to be much the more important. Disintegration by frost action, however, is widespread and in cold regions it may predominate.

RESIDUAL SOIL AND THE SOIL PROFILE

The surface changes in a fairly uniform rock brought about by weathering are well illustrated in the Sierra Nevada of California. The canyon walls expose hard gray granodiorite, composed chiefly of plagioclase, with smaller amounts of quartz, orthoclase, biotite and hornblende. The rolling forestcovered uplands between the canyons have a red-brown soil. The surface layer, called by soil scientists the "A horizon," is a sandy loam (a mixture of quartz sand, silt, minute clay particles and decomposed plant residues). In the fine-grained reddish matrix are embedded many irregular quartz grains, alike in size and shape to those in the granodiorite. At about a foot depth, the clay content increases.

This sandy clay is the subsoil or "B horizon." At two to four feet, the soil becomes paler and sandier. The sand grains include not only quartz, but abundant feldspar, stained but readily recognized by its flashing cleavage faces and also many micaceous flakes.

These are not jet black, as is the mica of the granodiorite, but iron-leached pearly yellow scales sometimes mistaken for gold. Still deeper, this material grades imperceptibly into stained and crumbly material in which the texture and characteristic minerals of the granodiorite can be vaguely recognized. The transitional material below the zone of high clay content (B horizon) is called the "C horizon.".

The conclusion seems inescapable that the surface material is developed by weathering of the underlying granodiorite. Such a soil, still in place on its source rock, is called *residual*. The A horizon of the residual Sierra soil is looser and sandier than the compact B horizon. Apparently some clay formed by decomposition of feldspar near the surface has been washed downward from the A horizon, giving the B horizon an extra portion. In contrast with the residual soil of the Sierra granodiorite is that of certain parts of Kentucky where the bedrock is limestone instead of granodiorite and the climate, though also temperate, is more humid.

The A horizon consists of black humus-rich clay; the B horizon of light-gray limy clay which passes downward into white limestone. Most soils show characteristic "profiles" composed of two or more distinct layers or soil horizons, of which many kinds have been distinguished from various climatic environments by soil scientists. Comparison of various soils is facilitated by comparison of their respective soil horizons.

SOILS DEVELOPED ON TRANSPORTED MATERIAL

Many soils have developed not on bedrock, but on unconsolidated material deposited by streams, glaciers, or other agents of erosion. Such materials have generally been at least partially weathered before deposition. After deposition the loose material allows easy entry of air and water and so weathers quickly to form typical A and B soil horizons.

The soils of the Great Valley of California afford an example. The residual soil of the Sierra Nevada, developed from granodiorite, has been washed into streams and transported to the Great Valley. On debouching from the mountains, some streams have deposited hundreds of feet of soil and rock debris at the mountain base.

In places, such sediments, even some gravels, have been left undisturbed long enough to have rotted into brown soils with a characteristic soil profile. The brown silty A horizon, two or three feet thick, contains scattered quartz pebbles and numerous groups of closely spaced angular quartz grains that are probably relics from decayed granodiorite pebbles.

The compact B horizon, one or two feet thick, has been formed in part by the filtering down of minute clay grains from the A horizon. Such compact clayey B horizons, especially when cemented by iron oxide, calcium carbonate, or other bonding materials, are called *hardpan*. The C horizon consists of the relatively unaltered gravel or other parent material below the B horizon.

Another example of soil developed on transported material is that in eastern Massachusetts derived from rock flour (very finely ground fresh rock), boulders and clay transported by glaciers in the geologic past. Here, the A horizon consists of three or four inches of dark humus-rich soil overlying sterile rock flour and fragments.

EXFOLIATION

The splitting away of successive scalelike layers of rock from an exposed or soil-covered surface of massive rock (not a platy rock such as shale or schist) is called *exfoliation*. The separated sheets or plates may be flat or curved, paper thin or many feet thick, a fraction of an inch or hundreds of feet long. Two types of exfoliation may be distinguished—small thin flakes on massive rocks of all kinds; and giant plates, usually on granite or granodiorite.

Closely fitted blocks of dolerite, sandstone, granite and other mediumor fine-grained rocks frequently develop numerous thin concentric layers of partially weathered material. During weathering, the material expands as hydrated clay minerals are formed from the original minerals. Probably the outermost shell weathered and expanded first, pulling away from the fresher rock beneath.

The shells formed successively inward in similar fashion. In each shell, chemical weathering preceded the mechanical process of exfoliation. The giant curved plates of fresh rock that have split away in forming the Yosemite domes in California, Stone Mountain in Georgia and Sugar Loaf at Rio de Janeiro, indicate that exfoliation may take place without much weathering. These domes are all in plutonic rocks that originally consolidated beneath a thick cover of overlying rocks. Perhaps unloading by erosion of thousands of feet of the cover has permitted upward expansion and the formation of cracks parallel to the surface.

Colours of Soils

Most soil colours are due to iron minerals or organic matter. Iron exists in two states: ferrous and ferric. A ferric ion may be formed from a ferrous ion by the losing of one electron, a process called *oxidation*. When hydrated ferric oxide ("limonite") forms by chemical decomposition of a ferrous silicate such as biotite, the reaction is made possible by the addition of oxygen from the air. The finely divided "limonite" stains the resulting soil yellow. If the soil is repeatedly dried out, as in a climate with a warm dry season, the "limonite" may lose its water and change to red hematite; the soil then becomes red.

Iron is reduced to the ferrous state as easily as it is oxidized. A pale greenish or a dark gray soil usually indicates that the iron has been reduced by plant residues or other reactions with organic matter. Ordinarily this occurs where the air is excluded because pores have been filled with water. Thus, a black soil may indicate swamp conditions; the iron has been reduced and the soil is rich in humus. Colours of soils give clues to the conditions of weathering and to the abundance of organic matter.

CLIMATIC FACTORS IN WEATHERING

In Moist Temperate Climates

The weathering of limestone and granodiorite produces mainly minute grains of one or more of the clay minerals, with more or less quartz sand, iron oxide, flakes of altered mica and perhaps some partially decomposed feldspar. For this climate, a rough, but not precise, rule is that the more iron and the less silica in a rock, the more rapid its weathering. Thus, basalt weathers readily, but quartzite is almost immune. Limestone weathers rapidly in this climate.

In Dry and Cold Climates

From the relatively few investigations reported it appears that disintegration is the chief weathering process in the driest deserts, such as those of southwest Africa. It is also dominant in polar latitudes and on the high mountain peaks of the temperate and torrid zones. In arctic regions, frost action predominates; in the subtropical deserts, shattering of unknown origin prevails, perhaps caused chiefly by slight chemical changes, facilitated by expansion of the surface layer by decomposition and by solar heating.

Obvious chemical changes are minor and fine-grained soils rare. Vegetation is sparse; hence, carbonic and other soil acids are of less than usual importance. In subarctic regions such as Finland, however, chemical weathering of a special type, producing very siliceous soils, appears to predominate, especially where vegetation is fairly abundant and its decomposition slow and steady.

Chemical weathering of another sort, producing soils rich in calcium carbonate and high-silica clay minerals, is predominant in semi-arid regions such as the western interior of the United States. The calcium carbonate accumulates because, during most of the year, the soil water rises to the surface through small connected (capillary) pores and releases its dissolved carbonate by evaporation, rather than draining away through subterranean pores, cracks and channels. A striking contrast between moist and semi-arid weathering is furnished by limestone, which dissolves to leave pock-marked valleys in humid regions but stands out as bold ridges, almost like quartzite, in a semi-arid environment like that prevailing over most of Arizona and Nevada.

In the Tropics

The rain forests of the Congo and Amazon basins grow in soils which have not been adequately studied, but which appear similar to the aluminum-silicate clay soils of moist temperate regions.

The grass- and tree-covered savanna lands characterizing wide belts north and south of the tropical rain forests are largely underlaid by yellow and red-brown soils called *laterites*. The characteristic dark-brown upper portions of laterites dry to bricklike hardness and are sometimes used as building materials. Typical laterite areas are found in India, Nigeria, Central America and Brazil.

Because laterites offer very different problems than do the products of weathering in temperate climates and because they illustrate one way in which a chemical element such as iron or aluminum may be concentrated into valuable ores.

Laterites vary considerably in mineral composition, but are typically aluminum and iron hydroxides and oxides mixed with some residual quartz. A rare variety called *bauxite* is almost pure $Al_2O_3 \cdot nH_20$ and, hence, valuable as the ore of aluminum. In laterites, practically all the silicon of the original silicates has been leached out by rain waters, along with the easily soluble sodium and potassium and the acid-soluble calcium and magnesium.

The chemical aspects of this are difficult to understand. Silica is normally soluble only in water containing many more OH- (hydroxyl) ions than hydrogen ions (an alkaline solution).

However, calcium and magnesium are not soluble in such a solution, but in those with excess hydrogen over hydroxyl ions (acid solutions). Yet all three of these elements have clearly been leached from the soils.

Most laterite regions have marked wet and dry seasons. It has been suggested that during the warm dry seasons organic acids are so completely oxidized (to carbon dioxide and water) that the dry soil contains no acidproducing materials because the carbonic acid goes into the atmosphere as a gas. But the soil may retain a very small amount of material which would, when dissolved, give an alkaline solution. Then, at the onset of the first rains, silica is carried off in the temporarily alkaline soil solution before the decay of new vegetation again restores the carbonic acid supply.

Residual laterite soils are characterized by a pale *zone of leaching* just above the parent rock and a dark-brown *concretionary zone* at or near the surface. Each zone is usually a few feet or a few tens of feet thick, locally deepening to hundreds of feet.

The concretionary zone is a concretelike mass, composed chiefly of either dark-brown "limonite" or of numerous limonitic nodules (*concretions*) of pea or marble size, more or less well cemented into a solid mass. Possibly the concretionary zone should be called the "B_1 horizon," and the zone of leaching the "B_2." The uppermost parts of some lateritic soils are crumbly or even sandy (because of unaltered quartz) and perhaps represent the A horizon.

In laterites some elements needed by plants have been leached away and others, such as phosphorus and iron, are precipitated as iron phosphates or other compounds so insoluble as to be practically unavailable to plants. Therefore, pineapple plants grown on the lateritic soil of the Hawaiian Islands are sprayed with soluble iron salts and small quantities of soluble phosphates are applied frequently.

What happens to the silica leached out during lateritization? No definite answer can yet be given because the subtropical distribution of even so abundant a substance as silica is little known. Extensive siliceous crusts have been reported from Angola, just south of the Congo Basin and from other parts of tropical and subtropical Africa, but laterite is not definitely known near them.

When the total amount of silica in certain rocks is compared to the amount in the streams that drain them, it appears probable that silica is more rapidly removed in tropical climates. It has been shown, for instance, that the relative solubility of the silica in certain igneous rocks in tropical British Guiana is about twice as great as the world average for all rocks in all climates.

RATES OF WEATHERING

The rate of weathering varies greatly with different rocks and different climates. In less than eighty years the Edinburgh inscription on marble in memory of Joseph Black, the discoverer of carbon dioxide, was rendered illegible by the action of carbon dioxide and other acid-forming substances. On the average, about nine millimeters of rock per hundred years has dissolved from faced limestone in Edinburgh. Furthermore, there has been marked disintegration of the limestone by frost action.

In small Scotch towns, with less coal smoke, the marble in tombstones has been dissolved more slowly, but the effects are notably greater on limestone than the barely perceptible roughening of slate slabs in one or two hundred years.

Europe, southeast Asia and Central America are dotted with ruined stone castles or temples. For many of these structures the dates of building and bandonment are known quite accurately. A few ruins have been studied by soil scientists or archeologists.

Kamenetz fortress in southern Russia was built in 1362 and remained in use until 1699. Thereafter, through neglect, the stone blocks began to disintegrate and decompose. The top of one limestone tower developed a soil 4 to 16 inches thick in the 231 years between 1699 and 1930.

This soil was much like that on undisturbed, similar limestone in the vicinity. Other towers and castles under the European climatic environment have weathered more slowly, especially if constructed of sandstone, granite, or other high-silica rocks. Most observers consider that the rate of chemical weathering is probably most rapid in moist tropical climates, even though the evidence from dated structures is inconclusive. The arched roofs of Angkor in Cambodia, Indo-China, stand intact after seven centuries of neglect.

Though the jungle crowds close and clumps of green plants grow in the crannies, the endless vistas of sculptured walls indicate only slight weathering of the rock. In a somewhat less luxuriant coastal jungle southeast of Vera Cruz, Mexico, a slab of basalt, definitely dated by archeologists, lay in moist soil for at least 400 years without blurring of the Mayan inscription, whereas other inscriptions at the same place, with an almost identical history, have become illegible.Weathering proceeds most slowly in a dry warm climate. Inscriptions carved between 3 B.C. and 79 A.D. above the doors of sepulchers in the crumbly sandstone cliffs of northwest Arabia could still be read in 1876. Forty centuries ago, not far from the site of Assuan Dam in Upper Egypt, a surface of Syene red granite was smoothed by men's tools and dated by an inscription.

This surface, though exposed to direct sunlight, is still firm. At various dates, ranging from 2850 to 313 B.C., colossal statues from this quarry were set up at Luxor and elsewhere in the somewhat moister atmosphere of Middle and Lower Egypt and blocks of the same rock were used to face pyramids near Cairo.

The average rate of exfoliation for all the Syene cut granite studied by D. C. Barton, an American visitor to Middle and Lower Egypt in 1916, was about 0.5 to 1.0 centimeter per 5,000 years. In general, the older monuments showed the greater effects, but the maximum measured exfoliation was 1.0 centimeter in the youngest blocks studied (313 B.C.).Two obelisks of the Syene granite, each bearing many deep-cut hieroglyphics and each now called Cleopatra's Needle, stood about 3,500 years at Heliopolis and Alexandria in Egypt with only slight weathering. One, removed to London, has weathered appreciably but not disastrously.

The other needle, brought to New York about 1880 and set up in Central Park where it is exposed to frost as well as frequent wetting, exfoliated so extensively that by 1950 (despite the application of shellaclike preservatives in the nineteen-twenties and thirties) part of the pictured story was completely erased. More weathering had taken place in 70 New York winters than in 50 times 70 Egyptian years.

Chapter 6

Volcanic Phenomena

VOLCANO

Volcano, an opening in the earth's crust, through which heated matter is brought, permanently or temporarily, from the interior of the earth to the surface, where it usually forms a hill, more or less conical in shape and generally with a hollow or crater at the top. This hill, though not an essential part of the volcanic mechanism, is what is commonly called the volcano. The name seems to have been applied originally to Etna and some of the Lipari Islands, which were regarded as the seats of Hephaestus, a Greek divinity identified with Vulcan, the god of fire in Roman mythology.

All the phenomena connected directly or indirectly with volcanic activity are comprised under the general designation of *vulcanism* or *vulcanicity - words* which are also written less familiarly as volcanism and volcanicity; whilst the study of the phenomena forms a department of natural knowledge known as *vulcanology*. Vulcanicity is the chief superficial expression of the earth's internal igneous activity. Vulcanicity is the chief superficial expression of the earth's internal igneous activity. It may happen that a volcano will remain for a long period in a state of moderate though variable activity, as illustrated by the normal condition of Stromboli, one of the Lipari Islands; but in most volcanoes the activity is more decidedly intermittent, paroxysms of greater or less violence occurring after intervals of comparative, or even complete, repose. If the period of quiescence has been very protracted, the renewed activity is apt to be exceptionally violent.

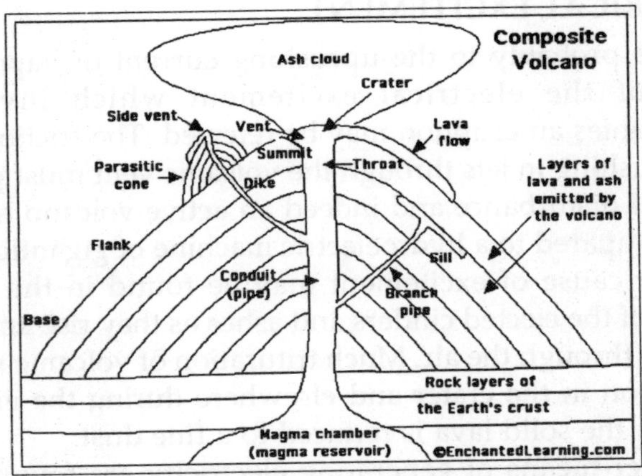

Fig. *Features of Volcano*

Thus, Krakatoa before the great eruption of 1883 had been dormant for something like two centuries and it is believed that the Japanese volcano Bandaisan previously to the gigantic outburst of 1888 had been silent for more than a thousand years. A volcano may indeed remain so long dormant as to be mistaken for one completely extinct.

It may happen that a volcano will remain for a long period in a state of moderate though variable activity, as illustrated by the normal condition of Stromboli, one of the Lipari Islands; but in most volcanoes the activity is more decidedly intermittent, paroxysms of greater or less violence occurring after intervals of comparative, or even complete, repose. If the period of quiescence has been very protracted, the renewed activity is apt to be exceptionally violent.

Thus, Krakatoa before the great eruption of 1883 had been dormant for something like two centuries and it is believed that the Japanese volcano Bandaisan previously to the gigantic outburst of 1888 had been silent for more than a thousand years. A volcano may indeed remain so long dormant as to be mistaken for one completely extinct. The volcanoes of central France are regarded as extinct, inasmuch as no authentic historical record of any eruption is known, but there are not wanting signs that in some parts of this volcanic region the subterranean forces may yet be slumbering rather than dead.

ELECTRICAL EXCITEMENT

It is probably to the uprushing current of vapour that much of the electrical excitement which invariably accompanies an eruption may be referred. The friction of the steam rushing in jets through the volcanic vent must produce electrical disturbance and indeed an active volcano has been aptly compared to a hydroelectric machine of gigantic power. Another cause of excitement may be found in the mutual friction of the ejected cinders and ashes as they rise and fall in showers through the air. Much trituration of volcanic material may go on in the crater and elsewhere during the eruption, whereby the solid lava is reduced to a fine dust.

Other means of generating electricity are found in the chemical reactions effected in the volcano and in the sudden condensation of the emitted vapour. L. Palmieri, in the course of his investigations at the observatory on Vesuvius, found that the vapours free from cinders carried a positive charge, whilst the cinders were negative.

The electrical phenomena attending an eruption are often of great intensity and splendour. The dark ash-laden clouds of vapour are shot through and through by volcanic lightning, sometimes in rapid horizontal flashes, then in oblique forked streaks, or again in tortuous lines compared to fiery serpents, whilst the borders of the cloud may be brilliant with electric scintillations, often forming balls and stars of fire.

During the great eruption of Krakatoa remarkable phenomena were observed by ships in the Strait of Sunda, luminous balls like " St Elmo's fire " appearing at the mast-heads and the yard-arms, whilst the volcanic mud which fell upon rigging and deck was strongly phosphorescent.

Quite distinct from any electrical phenomena is that intermittent reddish glare which is often seen at night in clouds hanging over an active crater and which is simply a glow due to reflection from,the incandescent lava and stones in the volcanic cauldron below.

VOLCANIC RAIN AND MUD

The condensation of the vast volumes of steam exhaled

during an eruption produces torrents of rain, which, mingling to a greater or less extent with the volcanic ashes, forms a hot muddy stream known in Italy as *lava d'acqua* and *lava di fango* and in South America as *moya.*

Deluges of such mud-lava may rush violently down the mountain-side and spread over the neighbouring country with terribly destructive effect, whence they are greatly dreaded by those who dwell at the base of a volcano. The solidified volcanic mud, often mingled with larger fragments of lava, is known as *duf f* or *tufa.*

Herculaneum was buried beneath a flood of mud swept down from Vesuvius during the Plinian eruption of 79 and the hard tufaceous crust which thus sealed up the illfated city came in turn to be covered by lava-flows from subsequent eruptions: hence the difficulty of excavating at Herculaneum compared with similar work at Pompeii, where there was probably much less mud, since the city, having been at a greater distance from the volcanic centre, was overwhelmed in great measure by loose ashes, capable of removal with comparative ease.

PREMONITORY SYMPTOMS

A volcanic eruption is usually preceded by certain symptoms, of which the most common are local earthquakes. The mountain, or other eruptive centre, may be thrown by internal activity into a state of tremor; the tremors perhaps continuing intermittently for months or even years and becoming more frequent and violent as the crisis approaches. At first they are usually confined to the volcano and its immediate neighbourhood, but may subsequently extend to a considerable distance, though probably never developing into earthquakes of the first magnitude.

The sudden opening of a subterranean crack, by rupture of a rock under strain, or the rapid injection of lava into such a fissure, will tend to produce a jar at the surface. For at least sixteen years before the first recorded eruption of Vesuvius in A.D. 79 earthquakes had been frequent in the Campania and had wrought havoc in the cities of Herculaneum and Pompeii.

Again, the formation of Monte Nuovo, near Pozzuoli, in 1538, was heralded by local earthquakes beginning several years in advance of the eruption.

So too in recent years many volcanic outbursts have been preceded by a succession of earthquakes; but as volcanoes are frequently situated in areas of marked seismic activity, the shocks antecedent to an eruption may not, unless exceptionally violent, receive much attention from local observers. It commonly happens that a volcanic outburst is announced by subterranean roaring and rumbling, often compared to thunder or the discharge of artillery underground. Other precursory symptoms may be afforded by neighbouring springs, which not unusually flow with diminished volume, or even fail altogether.

Possibly fissures open underground and drain off the water from the springs and wells in the immediate locality. Occasionally, however, an increased flow has been recorded. In some cases thermal springs make their appearance, whilst the temperature of any existing warm springs may be increased and perhaps carbon dioxide be evolved.

A disturbed state of the atmosphere is by no means a constant forerunner of an eruption, some of the greatest outbursts having occurred in a period of atmospheric stability: indeed the air is often felt to be close and still.

Immediately before a renewed outburst in an old volcano, the floor of the crater is generally upheaved to a greater or less extent, whilst the discharge of vapour from any fumaroles is increased. Where a crater has been occupied by water, forming a crater-lake, the water on the approach of an eruption becomes warm, evolves visible vapour and may even boil. In the case of cones which are capped with snow, the internal heat of the rising lava usually causes a rapid melting of the snow-cap, resulting perhaps in a disastrous deluge.

It seems probable that by attention to the premonitory symptoms a careful local observer might in many cases foretell an eruption. It generally happens that a great eruption is preceded by a preliminary phase of feeble activity. Thus, the gigantic catastrophe at Krakatoa on the 27th of

August 1883, so far from having been a sudden outburst, was the culmination of a state of excitement, sometimes moderate and sometimes violent, which had been in progress for several months.

Emission of Vapour

Of all volcanic phenomena the most constant is the emission of vapour. It is one of the earliest features of an eruption; it persists during the paroxysms, attaining often to prodigious volume; and it lingers as the last relic of an outburst, so that long after the ejection of ashes and lava has ceased an occasional puff of vapour may be the only memento of the disturbance.

By far the greatest proportion of the vapour is steam, which sometimes occurs almost to the exclusion of other gaseous products. Such, at least, is the usual and probably correct view, though it is opposed by A. Brun, who regards the volcanic vapours as chiefly composed of chlorides with steam in only subordinate amount. In the case of a mild eruption, like that occurring normally at Stromboli, the vapours may be discharged in periodical puffs, marking the explosion of bubbles rising more or less rhythmically from the seething lava in the volcanic cauldron. S. Wise observed at the volcano of Sangay, in Ecuador, no fewer than 267 explosions in the course of an hour, the vapour here being associated, as is so often the case, with ashes.

During a violent eruption the vapour may be suddenly shot upwards as a vertical column of enormous height, penetrating the passing clouds. For a short distance above the vent the superheated steam sometimes exists as a transparent vapour, but it soon suffers partial condensation, forming clouds, which, if not dispersed by winds, accumulate over the mountain. When the vapour is free from ash it forms rolling balls of fleecy cloud, but usually it carries in mechanical association more or less finely divided lava as volcanic dust and ashes, whereby it becomes yellow, brown, or even black, sometimes as foul as the densest smoke. In a calm atmosphere the dust-laden vapour may rise in immense rings with a rotatory movement, like that of vortex-rings. Frequently the vapours, emitted in a rapid succession of jets, form cumulus clouds, or are massed together in cauliflower-like forms. The well-known " pine-tree appendage " of Vesuvius, noted by the younger Pliny in his first letter to Tacitus on the eruption in the year 79, is a vertical shaft of vapour terminating upwards in a canopy of cloud and compared popularly with the trunk and spreading branches of the stonepine.

Fig. *Volcanic Mud*

Whilst in some cases the cloud resembles a gigantic expanded umbrella, in others it is more mushroom-shaped. In a great eruption, the height of the mountain itself may appear dwarfed by comparison with that of the column of vapour. During the eruption of Vesuvius in April 1906, the steam and dust rose to a height of between 6 and 8 m.

At Krakatoa in 1883 the column of vapour and ashes reached an altitude of nearly 20 m.; whilst it was estimated by some authorities that during the most violent explosions the finely divided matter must have been carried to an elevation of more than 30 m.

The emission of vast volumes of vapour at high tension naturally produces much atmospheric disturbance, often felt at great distances from the centre of eruption. It sometimes happens that volcanic mud is formed by the mingling of hot ashes not directly with rain but with water from streams and lakes, or even, as in Iceland, with melted snow.

A torrent of mud was one of the earliest symptoms of the violent eruption of Mont Pele in Martinique in 1902. This mud had its source in the Etang Sec, a crater-basin high up on the S.W. side of the mountain.

By the explosive discharge of ashes and vapours mingled with the water of the tarn there was produced a vast volume of hot muddy matter which on the 5th of May suddenly escaped from the basin, when a huge torrent of boiling black mud, charged with blocks of rock and moving with enormous rapidity, rolled like an avalanche down the gorge of the Riviere Blanche. If a stream of lava obstructs the drainage of a volcano, it may give rise to floods.

EJECTED BLOCKS

When a volcano after a long period of repose starts into fresh activity, the materials which have accumulated in the crater, including probably large blocks from the disintegration of the crater-walls, have to be ejected. If the lava from the last eruption has consolidated as a plug in the throat of the volcano, the conduit may be practically closed and hence the first effort of the renewed activity is to expel this obstruction.

The hard mass becomes shattered by the explosions and the angular fragments so formed are hurled forth by the outrushing stream of vapour. When the discharge is violent, the vapour, as it rushes impetuously up the volcanic duct, may tear fragments of rock from its walls and project them to a considerable distance from the vent. Such ejected blocks, by no means uncommon in the early stages of an eruption, are often of large size and naturally vary according to the character of the rocks through which the duct has been opened.

They may be irregular masses of igneous rocks, possibly lavas of earlier eruptions, or they may be stratified, sedimentary and fossiliferous rocks representing the platform on which the volcano has been built, or the yet more deeply seated fundamental rocks. By Dr H. J. Johnston-Lavis, who specially studied the ejected blocks of Vesuvius, the volcanic materials broken from the cone are termed " accessory " ejecta, whilst other fragmentary materials he conveniently calls " accidental " products, leaving the term " essential " ejecta for plastic lava, ashes, crystals, &c.

Masses of Cretaceous or Apennine limestone ejected from Somma are scattered through the tuffs on the slopes of Vesuvius; and objects carved in such altered limestone are sold to tourists as "lava" ornaments. Under the influence of volcanic heat and vapours, the ejected blocks suffer more or less alteration and may contain in their cavities many crystallized minerals. Certain blocks of sandstone ejected occasionally at Etna are composed of white granular quartz, permeated with vitreous matter and encased in a black scoriaceous crust of basic lava.

A rock consisting of an irregular aggregation of coarse ejected materials, including many large blocks, is known as a " volcanic agglomerate." Any fragmental matter discharged from a volcano may form rocks which are described as " pyroclastic." *Cinders, Ashes and* Dust.

After the throat of a volcano has been cleared out and a free exit established, the copious discharge of vapour is generally accompanied by the ejection of fresh lava in a fragmentary condition. If the ejected masses bear obvious resemblance to the products of the hearth and the furnace, they are known as " cinders " or " scoriae," whilst the small cinders not larger than walnuts often pass under their Italian name of "lapilli". When of globular or ellipsoidal form, the ejected masses are known as " bombs " or " volcanic tears."

Other names are given to the smaller fragments. If the lava has become granulated it is termed " volcanic sand "; when in a finer state of division it is called ash, or if yet more highly comminuted it is classed as dust; but the latter terms are sometimes used interchangeably.

The pulverized material, consisting of lava which has been broken up by the explosion, or triturated in the crater, is often discharged in prodigious quantity, so that after an eruption the country for miles around the volcano may be covered with a coating of fine ash or dust, sometimes nearly white, like a fall of snow, but often of greyish colour, looking rather like Portland cement and in many cases becomirg reddish by oxidation of the ferruginous constituents.

Even when first ejected the ash is sometimes cocoa-coloured. This finely divided lava insinuates itself into every crack and cranny, reaching the interior of houses even when windows and doors are closed. A heavy fall of ash or cinders may cause great structural damage, crushing the roofs of buildings by sheer weight, as was markedly the case at Ottajano and San Guiseppe during the eruption of Vesuvius in April 1906. On this occasion the dry ashes slipped down the sides of the volcanic cone like an avalanche, forming great ashslides with ridges and furrows rather like barrancos, or ravines, caused by rain. The burial of Ottajano and San Ginseppe in 1906 by Vesuvian ejecta, mostly lapilli, has been compared with that of Pompeii in 79.

Deposits of volcanic sand and ashes retain their heat long after ejection, so that rain will cause them to evolve steam and if the rain be heavy and sudden it may produce explosions with emission of great clouds of vapour. The fall of ash is at first prejudicial to vegetation and is often accompanied or followed by acid rain; but ultimately the ash may prove beneficial to the soil, chiefly in consequence of the alkalis which it contains. The " May dust " of Barbados was a rain of volcanic ash which fell in May 1812 from the eruption of the Soufriere in St Vincent. It is estimated that the amount of dust which during this eruption fell on the surface of Barbados, Ioo m. distant from the eruptive centre, was about 3,000,000 tons. The distance to which ash is carried depends greatly on the atmospheric conditions at the time of the eruption. Ashes from Vesuvius in an eruption in the year 472 were carried, it is said, as far as Constantinople.

During an eruption of Cotopaxi, on the 3rd of July 1880, observed by E. Whymper, an enormous black column of dust-laden vapour was shot vertically upwards with such rapidity that in less than a minute it rose to a height estimated at 20,000 ft. above the crater-rim, or nearly 40,000 ft. above sea-level, when it was dispersed by the wind over a very wide area. It is believed that the amount of dust in this discharge must have been more than 2,000,000 tons.

Enormous quantities of dust ejected from Krakatoa in 1883 were carried to prodigious distances, samples having been collected at more than a thousand miles from the volcano; whilst the very fine material in ultramicroscopic grains which remained suspended for months in the higher regions of the atmosphere seems to have enjoyed an almost world-wide distribution and to have been responsible for the remarkable sunsets at that period.

The ash falling in the immediate vicinity of a volcanic vent will generally be coarser than that carried to a distance, since the particles as they are wafted through the air undergo a kind of sifting. Professor J. W. Judd, who made an exhaustive examination of the products of the eruption of Krakatoa, found that the dust near the volcano was comparatively coarse, dense and rather darkcoloured, in consequence of the presence of numerous fragments of heavy, dark, crystalline minerals, whilst the dust at a distance was excessively fine and perfectly white. According to this observer, the particles tended to fall in the following order: magnetite, pyroxenes, felspar, glass.

The finely comminuted material, carried to a great height in the atmosphere, consisted largely of delicate threads and attenuated plates of vitreous matter, in many cases hollow and containing air-bubbles. The greater part of the dust was formed by the mutual attrition of fragments of brittle pumice as they rose and fell in the crater, which thus became a powerful "dust-making mill." By this trituration of the pumiceous lava, carried on for a space of three months during which the eruption lasted, the quantity of finely pulverized material must have been enormous; yet the amount of ejected matter was probably very much less than that extruded during some other historical eruptions, such as that of Tomboro in Sumbawa, in 1815.

The explosions at Krakatoa were, however, exceptionally violent, having been sufficient to project some of the finely pulverized lava to an altitude estimated to have been at least

30 m. It is usually impossible during a great eruption to determine the height of the column of "smoke" since it hangs over the country as a pall of darkness. The great black cloud, which was so characteristic a feature in the terrible eruptions in the West Indies in 1902, was formed of steam with sulphur dioxide and other gases, very heavily charged with incandescent sand or dust, forming a dense mixture that in some respects behaved like a liquid.

Unlike the Krakatoa dust, which was derived from a vitreous pumice, the solid matter of the black cloud was largely composed of fragments of crystalline minerals. According to Drs Anderson and Flett it is not impossible that on the afternoon of the 17th of May 1902, the solid matter ejected from the Soufriere of St Vincent amounted to several billions of tons and that some of the dust fell at distances more than 2000 m. east of the centre of eruption.

LAVA

The volcanic cinders, sand, ashes and dust described above are but varied forms of solidified lava. Lava is indeed the most characteristic product of volcanic activity. It consists of mineral matter which is, or has been, in a molten state; but the liquidity is not due to simple dry fusion.

The magma, or subterranean molten matter, may be regarded as composed essentially of various silicates, or their constituents, in a state of mutual solution and heavily charged with certain vapours or gases, principally water-vapour, superheated and under pressure.

In consequence of the peculiar constitution of the magma, the order in which minerals separate and solidify from it on cooling does not necessarily correspond with the inverse order of their relative fusibility.

The lava differs from the magma before eruption, inasmuch as water and various volatile substances may be expelled on extrusion. The rapid escape of vapour from the lava contributes to the explosive phenomena of an eruption, whilst the rate at which the vapour is disengaged depends largely on the viscosity of the magma.

Fig. *Volcanic Lava*

The lava on its immediate issue from the volcanic vent is probably at a white heat, but the temperature is difficult of determination since the molten matter is usually not easy of approach, by reason of the enshrouding vapour. Determinations of temperature are generally made at a short distance from the exit, when the lava has undergone more or less cooling, or on a small stream from a subordinate vent. A.

Bartoli, using a platinum electric resistance pyrometer, found that a stream of lava near a *bocca*, or orifice of emission, on Etna, in the eruption of 1892, had at a depth of one foot a temperature of 1060° C. In the lavas of Vesuvius and Etna thin wires of silver and of copper have frequently been melted. Probably the lava at the surface of the stream has a temperature of something like 110o° C., but this must not be assumed to be its temperature at the volcanic focus. C. Doelter, in some experiments on the melting-point of lava by means of an electric furnace, found that a lava from Etna softened at from 962° to 970° C. and became fluid at 1010° to 1040°, whilst a Vesuvian lava softened at 1030° to 1060° and acquired fluidity at 1080° to 1090°.

These results were obtained at ordinary atmospheric pressure, but it has been assumed that the melting-point of lava at a great depth would, through pressure alone, exceed that obtained in the laboratory. On the other hand the presence of water and of certain volatile fluxes in the magma lowers the fusing-point and hence the extruded lava from which these have largely escaped may be much less fusible than the original magma. It should be noted that all determinations of the melting-points of minerals and rocks involving ocular inspection of the physical state of the material are liable to considerable error and the only accurate method seems to be that of determining the point at which absorption of heat abruptly occurs - the latent heat of fusion. It is believed that the temperature of lava in the volcanic conduit may be in some cases sufficiently high to fuse the neighbouring rocks and so melt out a passage through them in its ascent. The wallrock thus dissolved in the magma will not be without influence on the composition of the lava with which it becomes assimilated.

Many interesting observations are on record with regard to the heating effect of lava on metals and other objects with which it may have come in contact. Thus, after the destruction of Torre del Greco by a current of lava from Vesuvius in 1794, it was found that brass in the houses under the lava had suffered decomposition, tje copper having become crystallized; whilst silver had been not only fused but sublimed. This indicates a temperature of upwards of 1000° C. Panes of glass in the windows at Torre del Greco on the same occasion suffered devitrification.

Notwithstanding the high temperature of lava on emission, it cools so rapidly and the consolidated lava conducts heat so slowly, that vegetable structures may be involved in a lava-flow without being entirely destroyed. A stream of lava on entering a wood, as in the sylvan region on Etna, may burn up the undergrowth but leave many of the larger trees with their trunks merely carbonized. On Vesuvius a lava-flow has been observed to surround trees while the foliage has been apparently uninjured.

A vertical trunk of a coniferous tree partially enveloped in Tertiary basalt occurs at Gribon in the Isle of Mull, as described by Sir A. Geikie and others; plant-remains in basalt from the Bo'ness coalfield in Linlithgowshire have been noticed by H. M. Cadell; and attention has been called by B. Hobson to a specimen of scoriaceous basalt, from Mexico, which shows the impression of ears of maize and even relics of the actual grains. In consequence of the slow transmission of heat by solid lava, the crust on the surface of a stream may be crossed with impunity whilst the matter is still glowing at a short distance below.

Lichens may indeed grow on lava which remains highly heated in the interior. The solidified surface of a sheet of lava may be smooth and shining, sometimes quite satiny in sheen, though locally wrinkled and perhaps even ropy or hummocky, the irregularities being mainly due to superficial movement after partial solidification.

The " corded lava " has a surface similar to that often seen on blastfurnace slag and is suggestive of a tranquil flow. After a lava stream has become crusted over on cooling, the subjacent lava, still moving in a viscous condition, tends to tear the crust, forming irregular blocks, or clinkers, which are carried forward by the flow and ultimately left in. the form of confused heaps, perhaps of considerable magnitude.

The front of a stream may present a wall of scoriaceous fragments looking like a huge pile of coke. As the clinkers are carried along, on the surface of the lava, they produce by mutual friction a crunching noise; and the sluggish flow of the lava-stream laden with its burden has been compared with that of a glacier.

Since the upper part of the stream moves more rapidly than the lower, which is retarded by cooling in contact with the bed-rock, the superficial clinkers are carried forward and, rolling over the end, may become embedded in the lava as it advances. Scoriae formed on the top of a stream may thus find their way to the base.

Rockfragments or other detrital matter occurring in the path of the lava will be caught up by the flow and become involved in the lower part of the molten mass; whilst the rocks over which the lava travels may suffer more or less alteration by the heat of the stream. The rapidity of a lava flow is determined partly by the slope of the bed over which it moves and partly by the consistency of the lava, this being dependent on its chemical composition and on the conciitions of cooling. In an eruption of Mauna Loa, in Hawaii, in 1855, the lava was estimated to flow at a rate of 40 m. an hour; and at an eruption of Vesuvius in 1805 a velocity of more than 50 m. an hour, at the moment of emission, was recorded.

The rapidity of flow is, however, rapidly checked as the stream advances, the retardation being very marked in small flows. Where lava travels down a steep incline there is naturally a great tendency to form a rugged surface, whilst a quiet flow over a flat plane favours smoothness. If the lava meet a precipice it may form a cascade of great beauty, the clinkers rapidly rolling down with a clatter, as described by Sir W. Hamilton in the eruption of Vesuvius in 1771, when the fiery torrent had a perpendicular fall of 50 ft.

In Hawaii the smooth shining lava, often superficially waved and lobed, is known as *pahoehoe,* whilst the rugged clinker beds are termed aa. These terms are now used in general terminology, having been introduced by American geologists. The fields of aa often contain lava-balls and bombs. It may be said that the pahoehoe corresponds practically with the *Fladen lava* of German vulcanologists and the as with their *Schollen lava*. Rugged flows are known in Auvergne as *cheires.* The surface of a clinker-field has often a horribly jagged character, being covered with ragged blocks bristling with sharp points. In the case of an obsidian-flow a most dangerous surface is produced by the keen edges and points of the fragmentary volcanic glass.

Fig. *Lava Arch, Hawaii*

If, after a stream of lava has become crusted over, the underlying magma should flow away, a long cavern or tunnel may be formed. Should the flow be rapid the roof may collapse and the fragments, falling on to the stream, may be carried forward or become absorbed in the fused mass. The walls and roof of a lava-cave are occasionally adorned with stalactites, whilst the floor may be covered with stalagmitic deposits of lava. The volcanic stalactites are slender, tubular bodies, extremely fragile, often knotted and rippled.

Beautiful examples of lava stalactites from Hawaii have been described by Professor E. S. Dana. Caverns may also be formed in lava-flows by the presence of large bubbles, or by the union of several bubbles. It may happen, too, that certain monticules thrown up on the surface of the lava are hollow, of which a famous example is furnished by the Caverne de Rosemond, at the base of Piton Barry, in the Isle of Reunion.It is of great interest to determine whether molten lava contracts or expands on solidification, but the experimental evidence on this subject is rather conflicting.

According to some observers a piece of solid lava thrown on to the surface of the same lava in a liquid state will sink, while according to others it floats. It has often been observed that cakes formed by the natural fracture of the crust on the lava of Kilauea sink in the liquid mass, but it has been suggested that the fragments are drawn down by convection-currents. On the other hand a solid piece, though denser than the corresponding liquid, may be buoyed up for a time by the viscous condition of the molten lava.

Moreover, the presence of minute vesicles may lighten the mass. Although the minerals of a rock-magma may separately contract on crystallization it does not follow that the magma itself, in which they probably exist in a state of solution, will undergo on crystallization a similar change of volume. On the whole, however, there seems reason to believe that lava on solidifying almost always diminishes in volume and consequently increases in density.

According to the experiments of C. Doelter the specific gravity of molten lava is invariably less than that of the same lava when solid, though in some cases the difference is but slight. In a vitreous or isotropic condition the lava has a lower density than when crystalline.

The differences are illustrated by the following table, where the figures give the specific gravity: Experiments by Dr C. Barus showed that a diabase of specific gravity 3.017 formed a glass of sp. gr. 2.717 and melted to a liquid of sp. gr. 2.52. J. A. Douglas on examining various igneous rocks found that in all cases the rock in a vitreous state had a lower sp. gr. than in a crystalline condition, the difference being greatest in the acid plutonic rocks. A. Harker, however, has called attention to the fact that the glassy selvage of certain basic dykes in Scotland is denser than the same rock in a crystalline condition in the interior of the dykes.

CHEMICAL COMPOSITION OF LAVAS

Lavas are usually classified roughly, from a chemical point of view, in broad groups according to the proportion of silica which they contain. Those in which the proportion of silica reaches 66% or upwards are said to be acid or acidic, whilst those in which it falls to 55% or below are called basic lavas. The two series are connected by a group of intermediate composition, whilst a small number of igneous rocks of exceptional type are recognized as ultrabasic.

Professor F. W. Clarke has suggested a grouping of igneous rocks as per-silicic, medio-silicic and subsilicic, in which the proportion of silica is respectively more than 60, between 50 and 60, or less than 50%. By far the greater part of all lavas consists of various silicates, either crystallized as definite minerals or unindividualized as volcanic glass. In addition, however, to the mineral silicates, a volcanic rock may contain a limited amount of free acid and basic oxides, represented by such minerals as quartz and magnetite. Rhyolite may be cited as a typical example of an acid lava andesite as an intermediate and basalt as a basic lava.

The various volcanic rocks are described under their respective headings, so that it is needless to refer here to their chemical or mineralogical composition.

In the course of the life of a volcano, the lava which it emits may undergo changes, within moderate limits, being at one time more acid, at another more basic. Such changes are sometimes connected with a shifting of the axis of eruption. Thus at Etna the lavas from the old axis of Trifoglietto in the Valle del Bove were andesites, with about 55% of silica, but those rising in the present conduit are doleritic, with a silica-content of only about 50%.

It seems probable that, to a limited extent, changes in the character of a lava may sometimes be due to contact of the magma with different rocks underground: if these are rich in silica, the acidity of the lava will naturally increase; while if they are rich in calcareous and ferromagnesian constituents, the basicity will increase: the variation is consequently apt to be only local and probably always slight.

By von Richthofen and some others it has been held that during a long period of igneous activity a definite order in the succession of the erupted rocks is everywhere constant; but though some striking coincidences may be cited, it can hardly be said that this generalization has been satisfactorily established. It has, however, often been observed, as emphasized by Professor Iddings, that a volcanic centre will start with the emission of lavas of neutral or intermediate type, followed in the course of a geological period by acid and basic lavas and ending with those of extreme composition, indicating progressive change in the magma.

The old idea of a universal magma, or continuous pyrosphere, has been generally abandoned. Whatever may have been the case in a primitive condition of the interior of the earth, it seems necessary to admit that the magma must now exist in separate reservoirs.

The independent activity of neighbouring volcanoes strikingly illustrated in Kilauea and Mauna Loa in Hawaii, only 20 m. apart, suggests a want of communication between the conduits; and though the lavas are very similar at these two centres, it would seem that they can hardly be drawn from a common source. Again, the volcanoes of southern Italy and the neighbouring islands exhibit little or no sympathy in their action and emit lavas of diverse type. The lavas of Vulcano, one of the Lipari Isles, are rhyolitic, whilst those of Stromboli, another of the group, are basaltic.

It is believed that the magma in a subterranean reservoir, though originally homogeneous, may slowly undergo certain changes, whereby the more basic constituents migrate to one quarter whilst the acid segregate in another, so that the canal, at successive periods, may bring up material of different types. It has often been observed that all the rocks from a definite igneous centre have a general similarity in chemical and mineralogical characters.

This relationship is called, after Professor Iddings, " consanguinity," and appears to be due to the fact that the rocks are drawn from a common source. Professor Judd pointed out the existence of distinct "petrographical provinces," within which the eruptive rocks during a given geological period have a certain family likeness and have appeared in definite succession. Thus he recognized a Brito-Icelandic petrographical province of Tertiary and recent lavas. It has been shown by A. Harker that alkali igneous rocks are generally associated with the Atlantic type of coast-line and sub-alkali rocks with the Pacific type.

Although changes in the character of an erupted product from a given centre are usually brought about very slowly, it has often been supposed that even in the course of a single prolonged eruption, or series of eruptions, the character of the lava may vary to some extent.

That this is not, however, usually the case has been repeatedly proved. M. H. Arsandaux, for instance, analysed the bombs of augite-andesite thrown out from Santorin at the beginning of the eruption of 1866, others ejected in 1867 and others again at the close of the eruption in 1868; and he found no important variation in the composition of the magma during these successive stages.

Moreover, Professor A. Lacroix found that the material extruded from Vesuvius in 1906 remained practically of the same composition from the beginning to the end of the eruption and further, that it presented great analogy to that of 1872 and even to that of 1631. All the Vesuvian lavas are of the type of rock known as leucotephrite or leucitetephrite, or they pass, by the presence of a little olivine, into leucite-basanite. Leucite is characteristic of the lavas of Vesuvius, whilst it is excessively rare in those of Etna, where a normal doleritic type prevails.

Nepheline, a felspathoid related to leucite, is characteristic of certain lavas, such as those of the Canary Islands, which comprise nepheline-tephrites and nepheline-basanites. Most of the lavas from the volcanoes of South America consist of hypersthene-andesite and it is notable that the fragmental ejectamenta from the eruptions of St Vincent and Martinique in 1902 and from Krakatoa in 1883 were evidently derived from a magma of this Pacific type.

It commonly happens that acid lavas are paler in colour, less dense and less fusible than basic lavas and they are probably drawn in some cases from shallower depths. As a consequence of the ready fusibility of many basic lavas, they flow freely on emission, running to great distances and forming far-spreading sheets, whilst the more acid lavas rapidly become viscid and tend to consolidate nearer to their origin, often in hummocky masses.

The shape of a volcanic mountain is consequently determined to a large extent by the chemical character of the lavas which it emits. In the Hawaiian Islands, for instance, where the lavas are highly basic and fluent, they form mountains which, though lofty, are flat domes with very gently sloping sides.

Such is the fluidity of the lava on emission that it flows freely on a slope of less than one degree. In consequence, too, of this mobility, it is readily thrown into spray and even projected by the expansive force of vapour into jets, which may rise to the height of hundreds of feet and fall back still incandescent, producing the appearance of "fire fountains."

The emission is not usually accompanied, however, by violent explosions, such as are often associated with the eruption of magmas of less basic and more viscous nature. The viscosity of the lava at Kilauea was estimated by G. F. Becker to be about fifty times as great as that of water. It may be pointed out that the fusibility of a lava depends not on the mere fact that it is basic, but rather on the character of the bases. A lava from Etna or Vesuvius may be really as basic as one from Hawaii.

Capillary Lava

A filamentous form of lava well known at Kilauea, in Hawaii, is termed *Pele's hair,* after Pele, the reputed goddess of the Hawaiian volcanoes. It resembles the capillary slag much used in the arts under the name of " mineral wool " - a material formed by injecting steam into molten slag from an iron blast-furnace. It is commonly supposed that Pele's hair has been formed from drops of lava splashed into the air and drawn out by the wind into fine threads.

According, however, to Major C. E. Dutton, the filaments are formed on the eddying surface of the lava by the elongation of minute vesicles of water-vapour expelled from the magma. C. F. W. Krukenberg, who examined the hair microscopically, figured a large number of fibres, some of which showed the presence of minute vesicles and microscopic crystals, the former when drawn out rendering the thread tubular. In a spongy vitreous scoria from Hawaii, described as "thread-lace," a polygonal network of delicate fibres forms little skeleton cells. Capillary lava is not confined to the Hawaiian volcanoes: it is known, for example, in Reunion and may be formed even at Vesuvius.

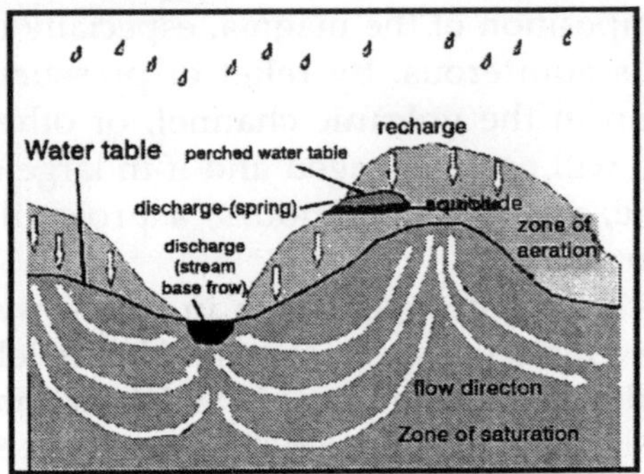

Fig. *Capillary Lava*

Pumiceous Lava

The copious disengagement of vapour in a glassy lava gives rise to the light cellular or spongy substance, full of microscopic pores, known as pumice. It is usually, though not invariably, produced from an acid lava and may sometimes be regarded as the solidified foam of an obsidian.

During the eruption of Krakatoa in 1883 enormous quantities of pumice were ejected and were carried by the sea to vast distances, until they ultimately became water-logged and sank.

Professor Judd found the pumice to consist of a vitreous lava greatly inflated by imprisoned vapours; the walls of the air-cells were formed of the lava drawn out into thin plates and threads, often with delicate fibres running across the cavities. Having been suddenly cooled, it was extremely brittle and its ready pulverization gave rise to much of the ash ejected during this eruption.

It has been shown by Dr JohnstonLavis that a bed of pumiceous lava, especially if basic, is generally vitreous towards the base, becoming denser, darker and more crystalline upwards, until it may pass superficially into scoria. The change is explicable by reduction in the temperature of the magma consequent on the conversion of water into steam.

Water in Lavas

Whether an eruption is of an explosive or a tranquil character must depend largely, though not wholly, on the chemical composition of the magma, especially on the extent to which it is aquiferous. By relief of pressure on the rise of the column in the volcanic channel, or otherwise, more or less steam will be disengaged and if in large quantity this must become, with other vapours, a projectile agency of enormous power.

The precise physical condition in which water exists in the magma is a matter of speculation and hence Johnston-Lavis proposed to designate it simply as H_2O. Water above its critical point, which is about 370° C. or 698°F., cannot exist as a liquid, whatever be the pressure, neither is it an ordinary vapour. It has been estimated that the critical point would probably be reached at a depth of about 7 m. At very high temperatures the elements of water may exist in a state of dissociation.

Fig. *Lava in The Ocean*

Much discussion has arisen as to the origin of the volcanic water, but probably it is not all attributable to a single source. Some may be of superficial origin, derived from rain, river or sea; whilst the upward passage of lava through moist strata must generate large volumes of steam. It has often been remarked that wet weather increases the activity of a volcano and that in certain mountains the eruptions are more frequent in winter. According, however, to Professor A. Ricco's prolonged study of Etna, rain has no apparent influence on the activity of this mountain and indeed the number of eruptions in winter, when rains are abundant, seems rather less than in summer.

The popular belief that explosive action is due to the admission of water to the volcanic focus is founded mainly on the topographic relation of volcanoes to large natural bodies of water, many being situated near the shore of a continent or on islands or even on the sea-floor. Salt water gaining access to heated rocks, through fissures or by capillary absorption, would give rise not only to watervapour but to the volatile chlorides so common in volcanic exhalations.

Yet it is notable that comparatively little chlorine is found among the products exhaled by the volcanoes of Hawaii, though these are typically insular. L. Palmieri, however, described certain sublimates on lava at Vesuvius after the eruption of 1872 as deposits of "sea-salt," to show that they were not simply sodium chloride, but contained other constituents found in sea-water.

It has been supposed that water on reaching the hot walls of a subterranean cavity would pass into the spheroidal state and on subsequent reduction of temperature might come into direct contact with the heated surface, when it would flash with explosive violence into steam.

Such catastrophes probably occur in certain cases. When, for example, a volcano becomes dormant, water commonly accumulates in the crater and on a renewal of activity this craterlake may be absorbed through fissures in the floor leading to the reopened duct and thus become rapidly, even suddenly, converted into vapour. But such incidents are accidental rather than normal and seem incompetent to account for volcanic activity in general.

When a stream of lava flows into the sea it no doubt immediately generates a prodigious volume of steam; but this is only a temporary phenomenon, for the lava rapidly becomes chilled by the cold water, with formation of a superficial solid layer, which by its low thermal conductivity allows the internal mass to cool slowly and quietly.

In the great eruption of Krakatoa in 1883 the sea-water gained occasional access to the molten lava and by its cooling effect checked the escape of vapour, thus temporarily diminishing the volcanic activity. But Judd compares this action to that of fastening down the safety-valve of a steam-boiler. The tension of the elastic fluids being increased by this repression would give rise subsequently to an explosion of greater violence; and hence the short violent paroxysms characteristic of the Krakatoa eruption were due to what he calls a " check and rally " of the subterranean forces. The action in the volcanic conduit has, indeed, been compared with that of a geyser.

The downward passage of water through fissures must be confined to the upper portion of the earth's crust known as the " zone of fracture," for it is there only that open channels can exist. Water might also percolate through the pores of the rocks, but even the pores are closed at great depths.

It was shown many years ago by G. A. Daubree that water could pass to a limited extent through a heated rock against the pressure of steam in the opposite direction. According to S. Arrhenius, water may pass inwards through the sea-bottom by osmotic pressure.

As the melting points of various silicates are lowered by admixture with water, it appears that the access of surface-waters to heated rocks must promote their fusibility. Judd has suggested that the proximity of large bodies of water may be favourable to volcanic manifestations, because the hydrated rocks become readily melted by internal heat and thus yield a supply of lava.

Whilst some of the water-vapour exhaled from a volcano is undoubtedly derived from superficial sources, notably in such insular volcanoes as Stromboli, the opinion has of late years been gaining ground, through the teaching of Professor E. Suess and others, that the volcanic water must be largely referred to a deep-seated subterranean origin - that it is, in a word, "Hypogene " or magmatic rather than meteoric.

It is held that the magma as it rises through the volcanic conduit brings up much water-vapour and other gaseous matters derived from original sources, perhaps a relic of what was present in the earth in its molten condition, having possibly been absorbed from a dense primordial atmosphere, or, as suggested by Professor T.C. Chamberlin, entrapped by the globe during its formation by accretion of planetesimal matter. Water brought from magmatic depths to the surface and appearing there for the first time, has been termed "Juvenile," and it has been assumed that such water may be seen in hot springs like those at Carlsbad. Professor J.W. Gregory has suggested that certain springs in the interior of Australia may derive part of their supply from juvenile or plutonic waters. According to A. Gautier, the origin of volcanic water may be found in the oxidation of hydrogen, developed from masses of crystalline rock, which by subsidence have been subjected to the action of subterranean heat.

Volcanic Vapours

It seems not unlikely that the vapours and gases exist in the volcanic magma in much the same way that they can exist in molten metal. It is a familiar fact that certain metals when melted can absorb large volumes of gases without entering into chemical combination with them. Molten silver, for example, is capable of absorbing from the atmosphere more than twenty times its volume. of oxygen, which it expels on solidification, thus producing what is called the " spitting of silver." Platinum again can absorb and retain when solid, or occlude, a large volume of hydrogen, that can be expelled by heating the metal in vacuo. In like manner molten rock under pressure can absorb much steam. It appears that many igneous rocks contain gases locked up in their pores, not set free by pulverization, yet capable of expulsion by strong heat.

Fig. *Volcanic Vapours*

Tilden has found that granite, gabbro, basalt and certain other igneous rocks enclose many times their volume of gases, chiefly hydrogen and carbon dioxide, with carbon monoxide, methane and nitrogen. Thus, the basalt of Antrim in Ireland, which is a Tertiary lava, yielded eight times its volume of gas having the following percentage composition: hydrogen 36.15, carbon dioxide 32.08, carbon monoxide 20.08, methane 10, nitrogen. No doubt some of the gases evolved on heating rocks may be generated by reactions during the experiment, as shown by M. W. Travers and also by Armand Gautier.

It has been pointed out by Gautier that the gas exhaled from Mont Pele during the eruption of 1902 had practically the same composition as that which he obtained on heating granite and certain other rocks. According to this authority a cubic kilometre of granite heated to redness would yield not less than 26,000,000 tons of water-vapour, besides other gases.

If then a mass of granite in the earth's crust were subject to a great local accession of heat it might evolve vast volumes of gaseous, matter, capable of producing an eruption of explosive type. Judd found that the little balls of Siberian obsidian called marekanite threw off, when strongly heated, clouds of finely divided particles formed by rupture of the distended mass through the escape of vapour. Pitchstone when ignited loses in some cases as much as to % of its weight, due to expulsion of water.

Much of the steam and other vapour brought up from below by the lava may be evolved on mere exposure to the air and hence a stream freshly extruded is generally beclouded with more or less vapour. Gaseous bubbles in the body of the lava render it vesicular,. especially in the upper part of a stream, where the pressure is relieved and the vesicles by the onward flow of the lava tend to become elongated in the direction of movement.

Vesiculation, being naturally resisted by cohesion, is not common in very viscid lavas of acid type, nor is it to be expected where the lava has been subject to great pressure, but it is seen to perfection in surface-flows of liquid lavas of basaltic character. A vesicular structure may sometimes be seen even in dykes, but the cavities are usually rounded rather than elongated and are often arranged in bands parallel to the walls of the dyke. A very small proportion of water in a lava. suffices to produce vesiculation. Secondary minerals developed in a cellular lava may be deposited in the steam-holes, thus producing an amygdaloidal rock.

After the surface of a lava-stream has become crusted over, vapour may still be evolved in the interior of the mass and in seeking release may elevate or even pierce the crust. Small cones may thus be thrown up on a lava-flow and when vapour escapes from terminal or lateral orifices they are known as " spiracles." The steam may issue with sufficient projectile force to toss up the lava in little fountains. When the lava is very liquid, as in the Hawaiian volcanoes, it may after projection from the blow-hole fall back in drops and plastic clots, which on consolidation form, by their union, small cones.

Vapour-vents on lava are often known as fumaroles. The character of the gaseous exhalations varies with the temperature and the following classification was suggested by C. Sainte-Claire Deville:

- Dry or white fumaroles having a temperature above 500° C. and evolving compounds of chlorine and perhaps fluorine.
- Acid fumaroles, exhaling much steam, with hydrocholoric acid and sulphur dioxide.
- Alkaline fumaroles, at a temperature of about 100°, with much steam and ammonium chloride and some sulphuretted hydrogen.
- Cold fumaroles, below100°, with aqueous vapour, carbon dioxide and sulphuretted hydrogen.
- Mofettes, indicating the expiring phase of vulcanism. A similar sequence of emanations, following progressive cooling of the lava, has been noted by other observers. During an eruption, the gaseous products may vary considerably. Johnston-Lavis found at Vesuvius that the vapour which first escaped from the boiling lava contained much sulphurous acid and that hydrochloric acid and other chlorides. appeared later.

If the vapours exhaled from volcanoes were derived originally from superficial sources, the lava would, of course, merely return to the surface of the earth what it had directly or indirectly absorbed. But if, as is now rather generally believed, much if not most of the volcanic vapour is derived from original subterranean sources, it must form a direct contribution from the interior of the earth to the atmosphere and hydrosphere and consequently becomes of extreme geological interest.

PHYSICAL STRUCTURE OF LAVAS

An amorphous vitreous mass may result from the rapid cooling of a lava on its extrusion from the volcanic vent. The common type of volcanic glass is known as obsidian. Microscopic examination usually shows that even in this glass some of the molecules of the magma have assumed definite orientation, forming the incipient crystalline bodies known as microlites, &c. By the increase of these minute enclosures, in number and magnitude, the lava may become devitrified and assume a lithoidal or stony structure. If the molten magma consolidate slowly, the various silicates in solution tend to separate by crystallization as their respective points of saturation are reached.

Should the process be arrested before the entire mass has crystallized, the crystals that have been developed will be embedded in the residual magma, which may, on consolidation, form a vitreous base. It is believed that in many cases the lava brings up, through its conduit, myriads of crystals that have been developed during slow solidification in the heart of the volcanic apparatus. Showers of crystals of leucite have occurred at Vesuvius, of labradorite at Etna and of pyroxene at Vesuvius, Etna and Stromboli. These "intratelluric crystals were probably floating in the molten magma and had they remained in suspension, this magma might on consolidation have enveloped them as a groundmass or base. A rock so formed is generally known as a " porphyry," and the structure as porphyritic. In such a lava the large crystals, or phenocrysts, evidently represent an early phase of consolidation and the minerals of the matrix a later stage. It is notable that the intratelluric crystals often lack sharpness of outline, as though they had suffered corrosion by attack of the molten magma, whilst they may contain vitreous enclosures, suggesting that the surrounding mass was liquid during their consolidation. It is believed that the more slowly consolidation has occurred, the larger generally are the crystals; and the higher the temperature of the magma the greater the corrosion or resorption. Possibly under certain conditions the phenocrysts and the ground-mass may have solidified simultaneously.

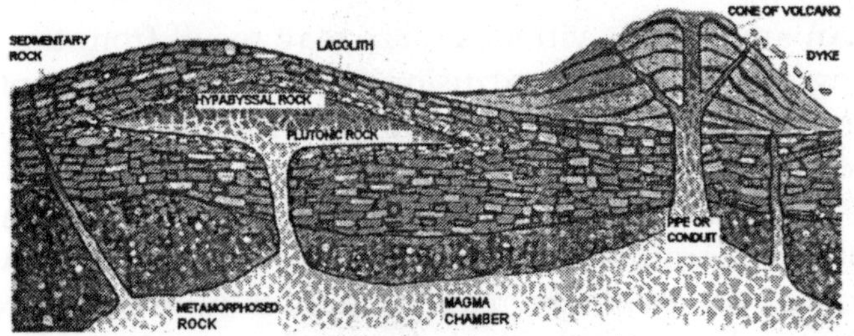

Fig. *Structure of Lavas*

In some cases the entire igneous mass assumes a crystalline structure, or becomes "holocrystalline." Such a structure is well displayed when the magma has consolidated at considerable depths, cooling slowly under great pressure and forming rocks which are termed "plutonic" or "abyssal" to distinguish them from rocks truly volcanic, or those which, if not effusive, like lava-flows, have at least solidified very near to the surface as dykes and sills.

Volcanic and plutonic rocks pass, however, into each other by gradual transition. The dyke-rocks, or intrusive masses, form an intermediate group sometimes distinguished under the name of "hypabyssal" rocks, as suggested by W.C. Brogger. Lavas extruded in submarine eruptions may have solidified under a great weight of seawater and therefore to that extent rather under plutonic conditions.

DESCRIPTION OF SPECIAL GASES AND VAPOURS

Hydrochloric acid, HCI, escapes abundantly from many vents, often accompanied with the vapours of certain metallic chlorides and is responsible for much of the acrid effects of volcanic exhalations. To avoid dangerous vapours an active volcano should be ascended on the windward side. Free hydrofluoric acid, HF, has sometimes been detected with the hydrochloric acid among Vesuvian vapours and silicon fluoride, SiF_4, has also been reported. Sulphuretted hydrogen, H_2S, is a frequent emanation and being combustible may contribute to the lambent flames seen in some eruptions. It readily suffers oxidation, giving rise to sulphur dioxide and water. By the interaction of hydrogen sulphide and carbon dioxide, water and carbon oxysulphide, COS, are formed; whilst by reaction with sulphur dioxide, water and free sulphur are produced, such being no doubt the origin of many deposits of volcanic sulphur.

Hydrogen sulphide may be formed by the decomposition of certain metallic sulphides, like that of calcium, in the presence of moisture, as suggested by Anderson and Flett with regard to certain muds at the Soufriere of St Vincent. Sulphur dioxide, SO_2, is one of the commonest exhalations, especially at acid fumaroles. It may be detected by its characteristic smell, that of burning brimstone, even when present in very small proportion and in the presence of an excess of hydrochloric acid. By hydration it readily forms sulphurous acid, which may be further oxidized to sulphuric acid. J. B. Boussingault found free sulphuric acid (with hydrochloric acid) in the water of the Rio Vinagre which issues from the volcano of Purace in the Andes of Colombia; and it occurs also in certain other volcanic waters.

Carbon dioxide, CO_2, is generally a product of the later stages of an eruption and is often evolved after all other gases have ceased to escape. Although it may sometimes be due to the decomposition of limestone, it seems to be mostly of true magmatic origin. At the well-known Grotta del Cane, at Lake Agnano, in the Phlegraean Fields near Naples, there has been for ages a copious discharge and analyses of the air of the cave by T. Graham Young showed the presence of from 61.5 to 7 1% of carbon dioxide.

Gautier, in 1907, found 96 to 97% of this gas in the vapours (excluding water-vapour) emitted from the Solfatara near Pozzuoli in the Bay of Naples. The gas by its density tends to accumulate in depressed areas, as in the Death Gulch in the Yellowstone Park and in the Upas Valley of Java. In the Eifel, in the Auvergne and in many other volcanic regions it is discharged at temperatures not above that of the atmosphere. This natural carbonic acid gas is now utilized industrially at many localities. In the gases of the fumaroles of Mont Pele, carbon monoxide, CO, was detected by H. Moissan.

Probably certain hydrocarbons, notably methane or marsh-gas, CH_4, often exist in volcanic gases. They might be formed by the action of water on natural carbides, such as that of magnesium, calcium, &c. Moissan found 5.46% of methane in vapour from a fumarole on Mont Pele in 1902.

Free hydrogen was detected by R. Bunsen as far back as 1846 in vapours from volcanoes in Iceland. In 1861 Deville and Fouque found it, with hydrocarbons, at Torre del Greco near Naples; and in 1866 Fouque discovered it at Santorin, where some of the vapour at the immediate focus of eruption contained as much as 30% of hydrogen.

It is notable that at Santorin free oxygen was also found. The elements of water may possibly exist, at the high temperature of the magma, in a state of dissociation and certain volcanic explosions have sometimes been attributed to the combination of these elements.

Oxygen is not infrequently found among volcanic emanations, but may perhaps be derived in most cases from superficial air and ground-water; and in like manner the nitrogen, often detected, may be sometimes of atmospheric origin, though in other cases derived from nitrides in the lava. In the vapours emitted by Mont Pele in 1902 argon was detected by H. Moissan, to the extent of 0.71%; and in those from Vesuvius in 1906 argon and neon were found by Gautier. The collection of volcanic vapours offers difficulty and it is not easy to avoid admixture with the atmosphere. F. A. Perret has successfully collected gases on,Vesuvius.

VOLCANIC FLAMES

Although the incandescence of the lava and stones projected during an eruption and the reflection from incandescent matter in the crater have often been mistaken for red flames, there can be no doubt that true combustion, though generally feeble, does occur during volcanic outbursts. Among the gases cited above, hydrogen, hydrogen sulphide and the hydrocarbons are inflammable. The flames seen in volcanoes are generally pale and of bluish, greenish or yellowish tint. The presence of hydrogen, sodium and hydrocarbons in the volcanic flames of Kilauea.

These, however, were referred to the kindling of sulphur deposited around the fumaroles, the flames being coloured by the presence of boric acid and arsenic sulphide. When a stream of lava flows over vegetation the combustion of the leaves and wood ma y be mistaken for flames issuing from the lava. In like manner brushwood may grow in the crater of a dormant volcano and be ignited by a fresh outburst of lava, thus producing flames which, from their position in the crater, may readily deceive an observer.

Volcanic Sublimates

Certain mineral substances occur as sublimates in and around the volcanic vents, forming incrustations on the lava. They are either deposited directly from the effluent vapours, which carry them in a volatile condition, or are produced by interaction of the vapours among themselves; whilst some of the incrustations, rather loosely called sublimates, are due to reaction of the vapours on the constituents of the lava. Possibly at the temperature of the magma-reservoirs even silica and various silicates may be volatilized and might thus yield sublimation products. Many of the volcanic sublimates occur at first as incandescent crusts on the lava. Being generally unstable they are difficult of preservation and are not usually well represented in collections.

Among the commonest sublimates is halite, or sodium chloride, NaCI, occurring as a white crystalline incrustation, sometimes accompanied, as at Vesuvius, by sylvite, or potassium chloride, KC1, which forms a similar sublimate. The two chlorides may be intimately associated. Sal ammoniac, or ammonium chloride, NH_4C1, is not uncommon, especially at Etna, as a white crystalline crust, probably formed in part by the reaction of hydrochloric acid with nitrogen and hydrogen in the vapours. Bunsen, on finding it in Iceland, regarded it as a product of the distillation of organic matter. At the Solfatara, near Pozzuoli, sal ammoniac was formerly collected as a sublimate on tiles placed round a bocca or vapourvent. Ferric chloride, FeCl 3, not infrequently occurs as a reddish or brownish yellow deliquescent incrustation and because it thus colours the lava it has received the name of molysite. The action of hydrochloric acid on the iron compounds in the lava may readily yield this chloride, which from its yellowish colour has sometimes been mistaken for sulphur. A crystalline sublimate from the fumaroles on Vesuvius, containing ferric and alkaline chlorides, KCl NH_4Cl 2FeC13 +$6H_2O$, is known as kremersite, after P. Kremers. From a scoriaceous lava found on Vesuvius after the eruption of 1906, Johnston-Lavis procured a yellow rhombohedral sublimate, which he proved to be a chloride of manganese and potassium, whence he proposed for it the name chlormanganokalite. It was studied by L. J.

Spencer and found to contain 4KC1. $MnC1_2$. Chlorocalcite, or native calcium chloride, $CaC1_2$, has been found in cubic crystals on Vesuvian lava. Fluorite, or calcium fluoride, CaF 2, is also known as a volcanic product. Lead chloride, PbC12, a rare Vesuvian mineral, was named cotunnite, after Dr Cotugno of Naples.

The action of hydrogen sulphide on this chloride may give rise to galena, PbS, found by A. Lacroix on Vesuvius in 1906. Atacamite, or cupric oxychloride, $CuC1_{2\cdot 3}Cu\ (OH)_2$, occurs as a green incrustation on certain Vesuvian lavas, notably those of 1631. Another green mineral from Vesuvius was found by A. Scacchi to be a sulphate containing copper, with potassium and sodium, which he named from its fine colour *euclorina* - a word which has been written in English as euchlorinite.

The copper in the sublimates on Vesuvius will sometimes plate the iron nails of a traveller's boots when crossing the newly erupted lava. Cupric oxide, CuO, occurs in delicate crystalline scales termed tenorite, after Professor G. Tenore of Naples; whilst cupric sulphide, CuS, forms a delicately reticulated incrustation known as covellite, after N. Covelli, its discoverer at Vesuvius. A sublimate not infrequently found in feathery crystalline deposits on lava at Vesuvius and formerly called " Vesuvian salt," is a potassium and sodium sulphate, (K. Na) $2SO_4$, known as aphthitalite (from Gr. a4B%Tos, imperishable and salt). A sulphate with the composition PbS04 (KNa)2 SO_4, found in the fumaroles at Vesuvius after the eruption of 1906, was named by A. Lacroix palmierite, after L. Palmieri, who was formerly director of the observatory on Vesuvius. Various sulphites are formed on lavas by the sulphurous acid of the vapours.

Ferric oxide, $Fe2O_3$, which occurs in beautiful metallic scales as specular iron-ore, or as an amorphous reddish incrustation on the lava, is probably formed in most cases by the interaction of vapour of ferric chloride and steam at a high temperature. Less frequently, magnetite, $Fe3O_4$ and magnesioferrite, $MgFe_2O_4$ are found in octahedral crystals on lava. An iron nitride ($Fe5N_2$) was detected thinly incrusting a lava erupted at Etna in 1874 and was named by 0. Silvestri, who examined it, siderazote.

Boric acid, H3B03, occurs in the crater of Vulcano so abundantly that it was at one time collected commercially. It has also led to the foundation of an industry in Tuscany, where it is obtained from the soffioni of the Maremma. From Sasso in Tuscany it has received the name of sassolin or sassolite. Realgar, or arsenic sulphide, occurs in certain volcanic exhalations and is deposited as an orange-red incrustation, often associated with sulphur, as at the Solfatara, where orpiment, As2S3, has also been found.

Of all volcanic products, sulphur is in some respects the most important. It may, in large quantity lining the walls of the crater, as at Popocatepetl in Mexico, where it was formerly worked by the Indian " volcaneros," or on the other hand it may be a rare product, as at Vesuvius. Sulphur appears generally to owe its origin in volcanic areas to the interaction of sulphur dioxide and sulphuretted hydrogen, or to the action of water on the latter.

A volcanic vent where sulphur is deposited is truly a solfatara (*solfo terra*) or a soufriere, but all volcanoes which have passed into that stage in which they emit merely heated vapours now pass under this name. The famous Solfatara, an old crater in the Phlegraean Fields, exhales sulphurous vapours, especially at the Bocca Grande, from which sulphur is deposited. In the orangecoloured sulphur of the Solfatara, realgar may be present to the extent of as much as 18%.

A brown seleniferous sulphur occurring at Vulcano, one of the Lipari Islands, was termed by W. Haidinger volcanite, but it should be noted that Professor W. H. Hobbs has applied this name to an anorthoclase-augite rock ejected as bombs at Vulcano. Sulphur containing selenium is known as a volcanic product in Hawaii, whilst in Japan not only selenium but tellurium occurs in certain kinds of sulphur.

At the Solfatara, near Pozzuoli, the hot sulphurous vapours attack the trachytic rocks from which they issue, giving rise to such products as alum, kaolin and gypsum. To some of these products, including alunogen and mendozite (soda-alum), the name solfatarite was given by C. W. Sheppard in 1835.

By prolonged action of the acid vapours on lava, the bases of the silicates may be removed, leaving the silica as a soft white chalk-like substance. The occurrence of kaolin and other white earthy alteration-products led to the hills around the Solfatara being known to the Romans as the *Colli leucogei.*

THE HOT DUST CLOUD AND AVALANCHE OF PELE

The terrific eruptions in the islands of Martinique and St Vincent in the West Indies in 1902, furnished examples of a type of activity not previously recognized by vulcanologists, though, as Professor A. Lacroix has pointed out, similar phenomena have no doubt occurred elsewhere, especially in the Azores. By Drs Tempest Anderson and J. S. Flett, who were commissioned by the Royal Society to report on the phenomena, this type of explosive eruption is distinguished as the " Pelean type."

Its distinctive character is found in the sudden emission of a dense black cloud of superheated and suffocating gases, heavily charged with incandescent dust, moving with great velocity and accompanied by the discharge of immense volumes of volcanic sand, which are not rained down in the normal manner, but descend like a hot avalanche. The cloud, with the avalanche, is called by Lacroix a *nuee Peleenne,* or *nuee ardente,* the latter term having been applied to the fatal cloud in the eruptions at San Jorge in the Azores in 1818.

In its typical form, the cloud seen at Pele appeared as a solid bank, opaque and impenetrable, but having the edge in places hanging like folds of a curtain and apparently of brown or purplish colour. Rolling along like an inky torrent, it produced in its passage intense darkness, relieved by vivid lightning. So much solid matter was suspended in the cloud, that it became too dense to surmount obstacles and behaved rather like a liquid.

It has, however, been suggested that its peculiar movement as it swept down the mountain was due not simply to its heavy charge of solids, but partly to the oblique direction of the initial explosion. After leaving the crater, it underwent enormous expansion and Anderson and Flett were led to suggest that possibly at the moment of emission it might have been partly in the form of liquid drops, which on solidifying evolved large volumes of gas held previously in occlusion.

The deadly effect of the blast seems to have been mostly due to the irritation of the mucous membrane of the respiratory passages by the fine hot dust, but suffocating gases, like sulphur dioxide and sulphuretted hydrogen, were associated with the water-vapour. Possibly the incandescent dust was even hotter than the surrounding vapour, since the latter might be cooled by expansion.

It is said that the black cloud as it swept along was accompanied by an indraught of air, not however sufficiently powerful to check its rapid advance. The current of air was likened by Anderson and Flett to the inrush of air at a railway station as an express train passes.

An attempt was made to determine the temperature of the fatal blast which destroyed St Pierre, but without very definite results. Thus it was assumed that as the telephone wires were not melted the temperature was below the fusing-point of copper: possibly, however, the blast may have passed too rapidly to produce the effects which might normally be due to its temperature.

SHAPE OF VOLCANIC CONES

Those volcanic products which are solid when ejected, or which solidify after extrusion, tend to form by their accumulation around the eruptive vent a hill, which, though generally more or less conical, is subject to much variation in shape. It occasionally happens that the hill is composed wholly of ejected blocks, not themselves of volcanic origin. In this case an explosion has rent the ground and the effluent vapours have hurled forth fragments of the shattered rock through which the vent was opened, but no ash or other fragmentary volcanic material has been ejected, nor has any lava been poured forth.

This exceptional type is represented in the Eifel by certain monticules which consist mainly of fragments of Devonian slate, more or less altered. In some cases the area within a ring of such rocky materials is occupied by a sheet of water, forming a crater-lake, known in the Eifel as a *rnaar*. Piles of fragmentary matter of this character, though containing neither cinders nor lava, may be fairly regarded as volcanic, inasmuch as they are due to the explosive action of hot subterranean vapours.

In the ordinary paroxysmal type of eruption, however, cinders and ashes are shot upwards by the explosion and then descend in showers, forming around the orifice a mound, in shape rather like the diminutive cone of sand in the lower lobe of an hour-glass. Little cindercones of this character may be formed within the crater of a large volcano during a single eruption; whilst large cones are built up by many successive discharges, each sheet of fragmentary material mantling more or less regularly round the preceding layer.

The symmetry of the hill is not infrequently affected by disturbing influences - a strong wind, for example, blowing the loose matter towards one side. The sides of a cinder cone have generally a steep slope, varying from 30 0 to 45°, depending on the angle of repose of the ejectamenta. Excellent examples of small scoria-cones are found among the puys of

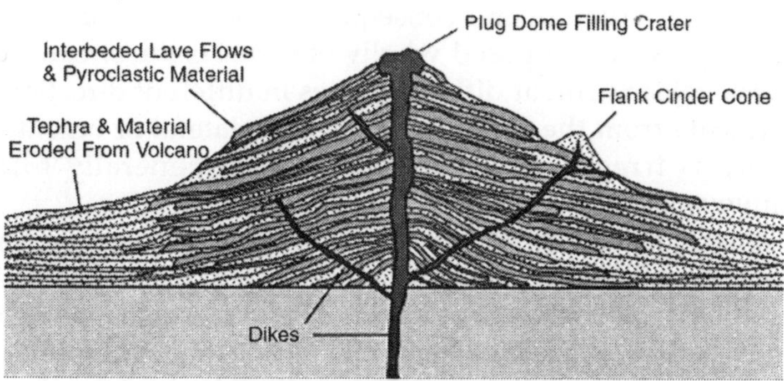

Fig. *Shape of Volcanic Cones*

Auvergne in central France, whilst a mag nificent illustration of this type of hill is furnished by Fusiyama, in Japan, which reaches an altitude of 12,000 ft. How such a cone may be rapidly built up was well shown by the formation of Monte Nuovo, near Pozzuoli - a hill 400 ft. high and a mile and a half in circumference, which is known from contemporary evidence to have been formed in the course of a few days in September 1538. The shape of a cinder cone may be retained for ages, since it is not liable to suffer greatly by denudation, as the rain soaks into the loose porous mass instead of running down the outside. If lava rises in the duct of a cinder cone, it may, on accumulation in the crater, break down the wall and thus effect its escape as a stream. Cones breached in this way are not uncommon in Auvergne.

It often happens that the cinders and ashes ejected from a volcano become mixed with water and so form a paste, which sets readily as a hard tufaceous mass. Such natural tuff is indeed similar to the hydraulic cement known as pozzolana, which is formed artificially from volcanic ashes and is renowned for durability. Although streams of volcanic mud are commonly associated with the ashes of a cinder-cone they may also form independent structures or tuffcones. These are generally broad-topped hills, having sides with an angle of slope as low in some cases as 15°. Lava-cones are built up of streams of lava which have consolidated around the funnel of escape. Associated with the lava, however, there is usually more or less fragmentary matter, so that the cones are composite in structure and consequently more acute in shape than if they were composed wholly of lava. As the streams of lava in a volcano run at different times in different directions, they radiate from the centre, or flow from lateral or eccentric orifices, as irregular tongues and do not generally form continuous sheets covering the mountain.

When lava is the sole or chief element in the cone, the shape of the hill is determined to a great extent by the chemical composition and viscosity of the lava, its copiousness and the rapidity of flow. If the lava be highly basic and very mobile, it may spread to a great distance before solidifying and thus form a hill covering a large area and rising perhaps to a great height, but remarkably flat in profile. Were the lava perfectly liquid, it would indeed form a sheet without any perceptible slope of surface. As a matter of fact, some lavas are so fluent as to run down an incline of 1 ° and flat cones of basalt have in some cases a slope of only 10° or even less.

The colossal mass of Mauna Loa, in Hawaii, forms a remarkably flat broad cone, spreading over a base of enormous area and rising to a height of 13,900 ft. Major Dutton, writing in 1883, said that "a; moderate eruption of Mauna Loa represents more material than Vesuvius has emitted since the days of Pompeii." Yet the lava is so mobile that it generally wells forth quietly, without explosive demonstration and therefore unaccompanied by fragmentary ejectamenta. Fluent lavas like those of Hawaii are also poured forth from the volcanoes and volcanic fissures of Iceland.

If the lava be less basic and less fusible, the hill formed by its accumulation instead of being a low dome will take the shape of a cone with sides of higher gradient: in the case of andesite cones, for instance, the slope may vary from 25° to 35°. Acid rocks, or those rich in silica, such as rhyolites and trachytes, may be emitted as very viscous lavas tending to form dome-shaped or bulbous masses.

Experiment shows that such lavas may persist for a considerable time in a semi-solid condition. It is possible for a dome to increase in size not by the lava running over the crater and down the sides but by injection of the pasty magma within the expanding bulb while still soft; or if solidified, the crust yields by cracking. Such a mode of growth, in which the dome consists of successive sheets that have been compared to the skins of an onion, has been illustrated by the experiments of Dr A. Reyer and the structure is typically represented by the mamelons or steep-sided domes of the Isle of Bourbon.

The Puy-de-Dome in Auvergne is an example of a cone formed of the trachytic rock called from its locality domite, whilst the Grand Sarcoui in the same region illustrates the broad domeshaped type of hill. Such domes may have no summit-crater and it is then usually assumed that the top with the crater has been removed by denudation, but possibly in some cases such a feature never existed. The "Dome volcano" of von Seebach is a dome of acid lava extruded as a homogeneous mass, without conspicuous chimney or crater. Although domes are usually composed of acid rocks, it seems possible that they may be formed also of basic lavas, if the magma be protruded slowly at a low temperature so as to be rapidly congealed.

The Spine of Pele

A peculiar volcanic structure appeared at Mont Pele in the course of the eruption of 1902 and was the subject of careful study by Professor A. Lacroix, Dr E. A. Hoovey, A. Heilprin and other observers. It appears that from fissures in the floor of the Etang Sec a viscous andesitic lava, partly quartziferous, was poured forth and rapidly solidified superficially, forming a domeshaped mass invested by a crust or carapace. According to Lacroix, the crust soon became fractured, partly by shrinkage on consolidation and partly by internal tension and the dome grew rapidly by injection of molten matter.

Then there gradually rose from the dome a huge monolith or needle, forming a terminal spine, which in the course of its existence varied in shape and height, having been at its maximum in July 1903, when its absolute height was about 5276 ft. above sea-level. The walls of the spine, inclined at from

75° to 90° to the horizon, were apparently slickensided, or polished and scratched by friction: masses were occasionally detached and vapours were continually escaping. Several smaller needles were also formed. Some observers regarded the great spine as a solidified plug of lava from a previous outburst, expelled on a renewal of activity. Lacroix, however, believed that it was formed by the extrusion of an enormous mass of highly viscid magma, perhaps partly solidified before emission and he compared the formation of the dome in the crater to the structure on Santorin in 1866, described by Fouque as a "cumulo-volcano."

Professor H. F. Cleland has suggested a comparison with the cone of andesite in the crater of the volcano of Toluca in Mexico and it is said that similar formations have been observed in the volcanoes of the Andes. Dr Tempest Anderson, on visiting Pele in 1907, found a stump of the spine, consisting of a kind of volcanic agglomerate, rising from a cone of talus formed of its ruins.

The Crater

The eruptive orifice in normal volcano - the *bocca* of Italian vulcanologists is usually situated at the bottom of a depression or cup, known as the crater. This hollow is formed and kept open by the explosive force of the elastic vapours and when the volcano becomes dormant or extinct it may be closed, partly by rock falling from its crumbling walls and partly by the solidification of the lava which it may contain. If a renewed outburst occurs, the floor of the old crater may reopen or a new outlet may be formed at some weak point on the side of the mountain: hence a crater may, with regard to position, be either terminal or lateral.

The position of the crater will evidently be also changed on any shifting of the general axis of eruption. In shape and size the crater varies from time to time, the walls being perhaps breached or even blown away during an outburst. Hence the height of a volcanic mountain in activity, measured to the rim of the crater or the terminal peak, is not constant. Vesuvius, for example, suffered a reduction of several hundred feet during the great eruption of 1906, the east side of the cone having lost, according to V. R. Matteucci, 120 metres. Whilst in many cases the crater is a comparatively small circular hollow around the orifice of discharge, it forms in others a large bowllike cavity, such as is termed in some localities a " caldera." In the Sandwich Islands the craters are wide pits bounded by nearly vertical walls, showing stratified and terraced lavas and floored by a great plain of black basalt, sometimes with lakes of molten lava.

Professor W. H. Pickering compares the lava-pits of Hawaii to the crater-rings in the moon. Some of the pit-craters in the Sandwich Islands are of great size, but none comparable with the greatest of the lunar craters. Dr G. K. Gilbert, however, has suggested that the ring-shaped pits on the moon are not of volcanic origin, but are depressions formed by the impact of meteorites. Similarly the " crater " of Coon Butte, near Canyon Diablo, in Arizona, which is 4000 ft. in diameter and 500 ft. deep, has been regarded as a vast pit due to collision of a meteorite of prodigious size.

Probably the largest terrestrial volcanic crater is that of Aso-san, in the isle of Kiushiu, which is a huge oval depression estimated by some observers to have an area of at least zoo sq. m. Some of the large pit-craters have probably been formed by subsidence, the cone of a volcano having been eviscerated by extravasation of lava and the roof of the cavity having then subsided by loss of support. The term caldera has sometimes been limited to craters formed by such collapse.

On the floor of the crater, ejected matter may accumulate as a conoidal pile; and if such action be repeated in the crater of the new cone, a succession of concentric cones will ultimately be formed. The walls of a perfect crater form a ring, giving the cone a truncated appearance, but the ring may suffer more or less destruction in the course of the history of the mountain. A familiar instance of such change is afforded by Vesuvius. The mountain now so called, using the term in a restricted sense, is a huge composite cone built up within an old crateral hollow, the walls of which still rise as an encircling rampart on the N. and N.E. sides and are known as Monte

Somma; but the S. and S.W. sides of the ancient crater have disappeared, having been blown away during some former outburst, probably the Plinian eruption of 79. In like manner the relics of an old crater form an amphitheatre partially engirdling the Soufriere in St Vincent and other examples of " Somma rings " are known to vulcanologists.

Much of the fragmental matter ejected from a volcano rolls down the inside of the crater, forming beds of tuff which incline towards the central axis, or have a centroclinal dip. On the contrary, the sheets of cinder and lava which form the bulk of the cone slope away from the axis, or have a dip that is sometimes described as pericentric or qua-qua-versal. According to the old "Crater-of-elevation theory," held especially by A. von Humboldt, L. von Buch and Elie de Beaumont, this inclination of the beds was regarded as mainly due to upheaval. It was contended that the volcanic cone owed its shape, for the most part, to local distension of the ground and was indeed comparable to a huge blister of the earth's crust, burst at the summit to form the " elevation crater." Palma, in the Canary Islands, was cited as a typical example of such a formation.

This view was opposed mainly by Poulett-Scrope, Sir Charles Lyell and Constant Prevost, who argued that the volcano, so far from being bladder-like, was practically a solid cone of erupted matter: hence this view came to be known as the " crater-of-eruption theory." Its general soundness has been demonstfated whenever an insight has been obtained into the internal structure of a volcano.

Thus, after the eruption of Krakatoa in 1883 a magnificent natural section of the great cone of Rakata, at the S. end of the island, was exposed - the northern half having been blown away and it was then evident that this mountain was practically a solid cone, built up of a great succession of irregular beds of tuff and lava, braced together by intersecting dykes. The internal architecture of a volcano is rarely so well displayed as in this case, but dissections of cones, more or less distinct, are often obtained by denudation. It should be mentioned that, in connexion with the structures called *laccoliths,* there may have been an elevation, or folding and even faulting, of the superficial rocks by subterranean intrusion of lava; but this is different from the local expansion and rupture of the ground required by the old theory. It may be noted, however, that in recent years the view of elevation, in a modified form, has not been without supporters. Where the growth of a volcanic mound takes place from within, as in certain steep-sided trachytic cones, there may be no perceptible crater or external outlet.

Again, there are many volcanoes which have no crater at the summit, because the eruptions always take place from lateral outlets. Even when a terminal pit is present, the lava may issue from the body of the mountain and in some cases it exudes from so many vents or cracks that the volcano has been described as " sweating fire."

PARASITIC CONES

In the case of a lofty volcano the column of lava may not have sufficient ascensional force to reach the crater at the summit, or at any rate it finds easier means of egress at some weak spot, often along radial cracks, on the flanks of the mountain. Thus at Etna, which rises to a height of more than io,800 ft., the eruptions usually proceed from lateral fissures, sometimes at least half-way down the mountain-side. When fragmental materials are ejected from a lateral vent a cinder-cone is formed and by frequent repetition of such ejections the flanks of Etna have become dotted over with hundreds of scoria-cones much like the puys of Auvergne, the largest (Monte Minardo) rising to a height of as much as 750 ft. Hills of this character, seated on the parent mountain, are known as parasitic cones, minor cones, lateral cones.

Such subordinate cones often show a tendency to a linear arrangement, rising from vents or *bocche* along the floor of a line of fissure. Thus in 1892 a chain of five cones arose from a rift on the S. side of Etna, running in a N. and S. direction and the hills became known as the Monti Silvestri, after Professor Orazio Silvestri of Catania. This rift, however,was but a continuation of a fissure from which there arose in 1886 the series of cones called the Monti Gemmellaro, while this in turn was a prolongation of a rent opened in 1883. The eruption on Etna in the spring of 1910 took place along the same general direction, but at a much higher elevation. The tendency for eruptions to be renewed along old lines of weakness, which can be readily opened afresh and extended, is a feature well known to vulcanologists.

The small cones which are frequently thrown up on lava streams were admirably exemplified on Vesuvius in the eruption of 1855 and figured by J. Schmidt. The name of " driblet cones " was given by J. D. Dana to the little cones and pillars formed by jets of lava projected from blowing holes at Kilauea, the drops of lava remaining plastic and cohering as they fell. Such clots may form columns and. pyramids, with almost vertical sides. Steep-sided cones more or less of this character occur elsewhere, but are usually built up around spiracles. Small cones formed by mere dabs of lava are known trivially as " spatter cones."

FISSURE ERUPTIONS

In certain parts of the world there are vast tracts of basaltic lava with little or no evidence of cones or of pyroclastic accompaniment. To explain their formation, Baron F. von Richthofen suggested that they represent great floods of lava which were poured forth not from ordinary volcanic craters with more or less explosive violence, but from great fissures in the earth's crust, whence they may have quietly welled forth and spread as a deluge over the surface of the country.

The eruptions were thus effusive rather than explosive. Such phenomena, constituting a distinct type of vulcanism, are distinguished as fissure eruptions or massive eruptions terms which suggest the mode of extrusion and the character of the extruded matter.

As the lava in such outflows must be very fusible, it is generally of basaltic type, like that of Hawaii: indeed, the Hawaiian volcanoes, with their quiet emission of highly fluent lavas, connect the fissure eruptions with the "central eruptions," which are usually regarded as representing the normal type of activity. At the present day true fissure eruptions seem to be of rather limited occurrence, but excellent examples are furnished by Iceland. Here there are vast fields of black basalt, formed of sheets of lava which have issued from long chasms, studded in most cases with rows of small cones, but these generally so insignificant that they make no scenic features and might be readily obliterated by denudation. Dr T. Thoroddsen enumerates 87 great rifts and lines of cones in Iceland and even the larger cones of Vesuvian type are situated on fissures.

It is believed that fissure eruptions must have played a far more important part in the history of the earth than eruptions of the familiar cone-and-crater type, the latter representing indeed only a declining phase of vulcanism. Sir Archibald Geikie, who has specially studied the subject of fissure eruptions, regards the Tertiary basaltic plateaus of N.E. Ireland and the Inner Hebrides as outflows from fissures, which may be represented by the gigantic system of dykes that form so marked a feature in the geological structure of the northern part of Britain and Ireland.

These dykes extend over an area of something like 40,000 sq. m., while the outflows form an aggregate of about 3000 ft. in thickness. In parts of Nevada, Idaho, Oregon and Washington, sheets of late Tertiary basalt from fissure eruptions occupy an area of about 200,000 sq. m. and constitute a pile at least 2000 ft. thick. In India the " Deccan traps " represent enormous masses of volcanic matter, probably of like origin but of Cretaceous date, whilst South Africa furnishes other examples of similar outflows. Professor J. W. Gregory recognized in the Kapte plains of East Africa evidence of a type of vulcanism, which he distinguished as that of " plateau eruptions. " According to him a number of vents opened at the points of intersection of lines of weakness in a high plateau, giving rise to many small cones and the simultaneous flows of lava from these cones united to form a far-spreading sheet.

Extrusive and Intrusive Magmas

When the molten magma in the interior of the earth makes

its way upwards and flows forth superficially as a stream of lava, the product is described as extrusive, effusive, effluent or eruptive; but if, failing to reach the surface, the magma solidifies in a fissure or other subterranean cavity, it is said to be intrusive or irruptive. Rocks of the former group only are sometimes recognized as strictly " volcanic, " but the term is conveniently extended, at least in certain cases, to igneous rocks of the latter type, including therefore certain hypabyssal and even plutonic rocks.

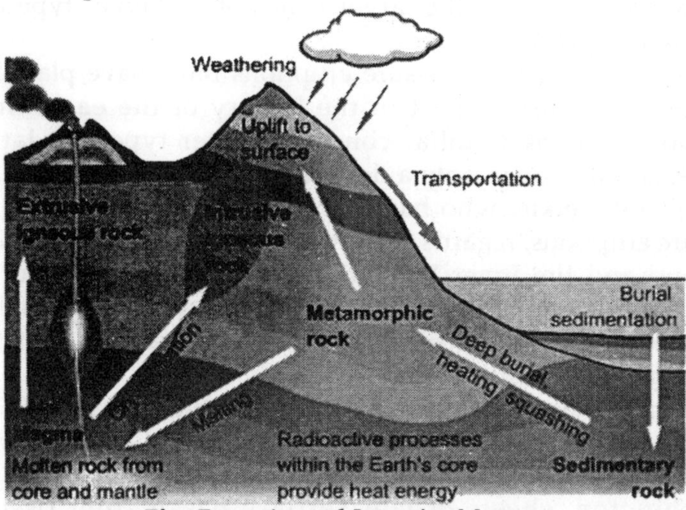

Fig. *Extrusive and Intrusive Magmas*

When the intrusive magma has been forced into narrow irregular crevices it forms " veins, " which may exhibit complex ramifications, especially marked in some acid rocks; but when injected into a regularly shaped fissure, more or less parallel-sided and cutting across the planes of bedding, it forms a wall-like mass of rock termed a " dyke. " Most dykes are approximately vertical, or at least highly inclined in position. The inclination of a dyke to a vertical plane is termed its " hade. " In a cinder-cone, the lava as it rises may force its way into cracks, formed by pressure of the magma and tension of the vapours and will thus form a system of veins and dykes, often radiating from the volcanic axis and strengthening the structure by binding the loose materials together.

Thus, in the Valle del Bove, a huge cavity on the east side of Etna, the walls exhibit numerous vertical dykes, which by their hardness stand out as rocky ribs, forming a marked feature in the scenery of the valley. In a similar way dykes traverse the walls of the old crater of Monte Somma at Vesuvius. Exceptionally a dyke may be hollow, the lava having solidified as a crust at the margin of the fissure but having escaped from the interior while still liquid.

When molten matter is thrust between beds of tuff or between successive lava-flows or even ordinary sedimentary strata, it forms an intrusive sheet of volcanic rock known as a " sill. " A sill may sometimes be traced to its connexion with a dyke, which represents the channel up which the lava rose, but instead of reaching the surface the fluid found an easier path between the strata or perhaps along a horizontal rent. Although a dyke may represent a conduit for the ascent of lava which has flowed out superficially, yet if the lava has been removed at the surface by denudation the dyke terminates abruptly, so that its function as the former feeder of a lava-current is not evident. In other cases a dyke may end bluntly because the crack which it occupies never reached the surface.

Lava which has insinuated itself between planes of stratification may, instead of spreading out as a sheet or sill, accumulate locally as a lenticular mass, known as a *laccolith* or *laccolite.* Such a mass, in many cases rather mushroom-shaped, may force the superincumbent rocks upwards as a dome and though at first concealed may be ultimately exposed by removal of the overlying burden by erosion. The term *phacolite* was introduced by A. Harker to denote a meniscus-shaped mass of lava intruded in folded strata, along a crest or a trough. The *bysmalith* of Professor Iddings is a *laccolith* of rather plug-like shape, with a faulted roof. An intrusive mass quite irregular in shape has been termed by R. A. Daly a *chonolith* whilst an intrusion of very great size and illdefined form is sometimes described as a *bathylith* or *batholite.*

STRUCTURAL PECULIARITIES IN LAVA

Many of the structures exhibited by lava are due to the

conditions under which solidification has been effected. A dyke, for example, may be vitreous at the margin where it has been rapidly chilled by contact with the walls of the fissure into which it was injected, whilst the main body may be lithoidal or crystalline: hence a basalt dyke will sometimes have a selvage formed of the basaltic glass known as tachylyte. A similar glass may form a thin crust on certain lava-flows.

In a homogeneous vitreous lava, contraction on solidification may develop curved fissures, well seen in the delicate "Perlitic" cracks of certain obsidians, indicating a tendency to assume a globular structure. This structure becomes very distinct by the development of "Spherulites," or globular masses with a radiating fibrous structure, sometimes well seen in devitrified glass.

Occasionally the spherulitic bodies in lava are hollow, when they are known as lithophyses, of which excellent examples occur at Obsidian Cliff in the Yellowstone National Park, as described by Professor Iddings. Globular structure on a large scale is sometimes displayed by lavas, expecially those of basic type, such as the basalt of Aci Castello in Sicily, which was probably formed, according to Professor Gaetano Platania, by flow of the lava into submarine silt, relics of which still occur between the spheroids.

Joints, or cracks formed by shrinkage on solidification, may divide a sheet of lava into columns, as familiarly seen in basalt, where the rock often consists of a close mass of regular polygonal prisms, mostly hexagonal. Each prism is divided at intervals by transverse joints, more or less curved, so that the portions are united by a slight ball-and-socket articulation. As the long axes of the columns lie at right angles to the cooling surface they are vertical in a horizontal sheet of lava, horizontal in a vertical dyke and inclined or curved in other cases.

It sometimes happens that in a basaltic dyke the formation of the prisms, having started from the opposite walls as chilling surfaces, has not been completed; and hence the prisms fail to meet in the middle. A spheroidal structure is often developed in basalt columns by weathering, the rock exfoliating in spherical shells, rather like the skins of an onion: such a structure is characteristically shown at the Kasekellar, known also as the Elfen Grotto, at Bertrich, near Alf on the Mosel, where the pillars of the lava are broken into short segments which suggest by their flattened globular shape a pile of Dutch cheeses. Although prismatic jointing, or columnar structure, is most common in basalt, it occurs also in other volcanic rocks. Fine columns of obsidian, for instance, are seen at Obsidian Cliff in the Yellowstone Park, where the pillars may be 50 ft. or more in height. Such an occurrence, however, is exceptional.

Vitreous lavas often show fluxion structure in the form of streaks, bands or trains of incipient crystals, indicating the flow of the mass when viscous. The character of this structure is related to the viscosity of the lava. Those structural peculiarities which depend mainly on the presence of vapour, such as vesiculation, have been already noticed and the porphyritic structure has likewise been described.

Submarine Volcanoes

Considering how large a proportion of the face of the earth is covered by the sea, it seems likely that 'volcanic eruptions must frequently occur on the ocean-floor. When, as occasionally though not often happens, the effects of a submarine eruption are observed during the disturbance, it is seen that the surface of the sea is violently agitated, with copious discharge of steam; the water passes into a state of ebullition, perhaps throwing up huge fountains; shoals of dead fishes, with volcanic cinders, bombs and fragments of pumice, float around the centre of eruption and ultimately a little island may appear above sea-level.

This new land is the peak of a volcanic cone which is based on the sea-floor and if in deep water the submarine mountain must evidently be of great magnitude. Christmas Island in the Indian Ocean. Andrews, appears to be a volcanic mountain, with Tertiary limestones,. standing in water more than 14,000 ft. deep. Many volcanic islands, such as those abundantly scattered over the Pacific, must have started as submarine volcanoes which reached the surface either by continued upward growth or by upheaval of the sea-bottom. Etna began its long geological history by submarine eruptions in a bay of the Mediterranean and Vesuvius in like manner represents what was originally a volcano on the sea-floor. As the ejectamenta from a submarine vent accumulate on the sea-bottom they become intermingled with relics of marine organisms and thus form fossiliferous volcanic tuffs. By the distribution of the ashes over the sea-floor, through the agency of waves and currents, these tuffs may pass insensibly into submarine deposits of normal sedimentary type.

Fig. *Submarine Volcanoes*

One of the best examples of a submarine eruption resulting in the formation of a temporary island occurred in 1831 in the Mediterranean between Sicily and the coast of Africa, where the water was known to have previously had a depth of too fathoms.

After the usual manifestations of volcanic activity an accumulation of black cinders and ashes formed an island which reached at one point a height of 200 ft., so that the pile of erupted matter had a thickness of about 800 ft. The new island, which was studied by Constant Prevost, became known in England as Graham's Island, in France as Ile Julie and in Italy by various names as Isola Ferdinandea. Being merely a loose pile of scoriae, it rapidly suffered erosion by the sea and in about three months was reduced to a shoal called Graham's Reef. In 1891 a submarine eruption occurred near the isle of Pantellaria in the same waters and the eruptive centre was termed by Professor H. S. Washington and Foerstner volcano, but it gave rise to no island.

A well-known instance of a temporary volcanic island was furnished by Sabrina - an islet of cinders thrown up by submarine eruptions in 1811, off the coast of St Michael's, one of the Azores. The island of Bogosloff, or Castle island, in Bering Sea, about 40 m. W. of Unalaska Island, is a volcanic mass which was first observed in 1796 after an eruption.

In 1883 another eruption in the neighbouring water threw up a new volcanic cone of black sand and ashes, known as New Bogosloff or Fire Island, situated about half a mile to the N.W. of Old Bogosloff, with which it was connected by a low beach. Another island, called Perry Island, larger than either of the others, made its appearance in the neighbourhood about the time of the great earthquake in California in 1906. It is reported that some of these islands have since disappeared.

Mud Volcanoes

Mud volcanoes are small conical hills of clay which discharge, more or less persistently, streams of fine mud, sometimes associated with naphtha or petroleum and usually with bubbles of gas. As the mud is generally saline, the hills are known also as "Salses."

The gases are chiefly hydrocarbons, often with more or less sulphuretted hydrogen and carbon dioxide and sometimes with nitrogen. Though generally less than a yard in height, the cones may in exceptional cases rise to an elevation of as much as Soo ft. The mud oozes from the top and spreads over the sides, or is spurted forth with the gases. Occasionally the discharge is vigorous, mud and stones being thrown up to a considerable height, sometimes accompanied by flames due to combustion of the hydrocarbons.

Mud volcanoes occur in groups and have a wide distribution. They are known in Iceland; in Modena; at Taman and Kertch, in the Crimea; at Baku on the Caspian; in Java and in Trinidad: Humboldt described those near Turbaco, in Colombia. In Sicily they occur near Girgenti and a group is known at Paterno on Etna.

Bythe Sicilians they are termed, *maccalube,* a word of Arabic origin. The " paint-pots " of the Yellowstone National Park are small mud volcanoes.

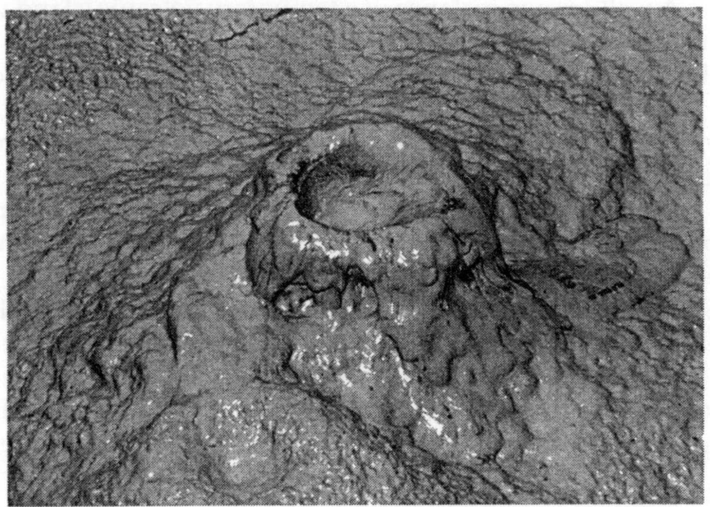

Fig. *Mud of Volcanoes*

Many, so-called mud volcanoes appear to be due to the derangement of subterranean water-flow or to landslips in connexion with earthquakes, whilst others may be referable to certain chemical reactions going on underground; but there are others again which seem to be truly of volcanic origin.

Hot water and steam escaping through clays, or crumbling tuffs reduced to a clayey condition, may form conical mounds of pasty material, through which mud oozes and water escapes.

Geysers are closely related to volcanoes, but in consequence of their special interest they are treated separately. For natural steam-holes and other phenomena connected with declining vulcanicity.

GEOGRAPHICAL DISTRIBUTION OF VOLCANOES

It is matter of frequent observation that volcanoes are most abundant in regions marked by great seismic activity. Although the volcano and the earthquake are not usually connected in the direct relation of cause and effect, yet in many cases they seem referable to a common origin.

Both volcanic extrusion and crustal movement may be the means of relieving local strains in the earth's crust and both are found to occur, as might reasonably be expected, in many parts of the earth where folding and fracture of the rocks have frequently happened and where mountain-making appears to be still in progress.

Thus, volcanoes may often be traced along zones of crustal deformation, or folded mountain-chains, especially where they run near the borders of the oceanic basins. They are frequently associated with the Pacific type of coast-line. The most conspicuous example of linear distribution is furnished by the great belt of volcanoes, coinciding for the most part with a band of seismic disturbance, which engirdles intermittently the huge basin of the Pacific; though here, as elsewhere in studying volcanic topography, regard must be paid to dormant and extinct centres as well as to those that are active at the present time.

As volcanoes are in many cases ranged along what are commonly regarded as lines of fracture, it is not surprising that the centres of most intense vulcanicity are in many cases situated at the intersection of two or more fracture-lines. On the eastern side of the Pacific Ocean the great volcanic ring may be traced, though with many and extensive interruptions, from Cape Horn to Alaska.

In South America the chain of the Andes between Corcovado in the south and Tolima in the north is studded at irregular intervals with volcanoes, some recent and many more extinct, including the loftiest volcanic mountains. in the world. The linear arrangement, often a marked feature in the distribution of volcanoes, is well exemplified in the general north-andsouth trend of the Andean ranges, the volcanoes being situated along the orographic axis.

These folded mountains with their volcanoes also illustrate the close relationship to the sea so frequently observed in volcanic topography, a relationship, however, not without many exceptions. The volcanic rock called andesite was so named by L. von Buch from its characteristic occurrence in the Andes. It is notable that the volcanic rocks throughout the great Pacific belt present much similarity in composition. The volcanoes of Ecuador have been described in detail by A. Stiibel and others.

Central America contains a large number of active volcanoes and solfataras, many of which are located in the mountains parallel to the western coast. Conseguina, on the south side of the Gulf of Fonseca, is remarkable for its eruption in 1835, when an enormous, volume of ash was ejected and the summit of the mountain blown away. Izalco, in San Salvador, came into existence in 1770 and is habitually active. In the centre of Lake Ilopango in Salvador,. which possibly occupies an ancient crater, a volcanic island arose in 1880 and attained a height of 160 ft. Guatemala is peculiarly rich in volcanoes, as described by Dr Tempest Anderson, who visited the country in 1907.

The Cerro Quemado, or the Volcano of Quezaltenango, was the scene of a great eruption in 1785. At the Volcano of Santa Maria there was an outburst in 1902 more violent than the simultaneous eruptions in the Lesser Antilles. The cones of Guatemala include the Volcan de Fuego and the Volcan de Agua, the former often active in historic times, whilst the latter is notable for the flood which in 1541 swept down from the mountain and destroyed Old Guatemala, but this flood was probably not of volcanic origin.

The plateau of Mexico is the seat of several active volcanoes which occur in a band stretching across the country from Colima in the west to Tuxtla near Vera Cruz. The highest of these volcanic mountains is Orizaba, or Cithaltepetl, rising to an altitude of 18,200 feet and known to have been active in the 16th century. Popocatepetl (" the smoking mountain ") reaches a height of about. 17,880 ft. and from its crater sulphur was at one time systematically collected. It came into existence rapidly during an eruption which began in September 1759, when it was said by unscientific observers that the ground became suddenly inflated from below.

The cone, though not of exceptional magnitude, is situated in an elevated district and its summit rises to about 4330 ft. above sea-level. In the neighbourhood of Jorullo there are three subordinate cones of similar character known as *volcancitos,* with great numbers of small mounds of cinder and ash formed around fumaroles on the lava and locally called *hornitos,* or " little ovens. " The streams of basaltic lava from Jorullo form rough barren surfaces, which pass under the name of *malpays,* or bad lands.

In the United States very few volcanoes are active at the present day, though many have become extinct only in times that are geologically recent. An eruption occurred in 1857 at Tres Virgines, in the south of California and the cinder cone on Lassen's Peak was also active in the middle of the 19th century. The Mono Valley craters and Mount Shasta, in California, are extinct. The Cascade Range contains numerous volcanic peaks, but only few show signs of activity. 1Vlount Hood, in Oregon, exhales vapour, as also does Mount Rainier in Washington. Mount St Helens was in eruption in 1841 and 1842; and Mount Baker the most northern of the volcanoes connected with the Cascade Range, is said to have been active in 1843. Few volcanic peaks occur in the Rocky Mountains, but evidence of lingering activity is very marked in the geysers and hot springs of the Yellowstone National Park. The earth's internal heat is also manifested at many points elsewhere, as at Steamboat Springs on the Virginia Range, an offshoot of the Sierra Nevada and in the Comstock Lode.

Volcanic activity is prominent in Alaska, along the Coast Range and in the neighbouring islands. The crater of Mount Edgecumbe, in Lazarus Island, is said to have been active in 1796, but this is doubtful. Mount Fairweather has probably been in recent activity and the lofty cone of Mount Wrangell, on Copper river, is reported to have been in eruption in 1819. In the neighbourhood of Cook's Inlet there are several volcanoes, including the island of St Augustine. Unimak Island has two volcanoes, which have supplied the natives, with sulphur and obsidian; one of these volcanoes being Mount Shishaldin, a cone rivalling Fusiyama in graceful contour. The.

Aleutian volcanic belt is a narrow, curved chain of islands, extending from Cook's Inlet westwards for nearly 1600 m. It is notable that the convexity of the curve faces the great ocean, as has been observed in other cases, the arcs following the direction of the rock-folds. According to Professor I. C. Russell, an authority on the volcanoes of N. America, there are in the Aleutian Islands and in the peninsula no fewer than 57 craters, either active or recently extinct.

From the Aleutian Islands the volcanic band of the Pacific changes its direction and passing to the peninsula of Kamschatka, where 14 volcanoes are said to be active, turns southwards and forms the festoon of the Kurile Islands. Here again the convexity of the insular arc is directed towards the ocean. This volcanic archipelago leads on to the great islands of Japan, where the volcanoes have been studied by Professor J. Milne, who also described those of the Kuriles. Of the 54 volcanoes recognized as now active or only recently extinct in Japan, the best known is the graceful cone of the sacred mountain Fusiyama, but others less pretentious are far more dangerous. The great eruption of Bandaisan, about 120 m. N. of Tokio, which occurred in 1888, blew off one side of the peak called Kobandai, removing, according to Professor Sekiya's estimate, about 2982 million tons of material. Aso-san in Kiushui, the southernmost large island of Japan, is notable for the enormous size of its crater. In the Bonin group of islands volcanic activity is indicated by such names as Volcano Island and Sulphur Island.

South of the Japanese archipelago the train of volcanoes passes through some small islands in or near the Loo Choo group and thence onwards by Formosa to the Philippine Islands, where subterranean activity finds abundant expression in earthquakes and volcanoes. After leaving this region the linear arrangement of the eruptive centres becomes less distinctly marked, for almost every island in the Moluccas and the Sunda Archipelago teems with volcanoes, solfataras and hot springs. Possibly, however, a broken zone may be traced from the Moluccas through New Guinea and thence to New Zealand, perhaps through eastern Australia (for though no active volcanoes are known there, relics of comparatively recent activity are abundant); or again by way of the Bismarck Archipelago, the Solomon Islands, the New Hebrides, the Fiji Islands and Kermodoc Island.

The great volcanic district in New Zealand is situated in the northern part of North Island, memorable for the eruption of Tarawera in 1886. This three-peaked mountain on the south side of Lake Tarawera, not previously known to have been active, suddenly burst into action; a huge rift opened and Lake Rotomahana subsided, with destruction of the famous sinter terraces. The crater of Tongariro is in the solfatara stage, whilst Mount Ruapehu is regarded as extinct. On White Island in the Bay of Plenty the cone of Wharkari is feebly active.

Far to the south, on Ross Island, off South Victoria Land, in Antarctica, are the volcanoes of Erebus and Terror, the former of which is active. These are often regarded as remotely related to the Pacific zone, but Dr G. T. Prior has shown that the Antarctic volcanic rocks which he examined belonged to the Atlantic and not the Pacific type.

Within the great basin of the Pacific, imperfectly surrounded by its broken girdle of volcanoes, there is a vast number of scattered islands and groups of islands of volcanic origin, rising from deep water and hiving in many cases active craters. The most important group is the Hawaiian Archipelago, where there is a chain of at least fifteen large volcanic mountains - all extinct, however, with the exception of three in Hawaii, namely Mauna Loa, Kilauea and Hualalai; and of these Hualalai has been dormant since 1811.

It is notable that the two present gigantic centres of activity, though within 20 m. of each other, appear to be independent in their eruptivity. Several of the Hawaiian Islands, as pointed out by J. D. Dana, who was a very high authority on this group, consist of two volcanoes united at the base, forming volcanic twins or doublets.

The volcanic regions of the Pacific are connected with

those of the Indian Ocean by a grand train of islands rich in volcanoes, stretching from the west of New Guinea through the Moluccas and the Sunda Islands, where they form a band extending axially through Java and Sumatra. Here is situated the principal theatre of terrestrial vulcanicity, apparently representing an enormous fissure, or system of fissures, in the earth's crust, sweeping in a bold curve, with its convexity towards the Indian Ocean.

Numerous volcanic peaks occur in the string of small islands to the east of Java - notably in Flores, Sumbawa, Lombok and Bali; and one of the most terrific eruptions on record in any part of the world occurred in the province of Tomboro, in the island of Sumbawa, in the year 1815. Java contains within its small area as many as 49 great volcanic mountains - active, dormant and extinct. The largest is Smerin, about 12,000 ft. high, but the most regularly active is said to be Gownong Lamongang, which is in almost uninterrupted activity, emitting usually only ashes and vapour, though in 1883 lava streamed forth. Many of the Javanese volcanoes present marked regularity of contour, with the sides of the cones rather symmetrically furrowed by tropical rains and probably ridged by ash-slides.

The radial furrows on volcanic cones are sometimes known as " barrancos." The little uninhabited island of Krakatoa in the Strait of Sunda appears to be situated at a volcanic node, or the intersection of two curved fissures and it is believed that the island itself represents part of the basal wreck of what was once a volcano of gigantic size. After two centuries of repose, a violent catastrophe occurred in 1883, whereby the greater part of the island was blown away. This eruption and its effects were made the subject of careful study by Verbeek, B g on and Judd.

Through the great island of Sumatra, a chain of volcanoes runs longitudinally and may possibly be continued northwards in the Bay of Bengal by Barren Island and Norcondam - the former an active and the latter an extinct volcano. On the western side of the Indian Ocean a small volcanic band may be traced in the islands of the Mascarene group, several craters in Reunion being still active. Far south in the Indian Ocean are the volcanic islands of New Amsterdam and St Paul.

The Comoro Islands in the channel of Mozambique exhibit volcanic activity, whilst in East and Central Africa there are several centres, mostly extinct but some partially active, associated with the Rift Valleys. The enormous cones of Kenia and Kilimanjaroo are extinct, but on Kibo, one of the summits of the latter, a crater is still preserved. The Mfumbiro volcanoes, S. of Lake Edward, rise to a height of more than 14,700 feet. Kirunga, N. of Lake Kivu, is still partially active. Elgon is an old volcanic peak, but Ruwenzori is not of volcanic origin. On the west side of Africa, the Cameroon Peak is a volcano which was active in 1909 and the island of Fernando Po is also volcanic. Along the Red Sea there are not wanting several examples of volcanoes, such as Jebel Teir. Aden is situated in an old crater.

Passing to the Atlantic, a broken band of volcanoes, recent and extinct, may be traced longitudinally through certain islands, some of which rise from the great submarine ridge that divides the ocean, in part of its length, into an eastern and a western trough. The northern extremity of the series is found in Jan Mayen, an island in the Arctic Ocean, where an eruption occurred in 1818. Iceland, however, with its wealth of volcanoes and geysers, is the most important of all the Atlantic centres.

According to Dr T. Thoroddsen there are in Iceland about 130 post-glacial volcanoes and it is known that from 25 to 30 have been in eruption during the historic period. Many of the Icelandic lava-flows, such as the immense flood from Laki in 1783, are referable to fissure eruptions, which are the characteristic though not the exclusive form of activity in this island. Probably this type was also responsible for the sheets of old lava in the terraced hills of the Faroe Islands, to which may have been related the Tertiary volcanoes of the west of Scotland and the north of Ireland.

An immense gap separates the old volcanic area of Britain from the volcanic archipelagoes of the Azores, the Canaries and the Cape Verd Islands. Palma a little island in the Canary group, with a caldera or large crater at its summit, from which fissures or barrancos radiate is famous in the history of vulcanology, in that it furnished L. von Buch with evidence on which he founded the " crater-ofelevation " theory. The remaining volcanic islands of the Atlantic chain, all now cold and silent, include Ascension, St Helena and Tristan da Cunha, whilst in the western part of the South Atlantic are the small volcanic isles of Trinidad and Ferdinando do Noronha. St Paul's rocks appear also to be of volcanic origin.

One of the most important volcanic regions of the world is found in the West Indies, where the Lesser Antilles - the scene of the great catastrophes of 1902 - form a string of islands, stretching in a regular arc that sweeps in a N. and S. direction across the eastern end of the Caribbean Sea. Subject to frequent seismic disturbance and rich in volcanoes, solfataras and hot springs, these islands seem to form the summit of a great earth-fold which, rising as a curved ridge from deep water, separates the Caribbean Sea from the Atlantic. The volcanoes are situated on the inner border of the curve. It is notable that the Antilles and the Sunda Islands, two of the grandest theatres of vulcanicity on the face of the earth, are situated at the antipodes of each other - one being apparently an eastern and the other a western offshoot of the great Pacific girdle.

The European volcanoes, recent and extinct, may be regarded as representing rather ill-defined branches thrown off eastwards from the Atlantic band. Vesuvius is the only active volcano on the mainland, but in the Mediterranean there are Etna on the coast of Sicily; the Lipari Islands, with Stromboli and Vulcano in chronic activity; and farther to the east the archipelago of Santorin, where new islands have appeared in historic times. Submarine eruptions have occurred also between Sicily and the coast of Africa; one in 1831 having given rise temporarily to Graham's Island and another in 1891 appearing near Pantellaria, itself a volcanic isle.

Of the extinct European volcanoes, some of the best known are in Auvergne, in the Eifel, in Bohemia and in Catalonia, whilst the volcanic land of Italy includes the

Euganean hills, the Alban hills, the Phlegraean Fields, &c. The great lakes of Bolsena and Bracciano occupy old craters and many smaller sheets of water are on similar sites. The volcanic islands no longer active include Ischia, with the great cone of Epomeo which was in a state of eruption in 1301; the Ponza Islands, Nisida, Vivera and others near Naples; and several in the Greek archipelago, such as Milos, Kimolos and Polinos.

From the eastern end of the Mediterranean evidence of former volcanic activity may be traced into Asia Minor and thence to Armenia and the Caucasus. East of Smyrna there is a great desolate tract which the ancients recognized as volcanic and termed the Catacecaumene. The volcanic districts of Lydia were studied by Professor H. S. Washington.

In the plateau of Armenia there are several extinct volcanic mountains, more or less destroyed, of which the best known is Ararat. Nimrud Dagh on the shore of Lake Van is said to have been in eruption in the year 1441. Dr F. Oswald has described the volcanoes of Armenia. Of the volcanoes in Persian territory not now active, Demavend, south of the Caspian, is an important example. Elburz is also described as an old volcano. It has been said that in Central Asia there are certain vents still active and recent volcanic rocks are known from the Przhevalsky chain and other localities.

The number of volcanoes known to be actually active on the earth is generally estimated at between 300 and 400, but there is reason to believe that this estimate is far too low. If account be taken of those volcanic cones which have not been active in historic time, the total will probably rise to several thousands. The distribution of volcanoes at various periods of the earth's history, as revealed by the local occurrence of volcanic rocks at different horizons in the crust of the earth. Periods of great earthmovement have been marked by exceptional volcanic activity.

Causes of Vulcanicity

The cause of vulcanicity two problems demand attention: first the origin of the heat necessary for the manifestation of volcanic phenomena and secondly the nature of the force by which the heated matter is raised to the surface and ejected. According to the old view, which assumed that the earth was a spheroid of molten matter invested by a comparatively thin crust of solid rock, the explanation of the phenomena appeared fairly simple.

The molten interior supplied the heated matter, while the shrinkage of the cooling crust produced fractures that formed the volcanic channels through which it was assumed the magma might be squeezed out in the process of contraction.

When physicists urged the necessity of assuming that the globe was practically solid, vulcanologists were constrained to modify their views. Following a suggestion of W. Hopkins of Cambridge, they supposed that the magma, instead of existing in a general central cavity, was located in comparatively small subterranean lakes. Some authorities again, like the Rev. O. Fisher, regarded the magma as constituting a liquid zone, intermediate between a solid core and a solid shell.

If solidification of the primitive molten globe proceeded from the centre outwards, so as to form a sphere practically solid, it is conceivable that portions of the original magma might nevertheless be retained in cavities and thus form "Residual lakes." Although the mass might be for the most part solid, the outer portion, or "Crust," could conceivably have a honeycombed structure and any magma retained in the cells might serve indirectly to feed the volcanoes. Neighbouring volcanoes seem in some cases to draw their supply of lava from independent sources, favouring the idea of local cisterns or "Intercrustal reservoirs."

It is probable, however, that subterranean reservoirs of magma, if they exist, do not represent relics of an original fluid condition of the earth, but the molten material may be merely rock which has become fused locally by a temporary development of heat or more likely, by a relief of pressure.

It should be noted that the quantity of magma required to supply the most copious lava-flows is comparatively small, the greatest recorded outflow ot having exceeded, it is said, six cubic miles; and even this estimate is probably too high.

Whilst in many cases the magma-cisterns may be comparatively small and temporary, it must be remembered that there are regions where the volcanic rocks are so similar throughout as to suggest a common origin, thus needing intercrustal reservoirs of great extent and capacity. It has been suggested that comparatively small basins, feeding individual volcanoes, may draw their supply from more extensive reservoirs at greater depths.

Much speculation has been rife as to the source of the heat required for the local melting of rock. Chemical action has naturally been suggested, especially that of superficial water, but its adequacy may be doubted. After Sir Humphry Davy's dis covery of the metals of the alkalis, he thought that their remarkable behaviour with water might explain the origin of subterranean heat; and in more recent years others have seen a local source of heat in the oxidation of large deposits of iron, such as that brought up in the basalt of Disco Island in Greenland. It has been assumed by Moissan and by Gautier that water might attack certain metallic carbides, if they occur as subterranean deposits and give rise to some of the products characteristic of volcanoes. But it seems that all such action must be very limited and utterly inadequate to the general explanation of volcanic phenomena.

At the same time it must be remembered that access of water to a rock already heated may have an important. physical effect by reducing its melting point and may thus greatly assist in the production of a supply of molten matter. The admission of surface-waters to heated rocks is naturally regarded as an important source of motive power in consequence of the sudden generation of vapour, but it is doubtful to what extent it may contribute, if at all, to the origin of volcanic heat.

According to Robert Mallet a competent source of subterranean heat for volcanic phenomena might be derived from the transformation of the mechanical work of compressing and crushing parts of the crust of the earth as a consequence of secular contraction. This view he worked out with much ingenuity, supporting it by mathematical reasoning and an appeal to experimental evidence. It was claimed for the theory that it explained the linear distribution of volcanoes, their relation to mountain chains, the shallow depth of the foci and the intermittence of eruptive activity.

A grave objection,. however, is the difficulty of conceiving that the heat, whether due to crushing or compression, could be concentrated locally so as to produce a sufficient elevation of temperature for melting the rocks. According to the calculations of Rev. O. Fisher, the crushing could not, under the most favourable circumstances, evolve heat enough to account for volcanic phenomena.

Since pressure raises the melting-point of any solid that expands on liquefaction, it has been conjectured that many deep-seated rocks, though actually solid, may be potentially liquid; that is, they are maintained in a solid state by pressure: only. Any local relief of pressure, such as might occur in the folding and faulting of rocks, would tend, without further accession of heat, to induce fusion. But although moderate pressure raises the fusing-point of most solids, it is believed,, from modern researches, that very great pressures may have a contrary effect.

It is held by Professor S. Arrhenius that at great depths in. the earth the molten rock, being above its critical point, can. exist only in the gaseous condition; but a gas under enormous pressure may behave, so far as compressibility is concerned, like a rigid solid. He concludes, from the high density of the earth as a whole and from other considerations, that the central part of our planet consists of gaseous iron (about 80% of the earth's diameter) followed by a zone of rock magma in a gaseous condition (about 15%), which passes insensibly outwards into liquid rock (4%), covered by a thin solid crust (less than i % of diameter).

If water from the crust penetrates by osmosis through the sea-floor to the molten interior, it acts, at the high temperature, as an acid and decomposes the silicates of the magma. The liquid rock, expanded and rendered more mobile by this water, rises in fissures, but in its ascent suffers cooling, so that the water then loses its power as an acid and is displaced by silicic acid, when the escaping steam gives rise to the explosive phenomena of the volcano. The earth was formed by the accretion of vast numbers of small cosmical bodies called planetesimals, the original heat of the earth's interior was due chiefly to the compression of the growing globe by its own gravity. The heat, proceeding from the centre outwards, caused local fusion of the rocks, though without forming distinct reservoirs of molten magma and the fused matter charged with gases rose in liquid threads or tongues, which worked their way upwards, some reaching the superficial part of the earth and escaping through fissures in the zone of fracture, thus giving rise to volcanic phenomena.

It is held that the explosive activity of a volcano is due to the presence of gases which have been brought up from the interior of the earth, whilst only a small and perhaps insignificant part is played by water of superficial origin.

Entirely new views of the origin of the earth's internal heat have resulted from the discovery of radioactivity. An ingenious hypothesis was enunciated by Major C. E. Dutton, who found in the radioactivity of the rocks a sufficient source of heat for the explanation of all volcanic phenomena. He believes that the development of heat arising from radioactivity may gradually bring about the local melting of the rocks so as to form large subterranean pools of magma, from which the volcanoes may be supplied.

The supply is usually drawn from shallow sources, probably, according to Dutton, from a depth of not more than three or rarely four miles and in some cases at not more than a mile from the surface. If the water in the local magma should attain sufficient expansive power, it will rupture the overlying rocks and thus give rise to a volcanic eruption. When the reservoir becomes exhausted the eruption ceases, but if more heat be generated by continued radioactivity further fusion may ensue and in time the eruption be repeated. According, however, to Professor Joly, it is improbable that sufficient heat for the manifestation of volcanic phenomena could be developed by the local radioactivity of the rocks in the upper part of the earth's crust.

Chapter 7

Sources of Ground Water

Ground Water is the water that saturates the pores and cracks in soil and rock beneath the land surface. It comes to the surface in springs and also swells the volume of many streams by seeping into them from their beds and banks. Ground water is also obtained for human use by sinking wells.

GROUND WATER

Water at the earth's surface has one obvious source—rain and snow—and, as Perrault showed long ago there are good reasons for considering these sources as practically the only ones. They are also the ultimate source of practically all ground water. Most soils and rocks are porous.

Small voids and openings occur between grains of soil, larger openings are formed by burrowing animals and cracks are formed by the shrinkage of drying soil clays. Even well-consolidated rocks may be riven by joint cracks or other openings into which water can penetrate. After a rain part of the water seeps into these openings and percolates beneath the land surface.

Some of the rain water that enters the soil remains near the surface, absorbed by the soil colloids, or held against gravity in the smaller voids by surface tension (*capillarity*, the force that pulls water up a very slender tube and holds it there against the action of gravity).

Deeper water may even be drawn back toward the surface of the ground by capillarity, thus helping to replace the losses from evaporation and transpiration by plants. At least a part of the rain water, however, percolates deeper and deeper through the larger soil openings until it ultimately reaches a zone beneath the surface where all the pores in the rock are completely filled with water.

THE WATER TABLE

The name *water table* is given to the upper surface of this zone below which all pores are filled with water. Except for the discontinuous films and irregular masses of water held by capillarity, or in process of transit downward, the pores above the water table are filled with air and so the zone between the water table and the surface is called the *zone of aeration*. Here weathering and other chemical changes go on under oxidizing conditions. Below the water table is the *zone of saturation*.

In most places the water table is a few feet, or a few tens of feet, below the ground surface, but in arid regions it may be hundreds of feet below. In marshes, on the other hand, the water table practically coincides with the ground surface, as it does at the edge of surface-water bodies such as lakes and streams. If many water wells have been dug, the form of the water table below the ground surface can be readily determined and accurate contour maps of it drawn from the elevation of the water surface in the various wells.

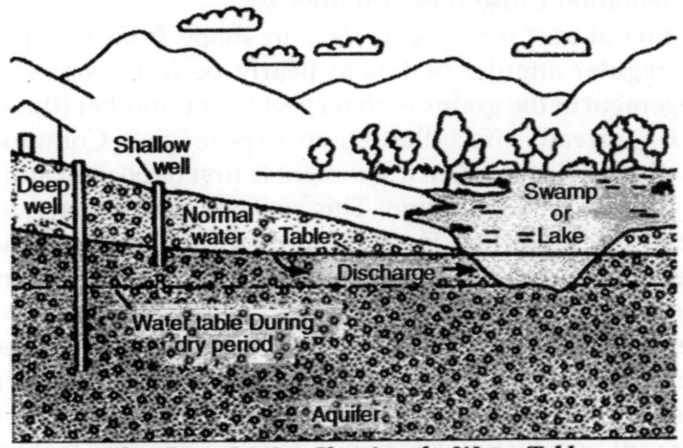

Fig. *Cross Section Showing the Water Table*

Such maps show that, in general, the water table is a subdued replica of the surface topography. It rises under the hills and sinks beneath the valleys, but is usually much smoother than the land surface. The level of the water table also fluctuates slightly with the seasons.

SUBSURFACE PORES, CRACKS AND CHANNELS

If subsurface pores are abundant and large below the water table, water in considerable volume can be pumped from wells. When wells are pumped, water from pores in the nearby rock flows rapidly into the well and replenishes the water being extracted. On the other hand, if the pores are small and widely spaced, only a little water can be recovered by pumping because replenishment is slow. The shapes, sizes and aggregate volume of the open spaces—pore spaces—in each kind of soil and rock are thus primary factors in determining the occurrence, amount and rate of movement of ground water.

Porosity and Permeability

Porosity is the ratio of pore volume to total volume, expressed as percentage. It is generally determined by testing samples of rock in the laboratory. Porosities of most uncemented elastic sediments range between 12 and 45 per cent. Variations in the *shapes* of the grains, their *sorting,* their *packing* and the *degree* of cementation cause these differences.

Mineral and rock grains vary in shape from thin plates and irregular angular bodies to nearly perfect spheres. The arrangement of the grains with respect to one another (the kind of *packing*) greatly affects the amount of pore space. Commonly, the packing is loose when a sediment is first deposited and the pore space is relatively high. This space is then progressively reduced by *compaction* from the pressure of new sediments subsequently deposited above. *Cementation* (deposition of mineral matter in the pores) further reduces the pore space. Closely packed spheres of uniform size have 26 per cent pore space. This is true whether the spheres are 1 millimeter in diameter, 5 feet in diameter, or any other size. Porosities above 26 per cent usually indicate nonuniform packing, or that some grains are themselves porous. Shape is also a factor, but the presence of nonspherical grains may either raise or lower the porosity. Porosities below 26 per cent usually result from poor sorting, or from compaction and cementation of the voids between grains angular bodies to nearly perfect spheres.

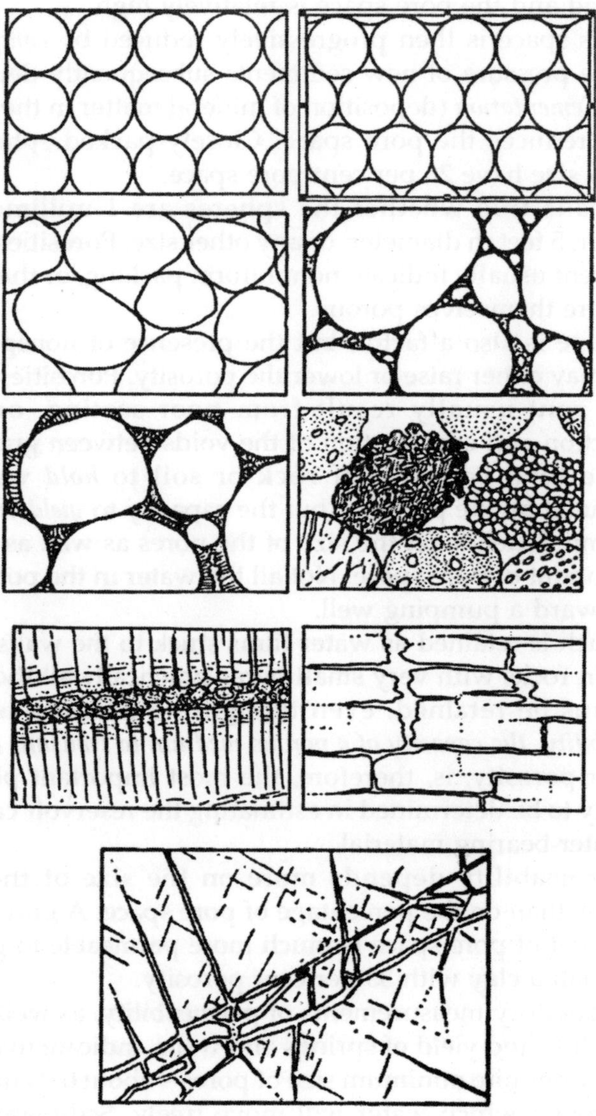

Fig. *Porosity in Tocks*

The arrangement of the grains with respect to one another (the kind of *packing*) greatly affects the amount of pore space. Commonly, the packing is loose when a sediment is first deposited and the pore space is relatively high.

This space is then progressively reduced by *compaction* from the pressure of new sediments subsequently deposited above. *Cementation* (deposition of mineral matter in the pores) further reduces the pore space. Closely packed spheres of uniform size have 26 per cent pore space.

This is true whether the spheres are 1 millimeter in diameter, 5 feet in diameter, or any other size. Porosities above 26 per cent usually indicate nonuniform packing, or that some grains are themselves porous.

Shape is also a factor, but the presence of nonspherical grains may either raise or lower the porosity. Porosities below 26 per cent usually result from poor sorting, or from compaction and cementation of the voids between grains.

The total capacity of rock or soil to *hold* water is determined by the porosity, but the capacity to *yield* water to the pump depends on the size of the pores as well as on the total amount of pore space. Not all the water in the pores will flow toward a pumping well.

Much is retained as water films stuck to the walls of the pores. In rocks with very small pores practically all the water may thus be retained, even though the porosity is high. *Permeability, the capacity of a porous medium to transmit a liquid* and not porosity, is, therefore, the most important physical property to be determined in estimating the reservoir capacity of a water-bearing material.

Permeability depends more on the size of the pore openings than on the percentage of pore space. A gravel with 20 per cent of pore space is much more permeable to ground water than a clay with 35 per cent porosity.

Laboratory measurements of permeability, as well as the distribution and yield of springs and wells, indicate that there is a fairly definite minimum size of pores—about 0.05 millimeters through which water will move freely. Sediments with smaller pores, such as clay, silt and shale, are impermeable even though their porosity may be high. Larger pores are present in sand and gravel and also in many incompletely cemented sandstones, conglomerates, jointed lava flows and porous limestones.

AQUIFERS

A body of rock or loose surface material that is permeable as well as porous and so can yield water to wells, is called an *aquifer*. Some aquifers may be empty of water, at least temporarily. Even in an aquifer, water movement is extremely slow, usually a fraction of a foot or a few feet per day, except in the immediate vicinity of a well or spring.

Most aquifers are sheets of sand or gravel, or beds of sandstone, limestone, or other permeable rock. A few aquifers are narrow sinuous bodies of gravel that fill former stream courses, but these are by no means as common as the popular expression "underground stream" indicates.

Water Witching

Before the Nineteenth Century men thought that ground water flowed in definite underground streams just as surface water does.

They reasoned that when one dug a well, if he were lucky, his well intersected one of these streams and produced a fine flow of water. A dry well, or one that produced only a little water, supposedly had failed to intercept any of the underground channels.

Since one cannot see beneath the surface of the ground, the drilling of wells was always an uncertain operation and in their doubt about where to dig for water farmers often consulted "water witches" or "dowsers".

Such people were supposedly endowed with supernatural powers that enabled them to discover the location of underground streams.

A dowser generally walks about with a forked stick, tightly held, which is supposed to dip violently when he crosses the position of an underground stream. His success, if any, has no known scientific basis.

Perched Water

An aquifer may rest on an impermeable substratum that overlies unsaturated material above the normal water table. The water in such an aquifer is *perched*. It is prevented from percolating downward to the normal water table by the impermeable material beneath. Bodies of perched water are especially common in arid or semi-arid regions.

Confined Water; Artesian Wells

A permeable sediment, such as an uncemented sandstone, may be overlain by an impermeable one, such as shale. If the rocks have been tilted and eroded, the permeable sediment may be exposed at the surface in hills or mountains that rise above the level of the overlying impermeable shale nearby. Under these conditions the ground water in the aquifer may be both *partially confined* (in the area where it is overlain by shale) and *easily replenished* (in the area where it appears at the surface). Water enters the aquifer at A where it is exposed at the surface—the intake area.

The water table (t-t2) within the aquifer of the intake area is higher than the ground surface at B, where the aquifer is confined beneath overlying shale. If a well (W) is put down to this aquifer in area B, the confined water in the aquifer will rise through the well under the pressure of the head of water from the intake area and flow out on the surface of the ground, making this an *artesian well*.

The name "artesian" was originally restricted to flowing wells. It is now applied, however, to any well in which the water rises above the elevation of the aquifer penetrated.

GROUND-WATER MOVEMENT

The water that fills the pores below the water table is not static. It moves slowly under the influence of gravity. The direction of movement is controlled by the form and slopes of the water table. From the higher points on the table the water moves laterally, vertically, or obliquely downward. Similarly, it flows from all directions toward troughs or low points. The overall tendency is to lower the high points on the water table and either to raise the low points or else discharge water from them. Points of surface discharge, located where the water table intersects the ground, are called *springs*. Most perennial streams also mark areas of discharge in the water table. They lie in troughs on the water table and the ground-water flow is toward and into them. Such streams are called *effluent* with respect to ground water. Streams that flow from mountains or other well-watered areas into deserts or semi-arid regions, however, may lose water to the water table. Such *influent streams* lie on ridges in the water table.

If permeability and other factors are uniform, movement between two points on the water table is determined by the *hydraulic gradient*, which is the ratio between the difference of elevation, or head (H) and the horizontal distance between the two points (L). Ground-water gradients are usually low, such as a fall of 1 foot per 1,000 feet (0.001) or 10 feet per 1,000 (0.01).

Effect of Variations of Intake on Water Levels

For any particular permeability, the hydraulic gradient adjusts itself to the supply of water. If the discharge, as into streams, is temporarily greater than the supply into the intake area, the water table flattens. Thus, in dry spells the water table sinks farther below the surface under the groundwater divides, reducing the hydraulic gradient and the discharge. In semi-arid western Texas, eastern New Mexico and adjacent parts of old Mexico the water table is relatively flat and, in many places, lies 500 to 1,000 feet below most of the land surface. Much of the region either lacks permanent surface streams or else the streams are on perched ground-water bodies.

DARCY'S LAW

The modern concepts of ground-water movement were discovered early in the Nineteenth Century. Almost all movements of ground water are so slow that they occur by laminar flow, whereas practically all movements in surface streams are turbulent. Flow lines in laminar flow are smooth, continuous and traceable. Investigation of the flow of water in pipes shows that for turbulent flow the velocity and discharge are approximately proportional to the square root of the hydraulic gradient, whereas *in laminar flow the velocity and discharge vary directly as the hydraulic gradient.* This fundamental law, as it applies to groundwater movement, was discovered and formulated in the 1850's by the French hydrologist Henry Darcy, during his study of the water supply of the city of Dijon. Darcy's law may be stated as:

$$V = PI$$

where: V = velocity of ground water P = coefficient of permeability—a constant determined by the character of the material through which the water moves—a measure of the ease with which water moves through a subsurface material I = hydraulic gradient—the slope of the water table

Geologists are usually more interested in the quantity of water moving than in its velocity. Hence, we usually replace the velocity term in Darcy's law by an equivalent term which states the quantity of water moving per unit of time (Q) through a given cross-sectional area (A).

and we write Darcy's law as or

$$Q = PIA$$

where: Q = quantity of water moving per unit of time (measured in gallons per day, cubic meters per day, etc.) A = cross-sectional area through which water moves

If A is expressed in square feet, if I is made 1 (100 per cent gradient) and if Q is measured in gallons per day, then P comes out in *Meinzer units*. Those aquifers that have been investigated in this way have permeabilities that range all the way from 10 to 90,000 Meinzer units. The approximate yield of a well may be predicted if the permeability of the aquifer is known, together with the hydraulic gradient under pumping conditions and the cross-sectional area (the area of perforated casing through which water is entering the well). Similarly, the amount of ground water passing through a cross section of an aquifer can be calculated.

RATES OF GROUND-WATER MOVEMENT

The movement of unconfined ground water through uniformly permeable material, a section showing both ground-water divides and effluent stream levels. Some flow lines go deep, but little water follows these. The maximum velocities (and hence the greatest volumes of water transferred) are near the streams where the hydraulic gradients are steepest.

The actual velocities of ground-water movement, though almost everywhere low as compared to stream velocities, are highly variable. They vary in different parts of a single body of water moving through rock of uniform permeability.

Mean or average rates of movement can be calculated most easily from Darcy's Law, if the coefficient of permeability is known. The late leader of American hydrology, O. E. Meinzer, considered the flow of 50 feet per year in the Carrizo sandstone of Texas to be typical of many aquifers. In places much more rapid movements have been demonstrated by use of dyes or salts, introduced at one observation well and recovered or measured indirectly at another. Movements of 10 or 20 feet per day are sometimes attained in highly permeable materials and one extreme velocity of 420 feet per day has been reported.

Balance Between Discharge, Permeability and Hydraulic Gradient

According to Darcy's equation, in materials of low permeability the gradient of the water table increases steeply as intake water is added locally to the mass of ground water. In materials of high permeability the water table is relatively flat and hydraulic gradients may be only a few feet or a few tens of feet per mile.

Discharge of Ground Water into the Ocean

The discharge of fresh ground water directly into the ocean may produce a body of fresh water extending far below sea level beneath an oceanic island or along a continental margin. It appears as if the lighter, higher column of fresh water were in static balance with the heavier sea water like a foreign mass floating within it. A column of sea water 1,000 feet high can balance one of fresh water about 1,025 feet high. Thus a column of fresh water rising to an elevation of 25 feet above sea level, near the ocean shore, might indicate that fresh ground water could be recovered to a depth of 1,000 feet below sea level. Such a water body is, however, not static; it is constantly discharging into the sea. The friction of flow through the pores retards the spreading out of the fresh water, but if it were not constantly replenished from rainfall, the vertical separation between salt and fresh water would soon disappear. The probable relationships and flow lines in such a water body may be deduced from Darcy's Law.

The body of fresh water under Oahu, one of the Hawaiian Islands, though hundreds of feet thick, is thinner than anticipated on the static flotation hypothesis. Furthermore, the fresh water is underlain by a thick transition zone of brackish water. Perhaps the mixing is due in part to oscillations caused by the intermittent pumping from many large wells. The original accumulation of fresh water extending far below sea level is presumably the result of flow following the lines induced by the hydraulic gradient of the fresh-water surface.

Drawdown by Pumping

A pumped well producing from an unconfined aquifer disturbs the water table by setting up a point of artificial discharge. We have already seen how the water table becomes adjusted to points and lines of natural discharge, such as springs and effluent streams. Similarly, removal of water through a well draws the water down adjacent to the well to produce a *cone of depression* in the water table that greatly increases the hydraulic gradient close to the well.

The pumping well removed water from uniformly permeable alluvium in the valley of the Platte River. The undisturbed water table sloped eastward 6 or 7 feet per mile. The lowering of the water levels was determined in more than 80 observation wells drilled in 8 radial lines extending out to about 1,200 feet from the pumping well is a section across the "cone" of depression in the direction of slope of the undisturbed water table. Note that a lowering, or *drawdown,* of 32 feet at the pumping well after 48 hours of pumping was accompanied by a lowering of 1 foot in the water table at a distance of 250 feet and that measurable lowering extended at least 1,200 feet from the pumping well. Flow lines toward the well must have extended equally far laterally and also far below the water table. Effects analogous to the cone of depression may also be observed in wells that tap a confined aquifer. If a group of artesian wells is capped so that no water escapes, the water in each well will exert a definite pressure against the top of the casing. If the cap of one well is opened so that the water begins to flow from it, a decline in pressure will soon be noted at all the other wells.

COMPOSITION OF GROUND WATER

Rain water, if it passes slowly through limestone, or even through decaying moderately calcium-rich rocks, such as granite, may pick up enough calcium ion—($C_aCO_3 + H^+ \rightarrow Ca^{++} + HCO_3$ -)—to become *hard water*. The amount of calcium ion in hard waters in humid regions is only a small fraction of one per cent. Such waters may contain as much sodium ion as calcium ion, but the sodium ion is not easily precipitated and ordinarily goes unnoticed.

In arid regions, however, water within a few feet of the earth's surface may be partially or completely evaporated after each rain, thus progressively concentrating the water-soluble salts in the soil water. The commonest salts thus concentrated are the carbonate, chloride and sulfate of sodium. The ground water in many arid regions is, therefore, more or less *alkaline*. Alkaline water is very toxic to plants and the "alkali soils" are nearly useless agriculturally. If such areas are drained and the alkaline water flushed downward by heavy irrigation, they may be reclaimed for agricultural use.

Still other ground waters are *salty*. Such waters contain enough sodium chloride to make their taste unpleasant and to make them injurious to plants. Some salty ground waters are in whole or part sea water, infiltrated directly from the ocean. Others have dissolved salt from salt beds in the rocks. Some water in deeply buried marine sedimentary rocks is supposedly sea water entrapped at the time of their deposition. Such entrapped sea water is called *connate water*. Connate waters rarely have exactly the composition of the present ocean. Variations are probably mostly due to dilution since burial, to concentration by evaporation previous to burial and to chemical reactions with the enclosing rocks.

GROUND WATER IN CARBONATE ROCKS

Solution Passages

Rain water is so effective a solvent of carbonate rocks that in limestone regions it enlarges cracks and pores and dissolves tunnels, irregular passages and even large caverns along joints or other openings in the rock. Solution of the rock may become so extensive that much of the surface drainage goes underground through vertical tubular passages called *sinks* and discharges through *caves*.

The development of sinks and caves is, of course, a very slow process. Water percolating down a crack in limestone enlarges the opening by solution. Weathering, rainwash and collapse of the walls by gravity widens the opening at the surface, thus enabling it to trap more and more surface water which, in turn, causes more rapid solution of the limestone walls. The vertical sink thus formed may extend completely through the limestone bed, discharging at its base into other caves and caverns formed by solution of the limestone. In time, the entire bed of limestone may thus become honeycombed with interconnected sinks and caverns. Within the caves a part of the dissolved calcium carbonate may be redeposited, often in bizarre iciclelike forms, producing masses of dripstone.

The easy movement of water through a cavernous limestone aquifer and the flatness of its water table have been often demonstrated by pumping tests in mines. At the Los Lamentos mine, in Mexico, about 100 miles southeast of El Paso, Texas, two years of constant pumping did not lower the water table appreciably and the mining of a rich ore body was necessarily stopped at the water table.

Karst Topography

In some limestone or dolomite regions there are few or no surface streams and the water table is flat and far below the surface. The rainfall passes underground at once through sinks and joints widened by solution and flows in underground streams with low gradients through large and small caverns, cascading at intervals to still lower levels, until, finally, it reappears at the surface as giant springs on the walls or floors of deep valleys.

Such an area of sinks and underground drainage is called a *karst* region, from the name of such an area in Yugoslavia. Other well-known karst regions are the Causses of southern France west of the Rhone River and parts of the Cumberland Plateau in Kentucky and Tennessee.

The topography of a karst region differs from that of an area with normal surface drainage. Instead of a system of slopes definitely integrated to surface streams, a karst region is pock-marked by a maze of large and small depressions. The smaller depressions are the upper ends of sinks and other solution passages, enlarged at the surface by weathering, rainwash and downslope movements.

Larger openings appear where the roofs of caverns have collapsed. Through such an area the main rivers flow at the bottoms of deep valleys, gaining most of their increase in volume from numerous large springs rather than from tributary streams. The largest springs arise from solution channels or even extensive caverns, which, in turn, are fed by surface drainage that goes underground. A stream may enter a sink in one valley and emerge through springs in a neighboring valley. As a result, many valleys are dry for short or long stretches.

Karst drainage has influenced men's actions for ages. Life is hard on the plateaus, easy in the well-watered valleys. Near the Causses of southern France great springs that emerge at the edge of the Rhone Valley have determined the sites of towns. Many of these, such as the ancient Roman city of Nîimes, have flourished since remote antiquity.

Ground Water in Volcanic Rocks

The cracks and cavities characteristic of many masses of volcanic rocks, as in the Hawaiian Islands or the Columbia

Plateau of the Pacific Northwest, make these rocks almost as permeable as cavernous limestone or dolomite. The most productive group of springs in the United States is in volcanic rocks. The springs are distributed along a 50-mile stretch of the Snake River, Idaho and discharge 14 million cubic meters of water per day, or 5,000 cubic feet per second, contributing to the Snake a volume of water two-thirds as great as the average discharge of the Mississippi River at St. Paul, Minnesota.

SOLUTION AND CEMENTATION BY GROUND WATER

In general, solution is more important above the water table; deposition and cementation below. The extreme effects of solution in carbonate rocks have already been described. Solution is not confined to carbonate rocks; in the long lapse of geologic time even minerals that the chemist regards as highly insoluble are appreciably dissolved.

Grains of garnet in sandstones may be pitted and etched by ground water and such highly insoluble minerals as augite and hornblende may be completely dissolved. Fossil shells composed of calcite are commonly leached out of shales and sandstones, leaving open cavities that may be filled to form "casts" which preserve faithfully the fossil form. Casts of soluble minerals such as halite or pyrite also are common.

The cementation of sand to form sandstone was mentioned. Calcite cements many sandstones but the reason it deposits in the sandstone pores is obscure. Evaporation of water and resultant deposition of calcite-the mechanism of dripstone formations can hardly operate below the water table.

Possibly one factor is the release of pressure on rising bicarbonaterich ground water allowing the escape of carbon dioxide and the resulting precipitation of calcite.

Around shell fragments, some sandstones are locally cemented, forming ball-like masses called *concretions*. This suggests that the shell fragments were nuclei upon which calcite from the saturated ground water precipitated. Ground water may also deposit other substances. Silica, as opal or quartz and iron oxide, as limonite or hematite, may fill or coat cavities in rocks. At depths of two or three miles temperatures are close to the boiling point of water at the earth's surface and many other minerals may form in the pores.

Ground-Water Supplies in the United States

Water supply is a primary concern of farmers and inhabitants of arid regions. Even in the humid but populous areas of western Europe and eastern North America industrial demands put a heavy strain on water supply. The total quantity of recoverable fresh water underground at any one time is a matter of interest, especially in regions dependent on ground water for their supplies.

All estimates indicate that the total amount of ground water is vastly less than the amount of water in the oceans, but much more than that in the atmosphere, or than that which falls as rain and snow in a single year. In almost any area some water can be obtained from wells, but in nonporous rocks, even where shattered, yields are relatively small and of course yields will be negligible from nonpermeable rocks even if they are porous. Abundant yields of ground water come chiefly from unconsolidated surface formations, mostly of Pleistocene and Recent age and secondarily from older consolidated, but still permeable, sediments or from lavas. The principal unconsolidated aquifers are:

- Alluvial gravels and sands that fill deep interior basins,
- Glacial outwash sands and gravels,
- The coarserparts of deltaic and other coastal plain deposits and
- Sands and silts beneath river floodplains.

Interior basins filled with unconsolidated sediments are common in the western third of the United States. Ground water from many of them has been extensively developed because of the general aridity of the region. They normally supply about one half the ground water used in the United States. Several large basins are in California, including the large central valley (basins of the Sacramento and San Joaquin rivers), which is 450 miles long and 35 miles wide; and the much smaller valley of southern California, which includes the coastal plain near Los Angeles. In 1948 the ground water extracted from these basins in California was enough to cover 9 to 10 million acres to a depth of 1 foot—over 35 per cent of all the ground water used in the United States.

Glacial outwash sands and gravels are interbedded with, or lie slightly south of, the relatively less permeable Pleistocene till sheets. Glacial gravels grade into river floodplain and deltaic gravels and sands, notably along the Mississippi River. Large floodplain aquifers are also found on the Great Plains and permeable coastal plain deposits extend from New Jersey to Texas. Among older, more consolidated rocks the chief aquifers are:

- Permeable sandstones,
- Well-jointed volcanic rocks,
- Cavernous limestone or dolomite and, more rarely,
- Fissured crystalline rocks such as quartzite or shattered granite. Sandstone aquifers supply large amounts of ground water in the Dakotas and elsewhere in the northern Mississippi Valley but are important also in Texas, Michigan and other regions. Basalt flows are important aquifers in the Pacific Northwest and in Hawaii. Cavernous limestones yield abundant ground water in Florida and in parts of the Cumberland Plateau.

The aquifers of three highly productive areas will be described in the following pages. These examples include the glacial outwash materials and Cretaceous sands of Long Island, the sands and gravels in the valley of southern California and the Dakota sandstone of the northwestern Mississippi Valley.

GROUND WATER OF LONG ISLAND

Long Island has no large streams, partly because of its size and partly because infiltration is so easy. The 40 to 50 inches of annual precipitation in west central Long Island, it has been estimated, is taken to surface runoff, ground water and transpiration 20, 40 and 40 per cent, respectively. The most productive Cretaceous sand is the basal bed, 100-250 feet thick. It is a clean quartz sand, practically uncemented and confined by an overlying shale.

The Cretaceous sands and shales are unconformably overlain by Pleistocene sediments as much as 300-400 feet thick. Interbedded Pleistocene till, sand and gravel form two ridges, one near the northwest shore and the other along the middle of the island. The outwash plains of sand and gravel lie between and south of the ridges.

In a small Brooklyn area excessive pumping lowered the water table until it was below sea level. The water table was 0-15 feet above sea level (0-50 feet below the ground surface) in 1903, but had been drawn down to a maximum depth of 34 feet below sea level by 1943. This reversed the hydraulic gradient and salt water from the sea invaded the margins of the aquifer. The State of New York thereafter has required water pumped from new cooling and air-conditioning wells to be returned to the aquifer from which it was withdrawn.

By 1946 more than 200 recharge wells were returning water underground in the urban portion of Long Island. In the rural areas several large surface recharge basins, where storm and industrial waste water could be stored until it would seep into the ground, had also been established. Total recharge in the summer of 1944 amounted to 60,000,000 gallons a day. The warmer recharge water raises the ground-water temperature a very few degrees Fahrenheit, making the water less valuable for cooling. The recharge wells and storm-water basins have served their main purpose, however; the water table is no longer falling and salt-water inflow has decreased.

Ground-Water Basin

In semi-arid southwestern California, several large interior basins have been filled with alluvium, across which the intermittent surface streams flow to the sea, charging or recharging the alluvium with ground water on the way. The basin is filled with moderately well-sorted alluvium, 500 to 1,200 feet thick.

Most of the alluvium came from the northeast and has been deposited in alluvial fans that have united to form a compound apron extending almost entirely across the whole basin. Coarse gravels are present at the heads of the fans, where the slope of the surface and initial dips of the strata are as high as 9°. Farther out on the fans the gravels grade into and are interbedded with, sands and silts.

Near the fan heads relatively impermeable soils have formed, several of which have been buried by upbuilding of the fans. The permeable gravels and coarse sands make a rather complex set of aquifers.

On the lower portions of the apron some wells were originally artesian, the aquifers being confined by finer sediments or soil zones and the hydraulic head being the result of the initial dip.

Many ground-water basins in California are broken by faults so recent that they cut through the alluvium. In this area the faults are effective barriers to ground-water movement because impervious clay gouge has been smeared along the fault planes. The water table drops 400 feet—a vital difference to those who must obtain their water from wells.

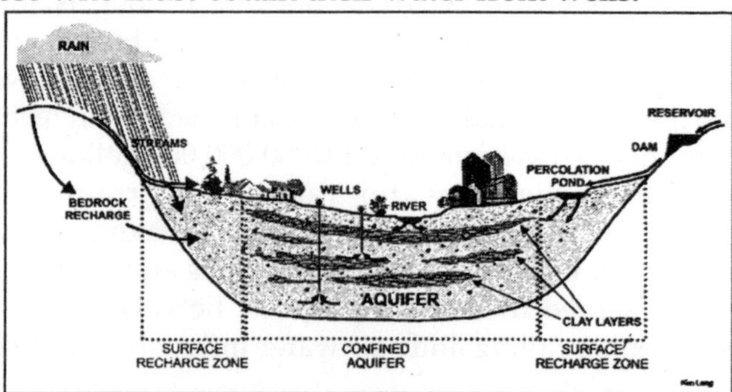

Fig. *Ground-Water Basin*

In this populous but arid region an attempt is made to keep the winter flood waters from reaching the sea by diverting them onto complex spreading grounds of coarse gravel which provide opportunity for the water to infiltrate. These spreading grounds serve the same purpose as the recharge wells and basins on Long Island.

It should be noted that lowering water levels in a basin by pumping is not wholly a misfortune. Surface or subsurface outflow from the basin is decreased or even stopped and the subsurface reservoir is partially emptied and so prepared to take in a much larger recharge.

DAKOTA SANDSTONE ARTESIAN AQUIFER

The great Dakota sandstone artesian basin is the largest and most important source of artesian water in the United States. It extends over much of North Dakota, South Dakota, Nebraska and parts of adjacent states. At least 15,000 wells have been drilled into this Cretaceous sandstone, which is generally somewhat less than 100 feet thick and is overlain by hundreds or even thousands of feet of other sediments, mostly impermeable shales.

The basin form resulted from folding. The principal intake areas are at the west, in the upturned zone along the edges of the Black Hills and Rocky Mountains, 2,000 or more feet higher than the sandstone on the east side of the basin. In general, there is a hydraulic gradient from west to east.

Pressures have decreased progressively since the first well was drilled in 1882. The sandstone is, however, not a simple aquifer. It includes a widespread shale bed near its eastern margin, dividing the sandstone into upper and lower portions and the compositions and pressures of the artesian waters are in detail rather complexly variable.

Movement of ground water from the intake area appears to be interrupted locally by relatively impervious portions of the formation. Despite these complicating details, the Dakota may be considered a single unit and an exceptionally important producer of ground water.

The Dakota sandstone is not the only productive aquifer in its area. One early forecast of production from a deeper sand is still remembered. The early railroads crossing this and region were large users of water.

Near the Black Hills the Dakota sandstone was an inadequate source. In 1905 N. H. Darton of the U. S. Geological Survey, who had just completed an investigation of ground water in the Great Plains region, recommended that the Burlington Railroad go deeper for water in the area near the Black Hills. The structure sections and geologic maps he had made indicated that by drilling at the town of Edgemont a probably productive Paleozoic aquifer would be encountered at a depth of about 3,000 feet.

After almost three years of old-fashioned impact drilling a well flowing more than 400,000 gallons a day was developed at a depth within 31 feet of that predicted. Forty years later a new well, completed in the same aquifer after 50 days of rotary drilling, had an initial flow of almost 1½ million gallons per day, an unusually large yield from a Paleozoic sand and very precious in this dry country.

Economic and Legal Aspects of Ground-Water Use

Where there is not enough surface and ground water to go around, disputes have arisen between individuals, communities and states and these disputes have been carried to the courts. Communizing the rule that the owner of the land surface also owns everything beneath the surface, all the owners of land above a ground-water basin are considered to have joint ownership of the basin.

Water may not be exported without good reason beyond the basin's surface area. In addition, the *principle of best use* has been established, as in a dispute between cattlemen, who wish to preserve feeble springs and truck gardeners, who wish to put the abundant ground water of the same area to more productive use.

In some states, such as New York and Maryland, the permission of state authorities is required for the drilling of large wells and return of used waters to the aquifer may be required. Ground water is an important public commodity and its use now requires regulation by well-informed public officials. Soil becoming saturated with faecal matter and specifically infected.

The ratio of cases to population living in Dublin on loose porous gravel soil for the ten years1881-1891was I in 94, while that of those living on stiff clay soil was but 1 in 145. " This is as we should expect, since the movements of ground air are much greater in loose porous soils than in stiff clay soils." A foul gravel soil is a most dangerous one on which to build.

For warmth, for dryness, for absence of fog and for facility of walking after rain, just when the air is at its purest and its best, there is nothing equal to gravel; but when gravel has been rendered foul by infiltration with organic matters it may easily become a very hotbed of disease.

Chapter 8

Erosion and Deposition

The loosening and transporting away of rock debris by moving agents operating at the earth's surface is called *erosion*. Weathering of rock is essentially a static process; if there were no natural agents to remove the clays, sands and other products of weathering.

All surface rocks would eventually be deeply buried beneath a thick mantle of their own decomposition products—a common situation on flat ground in areas with a humid climate. On sloping ground, however, the products of weathering are constantly being transported away by rainwash, rills and streams.

On gentle slopes the soil cover is generally a few feet deep, but on steeper slopes, weathering products are generally removed nearly as rapidly as they are formed. On many steep mountain slopes the erosional agents are so active that they even scour and remove fresh rock before it has been appreciably softened by weathering.

As the products of weathering are removed by erosion, new rock surfaces are exposed to attack by the weathering agents. Thus a cycle of change is set up: Weathering *prepares* the rock for transport by decomposing and disintegrating it. Erosional agencies *transport* the weathered material to a different locality (for example, to the ocean) and there deposit it in layers (strata) which may be ultimately *cemented* into new rock.

This rock may then be raised above sea level, exposed to the air and the rains and again weathered into soil, to start a new cycle.

OCEANIC WAVES AND CURRENTS

Turbulence in the water envelope of the earth (oceanic waves and currents), like turbulence in the air (winds), is an important agent of erosion. Anyone who has watched the waves of the open ocean crash against the coast need not be told that here is a powerful agent capable of picking up and transporting sand, gravel and other rock waste.

Houses, breakwaters and seaside roads may be destroyed overnight by the waves, cliffs are undermined, ships driven against the rocks and broken into matchwood and boulders and sand rolled about like ninepins in the swirling water along the shore. Sand drifted along the coast during a storm may completely block the mouth of a river, or may be swept far out into deep water.

Even on a relatively quiet day, the sand along an ocean beach is in constant agitation: each oncoming wave, as it strikes the beach, brings sand rolling along the bottom with it and as the wave ebbs the sand is rolled back seaward. During severe storms huge boulders may be dashed back and forth in the breakers and even hurled against a sloping rock beach with such violence as to catapult high in the air with the spray. Occasionally, shore-driven rocks are thrown over seawalls and come crashing through the roofs of buildings.

Obviously the shore of the open ocean is a zone of vigorous erosion. Most ocean coasts are cliffed because the waves rapidly undermine the shore. Along many shorelines, such striking changes take place during storms that valuable beach property ordinarily must be protected from erosion (or in some places from unwanted deposits of new sand or other beach materials) by breakwaters.

GLACIERS

Probably the Nisqually glacier, which is easily reached by highway in Rainier National Park, has been visited by more people than any other in North America. The Nisqually is a pygmy among glaciers; less than 5 miles long and less than 1,000 feet thick, it does not compare with the much larger Emmons and Winthrop glaciers in the more remote northeastern section of Rainier National Park, not to mention the 60-mile-long ice streams of Alaska or the vast moving ice sheets that blanket Greenland and Antarctica.

Dwarf though it is among its fellows, no visitor can examine the front of the Nisqually glacier without being impressed by the tremendous erosive power of moving ice masses. At the glacier front very little ice is visible-most of it is obscured by blocks of rock riding on the moving ice mass or pushed forward in front of it.

Some of these blocks are fragments of andesite, each weighing several tons. By matching the rock with outcrops far up the valley, it is obvious that they have been torn from parent rock ledges upstream and carried to their present resting place. Some have been rafted along on the back of the glacier, some were frozen into its body and dragged forward with it and some were pushed in front of it.

Smaller particles of rock and soil are mixed with the larger boulders. As the ice melts, they accumulate in hummocky piles in front of the glacier. Thick accumulations of loose rock also rest on top of the glacier, especially at its edges, because, here, boulders, loosened by frost action from the cliffs above, are constantly rolling down and snowslides and streams of meltwater also add debris to the glacier's surface.

The Nisqually glacier has been slowly shrinking in recent years. Downstream, below the front of the glacier, lie irregular piles of debris left by the ice as it retreated upstream. The rock floor, where bared by the melting ice, is scored and polished by the blocks of rock and finer material that has been dragged across it by the moving glacier. Meltwater pouring from tunnels in the Nisqually glacier is a dirty milky white. Examination shows that it is cloudy, not from clay, but from pulverized rock flour obviously made by the crushing and milling of rock particles against one another and against the bedrock floor as they were rolled beneath the tremendous weight of the moving glacier and dragged forward by it.

Relative Importance of Different Agents

Spectacular as the results of wind, ocean and glacier

appear, they are of relatively little importance as compared with the movement of rock debris by running water. In the years since the middle 1930's, rainwash has almost completely obliterated the drifts of silt and sand made during the dust storm of May 12, 1934.

Only in the driest deserts are features due to wind erosion conspicuous and even here the presence of numerous gullied slopes, dry stream beds and sheets of water-deposited detritus shows that the rare rainstorms are able to put their stamp upon the desert landscape. Oceans are limited in scope and effect.

The attack of the sea upon the land is chiefly confined to the shoreline and to the shallow submerged margin of the continents. Here they have undoubtedly played a part in extending the upper of the two dominant surface levels, but stream deposits have doubtless played an even greater role.

Again, compared to running water, glaciers are quantitatively unimportant as erosional agents. Streams are practically universal—there is scarcely any part of the continents that is not subject to at least some rainwash each year. Glaciers, on the other hand, are confined to high mountain ranges and to polar regions.

The combined area of all the glaciers constitutes only ten per cent of the land area. Their contribution to the denudation of the continents is insignificant when compared to the work of streams.

Streams, aided by downslope movements, are the great levelers. Although individual rivers during stage of flood may produce sudden and spectacular feats of erosion, the quantitative importance of streams is due to the fact that rain falls on all parts of the continental masses and in coursing off them to the sea it steadily removes rock debris from the surface. Let us turn our attention to an analysis of the way in which streams perform their work.

AGENTS OF EROSION

Erosion is a dynamic process. Only moving agents are capable of picking up rock debris and transporting it to a new location. Although many different agents—as varied in their nature as earthworms, lightning and avalanches of snow—can transport a minor amount of rock waste, the great bulk of erosional debris is transported by four main agents:

- The *wind;*
- *Waves and currents in the ocean* and, to a lesser extent, in lakes and other small water bodies;
- *Glaciers;*
- *Nunning water.* Another important process, because it modifies and increases the erosional effects of running water, is the slow, or occasionally rapid, *downslope movement,* under the action of *gravity,* of solid or semi-liquid masses of mud, soil and rock debris.

On hills and mountainsides such downslope movements deliver a steady increment of rock waste to the streams at the foot of the slopes. The streams then have material to transport.

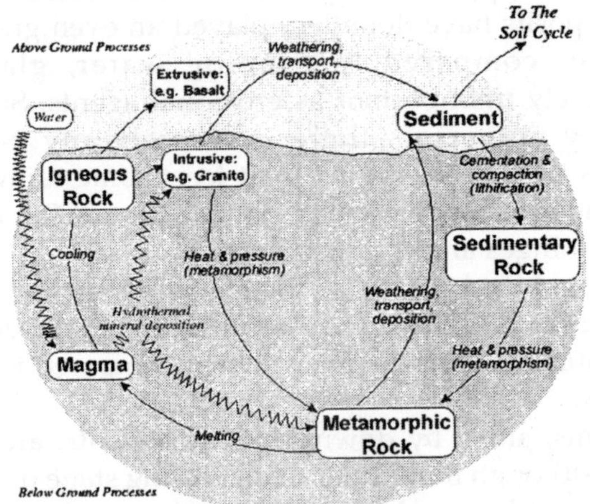

Fig. *The Cycle of Rock*

To evaluate the effectiveness of the different agents in eroding the land requires analysis of many factors. Because of the size of the earth, the complexity of the erosional processes and the difficulty of measuring accurately the amount of debris in transit at any one time, it is impossible to get quantitative figures on either the rate at which erosion is lowering the land masses (the *rate of denudation*), or on the amount of debris transported by any one particular agent of erosion. Yet, despite the incomplete data, fairly reliable estimates as to the relative importance of the different agents of erosion and a rough but useful approximation of the rate of denudation of the land areas can be made.

WIND

When the Great Plains area east of the Rocky Mountains was settled, the natural grass sod was plowed under and the light soils were excessively tilled in order to grow corn and wheat. Even before settlement, parts of this area had undergone considerable wind erosion. The farmers who raced madly into the "Cherokee strip" to stake out homesteads, when the Indian Territory of what is now Oklahoma was opened to settlement, were greeted by blinding dust storms.

As more and more land was put under the plow, these "black dusters" became bigger and more frequent. In the early 1930's a succession of dry seasons led to the terrible drought years of 1933 and 1934. Pulverized by tillage and parched by drought, the soil had lost not only its original protective grass cover, but much of its cohesiveness. Thus, the stage was set for one of the most spectacular and destructive events that has occurred in modern agriculture.

On May 12, 1934, strong winds lifted vast quantities of dust from the fields of Kansas, Oklahoma, Texas, Colourado and other Plains states high into the air and drove it swiftly to the east as a gigantic dust storm.

Sweeping swiftly out of the Plains as a blinding, choking mass of particles so dense as to blacken the sky and change day into darkness, the seething dust clouds rolled eastward across the well-watered lands east of the Mississippi River Here, where little dust could be gleaned from the forest and grass-covered landscape, the clouds lost some of their density, but they still retained enough solid material to blot out the still as they swirled around the skyscrapers of New York City and out over the Atlantic. Dense, dirty-brown clouds of dust engulfed ships 300 to 500 miles from shore and filtered even into well-protected holds, blanketing the cargoes.

Measurements of the amount of dust in the air, reported

by observers from widely scattered points over central and eastern North America, indicate that more than 100 tons of dust per square mile fell in the areas covered by the dust cloud. Since the area affected embraced approximately twothirds of the North American continent and much of the western Atlantic Ocean, it appears that more than 300,000,000 tons of soil was removed from the Great Plains during this storm and strewn over the lands and sea to the east.

Wind strongly winnows the material it transports. Only the finest particles are swirled high in the air and carried far. Most of the material moved by wind is not dust, but consists of coarser grains of silt and sand that roll or skip along the surface of the ground. Studies by soil conservationists indicate that in the Plains country about three-quarters of the soil moved by the wind does not rise into the air but drifts along the surface for a few yards, or at most a few miles and then accumulates in ditches, around clumps of vegetation and against fences, buildings, or other obstructions. Thus, for every ton of air-borne dust there are generally two or three tons of coarser debris piled in drifts of sand and silt within or near the source.

The dust storm of May 12, 1934, was followed by many others. How a vast area of the southern Great Plains was converted into a barren "Dust Bowl" with attendant economic disruption and enforced mass migration of people is now an episode of American history. Because of the great havoc wrought, many of the later dust storms were carefully investigated. One of the most completely studied storms was a much smaller one that swept out of the Dust Bowl on February, 7, 1937. Most of the land to the northeast was blanketed with snow so it was easy to collect and measure the dust that settled. At Ames, Iowa, 34.2 tons of material per square mile was deposited by this storm, 14.9 tons per square mile fell at Marquette, Michigan and 10 tons per square mile in southern New Hampshire.

More rain and better methods of farming have again brought prosperity to the Dust Bowl, but wind erosion on a minor scale still continues and grimly threatens a major revival if a series of drought years returns. Although the great dust storms of the 1930's were largely due to changes in the soil cover induced by man, wind erosion is constantly at work on all surfaces of the continent and is particularly potent in deserts. In the western Sahara, strong southward-blowing winds (the *Harmattan*, a local name for the Trade Winds) sweep across the desert for about six months of each year. In exceptionally stormy years they are reported to deposit as much as a foot of silt, dust and sand along the edge of the desert in northern Nigeria. At times, terrific dust storms also sweep northward across the Mediterranean into Europe. In 1859 such a dust storm covered most of Europe and deposited 80 to 100 tons of dust per square mile in southern Germany.

ANALYSIS OF STREAM EROSION

Source of the Runoff

"From whence springs the water of the rivers?" is a question to which reliable answers have been won only in comparatively modern times. The early philosophers held that rainfall was totally inadequate to account for the vast flow of water in rivers like the Seine, Rhône and Nile and that the earth's surface was too impervious to allow the percolation of rain water into the soil and rocks from which it could be returned in the form of springs. Although there were occasional dissensions, the general belief up to the middle of the. Seventeenth Century was that the water rising in springs and flowing to the ocean in streams could not possibly be the runoff from rain and snow.

The modern concept of hydrology (the science that treats of water) began with the work of a Frenchman, Pierre Perrault (1608-1680). Perrault made measurements of the rainfall in a part of the Seine River basin over a threeyear interval and also measured the amount of water flowing in the Seine. From the available maps, he then estimated the area of the river drainage basin above the point where he had measured its flow. These measurements indicated that the Seine carried off only one-sixth of the water that fell within its drainage basin as rain and snow. Perrault thus proved the fallacy of the old assumption that the streams carry more water to the sea than falls on their drainage basins.

In fact, the problem of the streams was now reversed—the question was no longer, What is the source of the water? but, What has become of the vast volume of water, equivalent to five times the flow of the Seine, that has been precipitated on the drainage basin but is not being carried out by stream flow? A few years after Perrault's measurements, Edmund Halley, an English astronomer, showed by experiment that the moisture evaporated from the surface of the oceans is entirely adequate to supply the water returned as runoff.

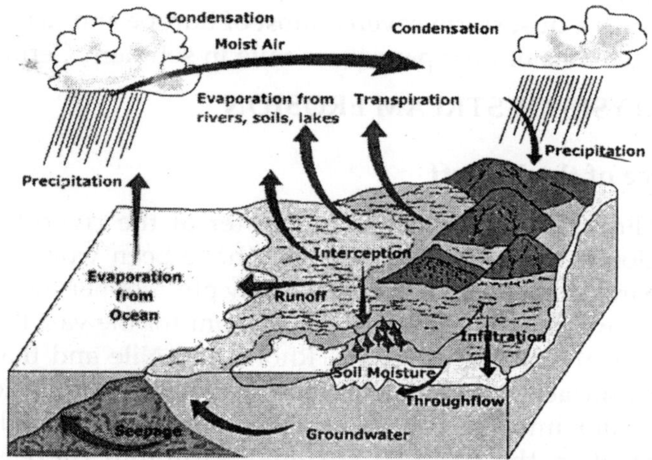

Fig. *The Hydrologic Cycle*

Since the time of these pioneers a wealth of hydrologic observations has become available. The flow of water in the major streams of practically all countries has been accurately determined by stream gaging. Such measurement, generally carried on by federal or state agencies, has many purposes. Precise data on river flow are valuable to navigation, irrigation, hydroelectric power, flood control and many other interests.

imilarly, the accurate recording of rainfall and snowpack has many uses; and data are available from thousands of stations scattered all over the world. Numerous tests have been made on the rate of evaporation from bodies of water under different climatic conditions and on the rate of loss of water into the atmosphere through transpiration by plants. Rates of underground percolation of water and measurements of the amount of water stored in underground reservoirs have been obtained from thousands of observations on springs and wells. From these measurements and observations we can fairly accurately deduce the relations between rainfall and runoff and can answer the question of what happens to the water that falls as rain.

The Hydrologic Cycle

Of the moisture precipitated on the land as rain and snow:

- Some evaporates at once from the surface of the ground and from the vegetation on which it falls;
- Some is absorbed by the roots of growing plants and is quickly transpired back into the atmosphere through their leaves, although a little enters into the plant tissues and is trapped there until the plant is dead and destroyed:
- Some seeps into the soil and rock where it may be temporarily stored underground until it is.
  - Brought to or near the surface by capillarity and evaporates;
  - Delivered to the roots of growing plants; or
  - Oozed to the surface at a lower elevation, reappearing either in springs or as ground-water runoff (*effluent seepage*) pouring into streams along their banks and bottoms;
- Some runs off in surface rills, brooks and rivers and is returned directly to the ocean. It is the last part, the *runoff,* that we are concerned with in the study of stream erosion.

The hydrologic cycle is depicted graphically. The sun is the source of energy that operates this great system of waterworks. Solar energy raises the temperature of the oceans and the lands and evaporates water from their surfaces. It also stimulates the growth of plants and leads to the transpiration of water vapour into the atmosphere from their pores.

Atmospheric moisture gained in these ways is wafted inland and raised to high altitudes by the winds. It thus acquires potential energy of position.

When the moisture falls as rain, this energy of position is converted to kinetic energy—the water has become a moving agent, capable of acquiring and moving a load. On reaching the surface of the earth, that part of the rain that constitutes the runoff flows downslope in response to gravity, moving obstacles in its path and sweeping soil and rock debris along in its currents.

In studies of the relation of runoff to precipitation it is convenient to define *runoff as the total discharge of water by surface streams*. Thus defined, the runoff includes not only that part of the precipitation that flows directly from the rains across the surface of the land, but also the increments of ground water that enter the streams from springs and effluent seepage.

It is also convenient to distinguish that part of the precipitation returned to the air by evaporation and transpiration as the *evapo-transpiration factor*.

Hydrologists, for convenience, usually simplify the relations within the hydrologic cycle by assuming that evapo-transpiration can be determined by subtracting runoff from precipitation. The equation is only approximately correct, however. It neglects such locally important (though generally unmeasurable) factors as the seepage of water through the soils and rocks directly into the ocean.

Precipitation = Runoff + Evapo-transpiration

For the land masses as a whole such seepage is not great, but in a few areas of highly permeable rocks, as in the Hawaiian Islands, it may locally dispose of a large part of the total rainfall. Other very minor factors are the amounts of water locked up in the tissues of plants and animals and the amounts chemically combined in minerals during weathering.

VARIABLE FACTORS AFFECTING THE RUNOFF

The ratio of runoff to precipitation is not everywhere 1:6 as Perrault determined for the basin of the Seine. Several variable factors affect the ratio.

Careful studies of many different hydrologic basins in the United States, made by the Water Resources Branch of the U. S. Geological Survey, have revealed great variations in the ratio of runoff to precipitation. From these measurements some of the factors may be summarized as follows:

Temperature

Evaporation is, of course, much greater in warm regions than in cold. High temperatures also favour growth of vegetation and hence increased loss by transpiration. In humid regions a close correlation has been found between the amount of runoff and the mean annual temperature. Even in arid and semi-arid regions, if large groups of data are considered, not simply those from a single drainage basin, there is also a reasonably close accord. The data relating temperature to runoff in the United States has been summarized by Langbein, a hydrologist of the U. S. Geological Survey. Thus, for an average annual precipitation of 40 inches, more than half the rainfall (21.5 inches) runs off where the mean annual temperature is 40° F.; but the runoff decreases to 10.2 inches in a mean annual temperature of 60° F. and to less than 3 inches in a mean annual temperature of 80° F.

Slope

Slope obviously exerts an important control on runoff. Steep slopes of bare rock in mountainous regions shed practically all of the rain that falls on them. On flat ground the rain may remain on the surface in shallow puddles until it is evaporated or absorbed by the soil and plants.

Permeability of the Ground

Different soils and rocks have highly different permeabilities. Most of the rain falling on the porous ash of a recent volcanic cone sinks directly into the ground and becomes a part of the ground-water circulation, but nearly all will run off on a landmass of similar slope that is underlain by relatively impermeable shale. The ability of a soil to absorb rainfall or snowmelt depends on other physical factors as well as permeability. Water-logged soil, although its permeability may be high, has reached its maximum capacity of absorption; frozen ground is similarly nonabsorptive. The absorption of a highly permeable soil also may be limited because of a nonpermeable subsoil or bedrock a few feet below.

Permeability, then, is not alone an accurate measure of the rate of infiltration. Hydrologists use the term *infiltration capacity* to define *the maximum rate at which the soil, in a given physical condition, can absorb falling rain*. Artificial tests of infiltration capacity made by playing sprinklers on enclosed plots of soil from which the runoff is caught and measured show, as would be expected, that the rate of absorption is high at the beginning of a rain, then diminishes rapidly until a fairly constant value is reached. The high initial rate usually changes to the slower uniform rate within a half hour after the rain begins to fall.

Amount and Duration of the Rainfall

Distribution of rainfall through the year greatly influences the runoff. Rain uniformly distributed in many small showers may be largely evaporated or absorbed by the ground and plants before it reaches flowing streams. During violent rainstorms, on the other hand, percolation is too slow to lead much of the water underground; most of it courses rapidly across the surface and into stream channels. Rapid melting of the winter snowpack by warm rains or winds is one of the most common causes of river flooding.

If the annual rainfall is high (80 inches or more), most of it is likely to run off, even though the rain may be uniformly distributed.Here, the underground reservoirs are constantly replenished and the pores within the soil and rock are filled with water almost to the surface of the ground. In deserts, on the other hand, most of the scanty rainfall is absorbed by the parched soil, or evaporated and there is little runoff except immediately after violent storms.

Vegetation

Vegetation retards the runoff. Tangled stalks of grass or

the mulch of decaying leaves and twigs in a forest absorb rain like a blotter. Earthworms and other burrowing animals that live in the plant-rich soil aid percolation by opening tunnels in the soil. Very striking increases in runoff have been observed where the forest cover has been burned off, or where the natural sod has been plowed under, as in our Dust Bowl. summarizes some of the data for a few such areas.

These variable factors result in marked differences in runoff in different stream basins. In the western United States runoff varies tremendously with differences in temperature, infiltration capacity and other local factors. The range is from less than 0.25 inch in the southwestern Arizona desert to more than 80 inches on the western slopes of the Olympic and Cascade mountains of Washington and Oregon.

AMOUNT OF WATER AVAILABLE FOR EROSION

The rivers of the United States, as determined by stream gaging, deliver water to the oceans at a rate of 1,800,000 cubic feet per second (about 330 cubic miles per year). Roughly one-third of this amount is carried by the Mississippi River.

The average annual rainfall of the United States, determined from thousands of rain-gaging stations, is 30 inches. The average runoff for the country as a whole is 8.6 inches. This gives a ratio of runoff to precipitation of approximately 1 to 3.5, as compared with the ratio of 1 to 6 determined by Perrault for the basin of the upper Seine.

Data for runoff and precipitation of other continents is less complete than for the United States. L'vovich, a Russian hydrologist, has summarized the available data on runoff for the various continents.

The table shows that Australia, with a runoff of only 3.0 inches, is the driest of the continents and that South America, with 17.7 inches of runoff, is the wettest.

Other hydrologists have reached slightly different and in general somewhat lower, for the average world runoff. Despite these uncertainties, we can make a fairly reliable estimate of the total amount of water available for erosion: The average annual precipitation over the land areas of the earth is about

40 inches per year. This means that each year approximately 35,000 cubic miles of water falls on the land in the form of rain and snow. According to different estimates, the rivers return to the oceans from 7.5 to 10.5 inches per year.

Using an average figure, this means that in round numbers approximately 8,000 cubic miles of water courses off the lands and into the seas each year. This is the amount of water that is available for the erosion of the lands.

ENERGY RELEASED BY THE RUNOFF

On the average, the lands stand about one-half mile above sea level. Therefore, the continental runoff descends from an average altitude of one-half mile to sea level in flowing from its source to the ocean. This is an average figure; runoff falling on a plain near sea level may descend only a few feet to reach the ocean; the meltwater from the snows on the summit of Mount Everest falls over five miles.

A tremendous amount of energy is developed by the world's streams during their descent to the sea. Some qualitative idea of the amount can be gained by imagining all the continental runoff to be concentrated in one gigantic river that is pouring over a huge waterfall equivalent in height to the average altitude of the landmasses. In other words, imagine 8,000 cubic miles of water per year cascading down a waterfall one-half mile high! One could easily compute the amount of power available at such a waterfall, but the resulting figure is so large in human terms as to be utterly meaningless.

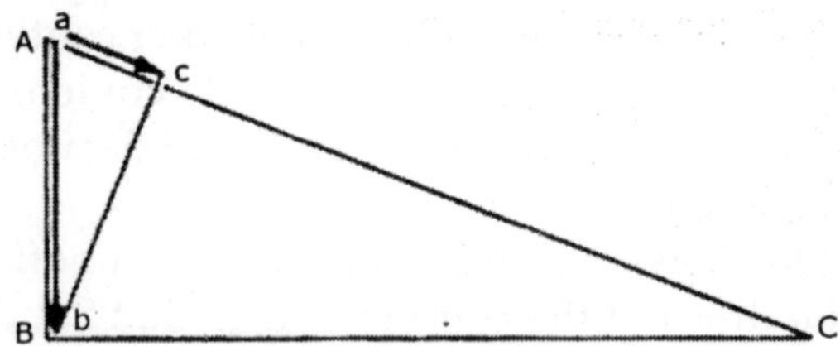

Fig. *Resolution of Gravity for a Stream of Uniform Slope*

Yet this is the amount of energy available for the erosion of the lands by streams. It is easy to see why running water is the great leveler—much more important than all the other agents of erosion combined. Fortunately for mankind, very few streams work at anywhere near their full erosional capacity. Because weathered material ready for transport is moved downslope to the streams only slowly, many rivers carry little load and dissipate most of their energy in work other than erosion. Otherwise, the continents would be largely denuded of soil.

MECHANICS OF STREAM EROSION

In dealing with influences on stream flow it is essential to keep in mind the *Law of Conservation of Energy* and the distinctions between potential energy (energy of position), kinetic energy (energy of motion) and heat energy. The *Law of Conservation of Energy* requires that, although one form of energy may be transformed into another, for example, kinetic energy into heat energy when a falling rock strikes the ground, the sum of all the different kinds of energy does not change during the transformation. In other words, the total amount of energy remains constant and no energy is gained or lost.

Flowing streams are powered by gravity. It is convenient to use a simple graphical representation, the "resolution of forces", to determine the component of the force of gravity effective down the particular slope on which a stream is flowing. Consider a mass of water poised at the top of a uniform slope.

Its total energy is measured by the potential energy of vertical fall and hence the final velocity after movement down the slope would, if there were no friction, be the same as for direct vertical fall, though the time required to attain the velocity would be greater because the component of gravity operating parallel with the slope is less.

The acceleration of gravity (an increase in velocity of 32 feet per second at the end of each second) is represented by the hypotenuse (*a-b*) of the right triangle *abc*. Under frictionless flow, the mass of water would move down the slope AC with an acceleration *ac* and would travel the distance *ac* in the same time that it would fall freely through the vertical distance *ab*. Even with the much flatter slopes characteristic of typical streams, the acceleration for frictionless flow would often be as much as 0.5 feet per second in a second. As there are 3,600 seconds in an hour, the stream, after an hour's frictionless flow down an ordinary slope, might attain the astounding velocity of 1,800 feet per second—more than 1,200 miles per hour!

No such velocities are even remotely approached in natural streams. Actual streams are slowed by friction of the moving particles against one another and against the land surface. They commonly attain and maintain velocities of only a few feet per second (generally less than 5 miles per hour); if their velocity changes appreciably, it usually decreases downstream. We conclude that the energy of fall is mostly converted to heat energy by friction.

Some of the energy, is also used tip in the transportation of eroded rock particles from upstream. The actual quantity of mud, sand and gravel eroded varies widely for different rivers. For example, compare the crystal-clear Columbia River at Wenatchee, Washington, with the Missouri River (often called the "Big Muddy") near Saint Louis, Missouri. The amount of water carried by each is about the same, but the Columbia, though a roily turbulent stream filled with violent rollers and eddies, contains very little silt and mud. It is flowing on smooth bedrock and coarse boulders, hence almost no fine debris is available in its bed to be picked up and carried.

By contrast, the Missouri is laden with mud and silt. It flows across the Great Plains, where the soil cover is deep and the vegetation scanty; consequently, abundant fine-grained rock waste is available for transport. Much of the energy of flow of the Columbia is used up in internal friction within its boils and eddies, whereas in the Missouri much of the available energy is used to transport the abundant rock waste.

But even the muddy Missouri is transporting far less material than it might. Only rarely does the Missouri carry more than 20,000 parts per million (2%) by weight of solid material. This is a high load for a big river, though it is exceeded by the Colourado, the Yellow River of China and several others. The Mississippi, generally considered a very muddy stream, carries only about 5,000 parts per million

(0.5%). Small streams in semi-arid regions with a thick soil mantle, however, may carry enormously greater loads—up to 30 per cent or more by weight—especially during floods. A sample from the San Juan River in Colourado when it was in flood contained over 75 per cent by weight of red silt and sand. There is, of course, a complete gradation from such highly loaded streams to mudflows and landslides.

The evidence indicates, however, that nearly all streams are working far below their potential capacity as erosional agents. If they used most of their energy in carrying rock waste, few slopes could retain a cover of soil.

RATE OF DENUDATION

What, then, are the actual figures on the rate of denudation? At the present time, measurements on the rate at which different drainage basins are being lowered are so few and incomplete and quantitative data so hard to obtain, that estimates can be only tentative.

The total load of debris being carried by a stream is difficult to measure. Although reliable figures covering a period of more than a year or two are available for only very few rivers, there are enough data on hand to indicate the general magnitude of erosion.

A carefully controlled series of measurements of the silt load of the Missouri River above Kansas City made during 1930 and 1931 indicates that this river is lowering its drainage basin about 1 inch in 650 years, or 1 foot in 7,800 years. The Colourado River, which, in the ten-year period preceding September 30, 1935, carried an average annual load of 250,000,000 tons of silt and sand past the Grand Canyon gaging station, is certainly lowering its drainage basin more rapidly than the Missouri; but the Columbia, at least above the junction with its muddy tributary, the Snake, is undoubtedly denuding its basin much more slowly than the Missouri.

Each year the Mississippi River brings to the Gulf of Mexico about 730,000,000 tons of dissolved and solid material. From such measurements it is estimated that *the rate of denudation for the entire United States is about 1 foot in 9,000 years.*

If this rate could be maintained and if there were no compensating upward movements of parts of the earth's crust, all the landmasses would be eroded to sea level in about 23 million years.

Changes in the Rate of Erosion

Because of certain variables not taken into account in the measurements, our figure of 1 foot in 9,000 years for the average rate of denudation must be modified considerably in dealing with specific erosional problems. For some kinds of problems the figure is too high; for others it is much too low. For example, can this figure be used as the average rate of erosion for all of geologic time?

There are many reasons for believing that present-day streams carry much higher loads than the average for the geologic past. The activities of civilized man contribute vastly to present stream loads. Rivers in populous areas carry a vastly increased load because they are polluted with industrial and municipal wastes.

Enormously increased amounts of silt and mud are also being washed into our streams from cultivated fields; before human settlement, the original forest and grass cover surely impeded erosion and greatly lessened the amount of soil washed to the sea. Such an increased rate of erosion has been proven recently by the Soil Conservation Service of the U.S. Department of Agriculture in making extensive tests of the amount of soil removed under varying conditions of tillage. The bureau caught and measured carefully the amount of soil and water running offfrom plots of ground with identical areas and slopes but with different vegetative covers..

These factors cannot be precisely evaluated and, thus, the rate of erosion during the whole of geologic time cannot be fixed quantitatively. Certainly, the rate of 1 foot in 9,000 years is much too high—perhaps two or three times too high. An ingenious attempt to measure the age of the oceans by their content of salt failed because the amount of dissolved salt carried to the sea by the rivers of the geologic past could not be accurately estimated—the age, as determined from the present rate, was many times too low. Can we apply the rate of 1 foot per 9,000 years to the problem of soil erosion? The soil conservationist wants to know how rapidly the topsoils of American farms are being gullied and rendered unusable by erosion. In many sections of the country productive topsoil is rapidly disappearing and once-prosperous farms are now barren wastes of deeply eroded gullies and exposed bedrock.

It is estimated that about 282 million acres of farmland in the United States have been so seriously eroded as to make farming unprofitable. Still another 775 million acres have lost enough topsoil to affect their productivity; and another billion acres of land show some evidence of damaging erosion. These figures do not include the large acreage of the Dust Bowl, in which the land has been damaged by wind erosion.

The soil conservationist has difficulty in reconciling this striking evidence of damaging erosion with the figure of 1 foot in 9,000 years for the rate of denudation. Most of our productive soils are at least 1 foot thick; how can so much damage have been accomplished in the relatively short time since the sod of the frontier was plowed into farmland? Only after intensive investigations of the kind represented thousands of soil plots scattered all over the United States did the answer become clear.

Most of the soil removed from the fields by rainwash is almost immediately redeposited; only a part is transported directly to the sea. In the words of Hugh Bennett, an American soil scientist with the Department of Agriculture:

The sediment entering the oceans represents merely a fraction of the soil washed out of the fields and pastures. The greater part is piled up or temporarily lodged along lower slopes, often damaging the soil beneath; or it is deposited over rich, alluvial stream bottoms or in channelways, harbors, reservoirs, irrigation ditches and drainage canals... Available measurements indicate that at least 3,000,000,000 tons of solid material is washed out of the fields and pastures of America every year. Bennett's figure for the tonnage lost from the fields is nearly three times the amount of soil delivered to the oceans by streams. Evidently, the figure of 1 foot per 9,000 years for the rate of denudation, though a reasonable estimate of total lowering of the continent, does not take into account the *local* erosion and redeposition of soils of primary interest to the soil conservationist.

GEOLOGIC EVIDENCE OF EROSION

Most of the data on which is based have been accumulated during the last one hundred years and much of it within the last fifteen or twenty. Since the earliest days of the science, however, geologists have been aware of the vast changes in the landscape brought about by running water. Their observations, that indicated an immense amount of erosion of the land surfaces, were not concerned with the amount of silt being carried to the sea in streams, nor even the spectacular effects occasionally produced by catastrophic river floods, but with the etched surface of the earth itself.

On the walls of the Grand Canyon of the Colourado River appear horizontal layers of limestone, shale and sandstone. The limestones, particularly the thick Redwall limestone which occurs 1,000 feet or more below the canyon rim, form great cliffs. These cliffs can be traced in and out of the short tributary canyons that score the walls of the Grand Canyon.

The cliffed limestone lies everywhere at the same level and is exposed on both walls of the canyon and its tributaries and in all the isolated buttes within the canyon that rise high enough to intersect it.

Clearly, the limestone must once have been continuous across the canyon; the Colourado River has cut down through the horizontal layers and eroded away the portions of them that formerly filled the area now occupied by the canyon.

At the bottom of the Grand Canyon the river is cutting into granite and metamorphic rocks. From their mode of origin, these rocks must have been formed far below the surface of the ground.

In many mountain ranges, strata of sandstone, shale and limestone have been tilted from their original horizontal position. The edges of these tilted beds, laid bare by erosion, can be followed in ridges. Obviously, the deeper and older beds in the sequence could only have been exposed to view through the removal by erosion of the younger overlying beds that once covered them. In several mountain chains such series of tilted beds are exposed in belts many miles wide; to strip away all of the overlying beds and to expose the deepest layers in some of these ranges have required the removal of 5, 10, or even more miles of overlying rock.

The study of this phase of erosion, however, can best be deferred until we have learned about geologic maps, for the most striking proofs of deep erosion have come from the mapping of the surface geology of mountain ranges.

OTHER IMPLICATIONS OF EROSION

There are other interesting implications that follow from our study of erosion. Streams, glaciers and waves all move material downhill; it ultimately comes to rest at or slightly below sea level. Can this be one factor in accounting for the upper of the two dominant levels of the earth's surface ?What isostatic response can we expect in the earth's crust as a result of the unloading of mountainous areas by erosion and the loading of the continental shelves and other low areas by deposition? Answers to these questions must be deferred, but the point emphasized here is that erosion is a process that is continuously operating to upset the condition of isostatic balance in the earth.

LAMINAR FLOW AND TURBULENT FLOW

If needle-thin streams of coloured fluid are admitted to a stream of water flowing very slowly through a glass tube, the coloured threads advance in parallel lines downstream from the points of contamination.

Where an obstruction is met, *left,* the flow lines move around the object in a "streamlined" way. This phenomenon, called *laminar flow,* is apparently produced by the sliding of thin layers of fluid over one another. It is characteristic of viscous liquids, such as cold lubricating oils or sticky molten lava. It can be produced in water only at velocities less than an inch or so per second (1/16 mile per hour).

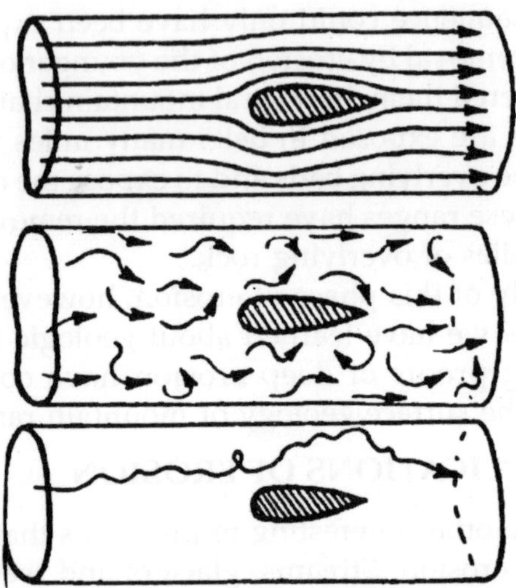

Fig. *Left, Laminar Flow; Centre and Right, Turbulent Flow*

If the flow is speeded up, the pattern of coloured threads abruptly changes. Instead of parallel lines they begin to trace irregular, tortuous paths, then swirl in all directions to the extent of losing their identity by mixing in the water stream. This is called *turbulent flow*. As the velocity of the stream increases, or if the walls of the tube are roughened, the turbulence is increased.

STREAM EROSION AND DEPOSITION

The sediment load of rivers and smaller streams provides undeniable evidence of the erosive power of running water. Surface runoff picks up and transports small amounts of mud and sand from its first confluence in rills and tiny brooks and eventually enlarges these miniature valleys into gullies. Rills join to form small streams and these, in turn, unite into rivers. The whole system erodes the surface of the land and transports the debris to the ocean. Knowledge of how streams erode, transport and deposit rock waste is essential to an understanding of the development of our present landscapes and to the interpretation of the origin of ancient sedimentary deposits.

Furthermore, the study of stream flow contributes to the

efficient use of rivers by man. Without a knowledge of stream mechanics we cannot hope to solve the problems of recurring flood disasters, erosion of rich valley soils, water storage and river navigation.

River Systems

Geologists and engineers often use the term *stream* in a general sense, that is, to designate bodies of running water of any size. For instance, the term *stream erosion* refers to the action of rivulets as well as rivers. A *river system* includes the various streams that join within a drainage area. A river or *trunk stream* is formed by the confluence of *tributary streams,* themselves formed by smaller tributaries and so on to the individual contributions of springs, rills and rainwash.

Several characteristics of river systems change fairly regularly from the smallest headwater tributary to the mouth of the main trunk river. The most obvious variables are:

- The *gradient* of the stream, commonly measured by *the number of feet of fall per mile of course*;
- The form of the stream valley;
- The amount of erosion or deposition along the course.

Most upper tributary streams flow with steep gradients (commonly over 50 feet per mile) through topography marked by valleys and ridges. They actively erode the bedrock along their courses. On the other hand, the gradient of a trunk stream, especially in its lower part, is ordinarily less than 2 feet per mile. Here deposition equals or may even exceed erosion and the river flows in a sinuous course upon a smooth plain of its own deposits.

It is along the lower, gently sloping parts of river systems that the problems of stream control are most keenly felt. It is here, also, that man uses rivers for navigation, builds great cities and tills the richest of farmlands.

The relations of rivers to the valleys in which they flow have long been recognized. In 1802 the Scotch mathematician and naturalist John Playfair observed the nice adjustments within a river system and concluded that rivers must surely have cut their valleys:

Every river appears to consist of a main trunk, fed from a variety of branches, each running in a valley proportional to its size and all of them together forming a system of vallies, communicating with one another and having such a nice adjustment of their declivities, that none of them joins the principal valley, either on too high or too low a level, a circumstance that would be infinitely improbable, if each of these vallies were not the work of the stream that flows in it.

Today, with a thousandfold more data than were available to Playfair, it is found that stream flow, erosion and deposition are in extremely sensitive balance. In evaluating this balance quantitatively, geologists and hydraulic engineers have recognized several variable factors. Among these are the gradient and shape of the channel; the velocity of the stream, including variations of velocity within the channel; and the nature of the material that is being eroded and transported by the stream. Before these variables even qualitatively, however, it is necessary to understand the nature of fluid motion itself.

VELOCITIES AND TURBULENCE OF NATURAL STREAMS

Many measurements show that nearly all natural streams, even the smoothest flowing, move rapidly enough to produce turbulent flow. The power of a stream to erode and transport apparently depends on this turbulence and also on the distribution of velocities and turbulence within its channel.

Range of Velocities

Stream velocities range widely, depending mainly on the *gradient*, the *shape of the channel* and the *discharge*, that is, the quantity of water passing a given point, usually measured in cubic feet per second. The effect of channel shape on velocity is well shown by the Rhine, which flows at a rate of 2 to 3 feet per second in the broad, flat channel between Basel and Bingen, whereas in the narrow, deep gorge guarded by the Lorelei, the velocity increases to 6 to 10 feet per second.

The effect of discharge on velocity is strikingly evident during floods. Mississippi River velocities measured near

Vicksburg in 1937 and 1938 were mostly between 2 and 7 feet per second, but increased to 10 and 14 feet per second during the February, 1937, flood. The Columbia River, about 170 miles from its mouth, had a mean velocity of 1.9 feet per second on March 12, 1945, when the discharge was only 78,000 cubic feet per second, but a mean velocity of 11 feet per second at flood flow of 1,018,000 cubic feet per second on May 3, 1948. Where stream gradients are very steep, as in mountain torrents, velocities in excess of 30 feet per second (more than 20 miles per hour) have been measured.

Distribution of Velocity and Turbulence

The velocity of a stream varies greatly in different parts of its cross section. Near the bed or banks a thin film of water, adhering closely to the solid surface, moves exceedingly slowly and probably entirely by laminar flow. Away from the banks the velocity increases rapidly through a foot or so of the stream. Several feet away from the banks or bed the current is nearly though there may be a slight increase to the maximum in the stream centre. Turbulent flow is most marked in the zone of sharp velocity change, bordering the stream's channel. Spiraling and rolling eddy currents are thrown from this layer of unbalanced flow into the main body of the stream. The regions of most prominent turbulence are shown in their relation to the velocity distribution.

HOW A STREAM CARRIES ITS LOAD

The Suspended Load

The turbulence in streams accounts largely for their ability to carry a load of silt and clay particles. Whether the particles are brought by rainwash and rills or are eroded directly from the stream's bed or banks, it is the *upward* currents of turbulence that allow their transport.

Once a grain has been lifted by an upsweeping eddy, it is kept from falling out again only by the force of other upward moving filaments. True, there are as many downward currents as upward, but both are randomly distributed, so a particle may long be sustained before striking a downward swirl. Even if the grain is eventually dropped, or carried down in an eddy, other grains will be picked up by currents acting in the opposite direction.

The important result is that, in the time interval during which the grain is picked up, jostled around and deposited again, it has moved downstream a distance that depends on the average forward velocity of the water. To an underwater observer moving with the current, the grains will appear to rise, gyrate, bob and fall; but to an observer on the bank the stream appears a homogeneous mixture of water and sediment, all flowing downstream. Because turbulent currents seem actually to suspend sediment in the stream, this part of the stream's load is called the *suspended* load.

The Bed Load

Even-flowing rivers seldom develop enough turbulence to lift particles larger than medium-grained sand from their beds. Coarser material may, however, be pushed or rolled along the bottom by the swirling currents of the boundary zone, thus becoming part of what is called the *bed load*.

Ex cept when stream velocities are high, the grains of the bed load do not move continuously but progress downstream by many stops and starts. Consequently, the bed load usually moves much more slowly than the suspended load. The motion of the bed load has been observed through windows in the walls of wooden flumes. Most of the grains roll, though some slide along the bottom and others bounce along or vault into suspension.

As the velocity of a stream flowing on a bed of sand is gradually increased, the motion of particles on the bed progresses through:

- Short transport of a few individual sand grains;
- Spasmodic movement and deposition of groups of grains;
- Smooth, general transport of many grains. Under some conditions the scour and deposition of grains produces ripples on the bed. At high velocities large numbers of grains may go into temporary suspension,

proceeding as such dense clouds that a distinction cannot be made between the immobile part of the bed and the moving, partially suspended load. A comparable situation is found in natural rivers, where, except at low velocities, the bed and suspended load apparently grade into one another.

The Dissolved Load

A third way that streams carry material is in solution. The dissolved particles are evenly distributed in the water by diffusion. Here transport does not depend in any way on the nature of stream flow. The chief source of the dissolved load is from inflow of ground water that has percolated slowly through weathering mantle.

Probably very little is dissolved from the channel walls except where streams flow on limestone. Thousands of chemical analyses show that few rivers carry more than 1,000 parts per million (0.1%) of dissolved materials. A general average for many American rivers is about 200 parts per million. It should be noted, though, that this may sometimes form a fourth or a third of a stream's total load.

Load Competence

A stream moving at a given velocity exerts a corresponding force upon the particles of its bed. For particles of the suspended load, the upward-moving currents of turbulence exert forces that depend on their velocities. The total force a current exerts against a grain depends on the grain's surface area, whereas the grain's weight is a measure of its resistance to movement (inertia).

Since the ratio of the surface area to weight increases with decreasing size, small grains are more easily moved than large ones. Thus for any stream, the current Will suffice to move particles only up to a certain limiting size. This maximum grain size gives us a simple measure of the stream's transporting power or *competence.*

Considerable experimental work indicates that the diameters of sand grains and coarser debris that can be moved by a stream increase approximately as the square of the velocity. That is, if the velocity is doubled, the diameter of grains that the stream can move is increased four-fold; if the velocity is tripled, the diameters increase nine-fold and so forth. The high velocities of steep streams, especially when in flood, make their currents almost irresistible. Some move boulders over 10 feet in diameter. When the St. Francis concrete dam in southern California broke in 1928 the great mass of water suddenly released carried blocks of concrete that weighed as much as 10,000 tons (63 × 54 × 30 feet) for half a mile downstream.

THE NATURE OF FLOW IN STEEP STREAMS

The flow of steep, tributary streams differs markedly from the even or tranquil flow of broad trunk rivers. Not only are their average gradients and hence their average velocities greater, but both gradient and velocity fluctuate more. Changes in gradient of the stream bed create what hydraulic engineers call *nonuniform* flow.

A typical case of nonuniform flow. Water from A passes over the uniform, gentle gradient AB with a moderate velocity. At B the steeper gradient sharply increases the velocity and hence the stream shallows from B to C. In the stretch from C to D the gradient and, therefore, the velocity are again moderate and the stream deepens. The alternating rapids, pools, riffles and waterfalls of most mountain streams illustrate such variations in depth and velocity.

The flow in the A to B stretch of the figure may be considered typical of all streams with low-to-moderate gradients. Turbulence disturbs the surface with mild boiling, rolling, or whirlpool motions. Surface waves produced by impact of the current against the bank or bed—or by dropping a stone into the stream—travel along the surface in all directions, disturbing the general downstream flow of the water. This kind of stream flow, because of its unhurried, though somewhat disturbed, nature, is called *tranquil* or *streaming* flow. Where such a stream suddenly quickens on a steeper slope its appearance changes abruptly. It breaks away from the surface of the tranquil stream in a curving lip and shoots swiftly down the steeper slope. If the bed is reasonably smooth, the stream surface is smooth and shimmering; where the stream flows over obstacles, there is a smooth roll or a boiling wave, but the wave's disturbance does not move upstream. Turbulent eddies are apparently flattened and attenuated, for disturbances of the surface are greatly reduced. This kind of flow is called *rapid* or *shooting* flow.

The critical velocity separating shooting from streaming flow cannot be simply defined, for it depends on many hydraulic variables. In general, shooting flow will be produced in a stream one foot deep and several feet wide if the slope is greater than about 3°; in larger streams shooting flow develops on lesser gradients.

Here the surface of the swift shooting stream must suddenly rise to the level required by the slower velocity below C. This abrupt rise, produced wherever the channel gradient sharply decreases, is appropriately called the *hydraulic jump*. The impact of the shooting water against the slower produces a strong rolling vertical eddy or a seething, air-charged mass of very high turbulence. Except for the waterfall, which is, in a sense, an extreme case of the hydraulic jump, the jump is the greatest dissipator of energy in a stream.

EROSION ON STEEP GRADIENTS

What mechanisms, inherent in the nonuniform flow of steep streams, allow them to erode their beds and banks? Whether a stream moves by shooting or streaming flow, most of its eroding power depends on the energy of large fast-whirling eddies.

These eddies are produced mainly by impact against, or deflection around, obstacles in the channel, such as boulders on the bed, bedrock projections, gravel bars, shallow depressions and deep holes.

During floods, swirling currents may be able to free and move jointbounded blocks of bedrock, a process called *hydraulic plucking*. Large nonuniform-flowing streams produce eddies of great competence. Below Great Falls, Maryland, where the Potomac River moves by streaming flow through a gorge, 15-inch boulders have been lifted as high as 60 feet above the channel bottom during floods.

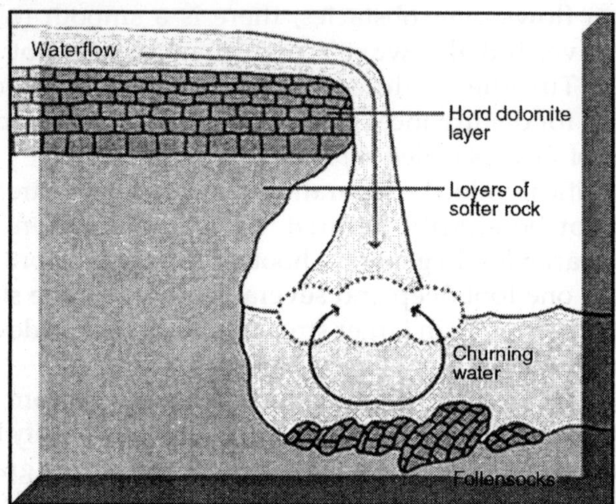

Fig. *Erosion at a Small Waterfall*

The impact of such missiles must effectively erode both bed and banks. In slower streams, swirling currents armed with coarse sand and small pebbles strongly abrade both moving rock fragments and exposed bedrock. Even fine sand may be an effective abrading agent if the currents are swift. Effects of eddy erosion can be seen along many streams during low-water stages.

Among them are rounding of the projecting rock, fresh angular scars where blocks have been torn from channel walls and cylindrical cavities (potholes) drilled in solid rock by stones caught in holes on the surface but rolled about by the swirling water.

The strong turbulence of the hydraulic jump causes relatively rapid erosion. Erosion at waterfalls illustrates an extreme in the competence of nonuniform flow. Water falling freely is accelerated by gravity, so that at the end of a second its velocity is about 32 feet per second (20 miles per hour).

The Niagara River falls at about 50 miles per hour at the base of the 150foot high Niagara Falls. Thus many waterfalls develop deep plunge pools at their bases which are cut back under the falls. Migration of waterfalls upstream by undercutting at the plunge pool and caving of the walls above may contribute greatly to the erosion of a stream valley. On the Niagara River, for instance, the Niagara Gorge between Queenstown and the Falls, a distance of about 7 miles, was cut principally by headward erosion of the waterfall, not by slow downward erosion along the entire channel.

THE NATURE OF SINUOUS RIVERS

The lower courses of most river systems show features that contrast strongly with the pools, rapids and waterfalls of steep streams. Trunk rivers commonly flow in broad, smooth valleys sloping so gently seaward as to appear flat. Apparently disdaining the shortest route to the sea, the rivers wind in intricate serpentine courses. The Mississippi River, from Cairo, Illinois, to the mouth of Red River, has a typically sinuous course. Such streams are often called *meandering rivers,* though, strictly speaking, bends called *meanders* should turn through 180° or more. Alternating with sinuous strctches are relatively straight reaches.

The broad flat on which such rivers flow is called the *river floodplain,* for it is more or less flooded when the river overflows its banks. Some floodplains are no wider than the meandering course of the stream—their bends impinge against the valley walls. Many floodplains, however, are far wider than the meanders and may broaden greatly where joined by sinuous tributaries, as does the Mississippi floodplain, shown by the dotted area.

Low indistinct ridges on floodplains, called *natural levees,* commonly border sinuous river channels. When streams overflow their floodplain, the great decrease in velocity and turbulence of the overbank flow results in deposition of even the finest suspended load. The coarsest part of the suspended load, dropping out just as the floodwaters overtop the banks, builds up the low natural levees.

The fine silt and clay are carried farther and deposited on the lowlands of the floodplain, behind the natural levees. These muddy floodplain accumulations are individually thin, but may build up to considerable total thicknesses over long periods of time. In the lower Nile Valley, silt has risen as high as 15 feet about the ancient Egyptian structures on the floodplain, indicating a depositional rate of 4½ inches per century. The floodplain behind the natural levees may be poorly drained and lakes, swamps, or indistinct sinuous channels are common.

THE CHANNELS OF SINUOUS STREAMS

Sounding of the Mississippi's bed reveals marked changes of the channel shape from bend to bend. The channel is deepest near the outer bank of each bend. The inner slope is gentle or even convex and often consists of shifting sand bars. In the short straight stretch between bends the river shallows considerably and the channel is more or less symmetrical. These shallow channels between bends, called *crossings*, troubled the old-time river pilots. Some are less than 10 feet deep at low water, as compared to "bendway" depths of 45 or more feet.

This variance in channel shape is directly related to the distribution of velocity and turbulence. In straight reaches or crossings the zone of maximum velocity, bordered by two zones of high turbulence, is midway in the channel. At a bend the conditions are different. The water moves forward into the bend with the direction given by the straight stretch above, so that the zones of maximum velocity and turbulence impinge against the outer bank of the bend. These competent currents contrast strongly with the sluggish flow over the convex slope on the inside of the bend.

EROSION AND DEPOSITION IN SINUOUS CHANNELS

The bank materials are generally unconsolidated sands, silts and clays; therefore the outer bank of a bend will be easily eroded by the strong currents directed against it. During a flood, when velocities and turbulence are greatest, the channel deepens, causing the outer banks to cave off rapidly. The fine-grained parts of the load thus acquired, the clays and silts, are readily carried away in suspension. Ordinarily, however, the coarser material is not moved far. Experiments show that almost all the sand and gravel is deposited on the next crossing and over the inside slope of the next bend.

Erosion of the outer bank and deposition on the inner slope of bends cause them to shift in position. This *migration* of the channel, or shifting of the meanders, allows a stream to wander widely over its floodplain.

If the floodplain materials are homogeneous, the meanders tend to sweep evenly downstream without lateral shifting. Over a long time the stream reworks the deposits left by previous meander courses and deposits material that is later cut away again as its channel continues to migrate. Thus the coarser sediments of floodplains move only slowly toward the sea. Erosional and depositional features on many floodplains clearly indicate former positions of sinuous channels, the low stripelike ridges that almost parallel the inside of the bends are beaches or bars left behind as the bends migrated. The crescent-shaped lakes are abandoned river bends. Such *oxbow lakes* develop either by *cutoff* or by *chute* formation. Cutoffs form when one meander is slowed in its downstream migration by a resistant bank, allowing the next bend upstream to catch up with it and cut through the narrowing neck between the two bends.

A chute, on the other hand, forms when a river in flood simply flows over the bar and beach deposits at a bend and then continues to use this shortcut channel. Chutes sometimes serve as alternate channels around bends, as illustrated in the middle distance of the enlarged map of a part of the Mississippi. Generally, the cutoff or chute channel is more efficient than the channel of the old bend, for, since the distance is shorter, the gradient is steeper and therefore the velocities and competence are higher. Velocity and turbulence in the cutoff bends become low or almost nil; consequently, sand is immediately deposited at both ends of the old bend, eventually damming it off from the new channel and converting it into an oxbow lake.

Significance of Sinuous Channels

Why should streams meander down their floodplains rather than flow in straight channels directly to the sea? In seeking an answer to this question geologists and hydraulic engineers not only have uncovered clues to the cause of meandering, but have come to understand many variables affecting river flow.

Observations of the discharge, load and velocity of natural rivers have contributed much to our knowledge of flow in sinuous channels, but many variables are too intricately interrelated to examine one at a time and hence their effects on sinuosity are difficult to evaluate. Because of the variability of natural rivers, investigators are turning more and more to stream models for help in solving both the theoretical and practical problems of stream flow.

Simple wooden troughs, used by early experimenters, have been replaced by carefully constructed scale models that not only exactly reproduce channel and valley shapes, but also have measuring and control devices for nearly all the variables of flow. Such models are used considerably in planning major projects of river regulation and construction and are thus the very intricate "test tubes" of the hydraulic engineer.

Many river model studies of meandering have been made at Vicksburg by the Mississippi River Commission. The models were not mere table-top arrangements, but were of considerable size. In one of the most revealing experiments a straight channel was carefully molded in uniform Mississippi River sand and water was allowed to run in it for three days.

The stream quickly developed a sinuous course, whose bends swept evenly downstream in the homogeneous, easily eroded floodplain material. The meandering began when the stream, cutting one of its banks more rapidly than it could transport all the debris, deposited a sand bar along that bank.

he sand bar at A forced the stream against the opposite bank, *B*, from which the current was deflected through the points *C*, *D*, *E* and so forth, like a ball oscillating against the sides of an inclined trough. These deflections made the currents at *B*, *C*, *D* and *E* more competent than at *F*, *G* and *H*.

The banks under lateral attack were eroded and the load thus acquired was deposited in the first downstream area of low turbulence. Deposits on one bank crowded the stream against the opposite bank, so that the angle of attack on this bank became greater and greater, gradually forming a series of migrating bends.

During the experiments discharge, valley slope, load and erodibility of bed material were individually varied, in an attempt to ascertain their effects on sinuosity. With increased discharge, the meanders widened in a regular and predictable way, verifying the general rule in nature that big rivers have big bends, little rivers little ones. For each set of flow conditions there was a certain maximum meander width. The chutes and cutoffs are apparently the mechanism—a sort of safety valve—that keeps this neat balance.

Likewise, the valley gradient directly and predictably affected the meander width: the steeper the slope, the wider the bends. A stream with steep gradient thus lengthens its course by forming wide bends, thereby decreasing its gradient.

When the streams were not given any load of sand at the head of the model channel, the meanders developed only *after* the stream was able to gain a load from its banks downstream. In experiments with channels molded in partially cemented sand, the streams could gain no load from their banks and did not meander at all.

Thus, as suggested by the mechanism of meander formation, in order for streams to develop sinuous courses, they must have easily erodible banks. This generalization is perhaps the most significant result of the Vicksburg experiments. It is strongly supported in nature by the common association of sinuous streams with wide valleys underlain by loose alluvial deposits.

BRAIDED STREAMS

A limiting case was reached in the experiments when streams were fed more sand than their gradients allowed them to carry away. Here the channel quickly choked with sand bars. The stream was then deflected in all directions into such a spread-out and intricately subdivided cross section that no continued attack could be made on a bank and no meanders developed. Such streams are called *braided streams*. They are common below glaciers where sediment chokes the stream they also abound in sandy regions where more material is washed to the streams than can be transported in low-water stages. The Platte River in Nebraska is an excellent example.

The Concepts of Base Level and Grade

All streams are limited in their ability to deepen their valleys by the level of the sea—or lake—into which they flow. If this limiting level, called the *base level* of the stream, remains unchanged, the stream tends to stabilize itself on that gradient which is just necessary to transport its load.

Sinuous streams like the lower Mississippi sidecut slowly through the alluvium of their valleys, but alter their gradients very little over long periods of time. Sand and gravel are moved from the outer banks of bends to crossings and thence to the growing bars on the convex slope of the next bend. Thus over a given stretch of the river, erosion and deposition practically cancel one another's effects, although load is being moved continuously downstream. Similarly, the average gradient remains the same over considerable stretches of the river, though the bed gradient varies from deep bendways to shallow crossings.

If the gradient and average channel shape remain the same, the hydraulic characteristics of the stream—the distribution of velocity and turbulence—do not change and thus the pattern of erosion and deposition are stable. Stream flow is just adequate to deliver at the lower end of a given stretch a load equivalent to that introduced at the upper end of the stretch. Streams, or parts of streams, that are thus balanced between erosion and deposition are said to be *graded* or to be flowing at *grade*.

If the slope of a graded stretch is reduced—as by slight movements of the earth's crust-the velocity and turbulence of the stream is reduced and the stream deposits sediment until its channel is built up to the slope required by grade. If the gradient is increased, the velocity and turbulence increase and the stream erodes its bed to the limiting slope of the graded condition. Changes in load or discharge similarly force streams out of the graded condition, until a new balance is attained. The steep tributaries of river systems are said to be above *grade* because they are obviously eroding downward into their beds. Rivers flowing *below grade,* characterized by braided courses, tend to deposit their load or *aggrade* their channels.

STREAM DEPOSITS

Besides the alluvial deposits of floodplains, stream deposition may produce deltas, alluvial fans and extensive alluvial plains. Some floodplains apparently result from long-continued stream erosion. A large number have been formed by the accumulation of stream deposits in pre-existing valleys. The aggradation of most of these valleys seems ultimately to have been caused by crustal warping, changes in sea level, or widespread glaciation.

Erosional Floodplains

Under ideal conditions, floodplains may develop by protracted valley cutting. This ideal process has been summarized by some geologists under the heading "Stages in the Evolution of a Valley." This theory is probably best explained by applying the principles of stream erosion and deposition thus far to a simple hypothetical example. Consider the stream. As long as it flows on a slope steeper than is required by the graded condition, it will cut into its bed, thereby tending to steepen its valley walls. This steepening initiates downslope movements on the sides of the valley. As long as the stream is competent to remove this debris and, at least intermittently, to attack its bed, its valley will have a V-shaped cross section (stage of youth).

Since the stream cannot cut below its base level, bed erosion gradually reduces the average gradient, thus diminishing its ability to cut down. Eventually the stream will become graded and, except during the strongest flood flows, its bed will be protected by a veneer of sand or gravel. Though such a stream can no longer cut downward, it may erode its banks by lateral attack against the valley walls. This will happen if the river is crooked at all. As in meandering streams, the mechanism of lateral erosion is simply the impingement of vigorous, turbulent currents against the outer banks of bends.

The weak currents along the inner banks of these bends will simultaneously drop their load as beach or bar deposits.As the stream slowly cuts into its valley walls, it leaves behind a planed-off bedrock surface covered by a thin veneer of sand and gravel (stage of maturity). Such a process can eventually produce a broad continuous floodplain. The attainment of grade separates the stage of youth from that of maturity.

This sort of lateral erosion must be much slower than that of streams meandering on alluvium, primarily because of the greater erosional resistance of most bedrock. Another factor is the enormously greater load acquired by lateral cutting and downslope movements.

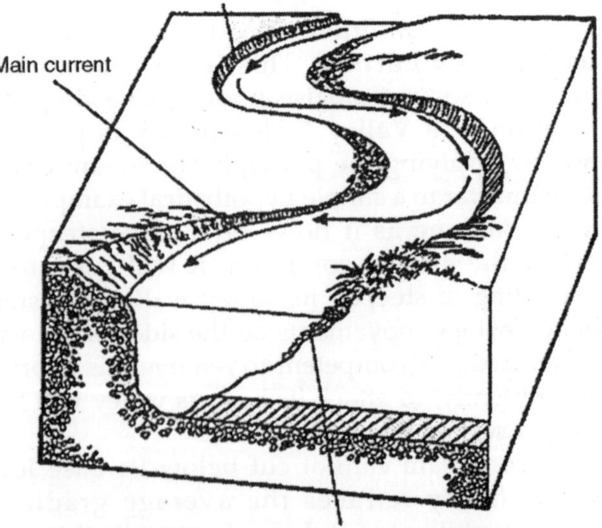

Fig. *Erosional Floodplains*

When a stream cuts even a foot into the wall of a deep valley, it makes it possible for downslope movement to add tons of debris to its load. Because of the great amount of time required to cut erosional floodplains, the probability that some change of conditions in or on the earth will upset the ideal progression of events is correspondingly great. It seems, indeed, that good examples of simple, laterally eroded floodplains are rare.

Most present-day streams have been sufficiently affected by crustal movements, sea level changes, or climatic changes to produce more complex valleys. Widespread floodplains seem to be chiefly of aggradational origin.

Aggradational Floodplains

Careful studies of the valleys of many sinuous streams show that their floodplains, instead of being alluvial veneers on a truncated bedrock surface, are underlain by deep deposits of sand, gravel and mud.

Among these streams are: the Lower Mississippi River; many stretches of the Missouri, Miami and Ohio rivers; the short sinuous stretches of the lower Colourado, Columbia and Snake rivers; and the SacramentoSan Joaquin River system of California. It is a notable fact that through the sinuous channels of these streams flows no less than two-thirds of the total runoff of the United States.

The evidence of aggradation is from drill holes and water wells in the floodplains. The Mississippi River Commission used several hundred drillhole and water-well records in its study of the deposits underlying the Lower Mississippi floodplain. These showed that the plain is underlain by alluvial debris ranging in depth from about 100 to over 400 feet. Near Natchez this debris is 260 feet thick, with its base 215 feet below sea level. Drill holes across the present floodplain indicate that the buried surface, the valley of a prehistoric Mississippi, is not a smooth plain but a steep-sided, though shallow, valley with many tributaries.

A detailed study of the Santa Ana River in southern California demonstrates similar aggradation. Here an ancient channel is buried beneath stream deposits 140 feet thick at the present mouth. This filled channel, traced upstream in drill holes, has about the same gradient as that of the modern stream. What has caused the marked filling in of such valleys as these? The deep filling of the Mississippi River Valley was probably induced in part by crustal downwarping (perhaps from isostatic sinking due to the load of the Mississippi Delta) and in part from a greatly increased stream load in the northern tributaries.

The aggradation of the Santa Ana River was probably caused by a rise in sea level since the ancient channel was cut. The ultimate cause of this change in sea level and in the load carried by the Santa Ana and other ancient rivers was apparently the growth and melting of very large glaciers during the recent geologic past.

These glaciers robbed the atmosphere and ultimately the ocean, of enough water so that sea level was distinctly lowered. As the glaciers melted, not only was this water returned to the ocean, so that the sea level rose, but also large quantities of rock waste released by the glaciers were dumped into the existing river systems.

Glaciers, affected the streams of North America in many ways, upsetting the ideal evolution of stream valleys toward erosional floodplains.

Sinuous Streams in Glacial Valleys

Some of the best examples of meandering streams occupy valleys that once contained tongues of moving ice. As noted at the Nisqually glacier of Rainier National Park, the load carried by glaciers is large; streams that flow from the snouts of glaciers commonly have very large loads supplied to them. As the glacier melts away, the load diminishes and the streams stabilize themselves on the flat, debris-filled valley floors.

In the process they take on a sinuous course and slowly rework the fluvio-glacial deposits which they have inherited. Literally thousands of streams in northern America, Europe and Asia owe their meandering courses directly to glacial erosion and deposition.

Deltas

Any laden stream must drop its load when it enters quiet water such as a lake or sea. Where the Nile emerges from its valley near Cairo, it splits into channels. These further subdivide and flow to the sea on a broad plain of river deposits. Because of its triangular shape, the name *delta* was applied by Herodotus to this flat depositional plain. Deltas may be triangular, often with a convexly curved border against the sea, or irregular with lobelike extensions like that of the Mississippi. The shapes and sizes of deltas are considerably affected by the wave and tidal characteristics of the water bodies in which they are formed.

Thus, the Mississippi and Colourado rivers, which empty into relatively calm and shallow gulfs, have prominent deltas, but the Columbia and Congo rivers have no subaerial deltas at all. The Columbia's load is distributed by ocean waves and currents for hundreds of miles along the sea coast; the Congo's is somehow washed down a long, deep submarine canyon into the depths of the Atlantic.

Where sand-laden streams flow into a deep, still body of water, the layers of deltaic sediment are not simply parallel to the lake or sea bottom, but show a characteristically discordant arrangement. By either bar deflection or overflow on the delta surface the stream tends to seek new channels, often subdividing into several *distributaries* over the delta surface. The stream deposits formed on the top surface of the delta are called *topset beds* and are commonly thin, because the distributaries must maintain certain minimum gradients across the delta. Thicker deposits, called the *foreset beds,* are laid down on the frontal slope of the delta.

The finest part of the load, kept in suspension for a long time by weak currents moving down the frontal slope, is deposited on the lake or sea bottom, in front of the advancing delta. These *bottomset beds* are eventually covered by later foreset layers, often with notable angular discordance.

Most large deltas are much more complex than this ideal example. The Mississippi delta, like that of most other large rivers, shows little or no discordance between topset, foreset and bottomset beds, probably because the load is fine grained and the sea bottom on which the delta is growing is only slightly inclined.

THE LONG PROFILES OF STREAMS

A fruitful way of finding the degree of balance between erosion and deposition in river systems is to make a graphic study of the stream gradient over long stretches of the stream's course. This is done by plotting the elevations of points on the stream's course against their distances from the mouth and connecting these points with a line. Such a line is called the longitudinal or long profile of the stream. In order to show variations in gradient it is necessary to exaggerate greatly the vertical scale of such a plot, for the length of streams is commonly very great compared to their vertical fall.

The long profile of the Arkansas River, which rises in the central Rocky Mountains and flows across the Great Plains, joining the Mississippi 440 miles above its mouth, shows slopes consistent with the general description of river systemsr. In the upper course the gradient is steep and irregular; here the river flows in high mountain valleys and, just above Canon City, through the chasmlike Royal Gorge. In contrast, the gradient of the lower 200 miles of the river is even and very low, nowhere exceeding 1 foot per mile. Here the river meanders in a tightly serpentine pattern on a nearly flat, broad floodplain bordered by low bluffs. The notable thing shown by the profile, however, is that between these two extremes the gradient varies regularly and so permits the profile to approximate a smooth concave curve, steepening markedly near the river's origin. Probably at least half of the thousands of long profiles that have been drawn show this general form, though really others show departures from it.

THE PROFILE OF EQUILIBRIUM

Examination of many streams shows that it is especially their sinuous portions on floodplains that have nearly smooth concave profiles. Since these sinuous courses are commonly the effect of a graded condition, it may be said that graded stretches of streams show nearly smooth profiles. The Arkansas River, for instance, shows a fairly smooth concave profile between Pueblo, Colourado and its confluence with the Mississippi; over nearly all this 1,350-mile course it flows on a floodplain, neither eroding nor aggrading its bed to any marked degree. The profiles of such graded rivers as the Mississippi, the Missouri, the Amazon, the Po and many others display nearly smooth concave curves everywhere except in their uppermost courses. Nearly smooth concave profiles, then, are the ultimate goal toward which all streams work as they approach grade. They are the commonest type of the *graded profile or the profile of equilibrium.*

The general concavity, itself, of graded profiles raises an important question. If such rivers as the Arkansas are, indeed, graded over certain long stretches, how is it that their gradient changes almost continuously? That is, if a river is at grade on one slope, how can it also be at grade on a steeper one upstream? The factors that apparently allow this are:

- The increase in discharge downstream,
- The change in the grain size and quantity of load downstream.

Increase in Discharge Downstream

There is a decrease in the ratio of friction surface to channel cross section as a stream grows. Consider the junction of tributaries. If the channels are all one foot deep with nearly vertical banks, the result of the union would be the removal of four cross-sectional feet of friction sur face, in the diagram a 30 per cent reduction. The energy previously dissipated in friction is available for erosion of the channel, resulting in a lowering of the gradient below the junction.

Can the actual increases in discharge be correlated with changes in the slope? Though data on the discharge of all the tributaries, particularly the amount of addition from ground water, are not available, a general assessment of the relation between slope and discharge is possible. The Arkansas River is initially formed by the confluence of dozens of mountain streams which are, generally, flowing above grade. From

Pueblo, where the river enters the plains, to Dodge City, contributions are restricted to small intermittent streams and seepage of underground water.

In summer much of this stretch is braided. Between Dodge City and Arkansas City the discharge is increased by many small tributaries and below Arkansas City by three major tributaries: the Cimarron, Neosho and Canadian rivers. The profile, the gradient flattens below the junction of these tributaries. A similar relation has been noted in the Lower Mississippi River, whose gradient and discharge have been carefully studied.

Some statistical studies have been made to determine the cause of the concavity of a stream's profile. For example, the discharges of rivers in the French Alps have been measured regularly, day after day, at 73 places. When the mean maximum discharge at each station was graphically plotted against its local gradient, the points approximated a smooth curve—strongly suggesting a causative relation between slope and discharge.

Effect of the Load

All graded streams do not show such a simple relation, probably because of the varying loads brought to the trunk stream by different tributaries. If the competence or capacity of a tributary is con siderably greater or less than that of the trunk stream, the confluence will force a steepening or flattening of the trunk stream's gradient. Such a situation may be illustrated by an extreme case on the Colourado River of Arizona.

Here tributaries are dry through most of the year, but during the seasonal floods their swift currents move great quantities of coarse debris into the main river. Much of this material is too coarse to be moved by the more sluggish Colourado and so it piles up as an apron of boulders, called a *debris dam*. Since these boulders can be transported only after they have been slowly ground down by abrasion, from the standpoint of human history, debris dams cause practically permanent convexities on the river's profile.

Decrease in Grain Size Downstream

Though the debris dams of the Colourado provide striking examples of the effect of the grain size of the load on a river profile, more important, though more subtle, examples may be found in graded, sinuous rivers. A careful study of the bed materials of the Mississippi River shows a marked decrease in the size of the particles downstream. The change in grain size of 600 samples that were collected.

The average composition of the delta sediments, which are 70 per cent silt and clay, is further evidence of this general change. On the Rhine River, between Basel and Bingen, there is a marked decrease in pebble size downstream that is closely correlated with a flattening of the river profile. It is noteworthy that the discharge increases only slightly over this stretch. Thus in some rivers the slow attrition of the bed load downstream appears to be a factor in allowing the stream to flow on an ever-decreasing slope.

DEPARTURES FROM THE IDEAL PROFILE

The smoothly concave curve of the ideal profile is closely approached but probably never attained by natural streams. Even in such rivers as the Lower Mississippi, probably as well-graded a stream as could be found, the profile shows slight but abrupt changes of slope between many individual stretches. This is not at all abnormal. For one thing, the inflow of most tributaries demands such changes in gradient. Note that the vertical scale of this profile has been exaggerated 2,100 times in order to show these slight changes. Other streams show much more marked changes in slope that must be considered to be actual departures from the ideal profile.

One of the principal values of the concept of the ideal profile is that marked departures from it call attention to exceptional circumstances. The debris dams on the Colourado are an example. Lava flows, landslides, or drifting sand dunes may upset the graded condition of a stream and impose an irregularity on its profile. Glaciation in the recent geologic past has greatly affected the slopes of many streams. Most, if not all, of the major streams of the northern United States have irregular profiles inherited from glacial erosion and deposition. Earth movements are also a factor, if they do not allow the stream quickly to adjust by degradation or aggradation. Some of the anticlines in the south central part of the state of Washington have apparently been uplifted rapidly and recently enough to outstrip the streams flowing down their flanks. Even in streams that have long been cutting toward grade in a relatively stable region, there may appear distinct changes in slope—called *nick points* —on the profile caused by local outcrops of especially resistant rocks. The profile both above and below such a nick point may show a smoothly concave shape.

Considerable time is required for the ideal profile to form in regions of resistant rocks. Therefore all kinds of structural or climatic "accidents" may intervene before the graded profile can be formed. It is a tribute to the eroding power of running water that many large streams flow on grades that approximate the profile of equilibrium.

SCULPTURE OF THE LAND BY STREAM EROSION AND DOWNSLOPE MOVEMENTS

Land forms result from the interplay between two opposing forces. On the one hand, earth movements and depositional processes raise certain parts of the crust; on the other, erosional forces constantly work to level them.

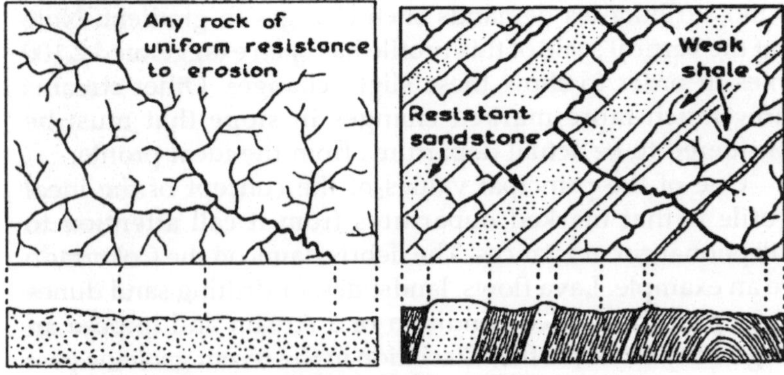

Fig. *Structural Control of Stream Patterns; Left, Dendritic Pattern; Right, Trellis Pattern.*

Streams and downslope movements, attacking uplifted areas, produce land forms whose shapes depend chiefly on the structure of the underlying rocks, the climate and the nature of the uplift that initiated erosion. The first two of these factors can ordinarily be determined and, if the processes of stream erosion are well understood, it is often possible to interpret the nature of the constructional force.

HEADWARD EROSION AND LANDSCAPE PATTERNS

Erosion in the steep headwater tributaries of a stream system not only deepens its valleys but also extends them by *headward erosion*. Stream systems thus tend to expand as they reduce the land and, coupled with downslope movements, create new ridge and valley systems or modify old ones. Headward erosion is guided by the arrangement of slopes and the structure of the rocks being dissected.

These two factors thus control the resulting pattern of stream courses and ridge lines. If the rock and its mantle offer uniform resistance to erosion and if the slopes of the land surface are more or less randomly arranged, tributaries will branch and erode headward in a random fashion, producing a *dendritic pattern*, resembling the branching limbs of a tree If the rocks being dissected are of unequal resistance, the extension and downcutting of tributaries will be most rapid on the weaker rocks, which will thus come to underlie valleys or lowlands between ridges or uplands of resistant rocks.

Thus in a stream system flowing across a region of steeply dipping parallel beds, tributaries tend to concentrate on the belts of weak rocks, forming a rectangular or trellis pattern. Other examples of the structural control of drainage patterns and ridge lines are the concentric patterns on eroded domes and basins and the strikingly linear stream courses along some faults. In temperate regions the structural control of land forms is usually more subtle than in the examples just cited.

The major ridges are underlain by resistant rocks and the larger valleys by soft rocks. Yet neither the drainage pattern nor the, arrangement of slopes gives much clue to the underlying structure. Even when on the ground, it is difficult to tell what resistant rock underlies a given ridge, because downslope creep of a layer of grass and soil has masked the rock contacts, forming smoothly rounded bills.

Structural Terraces and Plains

The structural control of stream erosion and downslope movement can go further than simply orienting individual valleys and ridges. In thick horizontal strata of unequal resistance to erosion, a stream will cut quickly through weak beds but will be arrested by a resistant layer.

Thus, a general, though temporary, base level is formed and headward erosion by tributaries may strip the weak layer over a large area, forming a *structural plain* on the resistant bed. In an arid or semi-arid climate the dissected upland border is eroded back as a steep embayed slope and temporary upland remnants called *mesas* and *buttes* are commonly left behind to attest to the stripping of the plain.

If a stream cuts its valley in a sequence of alternating weak and resistant beds, each resistant bed will form a step or *structural terrace* in the side slope of the valley. Such terraces are common in the Grand Canyon of the Colourado and where the Columbia and Snake rivers cut into the horizontal lavas of the Columbia Plateau.

STREAM TERRACES

More common than structural terraces are valley terraces that cannot be explained by differences in the resistance of flat-lying strata. Some such terraces are paired across a valley, others are single.

Some are made up entirely of river deposits, others are cut in bedrock but have veneers, channel fillings, or residual patches of river gravels on their flat surfaces. From such facts as these, it seems certain that these terraces are the remnants of old floodplains, now incised by the streams that once made them. Why a stream thus dissects its floodplain is in many cases difficult to understand.

In general, terraces are initiated by earth movements,

changes in sea level, or any factor that puts the stream above grade. Terraces obviously caused by earth movement were formed shortly after the Pleasant Valley earthquake of 1915. The 12-foot offset along the front of Sonoma Range produced a sharp fall in the profile of streams crossing the fault.

Erosion at this step was rapid, so that by 1930 it had retreated far upstream, leaving a 12-foot channel that broke the old floodplain into a pair of matched terraces. Because this stream now sidecuts at its new graded level, the terraces will be reduced in width, though probably parts of them, particularly downstream from projecting ridge spurs, will remain for a very long time.

It is of interest to note that even if the 1915 fault offset had not been observed, a clue to the movement could be found by examining the stream terraces, for they parallel the present stream profile in the Sonoma block, but end abruptly at the fault line.

Another example of terraces caused by faulting occurs on the west side of the Panamint Mountains, California. After faulting formed a series of small scarps across the large alluvial fans bordering the range, streams cut trenches in the uplifted portions of the fans and formed new fans below the scarps. Since, stream valleys within the mountains had no floodplains, the terraces extend only from tile new fault line to the mountain front.

Extensive Erosion Surfaces of Low Relief

An example of a well-marked erosion surface of fairly low relief, now being destroyed by erosion, but too extensive to be called a terrace, occurs in the central and southern Appalachian Mountains. This old surface is well developed in the vicinity of Harrisburg, Pennsylvania and so is called the "Harrisburg surface".

In some areas at or near the east edge of the range this surface is many miles wide, undulating and mostly underlaid by deep residual soils.

The surface has been formed on all but the most resistant rocks of the region, cutting across flat and tilted beds alike.

The Susquehanna and other rivers have cut 200- to 300-foot gorges into it and lesser streams have partially dissected it. The Harrisburg surface exists only as valley terraces in the parts of the Appalachians that are dominated by ridges of resistant sandstone.

The terraces are 200 to 300 feet above the streams in their lower courses, but the terrace gradients are commonly less steep than those of the present streams, so that upstream the terraces decrease in height.

Most river valleys in the southern Appalachians have similar terraces that can be approximately correlated with the Harrisburg surface. In some places broad passes between drainage systems, followed by roads and railways, are on the Harrisburg surface, though it is of interest to note that locally the surface is not at the same elevation in different valleys on the two sides of such passes.

Widespread erosion surfaces, commonly called *peneplains,* similar to the Harrisburg surface but generally with still lower ridges, occur in many parts of the world.

Some, as in the central part of the United States, from Missouri and Kansas south, lie near the graded level of through-going streams. How are such extensive, nearly flat surfaces formed? Careful study of the better preserved surfaces reveals that many are the result of erosion by streams flowing very near their base level.

Evidence for such an origin lies in the presence of a deep zone of thorough weathering, the fact that all but the hardest rocks-in all attitudes—are truncated, the presence of residual stream channels and stream gravels and the complete absence of surficial marine or glacial deposits.

Though these surfaces are similar to terraces in some respects, they differ from them in other ways besides their size. Floodplains may be formed by the sidecutting of a single stream, but surfaces such as that at Harrisburg are produced by downslope movements combined with the work of many streams. For this reason these surfaces are not flat, but undulating, their low crests and gentle slopes being the last remnants of hills and ridges.

Though it is tempting to suggest that such extensive plains are to be expected as the common end product of subaerial erosion, the present status of the earth's landscapes does not bear this out. All such surfaces of low relief that have been found by careful study to be the result of stream erosion are, like that at Harrisburg, variously uplifted, warped, dissected, or otherwise imperfect.

However, the existent remnants of these surfaces are evidence of more stable periods preceding the present and, like stream terraces, locally enable us to estimate the nature of crustal and climatic changes during the relatively recent geologic past.

Chapter 9

Displacement of Earth Crust

The miner deep in an Illinois coal mine who unearths a tree stump with the roots spreading out in the position in which they grew concludes without question that this tree, now 1,000 feet underground and 600 feet below sea level, once grew on the surface of the earth.

Similarly, the manager of a copper mine in northern Michigan whose ore comes from typical stream-deposited conglomerate or from the scoriaceous upper part of a basalt flow a mile underground reaches the same conclusion. The thought that land and sea are not always stationary but must locally have changed places has been entertained by nearly all peoples who have attempted to explain the occurrence of well-preserved marine fossils in the rock of plains, deserts and mountains. How else can we explain the marine fossils in the rocks of the Paris Basin and the deviation of the stratification in these rocks from the horizontal?

Spectacular but relatively small earth movements have accompanied many earthquakes Therein are described several destructive earthquakes that devastated densely populated areas. Most of these destructive earthquakes caused fissures to open in the ground and visible displacement of the earth's crust along the fissures. Such displacements are seen in the severed ends of roads, fences, strata, or other features that had extended uninterruptedly across the site of the break prior to the earthquake.

Breaks in the earth's crust along which slipping has occurred are called *faults*. Fault movement does not always produce an earthquake; slow, gradual movements along some faults have been observed without accompanying earthquake shocks. Sudden displacement along faults, however, appears to be the chief cause of earthquakes. The fault displacements observed during severe historic earthquakes are small—seldom more than a few feet.

In some, the movement was vertical, producing a small cliff along the fault; in others, the fault walls slipped laterally, offsetting roads or fences that formerly crossed the break; and in still others (for example, the Mino-Owari earthquake in Japan), the slipping was oblique, with both vertical and horizontal components.

Perhaps some historic earthquakes have produced even greater fault displacements on the sea floor. After an earthquake near Disenchantment Bay, Alaska, in 1899, the beaches stood 47 feet above the sea and a wide expanse of sea floor was made into dry land. This is the greatest well-authenticated single displacement known to have accompanied an earthquake. The great Japanese earthquake of September 1, 1923, which largely destroyed Yokohama and Tokio with a loss of over 140,000 lives, was at first believed to have been accompanied by a much greater displacement, but the evidence is inconclusive.

Comparison of soundings in Sagami Bay made before and after the earthquake showed local changes of more than 1,000 feet in depth of water. Detailed study suggests, however, that most and perhaps all of these great changes in the sea floor were not from fault displacement but more likely from large slides set up in the unconsolidated muds and silts on the floor of the Bay.

Thus, the historic record indicates that no single observed fault displacement accompanying an earthquake has been drastic enough to account for marine strata on mountain tops or rooted stumps in mines 1,000 feet or more below sea level. Is it possible, however, that these small displacements are but minor episodes in the growth of the fault on which they occur?

Examination of the structure of the rocks on either side of many earthquake fissures affirms this question. Although new faults with a few feet of displacement have formed during a few recorded earthquakes, most displacements have taken place by renewed movement along an old fault. The presence of such a fault is revealed by the impossibility of matching strata or other structures in the rocks on the two sides of the fault by merely restoring them to the positions they occupied prior to the earthquake.

The Pleasant Valley earthquake, which rocked the almost uninhabited desert of central Nevada in 1915, offers a good example. After the earthquake, the western base of the Sonoma Range was marked for 17 miles by a low cliff 1 to 16 feet high where none had existed before. The cliff was formed by renewed movement on an old fault which bounds the low-lying, partially alluvium-covered Pleasant Valley on the west from the rugged Sonoma Range to the east. During the earthquake, the Sonoma Range, relative to Pleasant Valley, moved directly up the steep fault at its base, forming the new cliff and increasing the height of the range up to 16 feet. Displacement is clearly seen in the vertical cliff offsetting the surface of the alluvium.

The small patch of alluvium clinging to the surface of the upthrown Sonoma Range block is 12 feet higher than the alluvial surface on the Pleasant Valley block. But the total displacement on the fault must be much greater; the dolomite of the upthrown block is against alluvium on the downthrown block for a vertical distance far greater than 12 feet. Furthermore, a short distance away, gullies cut in the Pleasant Valley alluvium show that it rests not on dolomite like that across the fault, but upon lava flows. Immediately across the fault, however, dolomite and other sedimentary rocks extend to the crest of the Sonoma Range over 2,000 feet higher.

Still farther north, the higher peaks of the Range are capped by lava flows like those on the Valley floor. Here, however, the lavas rest on the dolomite and associated sedimentary rocks. Obviously, therefore, the total vertical displacement on the fault has been at least 2,000 feet and perhaps it has been enough to account for the entire difference in level between valley and range, which is locally more than 3,500 feet. Apparently, the 1915 displacement was only the latest of many similar movements. Other historic ground displacements that have accompanied earthquakes show similar relations and warrant the conclusion that repeated small movements along a fault can ultimately displace the crust for thousands of feet.

GEOLOGIC EVIDENCE OF DISPLACEMENTS OF THE EARTH'S CRUST

Emergence and Submergence Along Coasts

Along most of the world's coastlines there is evidence of geologically recent (though prehistoric) crustal movement. The lobster fisherman who sets his traps in the water off the Maine coast now and then brings up fragments of the peat that floors the bottoms of some of the inlets and bays.

In this part of the United States, almost every stream ends in a tidal estuary. Dredgings from the bottom of these estuaries usually reveal river-deposited silts filled with decomposed grass roots, peat, or thinly stratified clay formed in freshwater lakes and marshes. All of these must have been deposited above sea level.

On a clear day, a person flying over southern San Francisco Bay can see the drainage channels of a former land on the shallow Bay floor. These channels, though interrupted by deltas or small wave-cut features at the shore, are clearly underwater continuations of the streams now entering the bay. They have apparently been drowned either by subsidence of the land on which they were flowing, or by a rise of the sea. Not far away, near.

Stockton, California, water wells drilled to as much as 1,000 feet below sea level penetrate buried soils, river silts with grass roots, peat accumulated in fresh-water marshes and other materials that must have been deposited above sea level. Off Denmark and Japan, dredging in shallow marine water has revealed kitchen middens (refuse heaps left by primitive man) and accumulations of charcoal and ashes marking the position of ancient campfires. High above the reach of present-day waves, along the shores of Alaska, Labrador, Newfoundland,

Chile, the East Indies, Scandinavia, Italy and many other countries, are abundant relics of former shorelines. Such features include barnacle shells and coral still attached to the rocks in the position in which they grew; rocks bored by marine clams (many with the shells still in the holes); sea cliffs, sea caves, stacks and other erosional features carved from solid rocks by former waves; and deposits of shell-strewn sand drifted along the ancient beaches by longshore currents. Conspicuous marine terraces border many sea coasts. Although locally masked by detritus washed down from the hills, or partly cut away by streams since their uplift, most of them preserve shellfilled sands or other clear proof that they are the uplifted floors of ancient seas.

Some of the most diverse examples of geologically recent changes of level are found in the East Indies, particularly in the region north of Sumatra, Java and the southern Moluccas.

Fringing coral reefs abound in the warm clear seas near these tropical lands. Growing reefs generally have relatively flat tops with surfaces less than 150 feet below sea level.

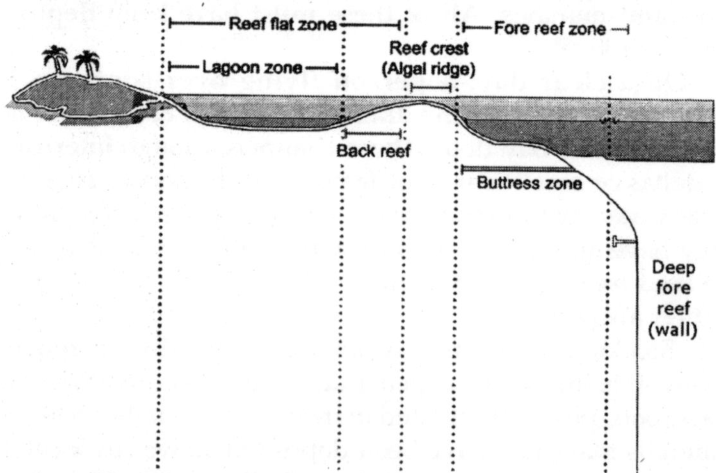

Fig. *Diagrammatic Sketch of a Typical Coral Reef*

The reef-building coral animals cannot stand depths greater than 200 feet and are also killed by a few hours' emergence. Coral limestone, largely built of reef corals and associated organisms, is, however, by no means confined to the present zone of coral growth. Far above the reach of the highest storm waves are great terraces of white coralline rock exactly like that accumulating off the present shore. Some of these uplifted reefs are only a few feet above the sea, others fringe mountain tops 3,000 feet high.

Some uplifted reefs form continuous level collars around the smaller islands. Others have been tilted, for they may be 100 feet or more above sea level on one side of an island, but when traced around to the other side they gradually decline to sea level, or even pass below it. Still other uplifted reefs have obviously been faulted and warped.

Each of these raised reefs, of course, testifies to uplift of the land with respect to the sea. But, as if to contradict this, we find equally clear evidence that some former land areas now lie depressed beneath the ocean. The chief evidence is from two sources: examination of samples dredged from the sea bottom and fairly detailed contour maps of the ocean floor prepared from sonic soundings.

Though coral reefs grow only in comparatively shallow water, at various points in the East Indies the dredge brings dead reef corals from depths of more than 1,000 feet. It also brings up land deposits such as muds that accumulated in fresh-water marshes and mangrove swamps or the poorly sorted silts and sands typical of river floodplains.

Sonic soundings also show indubitable evidence that some of the shallow East Indian seas were once dry land. One example is the Sunda Shelf between Borneo and Sumatra, just west of the area of uplifted coral reefs. The Sunda Shelf is a shallow sea, mostly less than 250 feet deep. The soundings show clearly that the Shelf is a submerged plain with three trunk rivers.

The headwaters of these rivers are the streams that still drain the northern parts of Sumatra and Java and the southern and western slopes of Borneo. The underwater extensions of the streams have been located from detailed sonic soundings. Further proof of this submerged river system lies in the identity of the fresh-water fishes and other stream-dwelling animals of southwestern Borneo to those of eastern Sumatra, although a wide salt sea now separates the two areas. Thus, a fairly small part of the East Indies yields clear geologic evidence of both submergence and emergence within comparatively recent time. Side by side and interfingering with one another, lie the drowned topography of the Sunda Shelf and the uplifted, tilted and faulted coral reefs of the southern Moluccas.

Elevated reefs are also found in Java and Sumatra, pointing to early submergence and then uplift prior to the drowning of the river system on the Sunda Shelf. Obviously, the movements that have changed the levels here have not been simple. They involve not merely vertical uplift and subsidence, but concurrent bending and folding of the earth's crust. Indeed, many of the tilted reefs fit into a regional pattern of folds in the basement rocks beneath the surface.

Study of the rock structure of the islands shows that most of the long curving island arcs are complex upfolds or arches in the rock, separated by basins and downfolds that still lie, in general, below the sea. The movements that produced these upfolds and basins are still going on, warping the fringing coral reefs above the sea at one point, drowning the reefs and the land-laid deposits behind them deeply at another and forming a complex but slowly growing system of folds in the crust of the earth.

Contemporaneous Folding, Erosion and Deposition

The folding of the earth's crust disclosed by the uplifts and submergences in the East Indies suggests that possibly such movements are related to the formation of mountains. Many mountain ranges show warped and bent strata that must have been deposited as nearly horizontal sheets.

They are now distorted into patterns more complex than the East Indian warps, but suggestive of them. Can we find any geological links between the broad warping of the East Indies and the more intense localized folding of mountain ranges such as the Alps or the Appalachians? Let us consider first the relatively simple warps and folds in the lava flows that make up the eastern foothills of the Cascade Mountains in northwestern United States. In central Washington and northern Oregon the walls of the canyons of Columbia River and its tributaries reveal flow upon flow of black basalt. The flows are 50 to 400 feet thick and, unlike those at Stolpen near Freiberg, have relatively little interbedded sedimentary material. Here and there, however, thin sheets of river gravel separate the flows, showing that streams advanced onto the barren volcanic plain after some eruptions only to be overwhelmed by new lava flows.

Elsewhere, narrow bands of red or black soil containing petrified (silicified) logs, or tree stumps with their roots still spreading out in the soil, appear between the flows. Obviously, time enough had elapsed for the scoriaceous flow tops to weather into soil and for a forest to grow on it before a new sheet of lava devastated the area. Elsewhere, layers of a brilliant white rock with paper-thin stratification contrast strikingly with the basalt.

Petrographic examination shows that these white rocks are made up almost entirely of the siliceous skeletons of minute onecelled plants called *diatoms*. Similar white diatomaceous deposits are being laid down today on the floors of shallow ponds and lakes nearby. In one part of central Washington, river gravels, floodplain silts and diatomaceous deposits continued to accumulate on the surface of the lava plain long after volcanism had ceased, until locally the black basalt flows were buried under several hundred feet of these unconsolidated river and lake deposits.

The point to be emphasized is that at many places these basalt flows and associated sedimentary rocks are no longer horizontal. They rise and fall in great arches and troughs. Here, on the flank of an arch, the scoriaceous top of a basalt flow and the diatomite strata above it dip at an angle of 20 degrees. There, beds of gravel and sand between basalt flows dip at 70 to 85 degrees. Obviously, no stream tumbling down a 70-degree slope could deposit gravel and sand upon it, nor could a molten basalt flow drape itself over a steep-sided arch and maintain a uniform thickness. Nor could the paper-thin sheets of diatomite remain on a 20degree slope on a lake bottom. They must originally have been nearly horizontal. Clearly, the flows of basalt and sheets of gravel and diatomite were folded into the present arches and troughs after they had been deposited. What is the relation of these arches and troughs to the present topography? Central Washington is rather exceptional among mountainous areas in that the major topographic ridges coincide with the arches in the lava.

The flanks of the arches are scored by many ravines, nearly parallel, extending straight down the slopes. Turbulent floods course down these ravines in the wet season, cutting deep canyons into the sides of the arches. Generally, these streams, aided by rillwash and downslope movements, have removed most of the unconsolidated river and lake deposits from the tops and flanks of the arches, but the eroded edges of these unconsolidated beds can still be seen farther down the flanks of the folds. In some places, a few of the much more resistant basalt flows have also been eroded and locally so many of them as to erase almost completely the fold as a topographic feature.

In general, however, the smooth curving surfaces of the arch, intact except for small ravines, indicate that erosion is only getting started in demolishing the uplift. Many of these ravines do not have the typical concave-upward profiles of normal streams. The stream canyons and the folds have grown synchronously and slow continuous arching of the lava sheets has modified the development of a normal stream regimen.

What is the relation of these topographic ridges to the troughs alongside? Conditions vary locally, but some of the partly dissected arches lie half buried in unconsolidated sediments deposited by streams and in lakes on the floors of the troughs. The uppermost beds of these trough deposits are nearly horizontal, abutting against the tilted rocks of the arches at high angles.

In a few places, younger lava flows are interbedded with these young sediments, but they do not extend onto the arches. They were erupted after the folding had started and poured down the troughs. In a few of the larger river canyons we can see and elsewhere from records of wells drilled in the troughs we can infer, that the older beds of even the younger strata have also been somewhat folded. Where they abut against the neighboring basalt arch, they dip with the basalt, though at a lower angle. Beds higher in the unconsolidated series dip similarly, but at still lower angles. In some troughs only the uppermost beds are nearly horizontal.

These relations confirm our conclusion from studying the erosion of the arches: While the troughs were being downfolded, sedimentation was also going on. In a few places, sediments were deposited on the trough floors almost as fast as they grew and have been progressively warped as the fold developed. Some of the sediments filling these troughs were eroded from the adjoining growing arches, but much was brought in by rivers from the rugged Cascade Mountains to the west, by the abundant quartz, pumice and andesite fragments—constituents foreign to the adjacent arches.

Such evidence of contemporaneous erosion and sedimentation must mean that the folds grew very slowly. Individual flows can be traced from the crest of an arch down its flank until they disappear beneath the unconsolidated sediments of the adjacent trough; but across the basin they reappear and can be seen to rise in the next arch.

Thus, originally horizontal flows have been folded until the surface of a particular flow on the crest of an arch is thousands of feet higher than it is at the bottom of an adjoining trough. But the folds did not grow overnight, for it is clear that folding, erosion and sedimentation were all occurring simultaneously. The first stage of the folding blocked out the respective areas of erosion (arches) and sedimentation (troughs). The rising folds dammed streams and impounded lakes in the troughs, interfering with normal stream development. Sedimentation also was locally interrupted by tilting of already deposited beds.

The concurrent erosion of the arches and deposition in the troughs as the folds grew shows that the movements were slow and gradual. There is meager but not wholly conclusive evidence from changes in the speed with which water is reported to flow in some of the irrigation canals built 10 to 50 years ago that the folding movements are still continuing and thus changing the gradient of the canals. We could cite many other examples to show that warping and folding go on slowly along with erosion and deposition. This has indeed been the rule in the young and growing mountain ranges fringing the Pacific Ocean and in the Alpine-Himalayan belt across southern Eurasia.

Measurable Slow Movements Along Faults

The Buena Vista oil field is on a rounded hill rising above the flat San Joaquin Valley in California. Soon after the first oil wells were drilled, a road was built across the hill and several pipelines were buried in shallow ditches along it. Within a few months, the roadbed cracked and a vertical displacement of less than an inch appeared on its surface. It was patched, but repeatedly broke at the same place. Within a few years, eight of the buried pipelines had buckled up out of the ground and they continued to arch upward slowly, a few inches each year.

The pipelines were thus shortened 9 to 19 inches in a period of 9 to 15 years. Cleaning tools lowered into the wells also revealed that 23 of the well casings were slowly being bent out of line. In some wells, bending continued until the casing col lapsed; in others, the casing, when pulled from the ground, was found to have been displaced about 15 inches horizontally within the 10 to 15 years after the well had been drilled. The casing failures in different wells occurred at depths ranging from 76 feet to 794 feet. What caused these curious phenomena? By plotting on maps and sections all of the points where such disruptions occurred, it was found that they lie along a smoothly curving surface. This surface slopes downward into the ground at a low angle and the rocks it cuts are offset against it.

Both the disturbances of man-made structures and the offsets in the rocks prove that the surface is a fault. Apparently the rocks above the fault surface are moving slowly upward and southward with respect to those below, breaking the roadbed, bending the well casings and buckling the pipelines out of the ground. Surveys show that the average rate of movement on the fault is 1 1/2 inches per year: No earthquake shocks arising from this fault have been detected, even by seismographs. Other examples of slow continuous movement along faults are known. In some parts of Japan, differential movements between different blocks of the earth's crust separated by faults have been detected by resurveys of the position and elevation of a series of points established by precise triangulation and leveling. For example, there were changes in level of a series of points along a 150-kilometre stretch of the northeast coast of the island of Honshu that occurred between leveling surveys made in 1900 and 1933. The changes indicate that throughout this area the earth's crust sank relative to sea level. The amount of displacement varies at different points, supposedly due to slow tilting and warping of fault-bounded blocks in the earth's crust.

Measurable Slow Movements Not Connected with Faults

The movements have taken place along faults, but there are also widespread and important movements in no way related to faults. Although the rate of movement is slow, it has been measured at several localities. In a few places such movements have changed the landscape appreciably within historic time. The classical and perhaps best known example of such movements is that of an ancient Roman ruin, the so-called "Temple of Jupiter Serapis" near Naples, Italy. Only three upright columns and a part of the floor are still in place. On the columns about 18 feet above the floor is a line. Above this the columns are smooth; just below it they have been bored full of holes by a marine rock-boring clam. Shells of this clam can still be seen in some of the holes. It is reasoned that the temple was built on dry land, that slow downward movements then brought sea level to a point 18 feet above the floor, allowing the marine clams to bore the columns and that, later, the land rose again.

Along the shores of the Baltic Sea in Sweden and Finland a still more striking movement has been observed for many years. The farmland and fresh-water marshes fringing the Baltic are littered with marine shells identical with those in the Baltic today. Over 150 years ago these conditions were correctly interpreted to mean that the land had risen out of the sea. In order to find out whether the movements are still going on, monuments were set up at high tide line along the shore. Today many of these monuments are a few feet above sea level and some as much as a mile inland.

The maximum rate of uplift—3 to 4 feet per century—is in the northern Baltic. In some parts of Scandinavia there is little or no movement and southern Denmark appears to be sinking about 2 feet per century. The rate of movement also appears to vary somewhat with time, as considerable differences from one decade to the next have been revealed by accurate tide gage records at many Baltic ports.

What causes this uplift? When the measured changes along the Scandinavian coast are compared with the obvious prehistoric evidence of uplift from raised marine beaches and terraces, an interesting relation appears. During the Pleistocene the Scandinavian area was covered by a huge sheet of ice such as mantles Greenland today. The Scandinavian uplift is greatest in the area where the Pleistocene glacier was thickest. The ice sheet was thickest in the northern Baltic and in the approximately 12,000 years since the ice melted (dated by varved clays in glacial lakes.

Because of the tendency of the earth's crust to reach isostatic equilibrium, uplift is what we would expect from unloading of a segment of the earth's crust by the melting of a large ice sheet and the greatest uplift should take place where the greatest load was removed.

Thus, this close coincidence between ice thickness and uplift is strong geologic evidence of the isostatic tendency. Since the uplift is still going on, we infer that Scandinavia has not yet risen to the elevation it must have had before the load of glacial ice began to accumulate and to depress the earth's crust under its weight.

Yet, there was no glacier or other large load upon the land at the Temple of Jupiter Serapis, which also sank and later rose. There must, then, be forces other than those tending to restore isostatic balance which are involved in crustal movements.

FOLDS THAT HAVE CEASED TO GROW

An older mountain range such as the Appalachian offers both contrasts and similarities to the structures of the East Indies and of the Cascade foothills cross section of the folded rocks of a part of the Appalachians of central Pennsylvania. This section depicts in detail the position and attitude of the rocks deep underground as compiled from surveys of the extensive coal mines and many exploratory drill holes.

The independence between the folds of this section and the topography suggests immediately that the Pennsylvania folds are not growing today, nor have they in the recent past, as in the East Indies and the Cascade foothills.

The folds have obviously long been dead, for the arches are no longer ridges nor are the troughs receiving sediments. In fact, throughout the folded Appalachians, it is common to find erosion-resistant rocks along the axis of a downfold forming a mountain summit, whereas adjacent upfolds with more easily erodible rock cores have been eroded to lowlands. In the Appalachians, the topography is indeed closely adjusted to the rock structure, but the adjustment is of a kind quite different from that of the East Indies and the Cascade foothills.

In the Appalachians, all of the ridges are underlain by erosion-resistant rocks and the lowlands by easily eroded rocks. Stream erosion, uninterrupted by folding and warping, has had time to etch out the easily eroded rocks and leave the resistant ones standing in relief. This contrasts with the East Indies and central Washington, where soft sedimentary rocks often cap the very summit of a mountain. Clearly, the arches and troughs in Pennsylvania are not growing today, nor have they grown in recent geologic time.

Can slow crustal movements that took place in the far-distant geologic past, long before the present streams developed their courses, explain the folded beds of the Pennsylvania coal mines? The evidence is certainly less clear. Obviously, erosion has been much deeper in Pennsylvania than in central Washington or the East Indies.

Gone are the sediments, if any, that were deposited in the troughs while the folds were growing. Any emerged shoreline features that may once have clung to the initially rising arches nave long since been eroded away. We can easily measure the fault displacements shown by the offset of the individual coal beds in the section, but we cannot tell whether they required a few seconds or millions of years to form.

However, nothing in the Appalachian structure differs fundamentally in form from that of the East Indian arches and troughs. Although the underground form of the East Indian folds is concealed, their surface form and the canyon exposures suggest that the folds extend to great depths. In the higher and more rugged parts of the Cascades to the northwest of the foothills area, erosion has already torn down some of the basalt arches and stripped so much sediment from the intervening troughs that the best evidence for their slow growth has here disappeared.

Thus, there is nothing about the Pennsylvania folds requiring different conditions of growth. Slow movements of the crust, like those now proceeding in the Dutch East Indies and Japan, are entirely adequate to explain the folds of Pennsylvania as well as those of the Cascade foothills.

"The present is the key to the past." Fossil marine shells at the tops of mountains and soils and fossil plants in deep mines are but what we should expect and predict from the evidence of slow but measurable changes of level that we can see taking place in different parts of the world today.

Furthermore, we can observe that, in some places and at certain times, the earth's crust has broken much like a brittle rock might be expected to break—along sharply defined faults. Elsewhere and at different times, in the very areas of faulting the rocks have folded like plastic dough. The reasons for this differing behaviour are obscure, for both kinds of deformation have taken place at the very surface of the ground. Nor is the ultimate cause of the deformation yet understood.

Chapter 10

Glaciers

Glaciers are slow-flowing, thick masses of ice. *Snowfields* are thinner and essentially motionless masses of permanent snow. Snowfields and glaciers are savings banks in the water economy of the earth. Each year, some of the water evaporated from the seas and lands falls as snow.

Most of the snow melts in summer, but in polar regions and at high altitudes part of it remains throughout the year in glaciers and snowfields.

If withdrawals by melting and evaporation exceed deposits as snowfall, the balance in the bank decreases and the glacier recedes, but if precipitation exceeds withdrawals, the glacier grows and spreads outward.

SNOWFIELDS

Above the snowline, on all except the steepest and windiest slopes, lie permanent snowfields. Excavations show that the beautiful geometric patterns of new-fallen snowflakes do not persist at depth.

Instead, the snowfield consists largely of small rounded granules of ice about the size of birdshot. This material, called *firn,* has been formed by compaction of the feathery snowflakes and by alternate thawing and refreezing of their edges.

This melting and refreezing is not due entirely to variations ill air temperature above the snowbank. Water expands 9 per cent when it freezes; hence, pressure tends to lower the melting point of ice.

Thus, even at temperatures slightly below the normal freezing point (0° Centigrade or 32° Farenheit), the snowflakes deep in a snowbank may be so tightly packed by the weight of later snowfalls that a thin film of water forms at the points of contact between them.

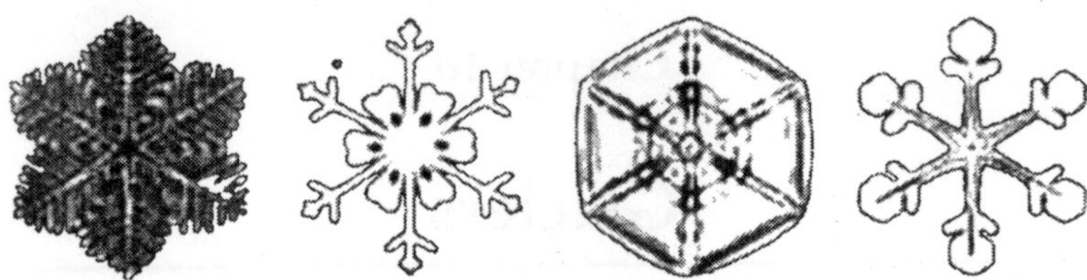

Fig. *Forms of Fresh Snowflakes*

This water film may then move to a nearby pore space, where the pressure is lower and refreeze into ice, even though there has been no change of temperature. Rains falling on the snowbank and meltwater formed on a warm day are absorbed as in a blotter by the granular firn. Perhaps the following night this water freezes, assisting in the compaction of the firn by excluding air spaces.

Thus, deep in the snowfield the firn gradually changes to interlocking granules of solid ice. Crevasses in glaciers commonly show all steps in the transformation: Snowflakes '! Firn '! Solid Ice.

If snow and firn merely continued to accumulate each winter, even the highest mountains in arctic regions would eventually be buried. Ice, however, is not a strong mineral and its strength limits the height to which it can accumulate. Before any great thickness is reached, the ice at the bottom begins to creep and flow outward under the overlying weight. Thus, from the snow and firn a *flowing ice mass*, or *glacier*, is formed.

GLACIER MOTION

Glaciers move so slowly that the motion cannot be seen, but the fact that the ice actually is moving can be readily proved by driving a straight row of stakes across the surface of the glacier. In a few days or months the straight line becomes a curve, bent downstream in the centre of the glacier where the speed is greatest. The rate varies in different glaciers and also with the season of the year. In a few of the Alaska and Greenland glaciers, velocities of more than 150 feet per day have been recorded, but such rates are exceptionally high. The lower parts of some glaciers on the coast of Alaska appear to be practically stagnant.

Indirect proof of glacier flow is given by the rocks frozen into the glacier. Most large glaciers contain abundant boulders which must have been carried from far upstream by the moving ice, because no similar rock is found in the valley walls nearby. The bodies of climbers who crashed to their death in the treacherous crevasses of the Bossons glacier in the Alps were released 41 years later at the terminus of the glacier several miles below.

Nature of Glacier Flow

The upper surface of glaciers is riven by crevasses and the ice appears to be hard and brittle. It shatters like glass under a sharp blow, just as does an ice cube from the refrigerator. It is difficult to imagine such a substance flowing. Actually, there is little or no evidence that the brittle, upper part of a glacier does flow. The abundant crevasses prove that fracture is the chief form of yielding. Deep within the glacier, however, deformation is almost entirely by flowage. Glaciers are perfect examples of the effect of scale on strength. Ice in small masses is rigid and brittle, but in glacial masses it flows under its own weight. The brittle, crevassed surface ice is rafted along on the flowing ice below.

The mechanism of glacier flow is complex. Microscopic studies of ice obtained from glaciers show that some of the ice crystals have been bent, others have glided along the sheets of atoms parallel to the base of the hexagonal crystals and still others have been granulated and sheared.

Furthermore, the crystals tend to be arranged in parallel planes, as in foliated metamorphic rocks, suggesting that they recrystallized during or after movement. Such recrystallization is demonstrated by the coarser crystals of glaciers as compared to those in snow and firn.

Crystals in the firn of Alpine snowfields average less than ¼ inch in diameter, but at the glacial snout the grains average about an inch in diameter. Crystals 6 to 8 inches in diameter may be collected from the edge of the Malaspina glacier in Alaska. The large crystals must have grown by recrystallization of smaller ones.

The ice at the front of a glacier, or in deep glacial crevasses, is generally layered. The layering superficially resembles stratification. It lies roughly parallel with the floor of the glacier and curves upward along its walls. Measurements at the glacier front have shown that the ice in adjacent layers moves at different speeds, proving that the layers are separated by shear surfaces. This layering is one manifestation of shear flow—the layers are due to the development of many closely spaced surfaces of shear within the glacier along which the ice and contained rock detritus are dragged forward and streaked out by movement.

The shear-banding of glaciers closely resembles the foliation of many gneisses and schists. Clearly, ice under a relatively small load is no longer brittle, but reacts to differential pressure by flowage and recrystallization. We have, in glacier ice, an example of a metamorphic rock, completely transformed from the original snow and firn into a wholly different streaky crystalline mass by movement and recrystallization under pressure.

THE SNOWLINE

The lowest altitude at which permanent snow rests is called the *snowline*. Obviously, it does not persist always in one position but is affected by variations in the annual snowfall, temperature, wind direction (which controls drifting) and topography (which controls both avalanching—snowsliding and shading from the sun).

Mean annual temperature decreases at higher altitudes and at higher latitudes, hence the snowline is higher near the equator than in polar regions. The amount of snowfall also affects its position.

Thus, the snowline stands over 5,000 feet higher (8,000 to 9,000 feet) on the dry eastern side of the St. Elias Mountains along the Alaska-Yukon border than on the wet western side (2,500 to 3,000 feet) and it is lower in well-watered Norway than on the colder but dry Taimyr peninsula. In much of Siberia, northern Alaska and Canada, the mean annual temperature is low enough, but the precipitation is too scanty to support permanent snowfields and glaciers. In such areas, the pore moisture in the soil and rock is frozen to considerable depths, forming a great sheet of *permanently frozen ground.*

KINDS OF GLACIERS

The form of a glacier is largely controlled by the topography over which it flows. *Valley glaciers* are streams of ice that flow down the steep-walled valleys of mountain ranges. Fed by snowfields on slopes and in catchment basins above the snowline, the glacier may extend far below the limit of perpetual snow.

All glaciers end at the point where the rate of melting at the front equals the rate of ice replenishment by flowage. Valley glaciers are typical of lofty mountain ranges the world over. Most of those in the United States, except for the extensive glacier system on Mount Rainier, Washington, are short blunt ice streams only a few hundred feet thick. It is difficult to distinguish many of them from snowfields; indeed there are all gradations.

The Rocky Mountains, the Cascade Range and the higher parts of the Sierra Nevada contain hundreds of small, rounded or irregular ice masses that lie in steep-walled clefts or else occupy amphitheater-like depressions opening out over steep slopes or cliffs.

Such masses are called *cliff glaciers* or *hanging glaciers.* At the foot of the St. Elias Mountains in Alaska several of these long systems of valley glaciers emerge from the mountain front and spread out on the plain below to form the Malaspina glacier, a lobate mass of ice covering 800 square miles. Such glaciers are called *piedmont glaciers.*

The largest glaciers are huge ice sheets, called *continental glaciers.* They are found today only in the polar regions. All of interior Greenland (about 637,000 square miles) is covered by ice, leaving only a narrow fringe of land along the coast. The Greenland glacier spreads outward in all directions from two high points in the interior. Locally, the coast is bordered by lofty mountains through which the glacier spills, splitting up between the peaks into valley glaciers which then descend the coastal valleys to the sea. Antarctica supports a much larger continental glacier, estimated to cover 5,000,000 square miles—an area larger than that of the United States and Mexico. The Antarctic glacier overrides the coast and projects into the "shelf ice" formed by the freezing of the sea. In places, the Antarctic glacier is held back by mountains through which the ice escapes in a series of large valley glaciers.

The famous Beardmore glacier, ascended by several of the early explorers in their quest for the South Pole, is one of these outlet rivers of ice from the high polar ice mass. The Beardmore glacier is 300 miles long, 12 miles wide and extends far out into the shelf ice even after it reaches the sea. Small masses of radially spreading ice are found on Iceland, Spitzbergen, in parts of Scandinavia and on the islands north of Canada. They are called *ice caps* to distinguish them from the huge continental glaciers. Glaciers are much more powerful agents of erosion than rivers, but, because ice is restricted to high altitudes and to polar regions, running water moves far more rock debris each year than does ice.

THE GLACIAL RASP

The head of a valley glacier is a zone of rapid accumulation of snow and ice heavily charged with rock debris. If accumulation has been rapid enough to produce a steep downstream gradient of the top of the glacier, outward movement in a thin ice mass may occur along curving shear planes similar to the movement in a landslide.

In a thicker and longer glacier, extrusion flow of plastic material may be less concentrated along shears, but it will follow similar paths. As the ice moves across the bedrock floor, rock debris frozen in its base is dragged across the glacier bed, abrading the bedrock. The glacier thus becomes an extremely effective cutting and polishing tool; it acts like a gigantic rasp, gouging into and scraping off irregularities on its bed and smoothing and grooving the rock beneath it.

Debris in Continental Glaciers

Practically the entire landmass occupied by a continental glacier is covered with ice; thus frost shattering and talus accumulation can take place only around relatively rare islands of rock that project through it. Also, little debris can be won by meltwater shattering of the rock floor at the bottom of crevasses. Yet, the ice at the margins of the Greenland glacier and other ice caps is heavily charged with rock debris, just as in valley glaciers.

Some of this debris is the eroded soil that formerly covered the land over which the glacier flowed. In many localities, where continental glaciers have melted back along their edges, a striking feature of the rock floor formerly covered by ice is the almost total absence of residual soils and of preglacial unconsolidated mantle such as stream gravels and floodplain deposits. The soil and loose rock have been frozen into the lower part of the glacier and dragged off bodily by the moving ice. As in large valley glaciers, rock floors under continental glaciers are smoothed and polished. On these polished floors are numerous scratches and even deep grooves. Obviously, the glacier has gouged and abraded material from its bed.

Rate of Erosion by Glaciers

The rapidity with which a glacier abrades its bed depends on four factors: the resistance to abrasion and plucking of the particular bedrock, the abundance of cutting tools (rock fragments) frozen into the base of the ice, the speed of flow and the weight (thickness) of the ice. Thus, thick continental glaciers flowing over weak or shattered rock are powerful agents of erosion. Similarly, on the floor of a cirque, glacial abrasion is generally great because here the ice is likely to be thickest and it is abundantly supplied with cutting tools.

ACQUIRING OF LOAD

Frost Weathering and Avalanching

Glaciers acquire rock debris in several ways. Valley glaciers commonly cover only a small part of the mountains in which they lie. On the gentle slopes between and above them lie extensive snowfields, but on steep windswept slopes even snow is unable to cling throughout the year. Hence, great expanses of craggy ridges, peaks and cliffs rise above the glaciers. These bare rock surfaces are strongly shattered by frost action. During the daytime, meltwater from snowbanks seeps into the rock crevices and then, perhaps that same night, freezes and breaks blocks loose from their parent ledges. Small grains, loose chips and even great blocks of rock thus freed tumble down the steep slopes and accumulate in talus piles on the edge of the glacier. Much debris is also carried down by avalanches (snowslides) during the spring thaw. Freshets from melting snow or summer rains also wash much debris onto valley glaciers.

Gradually, by being buried under new accumulations of snow or avalanche deposits and by falling into crevasses opened by the motion of the glacier, much of this rock debris becomes incorporated within the ice, especially along its edges. The edge of a valley glacier may contain so much debris that it shows up in photographs as a conspicuous dark streak. These stripes of dirty ice and loose rock along the edges of the glacier are called *lateral moraines*. Where two valley glaciers unite, the adjacent lateral moraines merge and form a *medial moraine*. If tributary glaciers are numerous, many medial moraines may streak the surface of the main glaciers. The moraines are not just surface features, they are wedgelike masses that extend deep into the glacier.

Meltwater Shattering

The exposed crags above a glacier are subjected to strong frost action but a glacier blankets its floor from extreme temperature changes. However, the rock floor of a thin glacier may be exposed to meltwater at the bottom of crevasses and here a change in temperature will cause some frost shattering. At the head of most valley glaciers is a deep arcuate crevasse, or series of closely spaced crevasses, called the *bergschrund*. The bergschrund gapes open in summer, but is filled or bridged over by snow in winter. It is inferred that the bergscbrund is formed by downstream movement of the plastic ice in the lower part of the glacier which breaks the brittle ice above and pulls it away from the rock wall.

Adventurous observers have descended into bergschrunds on ropes and found the lower parts to have a rocky wall on the upstream side and an ice wall on the other. The rock wall is shattered and riven by joints, some of which gape open and disclose slight displacements of the angular blocks of rock that they outline.

Some of the blocks are free but have moved only slightly; others lean out against the ice in precarious unbalance; and the rest, entirely free from their parent ledges, are now incorporated in the glacier. During the day, water from the melting snow pours over the face of the bergscbrund, filling the joints. The almost nightly frost loosens new blocks. Obviously, here is a zone of active glacial erosion, sapping the headwall by the plucking of frost-rifted blocks, steepening the cliffs behind and gradually extending the glacial channel headward. An open crevasse, however, cannot extend to depths of more than about 200 feet.

Experimental tests on ice under pressure show that at loads corresponding to greater depths ice will flow plastically and slowly close the deeper openings. Nevertheless, certain typical features of glacier heads indicate that shattering somehow does go on to greater depths.

One factor is that the meltwater itself opens channels in the ice at the bottom of the bergscbrund. On hot summer days much meltwater cascades into the bergschrund. A "chinook" wind or a warm summer rain may greatly increase the volume. Despite this great volume of meltwater, the bergschrund practically never fills with water to the point where streams emerge and flow over the ice. Apparently, the inpouring water melts its way down between the rock face and the glacier far below the bottom of the bergschrund, possibly, indeed, to the very base of the glacier.

Actually, local melting of the deep ice should be expected from physical principles; the compressed ice at depth melts at a slightly lower temperature than the ice and snow at the surface. Meltwater pouring into the crevasse also carries heat downward to melt more ice deep within the glacier. Plastic flow of the ice tends to close the meltwater tubes, but as long as heat is being continuously transferred downward, melting of the walls may offset the constriction by flowage. Furthermore, water in the tubes may be under high hydrostatic pressure. and the pressure opposes plastic closing.

Thus, we have a mechanism by which the rock wall at the head of a glacier may be wetted periodically to great depths. But, just as periodically, freezing cuts off the supply of water and the accumulated meltwater within the glacier and in the cracks in the rock wall may ultimately freeze. As a result, the rock wall is shattered and disintegrated to depths much greater than the bottoms of the crevasses.

Frozen meltwater also adds volume to the glacier. Such a glacier, now frozen tight to the headwall, moves slowly forward, drawing out the shattered blocks of rock with it and tending to open new cracks and crevasses to be filled by a further flow of meltwater.

It must be remembered, however, that meltwater in the deeper part of the glacier is protected from day and night changes in temperature. Deep within a glacier the temperature remains almost constant at the freezing point and ice and water are in physical equilibrium with one another. Change from liquid to solid may occur here chiefly in response to variations in pressure. A crack opened by the pull of the downsliding glacier will almost immediately fill with ice—the lowered pressure within the crack causes any moisture gaining access to it to freeze. But, in freezing, a gram of water gives up 80 calories of heat and, if this heat is not dissipated, freezing will stop. For these reasons, meltwater shattering must be less effective at great depths than on the wall of the bergschrund.

Cirques

Downstream flowage of the glacier removes the shattered debris and exposes new rock surfaces to attack. Where valley glaciers have melted away completely, exposing their headwalls, the glaciated valley generally ends in a semicircle of high cliffs that bound an arcuate rock basin generally occupied by a beautiful mountain lake. Such cliffed valley heads are called *cirques*. The curving cliffs at the head of the cirque are shattered clear to their bases and they generally meet the smoothed and polished floor of the valley at a sharp angle. The jagged and shattered cliff face—the product of frost action and meltwater shattering-contrasts strikingly with the smooth valley floor—the effect of rasping by glacial ice.

DEBRIS RELEASED BY GLACIERS

Till. That glaciers are effective eroding agents is well shown by the vast amount of debris released at the glacier front by melting. Piled in hummocky ridges (*moraines*) along the front of the ice are heaps of boulders, sand and silt, heterogeneously mixed without appreciable sorting or stratification. Such unsorted debris, deposited directly by the ice, is called *till.* The proportion of boulders to fine material in till ranges widely. Some till is largely coarse boulders, but that from thin ice caps eroding shale and limestone may be chiefly clay and silt, with only scattered boulders.

Rock fragments found in till are not shaped like those in stream or marine deposits. Stream and beach pebbles are commonly rounded, but most of the fragments in till are subrounded or sharply angular. Some have been cracked and broken by the crushing weight of the overriding glacier; many, especially in valley glaciers, are joint blocks loosened by frost action and little modified during transport. Some fragments show one or more nearly flat, grooved and polished surfaces. Obviously, these flat surfaces were faceted and smoothed as the boulder was dragged against the bedrock floor, or against other rock debris within the ice.

Rock Flour

The effectiveness of the glacial rasp is especially obvious from the large amount of fine material (silt and fine sand) released at the glacier front. Turbulent streams of milky water generally cascade from one or more tunnels in the front of a glacier. If we collect a beaker of the roily water and let it stand, most of the suspended material promptly settles to form a thick layer of silt and fine sand on the bottom of the beaker. A very small amount of fine clay particles may remain suspended for hours or days. The silt, examined with a microscope, differs strikingly from the mud brought down by freshets in an unglaciated region.

The minerals composing the silt of the glacial streams are not the ordinary soil minerals characteristic of chemical weathering. Instead, small sparkling cleavage fragments of unweathered feldspar and other undecomposed minerals are abundant and there is an almost complete absence of the yellow iron stain, the black and gray colours from humus and the slimy clays so characteristic of chemical weathering.

Clearly, most of this material is not an ancient soil removed by the glacier, but is *rock flour* made by the grinding and crushing of rock fragments against one another and against the bedrock.

Glaciers release an amazing quantity of rock flour. Glacial streams are almost invariably overloaded. They quickly deposit the coarsest material (rubble and coarse sand), building up a sloping alluvial plain over which they spread in braided courses. Most of the rock flour, however, is carried far beyond the front of the glacier and builds floodplain deposits farther downstream, or is spread into lakes as extensive delta and bottom deposits.

Such rock flour forms great *terraces of white silt* along the rivers of British Columbia and Alaska. The powdered rock is also widely scattered by the winds.

Extensive unstratified deposits of *loess* (a wind-deposited soil composed chiefly of undecomposed silt particles) are found just beyond the limit of glaciation in many areas. Loess characteristically, contains abundant small tubelike holes where grass roots have grown.

FORMS OF GLACIAL DEPOSITS

The debris dumped by glaciers and by the streams flowing from them is called *glacial drift*. The drift assumes characteristic and easily recognized topographic forms.

Moraines

Unstratified glacial drift (*till*) is deposited directly by the ice and occurs chiefly in the form of moraines. The term *moraine* is used both for material deposited by the ice and also for the material upon or within the moving ice. The largest and best developed moraines are generally at the front of the glacier. If the ice front is in equilibrium, so that the rate of flow equals the rate of melting, it will remain essentially fixed in position for a long time.

Here, the debris carried by the ice is released and accumulates in great hummocky morainal ridges. Because the ice front fluctuates with mild climatic changes, however, more than one morainal ridge is likely to form. The moraine that marks the farthest advance of the glacier is called the *terminal moraine*. Moraines that mark stages of halt as the glacier recedes are called *recessional moraines*.

Most terminal and recessional moraines of continental glaciers are broadly Iobate in plan and can be followed, with minor interruptions, for miles. The moraine at the front of a valley glacier is generally a high crescentic ridge curving around the front of the valley glacier and extending up its sides as lateral moraines. Small recessional moraines of valley glaciers may be almost buried in outwash debris.

During the advance and withdrawal of a continental glacier some till is deposited beneath the ice and overriden by it. Till may also be scattered over a glaciated area by melting out of the englacial (ice-enclosed) debris as the ice recedes. This irregularly scattered till, called *ground moraine,* is not concentrated into definite ridges like the terminal moraine. Some is found packed into hollows such as preglacial stream canyons; some is plastered around and upon low hills of the bedrock floor. Actually, the ground moraine is the most widespread deposit of a continental ice sheet. In most places it is spotty and only a few feet thick, but near the margin of the ice sheet it may cover hundreds of square miles in an almost continuous veneer.

In some areas, the ground moraine has been moulded into clusters of closely spaced hills, each shaped like the bowl of an inverted spoon, with the steeper slope facing the source of the ice. These streamlined hills are called *drumlins*. They vary in size, but are commonly about a mile long, ¼ mile wide and 50 to 150 feet high. Drumlins occur in clusters of scores or hundreds. Each drumlin in the cluster has its long axis roughly parallel to the direction of flow of the former glacier.

Although we cannot see drumlins forming in modern glaciers, they are doubtless masses of subglacial debris overriden and moulded into streamlined forms by an actively flowing ice sheet. Excavations show they are composed largely of till, commonly a sticky variety containing much clay. Some drumlins have cores of bedrock; others are composed entirely of glacial debris.

Stratified Glacial Drift

Meltwater from the glacier carries off much of the finer debris and deposits it either on the beds of the overloaded streams (*glacio-fluvial deposits*) or in lakes (*glacio-lacustrine deposits*) and in seas. In places, the deposit may consist of *pitted outwash,* characterized by numerous undrained depressions ranging in size from mere dimples a few feet wide up to great holes a mile long and more than 100 feet deep. These holes are called *kettles*. They are formed by the melting of ice blocks that were detached from the main glacier as it wasted away and were then partially or wholly buried in stream-deposited glacial debris.

The streams emerging from valley glaciers generally have somewhat steeper gradients than those from continental ice sheets. Nevertheless, they are so greatly overloaded that they immediately aggrade the valley floor to form a long narrow floodplain of coarse gravel and sand which grades downstream into finer sands and eventually into wide floodplains of silt. These aggraded valley bottoms, floored by glacio-fluvial debris, are called *valley trains*. When the glacier melts completely, the stream, being no longer overloaded, generally erodes the valley train into a series of terraces.

Stratified sand and gravel, containing here and there an irregular patch of till, forms long winding ridges in some areas from which continental glaciers have melted. Most of them are less than 100 feet high and a few hundred feet wide, but they may be several miles or even ten of miles long. They are called eskers, a term derived from the Gaelic meaning "ridge."

Eskers appear to be the deposits of aggrading streams that flowed in tunnels beneath the ice, or along the bottoms of crevasses, for some of them merge downstream into fans of outwash or into deltas. Most eskers must have formed when the ice sheet had wasted to a thin, almost stagnant mass, for, obviously, the strong thrust of an active glacier would close the tunnels and crevasses and scatter the esker deposit into ground moraine.

Where a glacier-fed stream enters a lake, a *delta* forms which may grow until it converts the lake to a swampy plain. The delta material is typically cross bedded and may range in grain size from coarse gravel to fine silt. On the floors of glacial lakes, mud and silt accumulate as thinly stratified layers. Each layer is commonly only a fraction of an inch thick, giving the deposit a conspicuous paperlike stratification.

Many glacial lakes are dammed on one side by the ice itself and the water may extend into crevasses and irregular-shaped holes in the wasting glacier. Such lakes are generally temporary and fluctuate in level as movement or melting of the ice diverts their outlets. Sediments accumulate in them. Because of fluctuations in lake level and movements of the glacier, these lake deposits are generally interlayered with stream deposits and with till. Upon melting of the glacier, patches of these sediments that filled former crevasses and holes in the ice are left in terraces and in isolated buttes called *kames* and *kame terraces*.

TOPOGRAPHY OF GLACIERS

It is apparent that glaciers differ greatly from rivers in their mechanism of flow and in their erosional and depositional features. The topography shaped by them contrasts sharply with the topography shaped by running water. Such differences are conspicuous in areas recently uncovered by glacial recession.

The polished and grooved bedrock floor is a feature not found in normal stream channels. Glaciated areas also contain swarms of lakes, ponds and marshes that fill either shallow rock basins scooped out by the ice, or depressions dammed by moraines and outwash. The streams that connect these abundant lakes are totally ungraded. An immature stream pattern with numerous waterfalls and lakes is typical of recently glaciated areas.

The upstream sides of peaks, hills, or small rock knobs overriden by ice are commonly rounded, polished and grooved; the downstream sides are irregularly jagged. It is inferred that the glacial rasp has sanded off the upstream side, but rock has been removed from the lee side, chiefly by quarrying of joint blocks caught by angular boulders dragged over them. Pressure relations may also be important. Pressure on the upstream face of the knob causes melting and the water freezes in the low-pressure area, causing plucking on the lee side of the knob.

The usual mountain stream valley has a V-shaped cross profile but in valleys from which glaciers have melted the lower part of the profile is characteristically U-shaped, though remnants of a former V-shape may be preserved on surfaces higher than the top of the vanished glacier. It is inferred that the spurs projecting between tributaries along the walls of a river canyon have been quarried away by the ice. A glacier cannot negotiate the sharp curves around spurs as easily as water; because of its bulk and brittleness it overrides the spurs and planes them into triangular facets forming a straight line of cliffs. At the same time, the floor of the valley is deepened by the glacial rasp. The widening, deepening and straightening of former stream canyons by ice transforms the original V-profile to a U.

The long profile of a valley formerly occupied by a glacier is irregular and is commonly interrupted by abrupt steps above and below which the smooth U-shaped valley may continue with flat gradients. Several such "cyclopean steps" may appear in a single glaciated valley, alternating with polished rock floors and lake-filled basins. Valley deepening by glaciers is not controlled by a smooth grade profile related to sea level as in streams, but by the thickness of the ice, the speed and duration of its flow and the structure of the rock in its bed. In restricted parts of its course, it is not unusual for a valley glacier to deepen its trough far below sea level.

Unlike stream valleys, glaciated valleys commonly head in cirques. Abandoned cirques are conspicuous features in recently deglaciated mountains. These gigantic semicircular basins may be so closely spaced that the divides between them are reduced to knife-edged comblike ridges, or to triangular mountain "horns". Stream valleys and glaciated valleys also differ in the relation of the tributaries to the main stream. The surface of a tributary glacier may accord with that of the main glacier—though, because of the rigidity of a thin ice mass, this is not always so—but the floor beneath the glacier and tributary is cut to depths depending on the volume and speed of the respective ice streams and the kind of rocks over which they move. The thin tributary glacier, unless favored by less-resistant bedrock, cannot keep pace with the scouring and plucking by the larger glacier. As a result, when the ice retreats, the tributary valleys are common left "hanging" high above the floor of the main valley and their streams descend into the main valley in high waterfalls or steep rapids.

Now that it is obvious glaciers leave clear marks of their former presence upon a landscape, let us see if there is evidence of widespread former glaciation in areas that today have a temperate or even a tropical climate.

Former Periods of Glaciation

The Iowa farmer, sweating under an August harvest sun, may doubt that "the present is the key to the past" if he is told that the soil he tills has come from deposits left by an ice sheet that once covered most of northern North America in the Pleistocene epoch. Similarly, the dark-skinned native of the Talchirs in India, who pauses to rest from the steaming tropical heat on a polished and striated ledge of rock, will doubtless consider the idea fantastic, though alluring, if told that his perch is the floor of an ancient glacier that spread widely over

India in Late Paleozoic time. Conclusions such as these so tax the imagination that, in spite of the clearcut evidence, they were not accepted even by geologists until every possible alternative—from the Biblical flood to some mysterious kind of volcanic activity—had been tested and found inadequate to explain the facts recorded in the rocks and in the drainage patterns upon them. Yet, evidence of the presence of former ice sheets can be readily gathered by any interested observer.

DATING PLEISTOCENE DEPOSITS BY VARVED LAKE SEDIMENTS

Lake deposits are abundant in glaciated regions. Deltas grow rapidly into glacial lakes where torrents of turbid meltwater pour into them. The finest silt and mud settles slowly and is spread throughout the lake. Many glacial lakes are deep green, owing to the dispersion of light by the abundant particles of suspended clay. In the winter, the lakes freeze over and the small tributary streams may freeze solid. During this quiet period, the suspended clay particles beneath the ice settle slowly to the bottom. By spring, most glacial lakes have nearly cleansed themselves. In summer, however, the surface ice melts and the streams become gushing torrents of debris-charged water which sweep coarse as well as fine material out into deep water. The coarse material settles promptly to the bottom, covering the fine winter deposit with a thin layer of coarse silt. Then, with the progress of the season, the silt is buried by successively finer deposits that gradually grade upward into the fine dark-coloured mud of the next winter. In the summer, microscopic organisms flourish in the warm surface layers. They extract calcareous matter from the water. The carbonaceous bodies of the organisms sink only slowly on death. Thus, the summer layer is richer in light-coloured silt, the winter layer in black organic matter. Such cycles have been observed to form in many existing lakes.

PRE-PLEISTOCENE PERIODS OF GLACIATION

Many ancient sedimentary formations are composed of unsorted debris showing all the characteristics of till except that they are tightly cemented. They contain striated and faceted stones and are associated with varved shales and slates or with sandstones and conglomerates showing the typical features of outwash deposits. These features and their local occurrence above polished and grooved rock floors, leave no doubt that they were formed during an ancient glacial epoch. Such cemented glacial tills have been called *tillites*.

Although small bodies of tillite have been found in rocks of many different ages, two periods in the earth's history before the Pleistocene are characterized by particularly widespread glacial deposits. During the Late Paleozoic, ice sheets spread widely over India, South Africa, Argentina and southern Brazil, southern Australia, the Boston Basin in the eastern United States and many other localities. In India, Australia and South Africa, grooved rock floors covered by the hard tillite may be seen at hundreds of localities. Their preservation today is due to the fact that they were buried under a succession of younger sediments. Deep burial protected them from weathering and erosion and the unconsolidated glacial deposits were slowly cemented into rock. Recent uplift and erosion of the overlying deposits has now exposed them to view. In the Late pre-Cambrian also, glacial climates appear to have been widespread. Ancient tillites of this age have been found on every continent except South America.

CAUSES OF GLACIAL CLIMATES

Geologists and climatologists have tried for more than a century to explain the recurrence of glaciation on a continental scale. Theory after theory has been suggested, but all explain too little or too much. None can be considered satisfactory, at least in its present form; yet they have an interest that justifies at least brief mention.

Facts to Be Explained

- Continental glaciers (in Greenland and Antarctica) occupy about 10 per cent of the land surface today. They formerly covered about three times as much area during four different epochs of Pleistocene time.

- Climatic zones during the Pleistocene were about parallel to their present positions but were displaced to the south during the glacial maxima and to the north during interglacial epochs. In low latitudes high rainfall was contemporaneous with glacial climates at higher latitudes.
- Estimates of the duration of the several glacial and interglacial stages of Pleistocene time do not suggest periodic recurrence; the climatic fluctuations seem to have been irregular.
- Evidence of continental glaciation on scales approaching that of the Pleistocene is also found in Late Paleozoic rocks (about 200 million years old) and in pre-Cambrian rocks (at least 500 million years old), but not in strata of intermediate age.

Suggested Explanations

The many theories that have been offered may be grouped under the following general categories: geographic, atmospheric and oceanic; geophysical; astronomical; and cosmic. The geographic theories attempt to account for glaciation by climatic changes that would be expected from changes in the areas of the continents with respect to the oceans, changes in oceanic and atmospheric circulation due to changes in the shape of ocean basins, mountain uplifts and so forth. These causes are inadequate because there is no evidence whatever of any notable shift in the distribution of landmasses or of violent fluctuations in their altitude since Pliocene time, yet we have conclusive evidence of four great glacial advances and of milder-than-present climates in some, at least, of the interglacial times. It has been suggested that changes in the amount of carbon dioxide and of volcanic dust in the atmosphere would bring about climatic variation. Carbon dioxide absorbs some of the heat radiated to space from the earth's surface. If its quantity were greater, more heat might be "blanketed in" and the temperature would rise, whereas, if it were less, the climate would become colder.

However, this mechanism is quantitatively inadequate,

especially because the effect of simultaneous change in amount of water vapour would practically compensate for the variations in carbon dioxide. Volcanic dust undoubtedly screens out some of the sun's radiation, but there is no sign that volcanoes were more active during the stages of ice advance than during interglacial times. Changes in salinity of sea water (and consequently modifications of the oceanic currents and their climatic influences) are likely results from glaciation but they can hardly have brought it about.

Geophysical theories that attribute glaciation to shifts in the positions of the continents are obviously not applicable to the Pleistocene climatic fluctuations, nor is there any strong evidence that these could ever have sufficed for earlier glaciations. Although it is conceivable that the amount of heat given off from the interior of the earth may have fluctuated from time to time, measured fluctuations are so small compared to the heat received from the sun (less than 1/10,000) that they seem negligible as a climatic factor.

The astronomical theories take account of the periodic variations of the earth's motion with respect to the sun. There are several such variations and there can be no doubt that they would have some climatic effect. Each affects the distance of a particular point on the earth's surface from the sun and hence the amount of solar radiation it receives.

They could not have alone produced the glacial periods, however, because their effects on northern and southern hemispheres, though not always exactly opposed, would not be identical, either. Yet, we now have recession of glaciers taking place simultaneously in both hemispheres and it is reasonable to think that earlier fluctuations were also simultaneous.

Furthermore, since all these astronomic motions have been going on through geologic time, we should have had continental glaciation every few hundred thousand years. Instead, we must go back to the Permian to find anything comparable to the Pleistocene glaciation in the geologic record. These geologic arguments seem conclusive, but several meteorologists contend, in addition, that the temperature changes produced by this mechanism would be entirely too small to produce glaciation even in one hemisphere. The cosmic theories are of two kinds:

- That the solar system from time to time encounters cosmic dust clouds or
- That the sun varies in the amount of heat it radiates. The cosmic dust clouds may either screen out the sun's radiation (if the component particles are fine enough) or blanket the because of lower precipitation; and, finally, when radiation fell to its present value we would be back to present conditions, with polar accumulations of relatively small size.

Thus, the somewhat paradoxical condition would arise, that a rise in the general temperature of the air would, by increasing precipitation and cloudiness, lead to continental glaciation of much of the present temperate zone. On this theory, one rise and fall of radiation would produce two glaciations in high latitudes, separated by a warm and wet interglacial time. The climate following the second glacial would be cold and dry like the present. In low latitudes there would be a single epoch of increased rains that would endure through both high-latitude glacials and the intervening warm, wet interglacial.

The four Pleistocene glaciations recognized in Europe and North America would, on this hypothesis, require two cycles of increased solar radiation, separated by a time during which the climate may have been much as it is today. Simpson computed that, during the maximum of solar radiation, the mean temperature was 5° to 10° C. warmer than at present and that the average cloudiness was increased from the present 54 per cent to between 70 and 80 per cent. Simpson computed that these changes are reasonably to be expected if the sun's radiation fluctuated about 20 per cent above its present value.

This hypothesis is attractive because it would account for the increased precipitation that would be necessary to feed ice sheets. The present line of mean annual temperature at the freezing point lies far closer to the equator than the limit of present ice sheets and this clearly indicates that low temperature alone is not enough to produce glaciation but that increased precipitation is needed. With the present configuration of land and sea, which is not greatly different from that during the Pleistocene, increased precipitation can only be brought about by increased evaporation from the oceans. The theory also accounts for the warm, moist climate suggested by plant fossils from some interglacial deposits.

On the other hand, there is no general evidence that the last interglacial period was generally warmer and moister than the second, as it should have been on this hypothesis. Nor is there general agreement that the increased rainfall in lower latitudes lasted throughout both the last two glacial epochs and the intervening interglacial epoch.

Finally, inasmuch as most astronomical processes are cyclic, if the variation of the sun's radiation is also cyclic, we should have had numerous other glacial episodes throughout geologic history. We must conclude that, even if this theory is satisfactory in some respects, an auxiliary hypothesis is needed to account for the long nonglacial periods of the geologic record. We thus see that no satisfactory hypothesis has yet been advanced to explain the widespread glaciation of the continents during the Pleistocene. It is impossible to say whether or not the present is simply another interglacial epoch. We cannot tell whether glaciers will again encroach on the sites of Chicago, Copenhagen and Warsaw within a few thousand years, or whether the climate will remain about the same as it is today. Much more work will have to be done to win an answer to this fascinating problem.

THE PLEISTOCENE GLACIATIONS

In most of Scandinavia, southern Canada, Labrador and parts of northern United States, the normal soil profile formed by weathering is either poorly developed or missing. Instead, rounded hilltops expose smoothly polished rock ledges like those beneath present glaciers. In places, the polish and smaller striations have been destroyed by weathering, but where protected by even a thin veneer of till, outwash, or peat, they may be as fresh and clear as beneath a modern ice sheet.

Boulders—many of huge size—are scattered erratically over the polished surface. Many boulders are of rocks entirely different from the local bedrock. In the fields of the Iowa farmer, boulders of gneiss and granite rest on flat-lying limestones and shales. The headwaters of streams in the area expose no such bedrock, so these boulders could not have been brought in by floods. In parts of Iowa, chunks of copper are occasionally found like the copper ore mined from rock ledges oil the Keeweenaw peninsula in Michigan or Isle Royale in Lake Superior, but not found elsewhere. Boulders of an unusual variety of granite called "Rapikivi," known in outcrops only north of the Gulf of Finland, are scattered widely across Estonia and even far into Poland.

The basalt plateau of eastern Washington is strewn with huge granitic boulders of kinds found only many miles to the north. To reach their present position they must have been transported directly across the Columbia River canyon, which is here 1,500 to 2,000 feet deep. A large nickel deposit in northern Finland was actually discovered by tracing its strewn ore fragments northward to their source near Petsamo.

Development of the Glacial Theory

The implications of such relations seem obvious to us now. Nevertheless, in the period from 1821-1835, when two European geologists, Venetz and Charpentier, showed that boulders of rocks characteristic of the central Alps were widely scattered across the broad Swiss plain to the north and correctly inferred a former extension of the present Swiss glaciers, their ideas met little but skepticism.

In 1836, however, Charpentier induced one of these skeptics, a young Swiss geologist named J. L. R. Agassiz (1807-1873), to accompany him on a visit to the active glaciers and, lower in the valley, the huge abandoned moraines of the Rhone glacier. Convinced on this expedition that the evidence for former more widespread glaciation was even stronger than

Charpentier had claimed, Agassiz set to work on the problem. He promptly saw and correctly interpreted the relation of the transported blocks to the polishing and grooving of the bedrock and showed that these smooth striated forms could not possibly have been produced by water. Agassiz's work, too, met with general disbelief. As his biographer relates: Men shut their eyes to the meaning of the unquestionable fact that... the former track of the glaciers could be followed, mile after mile, by the rocks they had scored and the blocks they had dropped. Agassiz, however, was not easily discouraged. In 1840, he visited much of Scotland, the north of England and parts of Ireland. There, he demonstrated glacial phenomena identical with those on the Swiss plain and he announced that not only had glaciers once existed in Britain, but that they formerly covered most of the country. At first this conclusion raised a furore of objections, but it also set geologists to observing and more carefully analyzing the evidence. Faced with the facts so easily gathered over much of the British countryside, it was not long until Agassiz's views prevailed.

Agassiz emigrated to America and began the studies of glaciation in New England since followed so fruitfully by many investigators. Studies since Agassiz's time have supplied a wealth of information. Modern maps portray in detail the erosional forms and deposits of the former ice sheets. Many abandoned moraines have been traced in the field, lobe by lobe and plotted on maps until their distribution in North America and Europe is well known. It is not difficult to map the moraines.

Although locally absent or poorly developed, most are conspicuous, easily identified, hummocky ridges of till that are readily traced for miles. Behind them, like signposts pointing the way, lie striated rock floors, locally strewn with till and foreign boulders. Here, too, are innumerable lakes and marshes where the vanished glacier gouged its floor or blocked older drainage by debris it left on melting. In front of the moraines are coalescing fans of outwash and, still farther beyond, sheets of loess, or terraces of silt. All assist in locating the position of the edge of the vanished glacier and all testify to its former existence. The landscapes of New York, New England, the Great Lakes states and the Northwest are nearly everywhere stamped with these glacial imprints.

In mountainous areas, such as Yosemite National Park, the Cascade Mountains, or the Rockies, former glaciers are recorded by the numerous U-shaped valleys; the fresh moraines; the abundant cirques; the superb waterfalls pouring from "hanging valleys"; the clear mountain lakes nestled in rock-scoured basins; and the terraces of outwash gravel, sand and silt extending downstream from the moraines.

Advance and Recession of Pleistocene Ice Sheets

Careful study of Pleistocene glacial deposits shows that they are not the result of a single advance and retreat of the ice. Instead, they record at least four stages of ice advance separated by long periods during which the glaciers withdrew to greater altitudes or latitudes than they occupy today, or even disappeared. At many places in North America and Europe, roadcuts or natural exposures reveal two tills, one on top of the other. The upper layer of till may contain boulders of almost fresh granite and gneiss, some with polished facets and striae. In the lower till, however, beneath a layer of thoroughly weathered clayey soil, the outlines of similar boulders can also be made out, but the rocks have weathered to the consistency of cheese—the feldspars are thoroughly rotted to clay and the ferromagnesian minerals have decomposed to limonite and other materials. Only chemically resistant rocks like quartzites are found intact in this lower till.

It is apparent that the older till had undergone long weathering and that typical A and B soil horizons had developed on it before the younger till covered it. The B-horizon of the weathered till is usually rich in clay that cements it into a tough sticky mass. Such tough clayey subsoils, whether of glacial origin or not, are popularly called *gumbo. Gumbotil* is the technical term used to designate deeply weathered clayey tills. Moraines with mantles of gumbotil, unlike younger unweathered moraines, have generally been considerably eroded since they were formed.

Careful study of the degree of weathering of different glacial deposits and mapping of their mutual relations have permitted the discrimination of four main stages of ice advance. Between the ice advances the climate appears to have been mild and warm, by the fossils found in stream and marsh deposits between the tills.

GLACIAL LOADS ON THE EARTH'S CRUST

There is strong evidence from deflection of the plumb line and from gravity measurements that large segments of the earth's crust are essentially in isostatic balance. Glaciers of continental dimensions also afford clear geologic evidence in support of this conclusion. The fact that former glaciers flowed over and eroded the ground on which they rested is proof that their surfaces sloped down from their feeding grounds to their termini. Wherever the glaciers spilled around mountainous topography, it is possible to get some measure of their thickness. In New England, for example, the highest peaks were unglaciated and a measure of more than 4,000 feet has been made for the Pleistocene ice in that region. In southern Canada the ice would certainly have been considerably thicker.

Although the density of ice is only a little more than one-third that of ordinary rock, a load of 3,000 feet of ice over an area as large as the Great Lakes and southern Canada would be the equivalent of a load of more than 1,000 feet of rock placed on the area and should bend the crust down.

Though the isostatic response by subcrustal flow might take considerable time, owing to the high viscosity of the subcrustal material, the lowering should be considerable and measurable, if isostasy really is general.Fortunately, the shorelines of ancient glacial lakes give us a means of testing the idea. They were level surfaces when formed.

After the ice load melted away the rise of the crust in response to unloading should tilt the shorelines, causing them to rise to the north. This is exactly what is found, not only in the Great Lakes region, but in the Baltic, in Labrador and even in the basin of Lake Bonneville, where the load was only 1,000 feet of water covering a much smaller area than that of the great continental glaciers. Suffice it to say here that the testing of the isostatic principle by glaciers is in full agreement with the geodetic evidence from which the theory was first announced.

[illegible]

ods, since between the ice advances the climate appears to have been mild and warm, by the fossils found in [illegible] and [illegible] deposits between the tills.

Encyclopaedia
of
GEOLOGICAL SCIENCE AND TECHNOLOGY

Encyclopaedia of Geological Science and Technology

Volume 2

Dr. Samayal Varadarajan

ANMOL PUBLICATIONS PVT. LTD.
NEW DELHI-110 002 (INDIA)

ANMOL PUBLICATIONS PVT. LTD.
Regd. Office: 4360/4, Ansari Road, Daryaganj,
New Delhi-110 002 (India)
Ph.: 23278000, 23261597
Branch Office: No. 1015, Ist Main Road, BSK IIIrd Stage
IIIrd Phase, IIIrd Block,
Bangalore-560 085 (India)
Tel.: 080-41723429
Visit us at: www.anmolpublications.com

Encyclopaedia of Geological Science and Technology

First Edition, 2009
ISBN 978-81-261-4125-8 (Set)

PRINTED IN INDIA

Printed at Mehra Offset Press, Delhi.

Contents

Volume 3

Preface

Geology is the science and study of the solid and liquid matter that constitute the Earth. The field of geology encompasses the study of the composition, structure, physical properties, dynamics, and history of Earth materials, and the processes by which they are formed, moved, and changed.

The field is important in academics, industry (due to mineral and hydrocarbon extraction), and for social issues such as geotechnical engineering, the mitigation of natural hazards, and knowledge about past climate and climate change.

The knowledge that geologists have known from the study of the earth a subject so large must be treated very briefly if it is to be presented between the covers of a single book; we have chosen to concentrate on the analysis of processes that are at work upon and within the earth, rather than to present a catalog of descriptive facts and terms.

We have felt, that the student is entitled to know something of the kind of evidence on which geologic conclusions are based, even though its presentation takes valuable pages that might be used to put forth more facts.

Author

Chapter 11

The Earth's Riddles

Since the dawn of civilization, men have been filled with curiosity about the earth on which they live. Why does a volcano erupt? What makes rain? What force causes the earthquake? What is the source of the water that bubbles up in a spring? As man's curiosity led him to seek solutions to such riddles, he often found that he was faced with new ones even more baffling. How did sea shells become entombed in the rocks of high mountain ranges?

Why does one stream have quicksand on one bank and solid rock on the other? What controls the beautiful geometric forms of snowflakes and other crystals?

Why does one well yield water in abundance, whereas another dug to the same depth is dry? It would be interesting to know how early man attempted to solve such riddles. By comparison, we of today have signposts along the way, for reasoning man has built up the method of investigation and the compilation of knowledge that is known as *geology*—the science of the earth.

To some of earth's riddles, geologists have won the final answers. To others, the suggested answers are still tentative; and to still others, only the faintest glimmerings of light that may ultimately illuminate the way to final solutions have, so far, been discovered. Progress in geology has not been at a uniform rate.

There have been periods, following fundamental discoveries, when outbursts of fruitful activity quickly revolutionized some of geology's theories and methods. At other periods there has been little advance. At times geologists even followed the wrong trail, and progress in some branches of the science came to a dead end. Then more information and new skills were acquired until, finally, the accumulated dogmas were overthrown and a new start made. The first roots of geologic knowledge are lost in antiquity.

The early Greeks and some of the peoples of other early civilizations made progress in geologic study, but their ideas were largely based on untested speculations, and little has survived. The modern science known as geology is of comparatively recent origin—the word itself is less than 200 years old.

Despite its youth, however, geology has already done much to stimulate and unshackle the thinking of mankind. The demonstration that sea shells and other fossils entombed in the rocks are but the remains of animals and plants that lived in the geologic past routed dogmas that had warped men's thinking for centuries. From the detailed study of the biological relationships of living and fossil organisms, coupled with geologic investigation of the sequence and changes of fossil assemblages with time, the doctrine of evolution emerged. This doctrine has profoundly influenced modern philosophical and scientific thought.

Evidence, well documented, that the landscapes about us are not static but are slowly changing, has not failed to stimulate the imagination of thinking men. The wheat farmer tilling the cold, wind-swept plains of Alberta is curious about the shells turned up by his plow, and becomes amazed when told that scientific comparison of these shells with living marine organisms shows that his farm was once the bottom of a warm, shallow sea.

Who can deny the thrill that comes with the realization that less than 20,000 years ago the site of Chicago lay under a sheet of ice such as enshrouds Antarctica today? Or that the green, well-watered hills of Scotland's Midland Valley were once the site of shifting sand dunes similar to those of the modern Sahara? Yet, preserved in the rocks and soils along Lake Michigan's shore and in sandstone quarries near the city of Glasgow are the proofs —as clearly recorded as are the deliberations of the Roman Senate preserved in the writings of Seneca and Cicero. The science of geology has brought to mankind new conceptions of time, just as astronomy has revolutionized ideas of space.

The rocks record events, some of which date back at least 1,800 million years, and throw into sharp perspective the short lapse of human history, as compared with that of the earth as a whole. There is a fascination in reading from the rocks the evidence on which we may reconstruct events of millions of years ago. It is this fascination that has led men to develop the science of geology—the attraction that will cause them to undertake deeper exploration of the earth's riddles.

MINERALS, WEALTH, AND POLITICS

Man's interest in the minerals and rocks of the earth's crust ceased long ago to be that of mere curiosity. There are sound practical reasons for his investigations. Our modern civilization makes many uses of the minerals and rocks that compose the earth's crust. Industry is almost wholly dependent on them. From minerals we obtain the iron, copper, aluminum, and other metals that make an industrial civilization possible. Our chief sources of power are the mineral fuels, coal and petroleum. In recent years, we have learned how to release stupendous amounts of energy from radioactive minerals.

Even our individual desires and needs are closely tied to the mineral industries. The bricks in our houses, the salt that seasons our food, the material that paves our highways, the gold and silver ornaments and precious stones with which we adorn ourselves—all have been won from mineral deposits in the earth's crust. Man's avid search for the gold and silver, the copper and gem stones that pleased his vanity and brought him security and wealth began early in the annals of civilization. With possession of the minerals, he sought to refine and improve them and to discover new uses to which they could be put.

As a result, the arts and crafts in metal and stone were born; and these, in turn, expanded into the vast industries we know today. On the international scene, the power and wealth of a nation is largely determined by its supplies of useful minerals, its authority over the areas that contain them, and its skill in discovering and utilizing them.

In this age of political readjustment between nations, we know that the vast accumulation of petroleum in Iran and Arabia is a potent force in world politics. We shall be wiser in world affairs if we know how petroleum occurs, how it is discovered, and how its quantity may be estimated.

Without the economic urge to find and exploit the mineral wealth hidden in the earth, many of the great forward steps in geology would never have been made; for geology is the science of the mine and the quarry, of the oil field and the placer.

STUDY OF GEOLOGY

Although geology is a complex and varied subject, it is also a stimulating and interesting one. Relatively few of its problems are so simple that they can be solved directly by one method of approach. Many even require supplementary investigations with the techniques of other sciences. Geologists are constantly taking over from chemistry, biology, physics, and engineering new methods, data, and theories that can be adapted to their needs. Geologists, in turn, have contributed data and ideas to these bordering sciences. Progress in one science advances all the others.

Because of the complexity of its problems, geology has not advanced so far as has physics or mathematics. The geologist cannot move a volcano to the laboratory to observe the growth of its cone, nor can he spread a bed of coal on the laboratory table to watch for millions of years its development. Yet, these are the simpler phenomena of geology. Factors of size and time make experimental study of many geologic processes difficult and often impossible.

It is not possible to put a lava flow in a calorimeter and measure its output of energy. Faced with these apparently insurmountable difficulties, geologists have had to devise ingenious, indirect methods for getting the answers to many of their questions. Despite these inherent difficulties in subject matter, however, geologists have been outstandingly successful in predicting where to drill for oil or other mineral deposits, and in arriving at verifiable solutions of complex scientific problems.

Geologists, if they would be successful, must develop resourcefulness and imagination. They must be able to make sound decisions on the basis of incomplete, and even partially conflicting, data. In deciding where to drill an oil well or where to develop a gold placer, the geologist must, in many instances, evaluate and coordinate several kinds of evidence.

His fundamental guides, of course, are the data from geologic mapping and other geologic techniques. He may also need to consider results from geophysical exploration, data regarding production of other wells or placers, and miscellaneous additional evidence drawn from engineering, economics, chemistry, physics, and many other sources.

These very factors of complexity and diversity, together with the newness of the science, combine to make geology a vigorous, rapidly expanding field. A student who selects geology as his profession has a wide choice of what he will learn and do. For his first two years of training, he will study more chemistry, physics, and engineering than geology. A sound elementary knowledge of these basic sciences is essential for many advanced geology courses.

The student will learn something of ordinary laboratory techniques, but will soon find that his main laboratory is not a building lined with bottle-filled shelves and machinery. The geologist's laboratory is the bold cliffs of high mountain peaks, the walls of deep canyons, and the slopes of desert ranges. A part of his education will be spent in strenuous hiking and climbing in some mountainous area, perhaps far from civilization, where he will map the rocks and their structures and collect other geological data. Such "field work" is essential to geological training.

Upon completion of his training, the geologist will find many opportunities open to him. He may work for an oil company and travel the earth in search of new petroleum deposits. He may direct exploration to find new bodies of ore in a mine. He may have the responsibility of estimating accurately the reserves of ore in the ground beneath a mining property or the amount of oil that can be recovered from a partially developed oil field. He may be called upon to decide which of several small mines is the best prospect for development and investment.

As an employee of a federal or state geological survey, the geologist may map rocks and mineral deposits, investigate conservation problems such as soil erosion and mineral depletion, or classify public lands as to their minerals, soils, water, and other natural resources.

Or the geologist may teach at a college or university, training future geologists, and at the same time, engaging in efforts to discover new principles or to unify and correlate old ones. Other opportunities for research are open in government work, in various research institutes, and in industrial laboratories.

In time of war, the geologist can serve by giving authentic information on problems of terrain, by discovering and assisting to develop critically short mineral supplies, and by selecting targets in enemy territory which, when demolished, will put an end to some vital industry of the enemy nation. He may sit at peace conferences and advise on the mineral resources of various nations and their resulting industrial potential.

There are also opportunities in commerce for the geologist. His geological knowledge can be used to good purpose in the development of a cement plant, or in the operation of a stone quarry, a brick yard, or a sand-andgravel pit. Whatever path the geologist takes, it is likely to lead to widespread travel, for the whole earth is the field for his investigations.

THE BRANCHES OF GEOLOGY

Geology is such a large and varied field that only the briefest résumé of its subject matter can be outlined here. Physical geology is concerned with the physical processes that operate on and within the earth—the processes that have given the rocks of the earth's crust their composition and structure, and the forces that have shaped the landscapes we see on its surface. Many separate geologic sciences contribute to the broad field of physical geology.

Among the more important are *mineralogy,* the science of minerals; *petrology,* the science of rocks; *geodesy,* which is concerned with measuring the form and size of the earth; *structural geology,* which seeks to interpret the structures to be seen in the rocks in terms of the dynamical forces that have produced them; and *geomorphology,* which deals with the origin of landscapes and with the changes that are constantly occurring in them.

The science that traces the evolution and development of the earth and of its animal and plant inhabitants with time. Historical geology is based, first of all, upon physical geology, but also draws extensively upon *paleontology,* the science that deals with the study of animals and plants of the geologic past, and on *stratigraphy,* the science that is concerned primarily with the order and sequence of the rocks that make up the earth's crust.

The application of the science of geology to the uses of man the finding and recovery of valuable mineral deposits from the rocks of the crust, the search for and recovery of oil and water from wells, or the study of the depletion of soils by erosion. The subdivisions we have enumerated are not independent sciences. For example, physical geology could never have attained its present development without the concurrent progress in paleontology. In a broad view of geologic science practically every branch contributes in some measure to all the others.

THE EARTH'S BROAD PATTERN

Some of the answers may be gained through observations made with the aid of an ordinary plumb bob. The plumb bob is one of the oldest and most useful instruments of civilization. The early Egyptians used it in recovering land boundaries after the annual floods of the Nile. Today, just as did our ancestors, we use plumb bobs in all surveying; in the construction of buildings, bridges, roads, and tunnels; in the making of all maps; and in measuring the size and shape of the earth.

THE EARTH'S GROSS SIZE AND SHAPE

The ancient Greeks noticed that the earth casts a round shadow upon the moon during eclipses. They also observed that the surface of any large body of water is curved, for only the upper part of a ship's mast is visible at a distance, and the ship appears to rise gradually out of the sea as it approaches. From these observations, the Greeks correctly inferred that the earth is, roughly, spherical. It is much easier for us to arrive at the same conclusion today. In modern airplanes, we can circle the globe in a tew days.

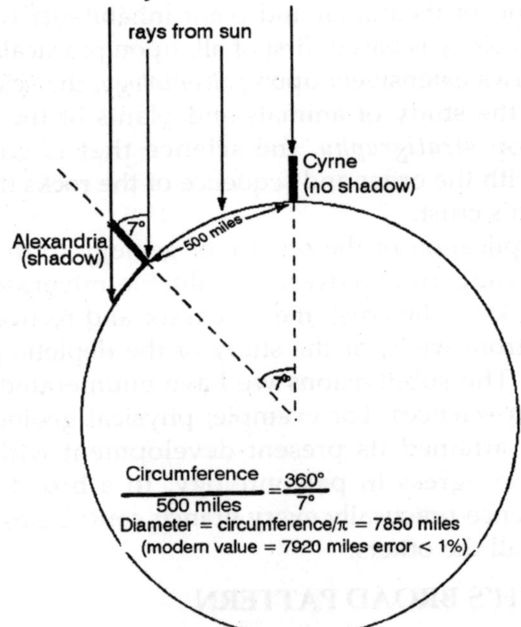

Fig. Eratosthenes' Method for Mefasuring Diameter of the Earth.

The curvature of the earth can be seen plainly in photographs taken from stratosphere balloons and rockets.

A few of the early Greeks made even more penetrating interpretations. More than 2,000 years ago, Eratosthenes, a

Greek geometer and astronomer, first measured the curvature of the earth's surface. Then, assuming the earth to be spherical, he computed its dimensions.

Though measuring techniques have been greatly refined, his reasoning is still used in modern geodesy. Eratosthenes learned that in southern Egypt, at Syene, the sun shines vertically down a well only at noon on the longest day of the year. On such a day, he measured the angle between the plumb line and the edge of the shadow cast by the sun at noon in a well at Alexandria. Alexandria lies 5,000 stades (the ancient Egyptian stade equals about 600 feet) north of Syene. On the premises

- That the sun is so distant that its rays to Syene and Alexandria are parallel;
- That Alexandria lies due north of Syene so that a plane through Alexandria, Syene, and the center of the earth also includes the noon sun;
- That the plumb line points directly toward the center of the earth; and
- That the earth is a sphere

the angle between the plumb line and the shadow at Alexandria is equal to the arc of the earth's curvature between the two points. On these assumptions, the circumference of the earth is given by solving the equation:

$$\text{circumference} = \frac{360°}{\text{Angle of sun's rays to vertical at Alexandria}} \times 5{,}000 \text{ stades}$$

Eratosthenes' earth was too large, but only 14 per cent larger than the figures now accepted. About a century later, Poseidonius the Greek philosopher, applied the same method to another are but was not so favored with compensating errors. The size of the earth deduced from his measurement was a quarter too small, and led to Columbus' error in mistaking America for India.

We shall see how, with refined modern measurements, the simple assumption that the earth is a sphere—which was satisfied by the early crude measurements—has had to be successively refined as succeeding measurements have become more and more accurate. It would, indeed, be difficult to find a better example of scientific method, and of successive changes in theory made necessary by improved observations, than is afforded by the history of investigation of the figure of the earth.

Modern Measurements

In the Seventeenth and Eighteenth centuries, as the expansion of navigation and the accurate surveying of land boundaries became more important, Eratosthenes' method came into wider use. Arcs of meridian (North-South lines), equivalent in length to one degree of geographic latitude or to some simple fraction thereof, were measured at many different localities.

It was found that a degree of latitude is longer near the poles than at the equator. Or, stated in another way, if we hold to the assumption that the earth is a sphere, its radius, as determined by observations near the pole, is notably greater than that deduced from observations near the equator.

Here the dashed circle, with the center P, the size of the earth based on observations made near the pole. The dotted circle, with the letter E at its center, the size as determined from observations made near the equator. Note the discrepancy in size. These measurements indicate that the earth is not a perfect sphere.

The inconsistency between the polar and equatorial measurements vanishes if we modify our assumption that the earth is a sphere and assume instead that it is slightly flattened at the poles—or, in technical words, that the earth is an ellipsoid. The solid line is an ellipse with its center at C. This line fits the data from both the polar and the equatorial observations.

Thus, the simplest model of the earth that conforms with modern measurements is an oblate ellipsoid—the solid figure that is obtained by revolving an ellipse about its shorter axis. In the ellipse indicated by a solid line, the short axis would be a straight line connecting the North Pole with the South

Pole. In the drawing, the flattening at the poles is greatly exaggerated, for the earth actually does not depart greatly from a sphere.

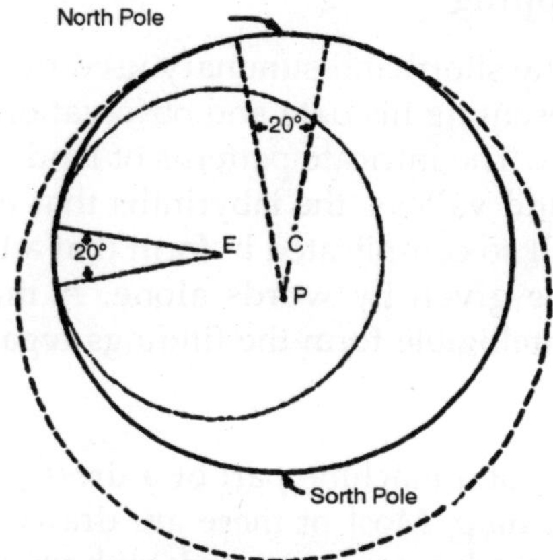

Fig. Circular Arcs Measured in Different Latitudes for the Earth.

The measurements now accepted and used internationally as the basis for official mapping, are:

Equatorial radius	6,378,388 meters	(3,963.5 miles)
Polar radius	6,356,912 meters	(3,950.2 miles)
Difference	21,476 meters	(13.3 miles)

In determining the size of the earth, Eratosthenes did not concern himself with mountains and other irregularities of its surface.

He regarded the earth as a smooth-surfaced sphere. Actually, this is a reasonably valid assumption, for the irregularities of the earth's surface in comparison to its size are not so great as those on the surface of a billiard ball. But to a Tibetan, living in the shadow of towering Mount Everest, such a comparison must seem a great over simplification.

How far does the earth's surface actually depart from perfect smoothness? To answer this question, we must find the difference in height between its highest and lowest points. This requires a kind of surveying different from that used by Eratosthenes, for now we are concerned with measuring the differences in altitude between various points on the earth's surface.

The results of such surveys have been assembled in the form of *topographic* and *hydrographic maps.*

Maps and Mapping

Maps are the shorthand summary used by the student of the earth in presenting his data and observations. The earth's crust is complex. The intricate patterns of land and water, the forms of hills and valleys, the labyrinths that men have dug in mining are all so complicated in form that a true picture of them cannot be given by words alone. A map, however, condenses in intelligible form the findings regarding them.

Map Scales

A blueprint of a machine part or a dress pattern may be thought of as a map. Most of these are drawn as *"full-scale" maps. An inch on such a map corresponds with an inch on the object it represents.*

Few geographic and geologic maps, however, are full size. Most of them are scale drawings. In such "reduced-scale" maps, an inch on the map may correspond to 10 inches, 1,000 inches, 1,000,000 inches, or whatever unit of reduction the map maker deems *desirable* to show the features he wishes to portray. If he decides to reduce the length of objects on the map to 1/10 of their true length on the ground, he plots on a "1/10 scale." The fraction is the ratio of reduction and simply means that 1 inch on the map equals 10 inches on the ground. Many of the newer maps of the *Topographic Atlas of the United States* are drawn on a scale of 1/24,000—1 inch on the map corresponds with 24,000 inches or 2,000 feet on the ground.

LIMITATIONS OF MAPS

On a full-scale drawing, it is possible to show, for example, the head of a nail 1/10 inch across in full size. If the nail were to be correctly represented on a 1/200 scale, however, it would have to be drawn as only 1/200 of 1/10, or 1/2,000 of an inch across. Such a point is too small to be visible. Thus, if such nails were to be shown at all on the 1/200 scale, they would have to be shown diagrammatically. Their positions might be indicated, but their size must be greatly exaggerated if they are to be seen. This limitation of reduced-scale maps must constantly be remembered by the map user.

All maps are generalizations, drawn to perform a particular service or function. Therefore, all represent selections of data chosen to the particular purpose and are commonly exaggerated with respect to other data. A navigator's chart emphasizes the features useful to navigation; a good road map stresses highway junctions and, in doing so, may distort the distances between them.

Whatever the scale, limitations in drawing and printing make it almost impossible for maps to be more accurate than 1/100 inch in the location of points. It is (difficult to make a legible pencil mark less than 1/100 inch across. A map of the United States reduced to a scale so that it could be printed on this page would be about 1/25,000,000; on it points less than 8 miles apart could not be shown as distinct without distortion. A thin line representing the Mississippi River on such a map would scale at least 4 miles in width.

Maps of the earth—the summaries of geographical knowledge—have still another limitation: They must depict, on a *flat* surface, the *curved* surface of the earth. It is impossible to do this without distorting the distances between points or the angles between intersecting lines. Most maps are compromises between these evils. Some of the distortions that result from various methods of plotting.

TOPOGRAPHIC MAPS

The maps described thus far may be called *planimetric maps;* they show the relative positions of points but do not indicate their elevations above or below sea level. The *relief* of an area is the difference in altitude between the highest and lowest points within it. Although, by means of skillful shading, a so-called *relief map* may give an impression of the relative steepness of the slopes in an area, such a map cannot be used to determine accurately the actual differences in elevation between any two points. It is impossible to read height accurately from such a map.

To meet this difficulty, geodesists have devised *topographic maps*, which are designed to show the elevations as well as the positions of points. They portray the three-dimensional form of the land surface, its *topography*.

A topographic map depicts a three-dimensional surface—one having length, breadth, and varying height above a reference plane or *datum* (usually *mean sea level*)—on a two-dimensional piece of paper. On such a map, lines called contours are drawn to portray the intersections of the ground surface with a series of horizontal planes at definite intervals above the datum plane.

A convenient way of visualizing contours is to regard each of them as the shoreline if the level of the ocean should rise to exactly the height that the contour represents. The 100-foot contour line on a map, therefore, shows the position of the shore if sea level should rise 100 feet. Every point on it is exactly 100 feet above the present sea level. Thus, each contour shows the course of a level line of definite elevation as it would appear if traced across the country shown on the map.

HYDROGRAPHIC MAPS

Hydrographic maps do for sea areas what topographic maps do for land areas. They not only depict the outlines of the water bodies but, by soundings (measurements of depth made by vessels at sea), they also show something of the topography of the bottom.

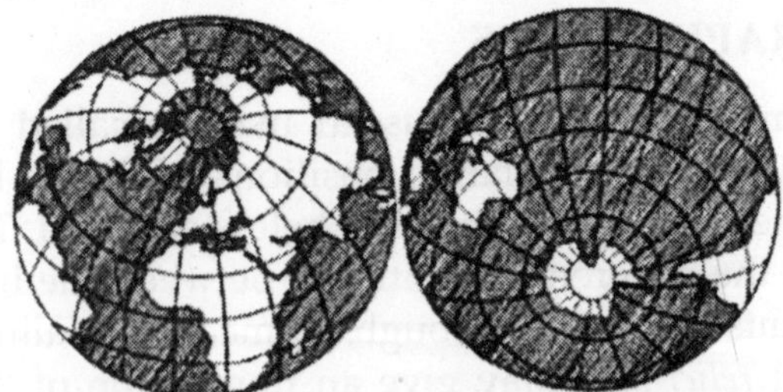

Fig. The Land and Water Hemispheres.

Although a few of the major irregularities of the ocean bottom became known during the Nineteenth Century, it was generally believed that the ocean floors were relatively much smoother than the surface of the land. With the advent of sonic methods of sounding however, it soon became evident that at least a few parts of the ocean floors are highly irregular. The change in the interpretation of the bottom configuration of an area in the South Atlantic with the added data from sonic sounding.

IRREGULARITIES OF THE EARTH'S SURFACE

Land and Sea Areas

The major irregularities of the earth's surface are, of course, the continents and the ocean basins. Perhaps the most obvious fact is that most of the earth's surface is covered by the ocean.

Using the best maps and charts available, the areas of land and sea have been carefully determined as follows:

Area of the sea	361,059,000 sq. km.	70.8%
Area of the land	148,892,000 sq. km.	29.2%
otal	509,951,000 sq. m.	100.0%

The land and sea are not uniformly distributed over the earth's surface. The surface of the earth can be divided into two hemispheres, one containing four-fifths of all the land, whereas the other is nearly nine-tenths covered by water. Let us now look at the topography of the earth's surface as it is recorded on maps of land and sea.

RELIEF FEATURES OF THE CONTINENTS

Relatively smooth *plains*, elevated and somewhat rougher surfaced *plateaus*, and rugged *mountain ranges* form the typical continental landscapes. Although details vary, several of the continents are characterized by lofty mountain ranges along one or more of their margins and by extensive plains in their interior.The highest and most extensive mountain systems on earth are in two great belts. One, the Circumpacific mountain system, is marginal to the Pacific Ocean; the other, the Alpine-Himalayan system, stretches across the southern part of the continent of Eurasia.

Surface Features of Eurasia

Eurasia, the largest continent, is also the most diversified topographically. Europe, the western part of this vast continent, can be divided roughly into three great topographic divisions:

- The northern highlands, which include rugged mountains rising 4,000 to 5,000 feet above sea level in the Scandinavian peninsula, and lower and more rounded ranges in the British Isles and northwestern France.
- The great central plain, which includes the vast plains of Russia, and which extends southwestward along the Baltic and North seas across Germany and into France.
- The Alpine system of mountain ranges and bordering highlands. This includes the lofty ranges of southern Europe, such as the Alps, Pyrenees, Apennines, Carpathians, and Caucasus, and also extensive areas of lower highlands that form the typical landscapes of Spain, central and northeastern France, Germany, Czechoslovakia, and Yugoslavia.

The mean altitude of Europe is slightly less than 1,000 feet. Its highest point is Mount Elbrus, which rises 18,465 feet above sea level; its lowest point is the Caspian Sea, whose surface is 86 feet below sea level.

Asia is much more diversified than Europe. It has both the highest elevation and the lowest depression of any continent. Mount Everest in the Himalayas towers 29,002 feet above the sea. The Dead Sea lies 1,294 feet below sea level.

Asia is wrinkled by scores of lofty mountain ranges. Many of them meet in the Pamir knot of northwestern India, which is often called the "Roof of the World." Among the great mountain chains that radiate from the Pamir knot are the Himalayas, Kuenlun, Hindu Kush, Karakorum, and Altyn Tagh ranges. Many other great mountain chains also diversify the face of Asia. Among the most prominent are the Elburz Mountains of Iran;

The Tian Shan, Great Altai, and Khingan ranges, which

stretch from west China northeastward into Mongolia and Siberia; and the prominent mountain chains that form the island arcs off the east coast of Asia, such as the Kurile, Japanese, Philippine, and East Indies archipelagos.

Between the great ranges of central Asia lie many high plateaus; among them are the Tibetan Plateau, the Tsaidan and Tarim desert basins, the great Plateau of Mongolia, and the Iranian Plateau.

Vast expanses of plain—the tundras and steppes of Asiatic Russia—occupy the area northwest of the central Asiatic plexus of mountain ranges. Other extensive plains form parts of China and India.

Surface Features of Africa

In general, Africa may be described as a vast plateau bordered in part by narrow coastal plains and diversified by numerous short, irregular mountain ranges. Unlike most of the continents.

Africa has no long continuous mountain systems, although the Atlas chain, along the northwestern coast, can be followed with minor interruptions for nearly 1,500 miles. Most of this chain consists of low mountains, but a few peaks rise more than 13,000 feet above sea level.

Extending from Abyssinia southward to Lake Nyassa is a mixed group of irregular ranges and elevated plateaus diversified by two linear depressions —the *Rift Valleys* of east Africa. Lakes Tanganyika, Albert, and Edward lie in the western Rift Valley; the eastern Rift contains Lake Rudolf and part of Lake Victoria.

The highest peaks in Africa are in this area. The extreme southern and southwestern edge of Africa is bordered by the relatively low Cape Mountains.

Surface Features of North America

North America is mainly a broad central plain lying between marginal mountain systems. Toward the north this plain widens to include the plains of the Canadian prairie provinces and, on the northeast, the low rolling surface broken by inconspicuous mountains known as the Canadian Shield. Far to the north, in the islands of the Arctic, Ellesmere Land and Grant Land, stand other mountains, higher than the Appalachians but little known.

The Appalachians, which border North America on the east, form a low but continuous series of small ranges and parallel ridges.

The much wider and higher Cordilleran mountain system lies west of the central plain. In the United States, the Cordilleran system is nearly 1,000 miles wide and includes the Rocky Mountains, the CascadeSierra Nevada chain, and, near the Pacific, a series of lower and somewhat discontinuous Coast Ranges. Between these mountain chains lie broad plateaus.

The principal ones are the Colourado Plateau west of the southern Rockies, the Columbia Plateau between the Cascades and the Rockies, and the Interior Plateau of British Columbia between the Rockies and the Coast Range.

In Mexico, a high plateau, the Mesa Central, lies between two mountain ranges which merge into a high volcanic table and south of Mexico City.

The highest point on the North American continent is Mount McKinley in Alaska, 20,300 feet, and the lowest is Death Valley in California, 276 feet below sea level. The mean elevation of the continent is approximately 1,300 feet.

Surface Features of South America

South America is similar in topography to North America. A long and high mountain system culminating in the Andes forms the continent's western margin. On the east, from southern Brazil northward, a low and irregular series of highlands lies near the Atlantic Coast. The interior of the continent between these two mountain systems is a vast plain which, in Argentina, extends unbroken to the shores of the Atlantic.

The Andes are surpassed in height only by the Himalayas. Several of the Andean peaks rise more than 21,000 feet above the sea; the highest is the Argentine volcano,

Aconcagua, 22,867 feet above sea level. At their northern end, the Andes split into three distinct chains. The easternmost extends into Venezuela and thence connects with the island arcs of the West Indies. Between the parallel ranges of the Andes lie high intermontane plateaus. The Pacific coastline is bordered, especially in the southern part of the continent, by a low and rather discontinuous Coast Range.

Surface Features of Australia

Australia is a broad, low plateau. It is diversified on the east by the Dividing Mountains, a low series of ranges extending from northern Queensland to Tasmania. Other low ranges are found north of Adelaide and in parts of western Australia, but most of the continent has little contrast in elevation. The highest peak is Mount Kosciusko in southeastern Australia, which rises only 7,323 feet above the sea.

Surface Features of Antarctica

The continent of Antarctica is only partially explored. It consists of a great interior dome of ice rising to elevations of 6,000 to 10,000 feet, through which project peaks of some of the higher mountain ranges. Most of the continent appears to be an irregular plateau buried beneath glaciers. Mount Erebus, a volcano overlooking the Ross Sea, rises to a height of over 13,000 feet.

RELIEF FEATURES OF THE OCEAN FLOORS

The outstanding relief features of the ocean floor are portrayed. Just as the continents are diversified with mountains, plateaus, and plains, the sea floor also has its larger topographic features. Among these are the *continental shelves* and *slopes*.

The continental shelves are the submerged edge of the continents. Off most coastlines the bottom descends gradually from the shore to a depth of about 200 meters, or perhaps somewhat less, before breaking off abruptly to the much steeper continental slope.

The shallow continental shelves include more than 7 per

cent of all the marine areas of the earth. They range in width from zero to as much as 800 miles (off the Siberian coast in the Arctic Ocean), and average about 30 miles in width for all coastlines. The surface of the continental shelves thus has an average seaward slope of about 20 feet per mile, though most of them are somewhat irregular and contain local ridges and depressions.

At depths that may be only a few meters or as many as 300 meters, the gently sloping surface of the continental shelf changes abruptly to the *very much steeper continental slope*. Off mountainous coasts, the continental slope may descend abruptly to the ocean floor at the rate of 300 feet or more per mile. Off wide coastal plains, its slope is generally about half this figure.

The continental slope off many shores is cut by so-called "submarine canyons". Some of these are deeper and more precipitous than the Grand Canyon of the Colourado, and form notches in the slopes recognizable to depths as great as 12,000 feet or more. Some head off the mouths of major streams such as the Congo and Hudson; others, equally impressive, head at the shoreline far from the mouth of any stream.

On the ocean floor itself, there is a diversity of topographic forms. Some are submarine mountains or other elevated tracts; some are depressions in the sea floor. Depressions may be grouped into the more-or-less rounded *basins*, the elongated *troughs* with gentle side slopes, and the similar but steeper-sided *trenches*.

Any one of these depressions that has a floor more than 7,000 meters (23,000 feet) below sea level is called a *deep*. Long, relatively narrow, elevated tracts are called *ridges*; or, if they are broader and larger, *rises*. Small, but significant submerged topographic eminences are the *sea mounts*, which are isolated steep-sided elevations; and the *banks*, which are shallower and broader, flat-topped elevations.

The Floor of the Atlantic Ocean

One of the greatest mountain ranges on earth, the *Mid-Atlantic Ridge*, almost bisects the Atlantic Basin from north to south. Its southern portion faithfully reflects the great bend of the coasts of South America and Africa.

Throughout its length, the summit of the Mid-Atlantic Ridge rises approximately 10,000 feet or more above the floor of the Atlantic on either side.

Most of its summit is less than 3,000 meters below sea level, and a few peaks—among them the Azores, Saint Paul's Rocks off the Brazilian coast, Ascension, Saint Helena, Tristan da Cunha, and Bouvet—project as islands above the surface of the sea.

The Walfisch Ridge, a range as high as the Alps and three times as long, branches from the Mid-Atlantic Ridge near its southern end and joins it to the west coast of Africa. The Rio Grande Rise, a lower and more gently sloping range, lies between the Mid-Atlantic Ridge and the coast of southern Brazil. In the North Atlantic, the Mid-Atlantic Ridge merges with the generally shallow bottom of the Norwegian Sea between Scandinavia and Greenland.

Two striking mountain arcs are present in the Atlantic. The Antillean Arc curves through the islands of the Lesser Antilles from the Venezuelan coast to the Florida shelf and separates the Caribbean Sea and the Gulf of Mexico from the Atlantic Basin proper.

In the South Atlantic, the are of the Southern Antilles is, similarly, convex eastward. It joins southern South America to the Antarctic continent through the South Georgia, South Hebrides, and South Sandwich islands. The only deeps in the Atlantic—the Puerto Rico Trough and the South Sandwich Trench, each more than 8,000 meters deep —lie on the convex sides of these arcs.

The Floor of the Indian Ocean

The Mid-Indian Rise, extending from India to Antarctica, resembles the Mid-Atlantic Ridge in lying medially in the ocean. It is broader and not so sharply marked off from the adjacent basins. Only about half its length reaches within 3,000 meters of the surface.

West of the Mid-Indian Rise lie several other ridges,

some of which branch from it. Others extend southeastward from the African coast; Madagascar lies on one of these. South of Australia, the Indian and Pacific basins are separated by a moderate rise that joins Australia to Antarctica west of the Ross Sea.

The area between Australia and southeastern Asia contains generally shallow seas, but the complex bottom topography of the East Indies includes several deep but relatively small basins and narrow trenches. This area also contains many curving island arcs, the archipelagos of the East Indies.

The Sunda Trench, south of Java, is the only deep in the Indian Ocean. Here, as in the Atlantic, the deep is on the convex side of an island arc.

The Floor of the Pacific Ocean

The Pacific Ocean floor is not well known. Apparently much of it is relatively flat. A sprawling, poorly defined rise extends from Antarctica to the coast of South America and across the southeastern Pacific to Central America. A considerably shallower rise between New Zealand and Australia reaches far to the northwest off the east coast of Australia and the north coast of New Guinea. This rise is notable for its steep eastern slope and for bordering several more-or-less isolated deep basins.

East of the Tonga and Kermadec islands, it drops off abruptly to the Tonga Deep (more than 9,000 meters). Another great deep (9,400 meters), south of New Britain, is exceptional in that it lies on the concave side of the arc that connects New Britain and Bougainville islands.

The Hawaiian Islands are perched on a rather straight ridge, trending northwest. A steep-sided ridge almost at right angles to the Hawaiian Ridge has been discovered recently. It crosses the Hawaiian Ridge northwest of the Hawaiian Islands. Northwest of Hawaii, the ocean floor appears highly irregular, though chiefly deeper than 5,000 meters.

The most striking features of Pacific Ocean topography are the great island arcs that festoon its northern and western sides. Among them are the Aleutian, Kurile, Japanese, Ryukyu, and Philippine arcs.

These islandcrowned arcuate ridges separate generally shallow seas—Bering, Okhotsk, Japan, Yellow, and East China seas—from the Pacific Basin proper. Several comparable arcs branch southward from Japan through the Bonin and Marianas islands. These are not bordered by shallow seas; the Philippine Basin, on their concave side, is as deep as much of the main Pacific.

As in the Atlantic and Indian oceans, great deeps are common on the convex sides of the Pacific arcs. Among them are the Aleutian, Kurile, Japan, Bonin, Philippine, Ryukyu, and Marianas trenches. The only other known deeps in the Pacific Ocean are next to the Andean arc off the Chilean coast and between New Zealand and Antarctica.

The Floor of the Arctic Ocean

The Arctic is, of course, the least known ocean. In a sense, it is a land-locked mass of water, for the greatest depth connecting it with the Atlantic is only 1,500 meters. The Bering Strait, its connection with the Pacific, is only 55 meters deep. The Arctic is notable for the very wide continental shelves that border Siberia, Europe, and northern Canada.

The sparse soundings that have been made in the Arctic have not yet revealed any great deeps. The greatest depth reported, 5,440 meters, north of Alaska, may even be in error, as indicated by later soundings in approximately the same area.

THE EARTH'S MAXIMUM RELIEF

Our brief review of the topography of the continents and ocean basins permits some generalizations about the earth's relief. It is in the great ocean deeps of the western Pacific that we find the maximum departure of the solid earth from sea level. Here, depths of more than 30,000 feet have been recorded by soundings.

The greatest depth so far known is in the Philippine Deep, which descends to more than 34,000 feet, or approximately 6.5 miles below sea level. The greatest height of land above sea level is Mount Everest, which rises 29,002 feet or nearly 5.5 miles. Great as these distances are by human standards, they dwindle into insignificance when considered in relation to the radius of the earth.

If the largest circle that call be drawn on this page represents the earth, a moderate weight pencil line would, on the same scale, include within its width all the irregularities of the earth's surface, ranging from Mount Everest to the Philippine Deep. Thus, we see that, in relation to its size, the earth actually is about as smooth as a billiard ball.

From the best maps and other data available, careful estimates have been made of the total areas of the land surface that lie between various altitude limits. These are summarized in the accompanying graph. This graph brings out two important facts:

- The highest mountain peaks and the extreme abyssal depths of the sea are very small in area;
- Within the extreme range represented by the Philippine Deep and Mount Everest, the altitudes are not evenly distributed by area but there are two altitude ranges especially prominent. One is the range between 4,000 and 5,000 meters (13,120 to 16,400 feet) below sea level, which embraces nearly one-fourth of the earth's surface.

  The other lies in the range from 200 meters below sea level to 500 meters above sea level and includes approximately one-fifth of the earth's surface. The submerged portion of this second level, extending from a depth of 200 meters to the sea coast, is the continental shelf. At its outer edge, the sea floor descends rapidly in the continental slope to the ocean floors proper, which lie at the first altitude range mentioned, approximately 4,000 to 5,000 meters below sea level.

What are the reasons for these two dominant levels in the architecture of the earth? Why is the boundary between continental shelf and ocean floor so marked? These questions are among the great problems of geology.

Chapter 12

Gravity, Isostasy, Strength

From our study of the earth's relief, we found that mount everest rises approximately twelve miles above the level of the philippine deep. Although twelve miles may seem insignificant in comparison with the 4,000-mile radius of the earth, nevertheless the irregularities of the earth's surface pose some fascinating problems. To one curious about the earth on which we live, several implications about the difference in level between its mountains and its ocean deeps are of lively interest.

An engineer knows that there are practical limits to the height of any building or tower. Even if we used the strongest rocks known, it would be quite impossible to construct a vertical tower as high as Mount Everest. The sheer weight of five and one-half miles of rock would crush the stone blocks at the base.

Long before the tower could be built to such a height, it would come crashing down under its own weight. How, then, is Mount Everest supported at such a high elevation? Why does not the floor of the Pacific Ocean crush in and fill the Philippine Deep? To gain some insight into how the great irregularities of the earth's surface are sustained, we must begin by studying the force that draws all objects toward the earth's center. This force is *gravity,* one manifestation of the universal force of gravitation.

GRAVITY

Water in a pond and coffee in a cup both assume a *level surface.* If we hang a plumb bob above either liquid, we find *that the plumb line is exactly perpendicular to the liquid's surface.*

Both the plumb bob and the liquid are responding to the force of gravity. Each has approached as close to the center of the earth's mass as its nature will permit—the liquid by flattening out so that it is as near the earth as possible, the plumb bob by bringing its mass as close to the earth as the string from which it hangs permits. *The earth evidently has an attraction for each particle of the liquid, and for each particle of the plumb bob.*

THE LAW OF GRAVITATION

The attraction of the earth for a plumb bob, for water, and for all other objects in the universe *Law of Gravitation,* first formulated by the great English scientist Isaac Newton (1642-1727). This generalization is usually stated: Every particle in the universe attracts every other particle with a force that is directly proportional to the product of their masses, and inversely proportional to the square of the distance between them.

The mathematical statement of this law is as follows:

F μ (read "varies as")

$$\frac{M' \times M''}{D^2}.$$

In this expression, F is the force of attraction, M' is the mass of one body, M'' that of a second body, and D the distance between the bodies.

As a numerical example of the way in which the law operates, consider the earth's attraction for a plumb bob that weighs one pound at sea level. At sea level, the distance of the plumb bob from the earth's center is approximately 4,000 miles. If this distance were doubled, the attractive force would be only one-fourth as great, for the denominator would be $8{,}000^2$ instead of $4{,}000^2$. In other words, if it were possible to suspend the plumb bob in space 8,000 miles from the earth's center (4,000 miles above the earth's surface), it would weigh only 1/4 pound.

Gravity and a Level Surface

In making a topographic map, or in setting the floor joists

of a building, what do we mean when we speak of a level surface? The surface of a pond is level, and it appears to be a plane but, as we know, in large bodies of water such surfaces are curved.

A level surface is thus not a flat plane, but a surface that is, at every point, exactly perpendicular to the direction of the plumb line at that point. The direction of the plumb line is the vertical—it defines the *zenith*, the point directly overhead.

What Eratosthenes did in measuring the size of the earth was to determine the curvature of the level surface of the earth by measuring the angle between two such vertical lines; and then, knowing the distance between the vertical lines at the surface, he calculated the size of the hypothetical, perfect sphere he considered the level surface to bound.

There is another method, involving careful *measurement of the exact force of gravity at various points on the earth's surface*, by which we can determine some interesting things about the size and form of the earth.

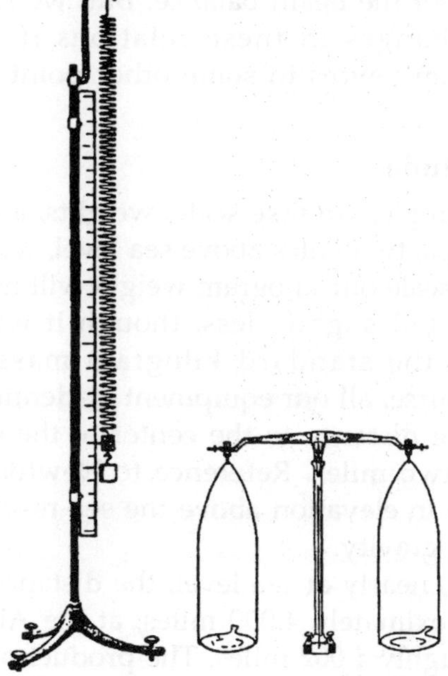

Fig. Two Common Weighing Instruments: Spring Scale.

Weighing a Mass

When we measure the force of gravity, we are, in reality, weighing a mass, for the *weight of any mass is the measure of the force of attraction between the earth and itself.* There are two common methods of weighing—the spring scale and the beam balance.

A spring scale measures weight by the amount of stretching of its spring; it is calibrated by marking the stretching produced by the earth's pull on a series of standard weights. By international agreement, such calibrations are based on *the standard kilogram mass* preserved at the International Bureau of Weights and Measures at Sèvres, France. At Sèvres, we calibrate our spring scale so that the pointer reads exactly one kilogram when the standard kilogram mass is placed on the scale.

A second weight that brings the pointer to the same mark weighs, of course, exactly one kilogram and will counterpoise the standard kilogram mass when the two are placed in opposite pans of the beam balance. But, we shall find some interesting changes in these relations if we move our equipment from Sèvres to some other point on the earth's surface.

Effect of Altitude

If, for example, we take scale, weights, and balance to a pass in the Alps, two miles above sea level, we shall find that on the spring scale our kilogram weight will no longer weigh one kilogram but slightly less, though it will still exactly counterpoise the standard kilogram mass in the beam balance. Of course, all our equipment is identical with that at Sèvres, but the distance to the center of the earth has been increased by two miles. Reference to Newton's Law shows why a change in elevation above the sea results in a change in the force of gravity.

At Sèvres, nearly at sea level, the distance to the earth's center is approximately 4,000 miles; at the Alpine pass, this distance is roughly 4,002 miles. The product of the masses of the earth and our kilogram weight must be divided, at Sèvres, by $4{,}000^2$ and, at the Alpine pass, by $4{,}002^2$. This is a difference of nearly 0.1 per cent. We see immediately that the distance from the earth's center must be considered when we weigh on a spring scale. It affects the force of attraction because it affects the denominator in the expression of Newton's Law.

Effect of the Earth's Rotation

Factors other than altitude also affect the force of gravity on objects at the earth's surface. Because the earth rotates on its polar axis, there is a tendency for objects near the equator to be thrown off, just as mud is thrown from a spinning automobile wheel, or sparks from an emery grinder. The force causing this tendency is the *centrifugal force of rotation*. It is very small compared to the force of gravity; hence, objects do not leave the equator and fly off into space.

Nevertheless, the centrifugal force of rotation is not negligible and must be considered in our studies of the form of the earth. As Newton showed long ago, this force should produce an equatorial bulge and a polar flattening in the figure of the earth. Thus is explained the difference between the polar and equatorial diameters actually found by measuring arcs of meridian-Eratosthenes' method.

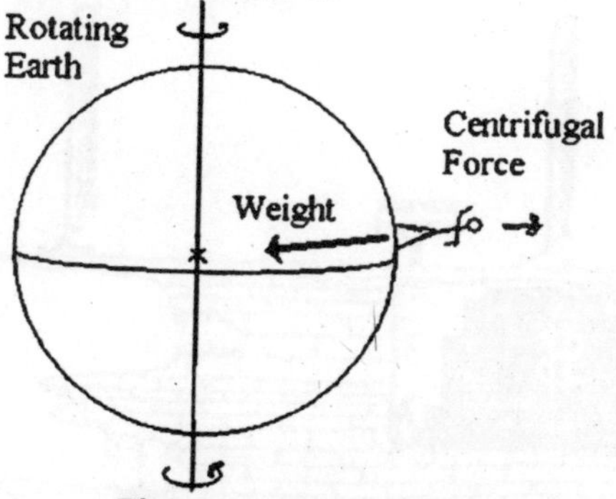

Fig. Centrifugal Force

The force of gravity is greater at the poles than at the equator because the earth is not a perfect sphere; the surface at the poles is 13 miles closer to the earth's center than at the equator. Even if the earth were a perfect sphere, an object at the equator would weigh less than the same object transported to one of the poles. The centrifugal force of rotation is greatest at the equator and steadily diminishes to zero at the pole.

The actual difference in the force that results from both of these factors is approximately 0.5 per cent. Thus, if a polar bear weighing 1,000 pounds in his natural habitat were transported to an equatorial zoo, he would there weigh only 995 pounds.

Effect of Local Variations in Rock Density

Still another factor produces variations in the weight of an object at different points on the earth's surface. This factor is the local distribution of rock masses of different density. Let us imagine that the gold buried at Fort Knox, Kentucky, as part of the United States Treasury reserve, formed a single block of pure metal.

A cubic foot of gold weighs more than 1,200 pounds. A moment's consideration of the inverse-square relation tells us that a sensitive device for weighing should show a greater weight for an object immediately on top of the gold than it would if the object were weighed on the roof of one of the great limestone caverns not far away, where much empty space is immediately beneath.

Fig. The Effect of Differences in Density of Nearby Masses Upon Weight.

Thus, even though the latitude and altitude of two localities are identical, the force of gravity at each point must vary with the *density* (mass per unit volume) of the immediately underlying rocks. The rocks just below the earth's surface do vary considerably in density from place to place, though, of course, not to the extreme degree of our artificial example at Fort Knox.

THE GRAVITY PENDULUM

In summary, we can see that three factors:

- Altitude,
- Latitude,
- Vvariations in density of the nearby rocks—affect the force of gravity at any point on the earth's surface. But the effect of each of these factors is small.

In order to learn much of value from studies of the variation in the force of gravity from one place to another we need a very sensitive instrument for detecting small variations in weight—far more sensitive than the crude spring scale of which we have spoken. Fortunately, we have precisely this kind of instrument in the *gravity pendulum.*

A free-swinging pendulum—that is, one which is not driven by clockwork or other external means—oscillates to and fro because of the force of gravity. If we suspend a heavy weight from a string, pull back the weight and release it, the weight falls toward the center of the earth following the arc that the suspending string allows. It is being pulled along its path by the force of gravity.

But the *inertia*—the resistance that an object offers to any change in its motion—of the weight carries it past the lowest point on its swing, and causes it to rise against the force of gravity along the arc that the suspending string allows. It continues to rise until its inertia is counterbalanced and eventually overcome by gravity.

Then it falls back again toward the lowest point, repeating the process again and again. The pendulum oscillates back and forth through the lowest point, rising in each oscillation to a little less height because of friction with the air.

Gradually, the oscillations die down until, ultimately, the pendulum loses its motion and hangs straight down perpendicular to a level surface. It has become a plumb bob.

We borrow from physics the law that governs the *period of oscillation* (the time for one complete to-and-fro movement) of a pendulum: The period varies inversely with the square root of the local acceleration of gravity, and directly with the square root of the length of the pendulum.

Newton demonstrated that this relationship explains why even good pendulum clocks show systematic gains or losses in time when moved about from place to place. We have seen that the *force of gravity* (the local acceleration of a freely falling body) varies from place to place on the earth's surface.

A pendulum clock that keeps good time in Paris would systematically lose time when taken to a point high in the Alps because the force of gravity (and hence the acceleration of the falling pendulum) is less at high altitudes than at low. From this effect of the force of gravity upon the period of a pendulum, we can measure gravity by means of a pendulum of known weight and length. To determine the force of gravity, we count the number of oscillations of the pendulum in a given time and calculate the force from the law for the period of oscillation.

Modern gravity pendulums are so constructed that they are almost frictionless. They are hung on knife-edge, jeweled bearings and are swung in chambers from which most of the air has been evacuated. The number of oscillations is counted by precision chronometers that can time the swing to less than 1/10,000 second. With such equipment, geodesists are able to measure the force of gravity with an accuracy of a few parts per million.

Ordinary gravity pendulums cannot be used on a ship at sea because of the rocking motion of the waves, but the Dutch geodesist, F. A. VeningMeinesz, has devised an adaptation of the gravity pendulum that can be used in a submarine submerged beneath the zone of strong waves. Thus, in recent years, we have obtained thousands of measurements from the land and sea areas of the earth.

From these measurements and from study of the angles between plumb lines at different points, geologists and geodesists have been able to make important deductions concerning the interrelations of gravity, topography, and the density of materials within the earth's crust.

Some of the most interesting of these interrelationships were discovered as a result of attempts to explain what appeared to be systematic errors in the location by triangulation of points on the earth's surface. Triangulation is the method of locating a third point by sighting on it from each of two points of known position.

But a point can also be located without recourse to triangulation by determining its latitude and longitude astronomically, as is done in navigation. Astronomic determinations of latitude and longitude are, of course, made with reference to the "vertical through the point," that is, the plumb line.

They are, in essence, determinations of the local zenith. If the plumb line invariably pointed directly to the earth's center, determinations of position made by triangulation should coincide exactly with determinations made by astronomic methods. The plumb line, however, does not everywhere point directly to the earth's center. According to the law of gravitation, a plumb bob is attracted by a given mass one mile away with a force one hundred times greater than it is attracted by an equivalent mass ten miles away. We ought, then, to find the plumb line deflected toward a nearby mountain. This is, indeed, the fact.

Consider the situation in a deep Norwegian fjord. A narrow arm of the sea lies between massive cliffs 4,000 feet high. A plumb line suspended near one side of the fjord will be deflected toward the nearby mountain mass. Also, the sea's surface will be tilted slightly upward toward the mountain mass and downward toward the center of the fjord.

Similarly, at the edge of a continent, the surface of the sea must be tilted upward slightly toward the continent by the attraction of the continental mass. Such tilts are very small, generally only a few seconds of arc, but they lead to appreciable errors in determining the position of a point by astronomic methods. Also, from consideration of such distortions of the water surface, we see that the concept of an earth with the figure of an oblate ellipsoid, though a closer approximation to the truth than the concept of a sphere, does not precisely represent, in detail, the shape of the earth's levelsurface—that is, the surface which is everywhere at right angles to the plumb line.

Isostasy

Even the largest mountains, however, have very small masses compared with the mass of the earth as a whole. Despite their being close at hand, we can expect their effect in deflecting the plumb line to be small. To calculate the theoretical deviations operating at a given station by the law of gravitation requires considerable mathematical labour, for usually one has to calculate the effects of many irregular masses lying in different directions and at different distances from the observing station.

When this is done for a large series of stations, however, an exceptionally interesting relation appears, showing that mountain chains are not mere loads of rock heaped on the surface of the earth, but that they are sustained in some other way. From such studies it has been discerned that mountain masses do not actually deflect the plumb line sideways as much as it would be deflected if the mountain were actually a load on top of the earth's crust. The same conclusion is reached from consideration of measurements of the force of gravity with the gravity pendulum.

If a mountain were really a load heaped on the crust of the earth, the force of gravity (after correction for the effect of added altitude) should be greater on the mountain than on an adjacent plain because of the added gravitational pull of the mass of the mountain. The results of many investigations with the gravity pendulum show, however, that over large areas there is little relationship between topography and the force of gravity, when this is corrected for elevation and masses above or below sea level.

From these relations, geologists and geodesists have inferred that the major irregularities of the earth's crust are sustained, not as loads borne up by the strength of the earth's crust, but by a process of flotation upon a heavier, plastic interior. Everyone knows that an oak plank will float lower ill water than a pine plank; it is denser and must sink deeper to displace its own weight of water. The idea is that *the excess mass of high areas,* such as mountain chains and continents, *is compensated for by a deficiency in density of the material of which the elevated tracts and their roots are composed.*

Mountain chains and continents float high because of their lower density. Plains and sea floors stand lower because either they are composed of rocks of higher density than the mountains, or else, if they are made of the same stuff as the mountains, this material is thicker under the mountains than under the adjacent lowlands. Such a condition of flotational equilibrium between blocks of the earth's crust is called *isostasy.* The word isostasy is also used to indicate a *tendency* for blocks of the earth to approach a condition of balance (flotational equilibrium) with adjacent blocks; that is, a tendency for heavy blocks to sink and light ones to rise, in the earth's plastic interior.

The theory of isostasy is an important concept that we shall refer to again and again. It will be well, therefore, to explain more fully the kind of geodetic evidence on which the theory is based by describing an actual example. There is, perhaps, no better example than the first study of this kind ever made—the analysis of the deflections of the plumb line discovered in the course of the survey of the land mass of India. It was the investigation of these deflections that first led to the discovery of the condition of isostatic equilibrium in the earth's crust.

THE TRIGONOMETRICAL SURVEY OF INDIA

In the middle of the Nineteenth Century, the Trigonometrical Survey of India was organized under Sir George Everest (for whom Mount Everest is named) to carry out an extensive plan of precise surveying in order to establish "control points" for mapping that great subcontinent. The method used for determining the relative position of the control points was triangulation. The triangulation was done very carefully and gave highly accurate distances between stations. Scores of points in India were located by this method, and their latitudes and longitudes were also carefully determined by astronomical observations.

If the latitude and longitude of one point in a triangulation net is known, it is, of course, a relatively simple matter to calculate the latitudes and longitudes of all the other points in the net with respect to the known one.

We know the amount of curvature of the earth, and, hence, the distance along the meridian that will be subtended by an arc of one degree. By the process of triangulation, we can determine the distance between two stations with a high degree of accuracy, and also determine the exact compass direction between the stations.

It is, therefore, easy to translate measured distances, determined by triangulation, into terms of degrees of arc, and hence to calculate the latitude and longitude of each point in the net. Such calculations are, of course, not dependent on astronomic observations at the new station. But if an astronomic determination of the latitude and longitude of the new station is made independently, it can be used as a check on the triangulation.

In the survey of India, it soon became apparent that, for some stations, determinations of their relative positions made by triangulation did not agree with those made by astronomic methods. At first, it was thought that perhaps errors had crept into the triangulation, but when checked the triangulation results were consistent. Also, several of the apparent "errors" were far too large to be accounted for by inaccuracies in surveying.

Two of the stations, Kaliana and Kalianpur, are among the stations discussed by Archdeacon Pratt, a British cleric who interested himself in the problem and who, in seeking an explanation for the apparent "errors," discovered the isostatic relationship. Now Kaliana is on the IndoGangetic plain, immediately beside the towering Himalayas. Kalianpur lies far to the south, near the center of the Indian peninsula.

Archdeacon Pratt expected that, at Kaliana, the plumb line

would be appreciably deflected northward toward the Himalayas, and, therefore, that the distance between the two stations, in terms of degrees of curvature of the earth's surface, would prove to be greater when calculated from the triangulation results than when based on astronomic determinations. This expectation was borne out.

Archdeacon Pratt was interested to find out whether the sideways gravitational pull of the Himalayas on the plumb bob accounted entirely for this discrepancy. Although no accurate topographic maps of the Himalayas were available, enough was known about the height and position of their principal peaks for Archdeacon Pratt to compute the approximate mass of the Himalayas above sea level and their average distance northward from each station.

Assuming that the mountains constitute an additional load on an otherwise uniform crust, he could then compute how much the plumb line should have been deflected northward at each station and the difference in astronomic latitude that should result from the deflection.

He found that, on these assumptions, the plumb line at Kaliana should have been deflected 27.853 seconds to the north, and at Kalianpur it should have been deflected 11.968 seconds. The difference between these is 15.885 seconds, over three times the 5.23 seconds difference actually found by the Trigonometrical Survey.

The difference between Pratt's calculated results and the figure actually found was far too great to be explained by errors in triangulation, or by errors in Pratt's estimate of the volume and mass of the Himalayas.

Pratt's Theory of Isostasy

Pratt saw that some of his assumptions were not in conformity with fact. One of the assumptions implicitly accepted was that the density of material below sea level was identical, whether the material lay beneath the plain of India or beneath the Himalayas.

He had assumed that the mass of the Himalayas, from their highest peaks to sea level, was a load heaped on a crust considered to be uniform in density below sea level. Pratt saw that the discrepancy might be explained if the rock below the Himalayas were of less mass (lower density) than that to the same depth beneath an equivalent area of the plain of peninsular India.

He suggested that both the Himalayas and the peninsula are "floating" on a deeper layer of still denser earth material, and that the heights to which their surfaces rise above the surface of this deeper layer are inversely proportional to the average densities of the material composing the two "blocks". In other words, the highstanding mass of the Himalayas is "compensated" or balanced by a corresponding deficiency of mass extending far below sea level beneath the Himalayas.

A simple illustration of Pratt's idea. In the upper diagram, blocks of four different metals, each of which weighs exactly the same amount and each of which has the same cross section, are shown floating in a pan of mercury, a liquid of very high density (13.6 gr./cm.3). The metal blocks have different lengths because their densities vary. The antimony block (density 6.6 gr./cm.3) is longer than the lead block (density 11.4 gr./cm.^{3}) because its volume must be about twice that of the lead block in order to weigh the same amount.

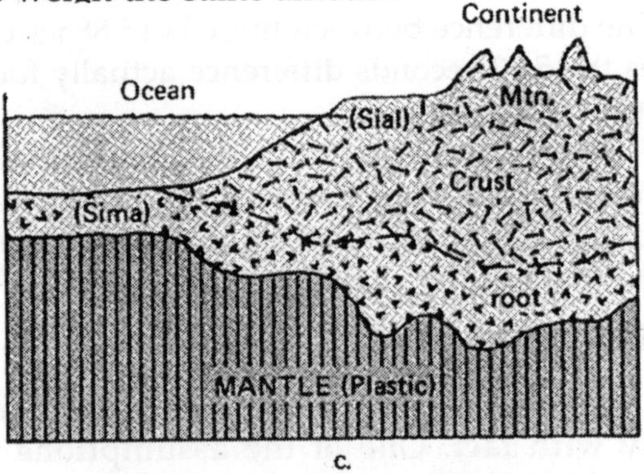

Fig. Pratt's Hypothesis of Isostasy

The blocks sink in the fluid until they displace enough mercury to equal their weight. Therefore, the antimony and lead blocks will sink to the same depth in the mercury, but, since the antimony block has nearly twice the volume of the lead block, it will project much farther above the surface than the lead. Similarly, zinc rises higher above the surface than iron, but all four metal blocks project into the mercury to the same depth—exactly deep enough to displace their weight of that fluid.

Pratt assumed that mountains, plains, and ocean floors showed essentially the same relations. Mountains are like the antimony block. They project higher above the surface of the "fluid" substratum because they are composed of material less dense than that which underlies a plain.

Pratt's suggestion was the first formulation of the theory of isostasy, and his scheme of "compensating" for differences in elevation above sea level by variation in density of the blocks below sea level is known today as the *Pratt Theory of Isostasy.* Pratt did not use the word isostasy; the condition of equilibrium which he discovered was named by Dutton some thirty-four years later.

Airy's Theory of Isostasy

The same volume of the *Transactions of the Royal Society* that contains Pratt's discussion also contains a brief contribution by G. B. Airy, the Astronomer Royal of Great Britain. Airy accepted most of Pratt's reasoning, saying that this conclusion should have been anticipated because it could be shown that if masses as great as plateaus and high mountains were loads on a solid earth, no rocks could be strong enough to sustain them.

The rocks beneath would break and flow out laterally until a state of balance was restored. The only possibility, then, is to regard mountain chains as floating masses, as Pratt did.

Airy, however, suggested a different mechanism of flotation. He saw no reason to believe that the density of the material immediately beneath a mountain is any different from that directly beneath a plain. If both blocks are of the same density, but of unequal thickness, the difference in surface elevation may still be readily explained.

The thick block (mountain block) will float higher above the surface, but it will also sink deeper into the heavier "fluid" below. The height of the mountain block is compensated by a "root" of light material that projects deeper into the underlying "fluid" layer and displaces it, just as an iceberg one hundred feet thick will float higher in water than one fifty feet thick, although it also projects deeper into the water.

Airy's view is called the *Roots of Mountains Theory of Isostasy*. This theory accounts for the "errors" in the India survey just as well as Pratt's theory, and accords much better with what we know and can infer about the composition of the materials beneath the surface.

Therefore it is more widely accepted by geologists.A simplified view of the "Roots of Mountains" theory. In the upper diagram, several blocks of copper, all with the same density but with different weights because they are of different lengths, are seen floating in a pan of mercury. The longest blocks rise higher above the surface and also sink deeper into the mercury.

The lower diagram shows how blocks of the earth's crust, all composed of the same kind of material, may, nevertheless, float to varying heights above sea level provided the blocks are of different thicknesses.

Good evidence will be developed to show that the earth's crust is not divided into simple blocks free to move past one another along frictionless boundaries. Also, the substratum is not a fluid, although it may respond by plastic flow to heavy loads of long duration and so act essentially like a very viscous fluid to these large stresses.

We know, furthermore, that the strength of the earth's crust is considerable; loading must reach a certain amount before this strength is overcome and the rock beneath gives way and begins to flow.

The point to be emphasized here is that both Pratt and Airy concluded that the Himalayas do not stand at their present height because their huge pile is a load supported by the strength of the earth's crust, but because it is, in some way, sustained by flotation. The general validity of this idea also finds strong support in studies of the intensity of the force of gravity as determined from observations with the gravity pendulum.

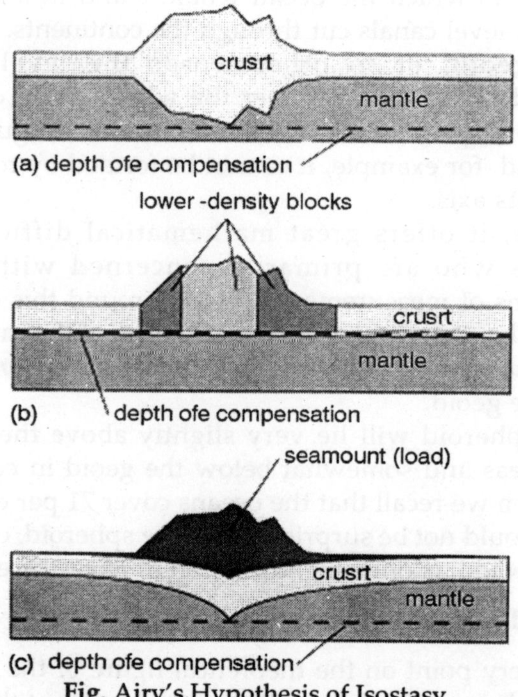

Fig. Airy's Hypothesis of Isostasy.

THE GEOID AND THE SPHEROID

We have seen how the modern measurements of arcs of meridian at different latitudes forced the modification of Eratosthenes' "Figure of the Earth" from a sphere to an oblate ellipsoid. But a moment's thought over the facts just discussed must show that still further refinement is necessary if we are to deal with the question in detail.

Though nearby mountain ranges do not deflect the plumb line as much as one might expect—a fact that we have already explained by the concept of isostasy—they nevertheless do deflect it. Thus, even the surface of the sea cannot be an oblate ellipsoid, at least near to the shores, for it is warped upward by the gravitational pull of the adjacent lands.

This warped surface of the sea, which is, of course, everywhere at right angles to the plumb line, is called the geoid. It may be thought of as the surface of the ocean, and as the surface to which the ocean would stand in a system of narrow sea-level canals cut through the continents.

The geoid, or in other words the level surface corresponding to sea level over the whole earth, obviously does not correspond to any regular mathematical figure; unlike the ellipsoid, for example, it cannot be formed by rotating an ellipse on its axis.

Hence, it offers great mathematical difficulties to geodesists who are primarily concerned with precise comparisons of measurements. To get around this difficulty, geodesists have devised a theoretical figure of the earth which is called the *spheroid*. The spheroid is really the *world-average* form of the geoid.

The spheroid will lie very slightly above the geoid in oceanic areas and somewhat below the geoid in continental areas. When we recall that the oceans cover 71 per cent of the earth, it should not be surprising that the spheroid, or "world-average geoid," is only slightly different from the ellipsoid.

Gravity Measurements and Isostasy

To every point on the theoretical figure of the earth, the spheroid, there corresponds a theoretical value of the force of gravity which depends only on the latitude of the station. That is, the theory takes account of the centrifugal force of the earth's rotation. By swinging a pendulum it is possible to measure the actual force of gravity. It is clear that, in order to compare the measured and the theoretical values of gravity on a common basis, we must allow for several factors. In the first place, nearly every continental station lies higher than sea level and hence is above the spheroid.

It is farther from the earth's center of mass than the theoretical point on the spheroid beneath. The effect of this is to make the measured value less than the theoretical. On the other hand, the material between the ground surface and the spheroid adds its local attraction to the measured value. This should make it too great as compared with the theoretical value. Also, there are the disturbing effects of nearby topographic features.

Obviously, many complex measurements and computations are necessary before the measured and theoretical values can be compared. When this work is done, however, as it has been for many thousands of stations on land and many hundreds at sea, a most significant fact is disclosed. By convention the theoretical value is subtracted from the measured value as corrected for all these disturbing influences. The difference is called *the gravity anomaly*.

Now, gravity anomalies that are computed after correcting for elevation of the station and for the mass between the station and sea level (called *Bouguer anomalies* after the French geodesist who first suggested this method) are chiefly negative on land areas and in general more highly negative the higher the station above sea. Conversely, the sea stations have slightly positive anomalies as a general rule. —

What this means can be easily seen. When we allow for the gravitational pull of the masses between sea level and the station, our corrected value is too low. But in making this "correction" we have assumed *uniform density of material below the spheroid*. As the anomalies are in general greater the higher the station, it is clear that one of two factors has influenced our result:

- The material *between the station and the spheroid* is not so dense and hence has not so great an attraction as we assumed it would have, or
- The material *below the spheroid* is less dense beneath high land than beneath low. But the material between the station and sea level cannot be far from the average rock in density. It must, then, be the second factor that has brought about our anomalies. But this is precisely the conclusion reached from plumb-line observations: *The excess mass of the rocks above sea level is compensated for by a deficiency of mass below sea level.*

Conversely, the positive anomalies at sea stations must be explained by the error involved in assuming that the oceans are underlain by material of the same density as that beneath the continents. If we replace this assumption with another—that the oceanic segments of the earth are underlain by *more dense material than the continents*—the anomalies are greatly reduced.

Thus, both the deflections of the plumb line and the relations between gravity measurements and the earth's relief indicate the same thing: The larger segments of the earth are roughly in floating equilibrium (isostatic balance), one with another. Large areas of highland stand above large areas of lowland because either:

- They are composed of less dense materials than underlie the lowlands; or
- If they are made of the same material, it is thicker beneath the highlands. We shall also show that, for some areas of limited size, isostatic equilibrium is far from complete—locally, indeed, there is good evidence indicating that masses of rock of the size of small mountain ranges may actually be held lower or higher than their equilibrium position by the strength of the crust.
- The maximum size of the masses that can be so held out of their equilibrium position is still vigorously debated by geologists. But there must indeed be a maximum limit, for considerations of strength and scale compel us to conclude that the great earth blocks comprising the continents and the ocean basins, and even blocks of the size of the major mountain chains, are incapable of supporting themselves except essentially in flotational equilibrium, one with another.

Strength

The preceding discussion of isostasy assumes that the earth has a plastic core capable of buoying up the crust by flotation. Airy emphasized the point that a "solid" or rigid core, no matter how "strong" we think the hardest rocks are, could not be strong enough to support the earth's crust; therefore, the core must behave as though it were essentially a fluid mass. What does this observation mean?

A piece of rock we pick up is strong and does not behave

at all like a fluid. We have also seen that the earth's surface is irregular in detail, with great mountain ranges towering thousands of feet above the sea and deep troughs submerged thousands of feet beneath it. The great mountain cliffs do not seem to be flattening out of their own weight. As far as we can see, they are made up of rigid and strong rocks capable of maintaining the topographic relief.

How can we reconcile this paradox between what we observe and the evidence that large blocks of the crust rest in flotational (isostatic) equilibrium upon a plastic interior? Must we conclude from these evidences of surface strength and of bodily weakness that the earth has a strong solid crust floating on an interior that is liquid? There are cogent reasons to doubt that the subcrust is really liquid. Perhaps a part of the answer may be found if we consider carefully just what is meant by "strength."

The strength of a body is defined as the force (load) per unit area that is required to deform that body permanently-that is, to break it or make it yield continuously. Thus, we say that a block of granite an inch square that breaks under a load of $30,000 pounds has a *compressive strength* of 30,000 pounds per square inch.

A steel cable with a cross section of one square inch that breaks when a weight of 150,000 pounds is suspended from its end has a *tensile strength* of 150,000 pounds per square inch. Each solid substance requires a definite force per square inch before it will rupture or flow.

Fluids have no strength. Water has no strength; it will flow into, and fill, all available spaces of a vessel into which it is poured. Tar has no strength; a piece of iron placed on it will eventually sink to the bottom. It is merely more viscous than water.

Indeed, part of the technical definition of a fluid is that it yields continuously under the slightest load or stress. Solids, too, can be made to flow—that is, yield continuously without rupture-under particular conditions of temperature and pressure.

But solids, unlike liquids, require that a definite threshold force per square inch must be applied before the strength of the material is overcome and continuous yielding begins.

Effect of Temperature

Whether a particular solid will break or will flow under an applied force often depends on the temperature prevailing while the force is applied. At red heat, a bar of iron is still solid, but it will flow under relatively small stresses that would not cause it to flow at room temperature—a fact that is utilized in forging iron.

Fig. Quarrying the State of Texas.

There is good reason to believe that many solid rocks buried deep within the earth's crust have yielded continuously by flowage in response to the heat and pressure of the earth's interior. The same rocks at the surface deform only by breaking. Such flowage of solids should not be confused with liquid flow; the rock does not melt, and a definite *stress* (equivalent to the strength of the material under the prevailing environment) must be applied before continuous yielding begins. The general rule is that *any particular solid is weaker at high temperature than at low.*

Effect of Size or Scale

In everyday life, we seldom think of the effect of scale upon the strength of materials, although, as we observed earlier, engineers have to consider scale in providing strength for massive structures. The earth is a large structure.

How are we to think of rock strength in huge masses? Suppose we are to quarry a single block of granite the size of the state of Texas. It is roughly 1,200 kilometers across, and we want to hoist a piece one-quarter as thick as it is broad—that is, 300 kilometers thick.

Grant that we have a quarry crane capable of hoisting it. The rock is assumed to be flawless and to have the average crushing strength of granite, about 30,000 pounds per square inch. Will it be strong enough to allow itself to be hoisted without disintegrating? Obviously, it is impossible to test such a problem directly, but we can investigate the properties of such a block by using a scale model. What would be the properties of a model reduced to a size suitable for our experiment?

To obtain a convenient size for our model of the state of Texas, we could reduce the true length of 1,200 kilometers to 60 cm (about 2 feet). Our model would then be on a scale of 5/ 10 millionths of the original. On this scale, the 300-kilometre thickness would be reduced in the model to 15 cm. (about 6 inches). The force of gravity does not concern us because both the original and the model are at the earth's surface. We can also use material of the same density as the original, about 3 gm/cm3. What other factors must be considered?

If the model is to act the same as the original—that is, if its mechanical behaviour is to be identical—there must be the same ratio of strength to size in the model as in the original. In order to obtain the same ratio of load in our model, we must use material whose strength is 5/10 millionths that of granite. The strength of the material in our model, then, will be pounds

$$\frac{5}{10,000,000} \times 30,000$$

per square inch, or.015 pounds per square inch.

In Hubbert's words: It is difficult to envisage a solid of this weakness. A crushing strength of.015 pounds per square inch is the same as 1 gram per square cm, which, for the density of 3 grams per cubic cm, would be the pressure at the base of a column 1/3 of a cm high. Any column higher than this would collapse of its own weight. Yet the size of the reduced block would be such that its thickness would be about 15 cm (6 inches) and its total weight about 180 pounds.

The pressure at its base would be about 45 grams per square cm or 45 times the crushing strength of the materials. Consequently, if we tried to lift such a block in the manner the eyebolts would pull out; if we should support it on a pair of sawhorses, its middle would collapse; were we to place it on a horizontal table, its sides would fall off. In fact, to lift it at all would require the use of a scoop shovel. That this is not an unreasonable result can easily be verified by direct calculation upon the original block. For it, too, the pressure at its base would exceed the crushing strength of its assumed material by a factor of 45. The inescapable conclusion, therefore, is that the good State of Texas is utterly incapable of self-support!

Effect of Confining Pressure

The strength of a substance is increased when it is placed under pressure from all directions simultaneously—that is, when it is under *confining* (hydrostatic) *pressure*. Laboratory experiments show that the crushing strength of the limestone from Solenhofen, Germany, is six times as great under a confining (hydrostatic) pressure of 10,000 atmospheres as it is under ordinary laboratory conditions. Its crushing strength under ordinary conditions is 25,000 pounds per square inch. Under a confining pressure equivalent to 10,000 atmospheres (147,000 pounds per squamre inch), its strength is 150,000 pounds per square inch.

Such a pressure is equal to the weight of a column of granite about 22 miles high. Presumably, then, it is also equal to the hydrostatic pressure at a depth of 22 miles within the earth.We see, therefore, that pressure due to gravity is a great factor in the earth's strength. In short, there is nothing really inconsistent in the apparent incongruity of an earth with "rigid and strong" rocks at its surface capable of maintaining relief features of considerable size yet so "weak" that it reacts almost like a liquid to large differential loads equivalent to the weight of a mountain chain. Our chief difficulty lies in visualizing the way in which a substance must act in very large masses or under different pressure-temperature conditions from those familiar to us.

Chapter 13

Minerals

The Earth's crust is not a homogeneous substance. If we examine its surface during a walk along a stream or a drive along the highway, we seldom fail to find a variety of rocks and soils which differ in colour, in coherence, in density, and in other characteristics. If we pick up a fragment of rock or a handful of soil and examine it more closely, we find that it, too, is a mixture of different substances. However, the individual particles that make up the rocks and soils are not mixtures.

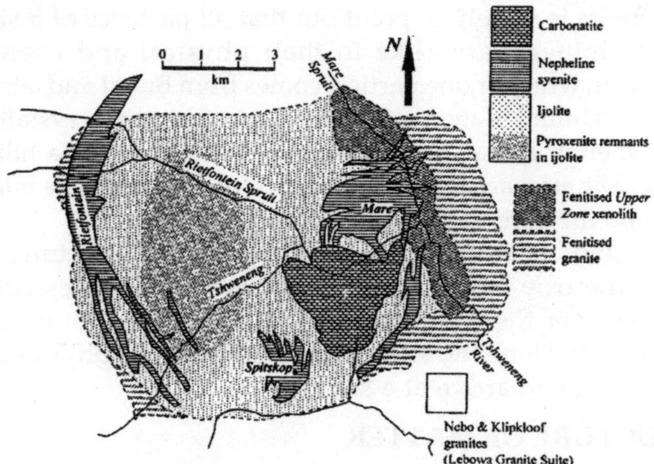

Fig. The Mineral Constituents of the Common Rock Granite.

Each is a distinct, homogeneous substance with definite chemical and physical characteristics. Some may be hard, transparent particles that resemble bits of broken glass; some may be dull, earthy grains; some may be tiny, elastic flakes that flash brilliantly in the sun. Each of these distinct, homogeneous substances is a mineral. *Rocks and soils are aggregates of minerals*. Hence, if we are to understand the origin and classification of rocks, we must learn something about the various minerals that compose them.

A mineral is a naturally occurring substance with a characteristic internal structure, and with a chemical composition and physical properties that are either fixed or vary within a definite range. Minerals, then, are natural substances, found ready-made out-of-doors.

The synthetic products that a chemist makes in a laboratory are not minerals, despite the corruption of the word in advertising. A druggist who tells you that a certain pharmaceutical preparation is "rich in vitamins and minerals" is using the term "mineral" in an entirely different way than it is used by a geologist. He is not prescribing a diet of mud and rocks.

To say that minerals have definite chemical and physical properties or properties that vary within certain definitely fixed limits is merely to point out that all particles of a single kind of mineral are alike in their physical and chemical characters, whether one particle comes from Brazil and another from the United States, or whether one may have crystallized in the shell of a snail and another from the water of a hillside spring. *The most definitive characteristic of a mineral is its internal structure,* the core of our definition.

To understand what we mean by internal structure, and, hence, the true nature of minerals, we must digress into a discussion of the fundamental structure and properties of matter itself. Here, the sciences of chemistry and physics unite with geology to arrive at a solution.

STRUCTURE OF MATTER

Atoms

The English chemist, John Dalton, in 1805 advanced the hypothesis that all matter is made up of individual particles which he called *atoms*. Later work of many physicists and chemists has verified Dalton's prediction. All solids, liquids, and gases are made up of atoms. Atoms are extremely minute particles. One hundred million of them placed side by side would make a row only an inch long. Yet, a century after Dalton's work, it was discovered that the atom, small as it is, is composed of particles still smaller.

Only three of the various subatomic particles that have been discovered are important in discussing the chemical behaviour of minerals—the proton, the neutron, and the electron. Every atom has a nucleus that is very small and dense. The nucleus contains one or more protons and, nearly always, one or more neutrons. There is an important difference between these two kinds of particles: The proton has a unit of positive electrical charge, whereas the neutron is electrically neutral.

The part of the atom that surrounds the nucleus is largely empty space, but within it the electrons of the atom revolve in orbits. Compared with the diameter of the nucleus, this outer part of the atom is very large—about 10,000 times as great. Each electron has a unit of negative electrical charge. Each atom contains the same number of electrons as protons and, therefore, is electrically balanced or neutral. Electrons are very much lighter than protons or neutrons; they weigh only 1/1,845 times as much.

The orbits along which the electrons revolve about the nucleus are arranged in successive shells. The simplest atom, that of hydrogen, consists of a nucleus containing a single proton with a single electron revolving about it. Helium has two protons and two neutrons in the nucleus, and it contains two electrons revolving in an electron shell close to the nucleus. In more complex atoms, the inner electron shell containing two electrons—the helium structure—is retained, but additional electrons, sufficient to match the number of protons contained in the nucleus of the atom, are added in one or more successive shells farther out from the nucleus.

The size of an atom, however, is not determined solely by the number of electron shells it contains. The calcium atom, which has 20 electrons distributed in 4 electron shells, is nearly identical in size to the sodium atom, which contains 11 electrons distributed in 3 shells. Both are very much smaller than the potassium atom which has 19 electrons distributed in 4 shells. This matter of atomic diameter is highly significant in minerals one kind of atom can substitute for, or replace, another atom within the atomic latticework that makes up a crystal.

Elements and Compounds

From the chemist's point of view, the number of protons contained in the nucleus of an atom is fundamental, because the chemical characteristics of an atom are determined entirely by the number of positive electrical charges in its nucleus. The number of these electrical charges, of course, equals the number of protons in the nucleus.

All atoms having the same number of protons in the nucleus belong to one species of matter, called an *element*, or elementary substance. About ninety elements have been recognized in nature, and the atomic physicists and chemists have recently produced a few new ones by artificial means.

Each element has been assigned a definite *symbol*, such as H for hydrogen and Pb for lead—a convenient shorthand in writing chemical formulas and equations. The elements, together with their symbols, atomic numbers (number of protons in the nucleus), and atomic weights.

An elementary substance, such as hydrogen, oxygen, or iron, is made up entirely of atoms of a single kind, whereas a *chemical compound*, such as water, salt, or mica, is composed of two or more kinds of atoms bound together by electrical charges. We shall now consider the nature of these electrical bonds that hold diverse elements together in compounds.

For some reason, those elements are most stable whose outermost electron shell contains 8 electrons. Such elements—for illustration, the gases argon, neon, and xenon—are never found combined with other elements in the form of chemical compounds and, hence, are called the *inert gases*. The atoms of other elements may contain from 1 to 7 electrons in their outer shell.

The number of electrons in the outer shell is a major factor in determining the relative ease with which elements combine to form compounds.

For example, the elements sodium (Na) and chlorine (Cl) are highly active chemically. Sodium has 1 electron in its outer shell, chlorine has 7. Sodium has only to lose an electron and chlorine only to gain one for both to attain the stable grouping of 8. This is just what does happen when sodium combines with chlorine to form *sodium chloride* (NaCl), the substance we know as table salt.

An electron is transferred from the sodium atom to the chlorine atom. But, the loss of an electron leaves the sodium atom no longer electrically neutral; it now has one more positively charged proton in the nucleus than it has negatively charged electrons in its electron shells. Hence, the atom has one unbalanced positive charge.

Similarly, the chlorine atom, by gaining an electron, acquires one unbalanced negative charge; there is one more electron in its electron shells than there are protons in its nucleus.

Such a charged atom, with the number of its protons either more or less than the number of its electrons, is called an *ion*. As unlike charges of electricity attract, and like charges repel, if the positively charged sodium ion is free to move, as in a solution or a gas, it is drawn to the negatively charged chlorine ion and the two may be held together to form a molecule of sodium chloride.

Molecules are distinct groups of two or more atoms tightly bound together. The compound, sodium chloride, has very different properties from either of the two elements that combined to form it.

Two atoms also combine to form stable outer shells of 8 electrons without actual transference of one electron to another. This happens by the "sharing" of one or more electrons in such a way that, if the shared electrons are counted as belonging to each of the bonded atoms, each atom achieves a stable shell of 8. Thus, two chlorine atoms with 7 electrons in their outer shells may unite to form a chlorine molecule by sharing two electrons

$$:\ddot{\underset{..}{Cl}}. + :\ddot{\underset{..}{Cl}}. \rightarrow :\ddot{\underset{..}{Cl}}:\ddot{\underset{..}{Cl}}$$

Chlorine atom chlorine atom chlorine molecule.

The chemist's concept of molecules thus helps to explain the formation of chemical compounds and also the structure of gases and liquids in which most chemical reactions take place. The molecular concept of matter is also useful in describing the structure of some solids.

With most solids, however, a somewhat different view of the structure of compounds, a view in which the idea of molecules is not strictly applicable, seems to be needed. Most minerals are solids of this kind, as has been proved by many studies within the past twenty-five years. Some of these studies throw much light on the question of how elements combine, as well as on the internal structure of minerals.

STRUCTURE OF CRYSTALS

Geometrical Form

Everyone is familiar with crystals of certain minerals: halite (or rock salt), garnet, quartz (also called rock crystal), ice. Some crystals occur in beautiful geometric forms with surfaces bounded by smooth planes called *crystal faces*. Even the early Greeks noticed that some minerals, such as garnet and quartz, commonly have characteristic crystal forms.

Perfect crystals are rare. Most snowflakes fall as beautiful, six-sided, perfect crystals; but frost on a windowpane shows much less perfect ones, and the granules of ice that form on the surface of a freezing pond may show few, if any, crystal faces.

This is also true for other minerals. Relatively few are completely bounded by plane faces and some show none at all. Nevertheless, study of the common imperfect crystals, together with the relatively rare perfect ones, permitted mineralogists (as geologists who specialize in minerals are called) to make sound deductions about the internal structure of minerals and the fundamental properties of matter long before physicists and chemists had proved the atomic theory.

The first important step in the analysis of minerals by their crystal faces was made by Nicolas Steno, a Danish physician who lived in Florence, Italy. Steno, one of the outstanding figures in the history of geology, also made fundamental contributions to our knowledge of the origin and structural relations of rock strata.

Fig. Common minerals Showing Good Crystal form: A, Epidote; b, Orthoclase; c, Garnet; d, Pyrite.

Constancy of Interfacial Angles

Steno showed, with the crude instruments at his disposal, that, if crystal faces are found on a specimen of quartz, they always meet in characteristic angles regardless of the size and gross shape of the crystal. An Italian student, Guglielmini; showed in 1688 and 1705 that similar relations held for other minerals. He also noted that the angles characteristic of one species of mineral differed from those of another.

Thus, in halite (NaCl) the angle between adjacent surfaces is always a right angle; this means the crystal may be a cube, a rectangular, boxlike figure, or any other right-angled parallelepiped. In quartz, as Steno had found, the angles between the long crystal faces that form the sides of the crystal are always 120°. Steno's and Guglielmini's methods were refined and extended by later workers until thousands of similar measurements on many kinds of mineral crystals had been made. From them, mineralogists long ago reached the conclusion that the internal structure of each kind of mineral is unique.

They reasoned that the constancy of the interfacial angles in different specimens of the same mineral, regardless of the size and shape of the crystals, could mean only that each mineral is built up of minute particles regularly packed together in a definite geometric pattern. The pattern of packing determines the angles between faces and is identical in all specimens of a particular mineral. The size of the specimen depends merely on the number of such particles it contains.

Optical Properties

Other studies of minerals from a very different viewpoint fortified this conclusion. Among the most important were the studies of the effects that crystals produce on light transmitted through them. The Dutch physicist, Christian Huygens discovered the phenomenon of "double refraction" while studying the transmission of light through the mineral calcite.

One can easily observe this phenomenon by placing a fragment of transparent calcite above a dot on a sheet of paper. Instead of one dot, two are seen. If the calcite is revolved slowly, one dot describes a circle about the other. More than 100 years later, physicists showed that this phenomenon could be explained in terms of a theory of light.

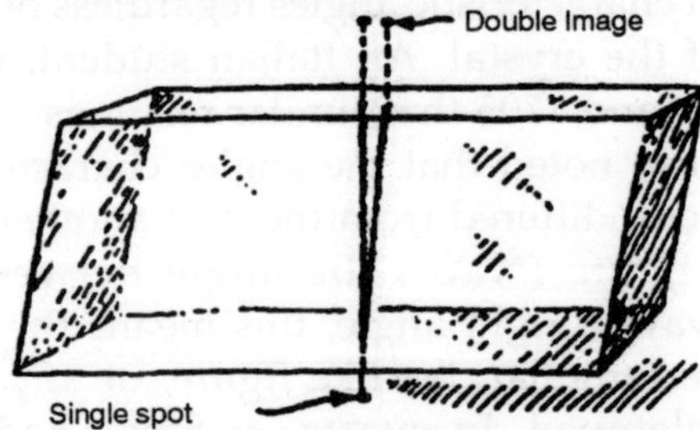

Fig. A Fragment of Clear Calcite Showing Double Refraction.

According to this theory, the light ray which penetrates

the crystal is broken into two rays which deviate slightly from each other as they travel through the crystal. Furthermore, in contrast to ordinary light which vibrates in all directions at right angles to the line of propagation, each of the two rays that pass through the crystal vibrates in only one plane. Light so modified that it vibrates in only one plane is said to be *polarized*.

These discoveries opened the way for an important new technique in studying and identifying minerals. William Nicol, who taught natural philosophy at Edinburgh, showed in 1829 that transparent fragments of calcite could be cut and glued together in such a way as to eliminate one of the two polarized rays.

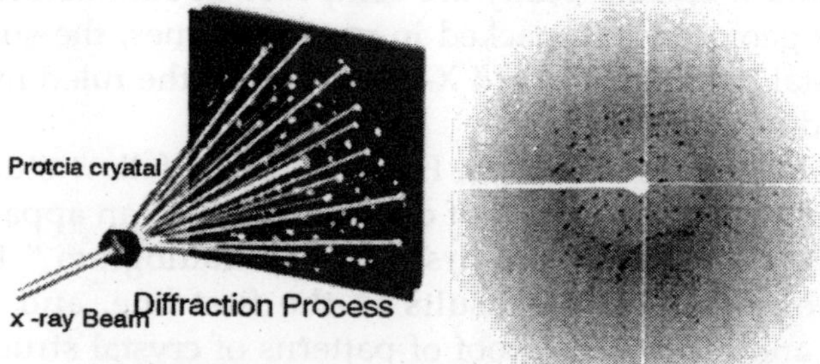

Diffraction Process Diffraction Pattem form NSLS

Fig. Diffraction Pattern of Crystal

Others adapted these Nicol prisms to microscopes, making it possible to study the effects that crystals produce on transmitted polarized light.but it should be emphasized that the *optical properties* of minerals, determined by means of the *petrographic microscope,* are precise and diagnostic. By them, most minerals can be quickly identified. The petrographic microscope is the geologist's most efficient instrument for mineral and rock study.

But more important for our present purpose is the fact that the petrographic microscope gives us clues to the fundamental nature of matter and light. The behaviour of light in minerals is systematically related to the angles between the crystal faces.

This suggests that light is influenced by very minute, systematically arranged particles within the crystal, and strengthens the conclusion from interfacial angles that minerals are made up of submicroscopic particles systematically packed together.

X-ray Studies

A young Munich student, W. Friedrich by name, was interested in the theory on wave properties of X-rays in relation to the structure of crystals, as expounded by M. Laue, a specialist in the physics of light.

When a series of closely spaced parallel lines are scratched on a mirror surface, the light reflected from the mirror is broken into the colors of the spectrum. Laue reasoned that if X-rays are like light but much shorter in wave length, and if crystals really are composed of submicroscopic particles geometrically packed in parallel planes, the surface of a crystal might act toward X-rays much as the ruled mirror surface does toward light.

In order to test this idea, Friedrich and a fellow student, P. Knipping, using a crystal of copper sulfate on an apparatus much like, developed the first "Laue X-radiogram." Later tests showed the same results as the first one, and gave definite and conclusive proof of patterns of crystal structure, confirming Laue's reasoning on the wave properties of X-rays. The experiments also proved beyond debate the inference of mineralogists regarding the internal structure of crystals. Also, a versatile new tool had been made available for the study of minerals.

By X-ray studies the geometrical arrangement of the atoms within a crystal, its *internal structure*, could now be worked out and

The technique could even be applied to grains so small that they can hardly be seen with the petrographic microscope. By X-ray studies, it also became possible to measure accurately the volume occupied by atoms of almost all the elements represented in the crystal, since the distances between similar planes in the geometrical pattern could now be measured.

From the spacing and arrangement of the different kinds

of atoms, it was possible to analyse more closely the way in which the atoms in crystals are held together.

THE CHEMICAL BOND IN CRYSTALS

X-ray studies show, for example, that halite (NaCl) has the structure. We have already seen that by the transference of the one lone electron in the outer shell of a sodium atom to the almost filled outer shell of a chlorine atom, both atoms can achieve the stable arrangement of 8 electrons in the outer shell. In acquiring this stable arrangement, each atom sacrifices its electrical neutrality and acquires a charge. It becomes an ion. In a liquid or gas, two ions with unlike charges, such as a sodium ion and a chlorine ion, might be drawn together to form a molecule.

In a crystal, the geometrical packing of the particles necessitates a fixed arrangement of the ions, an arrangement that must satisfy the electrical forces set up by the attraction of ions of unlike sign and the repulsion of those with the same sign. Each positively charged sodium ion is equidistant from, and at the center of, 6 symmetrically placed chlorine ions.

Fig. The Cubic form, Right, and Internal Structure of Halite.

Each negatively charged chlorine ion is similarly surrounded by 6 symmetrically placed sodium ions. Therefore, we say that halite is an *ionic crystal* held together by the electric charges on its symmetrically arranged ions. Most minerals are held together by similar chemical bonds, although, in general, the internal structure is far more complex and not so easily visualized as in halite. Some minerals—diamond, for example are held together in another way.

Diamond is composed entirely of carbon; it is one of the

crystalline forms of this element. Carbon atoms have 4 electrons in the outer shell. In crystals of diamond, each carbon atom is linked with four others. This linking allows each of the 4 outer electrons in each carbon atom to be "shared" with an adjacent carbon atom.

Thus, the carbon atoms in diamond achieve stability; each may be considered to have a complete outer shell of 8 electrons, though every electron is actually shared with a neighboring carbon atom. In terms of atomic structure, each electron may be thought of as oscillating between the orbits of the two neighboring carbon nuclei.

This close bonding of the atoms in diamond is very strong; hence, diamond is the hardest substance known. In this type of crystal, there are no ions—the atoms retain electrical neutrality but are selectively packed in positions so that electrons may be "shared."

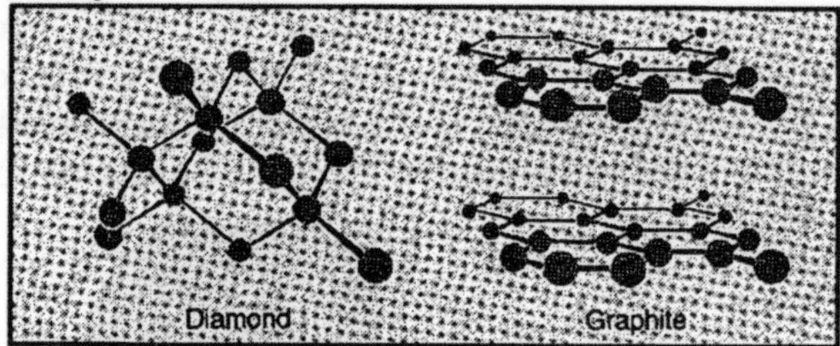

Fig. Structures of Diamond and Graphite.

It is readily seen that in either ionic crystals, such as halite, or in atomic crystals, such as diamond, there is no particular part of the crystal that can be considered a molecule. The chemist's concept of molecular association of atoms, applicable to gases and liquids, fails to apply to most minerals.

THE CHEMICAL COMPOSITION OF MINERALS

Some minerals, like diamond, sulfur, copper, and gold, are elements. Others, like ice (H_2O), quartz (SiO_2), calcite ($CaCO_3$), and kaolinite ($H_4Al_2Si_2O_9$), are compounds whose compositions can be expressed by simple chemical formulas. Nevertheless, as is indicated in the definition of mineral, many minerals vary within certain limits in the percentage of the various elements they contain, and the composition of such a one cannot be expressed by a simple formula. The fundamental distinction among minerals rests not on their chemical composition but on their differing internal structures.

Perhaps the most striking illustration of this is found in the contrast between diamond and graphite. Both are pure carbon; their chemical composition is identical but they occur in different kinds of crystals.

Diamond is the hardest known substance, graphite is soft and greasy. Most diamonds are transparent, graphite is opaque. Diamond is used as an abrasive and cutting tool, graphite as a lubricant because it yields fine flakes that glide readily over one another.

ISOMORPHISM

Certain elements may replace one another within a compound so that a range of chemical composition occurs within a single mineral species. Such replacement is called *isomorphism*. The elements that substitute for one another may be chemically similar, but in many of the commonest minerals, the plagioclase feldspars, for example, they are not. Perhaps as simple an isomorphous series as any is the olivine group of minerals.

The formula of this group is written $(Mg,Fe)_2\ SiO_4$, meaning that different specimens of olivine may have chemical compositions intermediate between the two "end members"; that is, they range from pure $Mg_2\ SiO_4$ to pure $Fe_2\ SiO_4$. It is only the iron (Fe) and the magnesium (Mg) that vary; the proportions of silicon and oxygen remain constant. The intermediate members of the series are regarded as *solid solutions* of the two end members because they are homogeneous crystals and not merely mixed aggregates of two minerals.

Mechanism of Substitution in Isomorphism

The application of X-ray studies to minerals has revealed much of the mechanism by which substitution of one element for another can take place in crystals. The controlling factor in isomorphism is not the number of electrons in the outer shell of the atoms of the two elements concerned, as might have been thought, but their atomic (or ionic) diameters. These are measured in units of length called *Ångströms*. An Ångström is 0.00000001 centimeter long. In olivine, Fe and Mg can readily substitute one for the other, for not only does each contain 2 electrons in its outer shell, but the ionic diameters are very nearly the same.

In many mineral groups, sodium readily substitutes for calcium (2 electrons in the outer shell) because their ionic diameters are very similar (0.98 and 1.06 Ångströms, respectively). But sodium cannot substitute to nearly the same extent for potassium, although each has 1 electron in its outer shell, because the potassium ion is much larger (1.33 Ångströms) than the sodium ion.

This discovery was at variance with earlier chemical theory because, since sodium and potassium are very similar in chemical properties, it had been thought they would substitute readily for one another, as indeed they do in aqueous solutions.

In crystals, however, a small amount of potassium may be substituted for sodium—the internal structure is warped to take care of the difference in ionic diameters—but, when the substitution exceeds a certain amount, the warping is evidently too great for the structure to remains stable; it breaks up into interlocking crystals of two distinct minerals.

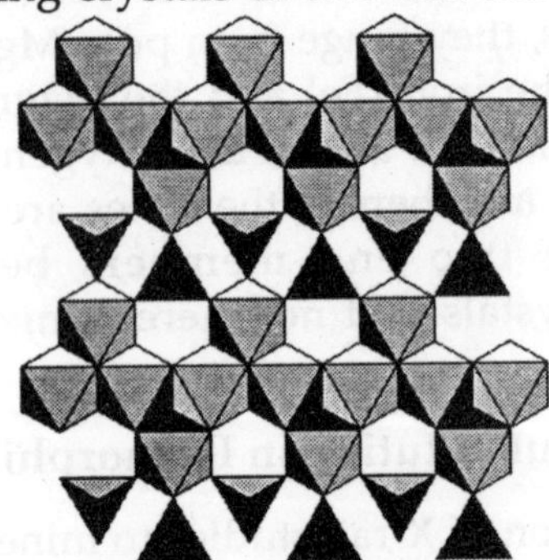

Fig. Structureof Olivine

Such warping of the internal structure during substitution, which shows the result of substituting calcium ions for about half of the iron and magnesium ions in the internal structure of olivine, giving the slightly different structure of monticellite, a mineral of closely related chemical composition.

Clearly, an element with one electron in its outer shell cannot be substituted for another with two electrons in its outer shell without destroying the electrical neutrality of the structure; therefore, a second, concurrent substitution is required to maintain neutrality.

For example, in the plagioclase series of solid solutions, the change from pure albite ($NaAlSi_3 O_8$) to pure anorthite ($CaAl_2 Si_2 O_8$) takes place by the simultaneous substitution of Ca (2 electrons in outer shell) and Al (3 electrons) for Na (1 electron) and Si (4 electrons). Hence, chemical neutrality is attained, since 3 + 2 = 4 + 1.

The ionic diameter of sodium is nearly the same as that of calcium; the ionic diameter of aluminum is sufficiently near to that of silicon so that the internal structure is not warped enough to become unstable. These slight changes in structure and composition, however, are sufficient to produce small variations in the optical properties.

In many minerals, similar substitutions of one element for another are numerous, and there may be several end members instead of only two. The composition of such minerals can seldom be expressed by a simple chemical formula in which the ratios of the elements are expressed in whole numbers.

Chapter 14

Rocks

Geology, like any science, has certain principles and generalizations that help to systematize and interpret the data collected by observation and experiment. The inquiring student should look critically into the validity of these generalizations. A basic principle in geology, often called the *Uniformitarian Principle*, was proposed by James Hutton of Edinburgh in 1785, and was popularized by the British geologist, Charles Lyell, in 1830. This principle of "uniformity in the order of nature" may be stated as follows:

"The present is the key to the past," or, applied more specifically to our present subject: Rocks formed long ago at the earth's surface may be understood and explained in accordance with processes now in operation.

The Uniformitarian Principle assumes the *uniform operation of physical laws throughout the geologic past*. It assumes, for example, that in the geologic past, just as today, water collected into streams and carried loads of mud and silt to the sea, that marine organisms lived and died in the ancient seas, and that their shells were buried in the sands and mud accumulating on the sea floor.

Hence, if features identical to those we now see in process of formation along streams and beaches can be recognized in and among the solid rocks, it is reasonable to infer that these features were produced by processes of the kind we now see in operation.

The Uniformitarian Principle, like any other scientific "law," rests on the circumstance that all the facts known conform to it. Long and careful searches have failed to find good evidence for ancient conditions totally unlike any existing today. Yet, like most scientific laws, this one must be interpreted carefully and rather broadly. In applying the principle that "the present is the key to the past," we must keep in mind that, although there is good evidence to believe geologic processes have always operated in the same way, they have not necessarily always operated at their present rate or intensity.

Climates of the world were colder some 15,000 years ago and that glaciers were much more widespread than they are now; but there is every reason to believe that the glaciers of that time formed, moved, eroded, and deposited in precisely the same way that glaciers do today.

In the following discussion it will be apparent that some applications of the Uniformitarian Principle may be made to the problems of the origins of rocks.

THREE GREAT CLASSES OF ROCKS

The walls of many deep roadcuts show that the surface soil and the unconsolidated mantle rock beneath it form only a thin veneer. At depths of a few feet, we pass through them and enter rock. The deepest wells that have been drilled extend downward into the rocks about 20,000 feet—less than one-tenth of 1 per cent of the earth's radius. Materials now at greater depths cannot be examined, that the rocky shell extends to depths of at least several miles, and that some rocks now at the surface were once buried deeply.

Let us briefly summarize a few of the conclusions geologists have reached about rocks, leaving the evidence upon which these generalizations are based until later. Rocks are aggregates of minerals. Three great classes are recognized. The *sedimentary rocks* have been formed at the surface of the earth, either by the accumulation and cementation of fragments of rocks, minerals, and organisms, or as precipitates from sea water and other surface solutions.

The *igneous rocks* have formed from molten material that solidified on cooling. The *metamorphic rocks* have been formed by the transformation, *while in the solid state,* of pre-existing rocks beneath the earth's surface through the agencies of heat, pressure, and chemically active fluids.

Rocks are difficult to classify because they grade one into the other. Even the three broad classes of rocks intergrade, and not every rock can be placed unequivocally in one class or the other. Nevertheless, we can classify most rocks into these three great classes and also into smaller divisions, just as it is possible to classify people into tall and short, even though we know there is every gradation between.

Sedimentary Rocks

Everyone has observed how rills formed on a hillside during a downpour of rain spread sheets of mud, sand, and gravel at the base of steep slopes. Similarly, every stream, whether it be a small brook or a great river, carries unconsolidated debris downstream.

Most of the rock waste is dropped in sand bars or in beds of silt and gravel in the slack-water parts of the stream course, but it is carried further by later floods and most of it eventually reaches the ocean and accumulates in a delta at the stream mouth or is scattered over the beaches and sea floor by ocean waves and currents.

So commonplace and easily observed are the transportation and deposition of debris by running water that even the Greek philosophers learned to recognize water-borne deposits. They reasoned that beds of gravel and sand clinging to the walls of steep valleys high above the reach of presentday floods were the deposits of former streams. They saw in the clam and oyster shells protruding from some sandstone ledges on inland mountain ridges evidence of former higher-standing seas.

It was a more difficult step—and one probably not made by the ancients —to conclude that a hard, well-consolidated sandstone containing scattered fossil shells and exposed in the gorges of a mountain range is merely the *cemented,* shell-strewn sand of an ancient beach or sea floor that has been warped upward to its present height. In the Middle Ages, churchmen concluded that such fossils, many of which are not closely similar to the shells of animals now living in the sea, were not organic remains but "sports of nature," perhaps put in the rocks by the devil to confuse mankind.

Even the accurate and relatively enlightened German scholar, Georg Agricola, in whose words and woodcuts the late medieval Saxon mines and miners are still preserved for us, described only the leaves, wood, bones, and fish skeletons embedded in the rocks as organic remains. To him the fossil shells were "solidified accumulations from water" (whatever that may mean).

The restraints of tradition and authority were not thrown off until the geological pioneers of the late Seventeenth and Eighteenth centuries collected and compared the shells in hard rocks with those in unconsolidated sands near the seashore and, in the years between 1790 and 1815, made maps showing the surface distribution of strata of sandstone, limestone, and other rocks, demonstrating that soft sandstones may grade laterally into hard rock and that in some localities both types contain the same varieties of fossil shells.

Today, by modern tools such as the petrographic microscope, it is only the work of minutes to trace the steps whereby loose sand like that on the floor of the sea has been transformed to solid rock. It is easy to see that the individual grains of the fossiliferous sandstone, when magnified by the microscope, are of the same shapes and composed of essentially the same minerals as the sand grains of a modern beach.

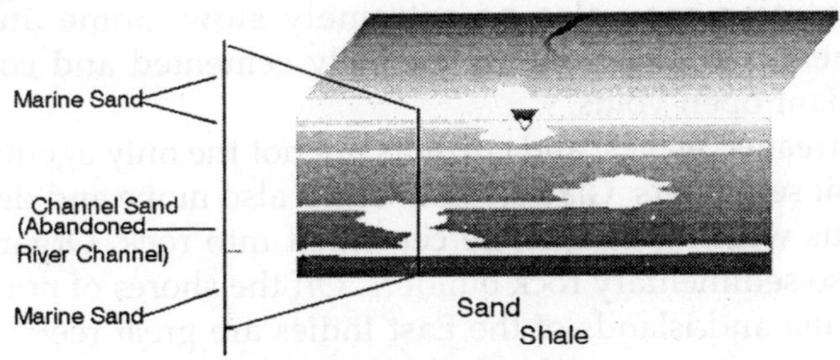

Fig. Cementation of Sand

The fossil shells of the sandstone, even though different from those of animals living today, show under the microscope structures so similar as to compel the belief that they are the remains of organisms that lived in the past. Thus, the loose beach sand and the firm sandstone differ only in cementation. The voids or pores between the grains of the fossiliferous sandstone have been almost completely filled with mineral matter. By the filling of these voids, unconsolidated sand has been made into solid rock.

The process of natural cementation is slow, but partially cemented rocks are common and, since all intermediate stages have been observed, the inference that sandstone is cemented sand is logical. By digging deeply into some modern beaches, we find that certain layers of sand below the surface, where they are not disturbed by wave action, are coated with a thin film of mineral substance that evidently has precipitated from the solution surrounding them.

Although this film may not have grown thick enough to bind the sand grains firmly, complete cementation will surely occur if the process keeps on for a sufficiently long time. More striking examples may be seen along certain stream beds in localities that have hot, dry summers.

Here, the newly deposited gravel brought down by a spring flood may, under exceptional conditions, be completely cemented in a single season by the calcite and other minerals precipitated out of the water as the stream dwindles and finally disappears under the hot summer sun. Thus, even modern sediments in process of accumulation show many stages in cementation from unconsolidated debris to solid rock. Yet cementation may also be extremely slow. Some ancient sedimentary rocks are only partially cemented and contain abundant open voids.

Streams, waves, and currents are not the only agents that deposit sediments. Glaciers and winds also move and deposit detritus which may then be cemented into rock. Organisms are also sedimentary rock builders. Off the shores of northern Australia and islands of the East Indies are great reefs many miles long, made up of the shells of corals, clams, marine snails, and a wide variety of other organisms which live in the shallow, clear water of these warm seas.

Even today, the living shells are being cemented together by lime-depositing plants and the interstices filled with microscopic shells and limy (calcite-rich) mud. Such accumulations of cemented shells form the common sedimentary rock called limestone.

Limestones that show all the characteristics of these underwater reefs are also found on land far above the reach of present seas. Even the New Guinea natives recognize that some of the limestone in hills near the shore has been formed in the same way as the reefs accumulating offshore today. Similar reef limestones, interstratified with other marine sedimentary rocks, are found in west.

Texas, the Balkans, and many other places thousands of feet above the sea all hundreds of miles from it. The conclusion that these rocks were once deposited in ancient seas as "coral reefs" of the kind that we now see in the southwest Pacific seems logically inescapable.

Coal is another organic sedimentary rock; the well-preserved cell structures and other plant characteristics visible under the microscope prove that it is made from accumulations of plant remains.

Still other sedimentary rocks, such as the salt deposits of the Bonneville flats west of Great Salt Lake, Utah, are residues from the evaporation of saline lakes—identical beds can be seen in process of forming in parts of the lake a few miles farther east.

Most sedimentary rocks form distinct layers or *strata.* This *stratification* (also called *bedding*) generally results from variations in the supply of sedimentary detritus during deposition, or from changes in the velocity of currents that are laying down the material, or from still other factors.

Even a few minutes' observation of a sand bar in a small stream shows that such variations have taken place recently. A smooth-walled pit dug a few inches into the sand bar usualy reveals stratification identical with that of many sandstones. Visits to an ocean beach during and after a storm reveal comparable changes in the coarseness of the beach material.

A pit dug by a child in the sand commonly shows layers of varying coarseness recording such variations in current and wave strength. The strong waves and currents of a heavy storm may partially erode an earlier deposited layer of sand and mud on the sea floor and sweep a sheet of coarse gravel over it, as has been proved by samples dredged from the same spot before and after storms.

Even the accumulation of shells in an offshore reef is sometimes interrupted by a fall of ash and pumice from a nearby volcano or by mud swept far out to sea during unusually heavy floods in the rivers of the adjacent land. By such interruptions and accidents during deposition, distinct sedimentary strata are formed; and because such changes vary in intensity and frequency, some strata are thin, others many feet in thickness.

Laws of Sedimentary Sequence

Observations of strata now accumulating make possible the following rather obvious generalizations, which are useful in interpreting ancient sedimentary rocks:

In any pile of sedimentary strata that has not been disturbed by folding or overturning since accumulation, the youngest stratum is at the top and the oldest at the base. In other words, the order of deposition is from the bottom upward. This is the *Law of Superposition,* first clearly stated in 1669 by Nicholas Steno.

Water-laid sediments are deposited in strata that are not far from horizontal, and parallel or nearly parallel to the surface on which they are accumulating. Though the bottom surface of a stratum may conform roughly to the irregularities of the base, its top must be nearly horizontal. This is the *Law of Origirial Horizontality,* also stated by Steno.

Many applications of these generalizations will appear in subsequent that they are not insignificant truisms may be realized from the fact that in most mountain ranges the strata of sandstone, limestone, and other sedimentary rocks are no longer horizontal. Instead, they are steeply tilted or even overturned. Because we know that these rocks were once sheets of sand, shells, and gravel deposited in nearly horizontal layers and then cemented together, their present distorted attitudes show that great forces must have been at work in the region. Thus, a simple structure like the stratification of a sediment gives us the clue to read the record of great changes of energy in the earth—changes that have deformed and mashed once horizontal sheets of rock into fantastically complex patterns.

CLASSIFICATION OF SEDIMENTARY ROCKS

Sedimentary rocks are classified and named on the basis of their texture, that is, the size and shape of their constituent particles; and on their composition, that is, the kinds of material that compose the particles and cements. Two general subdivisions may be recognized:

Clastic sedimentary rocks are composed mainly of fragments of rocks and minerals which have been transported to the site of deposition and cemented there; organic and chemical sedimentary rocks are composed dominantly of the cemented shells of organisms, or of precipitates from aqueous solutions such as sea water.

Hundreds of different kinds of sedimentary rocks have been described and named but most of them are comparatively rare. For an elementary knowledge of geology.

The descriptions should be studied with a specimen of the rock at hand, so that one can note and compare its properties with those listed in the description. In making such a comparison, do not be dismayed if specimen and description do not correspond exactly, for, as we remarked earlier, rocks vary widely and they grade into one another.

The brief notes on the origin of each of the rocks only to present summaries of conclusions reached by geologists from evidence outlined. To give all of this evidence here would involve lengthy presentation of indirect as well as observational data.

Igneous Rocks

At active volcanoes, also, we can see rocks being made.

Molten material rises to the surface and flows for a time in lava streams, but eventually cools and hardens into *igneous rock*.

Two subdivisions of the igneous rocks are recognized. The lavas and solid fragments erupted from volcanoes are called *volcanic rocks*. They are composed in large part of microscopic mineral crystals and glass. In a few places volcanic rocks can be seen to grade into rocks composed of much larger crystals.

These coarse-grained igneous rocks are called *plutonic rocks*, and are widespread at the earth's surface, but have not been observed in process of formation. There is good reason to believe that the plutonic rocks were not spewed out to the surface like the lavas, but solidified deep underground. Subsequently, the roof rock that covered many of these buried masses was eroded away, thus exposing the plutonic rock at the surface.

Volcanic Rocks

In January 1938, a white-hot stream of molten lava issued from a fissure near the base of the volcano Nyamlagira, in central Africa, and poured quietly downward into a forested plain below.

For two years and four months the lava continued to emerge until more than 500,000 cubic yards of molten rock had devastated an area of over twenty-five square miles. Finally the flow ceased, the fissure froze over, and the lava congealed into the black slaggy rock that we call basalt. Similar but smaller flows have been observed at many other volcanoes.

Some volcanoes erupt explosively, blowing vast quantities of volcanic "ash" and broken rock fragments into the air. Around volcanoes such as Mount Katmai in Alaska, Bandai-San in Japan, or Mount Pelée in the West Indies, large areas have more than once been buried under fifty feet or more of hot ash and rock. Such loose debris, erupted from Vesuvius in Italy, has been known to consolidate into firm, coherent rock within a generation.

That rocks were made by volcanic action was well known to the peoples of early civilizations because of the many active volcanoes in the Mediterranean countries and in Persia. Early writings contain a wealth of information on Vesuvius in Italy, Santorin in the Aegean Sea, Etna in Sicily, and Erebus in ancient Persia.

The eruption from Vesuvius that overwhelmed Herculaneum and Pompeii in 79 A.D. was but one of a series of volcanic disasters that impressed on early peoples both the awesome power of volcanic eruptions and the characteristics of the lavas and explosive products that come from them.

Nevertheless, the ancients did not recognize that volcanic rocks are widely distributed over the surface of the earth in areas far removed from volcanoes now active.

It is one thing to stand on the brink of a fissure and watch liquid lava emerge, roll down a slope, and congeal into a mass of basalt, and another to recognize a basalt flow that was extruded millions of years ago, and has since been detached from its parent cone or fissure by erosion, or buried under a later accumulation of sedimentary rock. From their manner of formation, it is common for volcanic and sedimentary rocks to be interstratified.

In the Samoan Islands, flows of basalt have been seen to spread over reefs in which limestone was forming; today, deposits of coral and shells are collecting on the upper surface of the congealed lava.

The great flow at the base of Nyamlagira covers older volcanic material, but earlier flows in the same region spread out over a plain underlain by lake and river deposits and, in turn, were partially buried under new deposits from the lakes and rivers. Scarcely any thick pile of sedimentary rocks is completely free from volcanic interlayers.

It is not surprising that flows of lava and beds of volcanic ash interstratified with sedimentary rocks were considered to be hardened sediments by the early geologists, just as they often are by the uninitiated today. It was not until after much careful field study that reliable criteria were put forth distinguishing buried lava flows from sedimentary deposits. Indeed, fifty years of violent controversy took place before the volcanic origin of basalt was definitely established.

THE CONTROVERSY OVER THE ORIGIN OF BASALT

The interpretation of geological phenomena is often influenced by the philosophy of the worker. The history of geology is replete with examples of unsuccessful attempts to fit the features seen in the field into the preconceived notions of the observer. One of the classical examples of the conflict between theoretical and field interpretations was the controversy concerning the origin of basalt, a controversy that raged from about 1775 to 1822.

Some of the hills of Saxony near the famous mining academy of Freiberg (Bergakademie) are composed chiefly of sedimentary rocks, but interstratified with them are a few layers of a hard, dark-colored rock long ago named basalt. The basalt is more resistant to erosion than the sedimentary rocks with which it is associated and appears in picturesque colonnaded cliffs at or near the summits of many of the hills.

In 1775, the Stolpen, one of the basalt-capped hills, was visited by Abraham Gottlob Werner, professor of mining and mineralogy at Freiberg, a scientist who was destined to wield great influence on the development of geology ("geognosy," as he called it). From his observations on this and later visits, Werner wrote, in 1787, that the hill showed "not a trace of volcanic action, nor the smallest proof of volcanic origin.

After further more matured research and consideration, I hold that no basalt is volcanic, but that all these rocks, as well as the other Primitive and Floetz * rocks, are of aqueous origin." Following up his idea that all basalts were precipitated from the ocean.

Werner developed a "system." With all the precision that he applied to the classification and organization of the mineral specimens in the laboratory collections at Freiberg, he proceeded to divide the crust of the earth into a series of "Universal Formations." These, he taught, were all precipitated from a primeval ocean, and could be definitely recognized all over the world, each formation having the same character and occurring in the same order no matter in what country it might be found.

Werner's Personal charm was great, and he attracted large numbers of able students whom he fired with great zeal. They came to believe that the "Universal System" of the Master would unlock the geologic history of every country.

Some of them were rudely awakened; in fact, Wernerism received its first really serious setback from two of Werner's own students, D'Aubuisson and von Buch. Both of these young men became disillusioned with Werner's "system" after visiting the Auvergne region in central France. The Auvergne has had no eruptions within historic times but it contains almost perfectly preserved craters, lava flows, and other volcanic phenomena.

The Auvergne had been studied several years before by Nicholas Desmarest (1725-1815), a hard-working government official who in his spare time published several excellent papers on geology. Among these, his treatises on the Auvergne volcanoes have won for him the title of "Father of Volcanology."

How different was the approach of this thorough observer from that of the dogmatic Werner to essentially the same problem! In his first journey into the Auvergne in 1763, Desmarest found a cliff of basalt. Searching at the base of the cliff, he noticed that the soil beneath the flow had apparently been burned and hardened.

He also noticed that the basalt grades into and contains masses of *scoria,* a coarsely frothy basaltic rock filled with small, spherical holes. Scoria is common at the base, and even more abundant in the upper part, of basalt flows; it has been observed to form when rising bubbles of steam are caught by the congealing of the sticky lava around them.

When he first visited the Auvergne, Desmarest had never seen moving lava from an active volcano. By careful observation and by reasoning, bowever, he established two of the criteria now generally used in the recognition of ancient basalt flows—the baking of the ground beneath the flow and the presence of scoria formed by the congealing of bubble-filled lava. Desmarest, however, did not proceed forthwith to establish a "system."

His curiosity was aroused, but he drew no certain conclusions. He decided to examine and map the outlines of the entire flow. This work disclosed similar features at many other points along the base of the flow. He also traced the flow to its source at the base of a round, steep-sided bill which still retains the characteristic form of a volcanic cone.

Still not entirely satisfied, Desmarest decided to plot on a map the distribution of all the different kinds of rocks of the Auvergne. By carefully following the junctions between the lavas and other kinds of rocks, and by plotting these junctions or *contacts* on maps, he proved that the volcanic history of the Auvergne was very long.

Some eruptions had been followed by long quiet periods during which streams cut valleys in the flows and removed much of the ash from the cones. Then, these newly cut valleys were filled and obliterated by renewed volcanic activity. Eventually he traced out three main cycles in the volcanic history. His map, one of the first geologic maps produced, is a monument to his good judgment and ability to interpret field relations. Desmarest's maps will stand comparison with modern maps of similar volcanic districts.

Thus, nearly 200 years ago, the failure of Werner's speculations to withstand the test of Desmarest's rigorous field observations showed geologists that the ultimate worth of a theory can only be found in the field. Ironically, although most of Desmarest's reports were published before Werner tried to compress all the vagaries of volcanic action, sedimentation, and erosion into a "Universal System," they remained almost unnoticed for many years. While Werner's teachings and ideas were sweeping over Europe, Desmarest took no part in the controversy. When asked for his opinion on the origin of basalt, he would reply:

Eventually D'Aubuisson and von Buch, eager to establish the "Universal System" of their teacher in other countries, visited the Auvergne. How great was their disillusionment as they followed, step by step, the same evidence and lines of reasoning that had led Desmarest to his conclusion! Here in the Auvergne, D'Aubuisson and von Buch saw the burned and hardened ground on which the lava had flowed, and the scoriaceous tops and bottoms of the flows where rising gas bubbles had been trapped in the congealing lava.

All these features could have been seen at Stolpen had Werner taken his students there on field trips instead of merely lecturing about the "Universal System" and giving them specimens of basalt to study in the laboratory. Even more devastating for the Wernerian but nearby they also rested on granite, which supposedly occupied quite a different position in the "Universal System."

The reports of D'Aubuisson and von Buch did much to overthrow the theory that basalt was a precipitate in the "Universal Ocean," although other students of the great teacher, still content to work in the laboratory instead of examining rocks in the field, continued to promulgate the Wernerian doctrines.

The controversy over the origin of basalt had happy results for the development of volcanic geology. Interest in ancient volcanoes spread wide and far and bore abundant fruit, particularly in the British Isles, an area uniquely rich in varied and spectacular volcanic phenomena, notwithstanding the fact that none of the British volcanoes has been active in historic time.

Plutonic Rocks

As the details of lava streams and deposits of volcanic ash became better understood, the British geologists turned their attention to the conduits through which the molten rock, or *magma,* as such natural silicate melts are called, had reached the surface. In the Midland Valley of Scotland and in parts of the Hebrides, erosion has swept away most of the lava streams and cones and has cut deeply into the foundation beneath.

Here, as in northwestern New Mexico and in central Oregon, erosion has displayed in clear and diagrammatic fashion the conduits that fed the volcanoes. Many of these feeders are simple, pipelike masses (*volcanic necks*) or filled fissures (*dikes*); but some connect with other subterranean igneous masses much more complex in form and origin.

Soon a wide variety of different kinds of igneous bodies formed by the solidification of magma underground had been mapped and described. These are called *intrusive igneous bodies* in contrast to the *extrusive bodies* formed by magma erupted to the surface. The rocks that have solidified within them are called *plutonic rocks*, in contrast to the *volcanic rocks* that have formed at the surface.

Observations of hundreds of different intrusive masses have shown that plutonic rocks are characterized by coarser mineral grains than comparable lava flows. This has led to the inference that the slower-cooling underground affords time for the mineral grains to grow larger. In large intrusions, some of which are many miles in diameter, the grains are easily visible, and the minerals can be readily identified without the microscope.

Laboratory investigations prove that heat is transmitted by solid rock less readily than it is carried off by rising air (convection currents) above a molten lava flow. Hence, the roof rock acts as a blanket which permits only slow cooling of a magma emplaced below the earth's surface.

Rounded gas bubbles are absent or at least very rare in plutonic rocks, apparently because of the considerable pressure of the roof which keeps the gas in solution in the magma. The gas collects into bubbles only when the pressure is released, as when magma is suddenly extruded from depth to a point upon or near the surface. Although some intrusive bodies can be traced continuously into recognizable lava flows so that their origin from a common magma is obvious, most are separate bodies. These may never have had any direct outlet to the earth's surface.

That these masses were likewise formed by crystallization from a magma is proved by several lines of evidence: They commonly harden, or even completely recrystallize, the rocks along their contacts; tongues and stringers from them penetrate into cracks in the adjacent rock in the manner of a liquid; they commonly show chilled borders of finetextured rock similar to that of lava flows; and they display, with minor exceptions, the same mineral and chemical compositions as surface flows. Most intrusive bodies cut across the bedding of the enclosing sedimentary rocks, hence they are called *discordant,* but a sill is a rock mass congealed from magma forced *concordantly* between sedimentary strata. The magma has spread out like the grease squirted between metal surfaces by a grease gun.

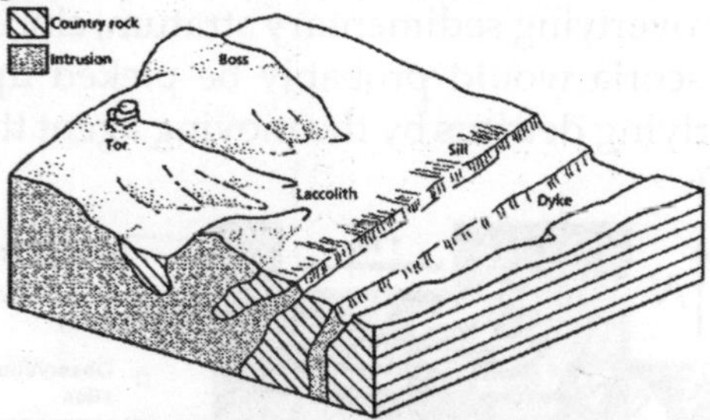

Fig. Discordant Intrusions.

A sill might thus occur low in the sequence and yet be much younger than strata hundreds of feet above it. Here, then, is a possible source of error in applying our generalization about the sequence of strata, for how can we tell an intrusive sill from a lava flow that has been buried tinder a later accumulation of sedimentary rocks?

CONTACTS OF ROCK MASSES

The key to this question, as to many others in geology, lies in the interpretation of the contacts between rock bodies. A mass of a single kind of rock, whether sandstone, basalt, or any other, (does not extend indefinitely; somewhere it must end against another. These common boundary surfaces of adjacent rock masses are called their *contacts*.

There are two general kinds: *abrupt*, with a definite surface of junction; and *graduational,* in which there is no sharp boundary but an intermediate zone of greater or less thickness in which the transition from one rock mass to the other takes place.

In interpreting the relative age of two rock masses in contact, the following generalizations are useful. They apply not only to problems of sequence among igneous rocks, but to all kinds of rock bodies:

Of two rock masses in contact, that one is younger which contains within it fragments or inclusions of the other.

Thus, a sill might be expected to enclose, at some places near its upper surface, fragments torn from the overlying stratum at the time of intrusion. On the other hand, fragments of a buried lava flow are likely to be found as inclusions in an immediately overlying sedimentary stratum, since loose pieces of lava and scoria would probably be picked up and mixed with the overlying detritus by the moving agent that deposited the sediment.

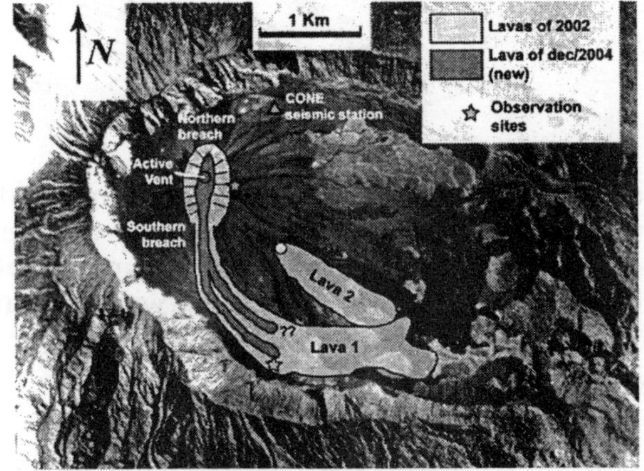

Fig. Buried Flow of lava

If a rock sends tongues and branches into another, it is younger than the rock it penetrates. The sedimentary strata above most sills, or for that matter the older rock in contact with any igneous intrusion, commonly contain tongues or (dikes of igneous rock formed by the magma penetrating fissures or other openings in the older rock and solidifying there.

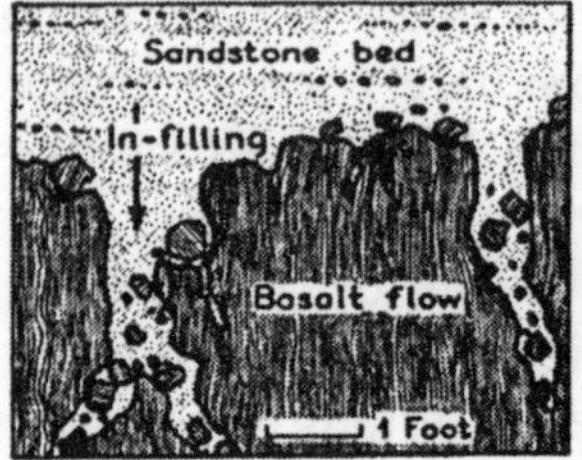

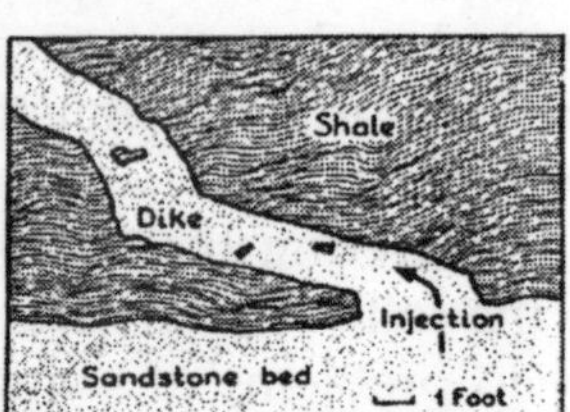

Fig. Sandstone Dikes

Although this rule applies mostly to igneous rocks, sediments may also, in places, penetrate adjacent rocks. Thus, cracks and openings in a lava flow may contain debris identical with that of an immediately overlying sedimentary rock. Presumably, some of the unconsolidated detritus filtered into the cracks during deposition of the sediment and cemented there into rock.

The generalization must be used with care, however, when applied to sedimentary rocks. For example, an underlying sandstone may send dikes upward into overlying shale. This is an exception to the rule as stated, though of course the rise of the sand into dikes actually occurred after the formation of the overlying shale.

What apparently happened was that underlying, unconsolidated sand was forced into cracks opened in overlying rock. Presumably cementation of the sandstone was delayed until after the shale had been laid down and consolidated.

If an igneous rock bakes or alters another rock with which it is in contact, it is younger than the rock it bakes. So stated, this is an obvious truism, but to recognize such contact alteration is not always easy. Some rock masses are exceptionally well cemented or impregnated with mineral matter along their contacts because of later percolating waters. Along other contacts, rocks have been discolored by water stains until they may look like chilled zones. Generally, careful observation or even microscopic study is needed to distinguish these spurious features from actual baking or chilling.

We have already seen how Desmarest noted the alteration of the rock beneath the Auvergne, lavas. A sill will alter not only the strata that form its floor, but also those of its roof. A moving lava flow has no roof, and lava congeals before any considerable sedimentary cover can be deposited upon it.

In general, the criterion of alteration a hot magma can be more readily applied to intrusive rocks than to lava flows. The interior of a lava flow commonly remains molten long after a thick rind of congealed lava has formed on its surface. Many flows have been seen to move forward beneath this rind; during movement the still-liquid interior breaks up the solidified crust and rolls blocks of it under the front of the flow.

Such blocks are commonly so cool that they fail to bake the material beneath—or at least baking is so slight that it is not readily recognized. But intrusive masses of igneous rock invariably alter their walls, at least to some degree.

THE INTRUSIVE NATURE OF GRANITE

As careful geologic mapping in many areas of volcanic and shallow plutonic rocks disclosed more and more about the contact relations of intrusive igneous rocks, a spirited discussion arose over the origin of granite, a common, coarse-grained rock composed chiefly of feldspar and quartz. The origin of granite is still a topic of lively discussion but many points have been clarified by careful study of granite contacts.

Geologic mapping in many different countries shows that granite and the similar rock, granodiorite, are among the most abundant rocks in the accessible part of the earth.

They are found in great monotonous bodies hundreds of square miles in extent. At many places, eroded granites are covered with sedimentary rocks that contain pebbles of the underlying granite and are therefore younger than the granite. Other bodies are covered by younger lavas and beds of volcanic ash. These relations persuaded some early geologists to regard granite as the earth's "original crust," formed when the earth solidified from a molten state, though, as we have seen, Werner considered it to be the first precipitate out of a universal ocean.

The idea that all granites are old, however, was not left unchallenged. In many localities, geologists began to note that granitic dikes, tongues, and complicated offshoots of considerable size penetrate the overlying rocks, indicating that the granite had injected the roof rocks as a molten mass. Only in scale of the injection features—in miles or thousands of feet rather than in feet or inches—were the contacts different from those of small intrusive masses. Furthermore, the invaded rocks at such contacts were altered, and, in places, so thoroughly recrystallized that it was difficult to tell whether they had originally been sedimentary or volcanic rocks.

Such changes in the wall rocks suggested the effect of high temperatures. More detailed field studies also showed that most of the great expanses of granite, far from being as monotonous as had been supposed, actually embrace several distinct intrusions of granite and related plutonic rocks, the younger ones penetrating the older. Also, along many contacts, fragments of the adjacent older rock are strewn through the margin of the granite in large numbers as though they had been pried off by the intruding granite magma.

Fig. Intrusive Granite

Today, no remnants of the hypothetical "original crust" have survived the results of detailed mapping. At some points along their contacts, most granite masses can be seen to have invaded and altered rocks of sedimentary or volcanic origin, although elsewhere younger sedimentary and volcanic rocks may have been deposited upon the eroded granite surface.

Thus, many granites were shown to be igneous rocks, congealed at considerable depth. Diked contacts and other injection phenomena also indicated that some masses were not single intrusions but had been formed by successive invasions separated by intervals of inaction. This recalls the Auvergne, where Desmarest demonstrated three distinct effusive periods separated by long quiet interludes.

Apparently the emplacement of large igneous bodies, whether at the surface or far beneath it, is a slow and somewhat intermittent process.

If we left the subject of the origin of granite here, however, we would be greatly oversimplifying a complex problem. Not all granite contacts are either clearly erosional or clearly intrusive. Many of them are gradational, the granite appearing to fade gradually into rocks of undoubted sedimentary or volcanic origin.

Also, certain rocks, definitely granites in composition and texture, contain faint, nebulous patterns that resemble stratification, outlines of pebbles, or other structures found in sedimentary or metamorphic rocks. The origin of these transitional contacts and of granites with ghostlike foreign structures is a fascinating problem that has not yet been completely solved.

Classification of Igneous Rocks

With the petrographic microscope, much information can be obtained about the minerals in an igneous rock, including their kinds, shapes, sizes, patterns of arrangement, order of crystallization, and the alterations brought about in them either by hot gases from the magma or by outside agencies such as air and atmospheric moisture. Great variation in so many different features yields infinite possibilities for classification, and so literally hundreds of kinds of igneous rocks have been discriminated and given separate names.

However, the great bulk (approximately 95 per cent) of the igneous rocks can be lumped into about 15 major groups. This simplified list suffices for our purpose; we shall not need names for the numerous varieties within each group, or for the rare groups not included. Such a simplified classification of the igneous rocks can be based on the minerals and textures visible to the eye, thus dispensing completely with the microscope.

Comparison of mineral composition and the bulk chemical composition of thousands of igneous rocks shows that the kinds and amounts of different minerals depend, in general, on the chemical composition of the original magma. Magmas rich in silica yield, on cooling, large amounts of feldspar and quartz; magma low in silica form rocks rich in the ferromagnesian minerals such as pyroxene, olivine, and amphibole.

Textures of Igneous Rocks

The size of the mineral grains in an igneous rock depends chiefly on the rate of cooling, although bulk chemical composition of the magma plays a part. It has been inferred from field observations, and confirmed by laboratory experiment, that a high content of water and other volatile substances promotes the growth of larger crystals. Field observations clearly indicate that when magma congeals slowly in a large subterranean mass, the minerals commonly grow large enough to be readily identified with the eye.

If the magma forms a lava flow, however, rapid cooling in contact with the air prevents the growth of large crystals, and, instead, the rock congeals to mixtures of microscopic mineral grains and glass. If cooling is even more rapid the magma may congeal into a glass containing almost no crystals.

These differences in degree of crystallinity and in size of crystals determine the texture of the igneous rock. A rock is said to have a granular texture if its mineral grains have grown to a size large enough to be seen and identified without the aid of lens or microscope.

In different rocks, the average size of the grains may vary from about 0.5 mm. to more than 1 cm. in diameter, but the common coarse granular rocks such as granite have grains averaging from about 3 mm. to 5 mm. in size. In rocks with aphanitic texture, the constituent minerals are less than 0.5 mm. in diameter; they are commonly mere specks, too small to identify without the microscope. In many aphanitic rocks, the microscope shows a considerable residue of uncrystallized glass between the minute crystals, but others are completely crystalline. Rocks with glassy texture are almost entirely volcanic glass, though even they may contain a few scattered crystals.

Pyroclastic texture is the texture of rocks formed of volcanic ash and rock fragments blown out of a volcano and cemented together in the manner of sedimentary rocks, or else welded together by heat from the chemical reactions that take place between the gases in the eruption cloud.

Many igneous rocks contain crystals of two widely different sizes and therefore appear spotted; these rocks are said to have a porphyritic texture and are often loosely called "porphyries". Because porphyritic texture is most common in small intrusive bodies or in lavas, it has been attributed to a change in the rate of cooling while the magma was crystallizing.

The process is hypothetically explained as follows: A large body of magma underground may cool to the temperature where one or more minerals begin to crystallize. Because cooling is slow, the crystals of these minerals grow to considerable size. If, when the magma is perhaps half crystallized, a fissure opens in the roof of the chamber, some of the magma with its suspended crystals may escape to form a lava flow at the surface.

The still-liquid portion of the magma quickly freezes and surrounds the large crystals, or phenocrysts, with a groundmass of crystals of aphanitic size. The phenocrysts were formed underground, the aphanitic groundmass at the surface. Such a lava has a porphyritic aphanitic texture.

Rocks with porphyritic granular texture—large crystals in a granular groundmass of finer grain—are common in some intrusive bodies. Porphyritic glassy texture is seen in some lava flows. Rarely, conditions other than a change in the rate of cooling may locally produce porphyritic rocks. With the five basic textural distinctions—granular, aphanitic, glassy, porphyritic, and pyroclastic—we can build a table of the major groups of igneous rock.

Metamorphic Rocks

We have seen that the sedimentary origin of many rocks can be confidently inferred from their stratification, water-worn pebbles, fossils, or other characteristic features. Other rocks display textures and mineral compositions that, by analogy with those of rock masses of known volcanic or plutonic origin, permit us to classify them just as confidently as igneous rocks. There remains a third great group of rocks in which diagnostic features of igneous or sedimentary origin either are absent or have been so obscured and altered by the growth of new minerals and new textures that they are scarcely recognizable.

Thus, in places, we find rocks showing rounded pebbles and stratification much like that of normal conglomerates, but the pebbles are ellipsoidal instead of spherical; the matrix between them consists not of sand and clay as in ordinary conglomerate, but of lustrous foils of mica and quartz; and, cutting across two or more adjacent pebbles and the intervening matrix, are long delicate needles of tourmaline.

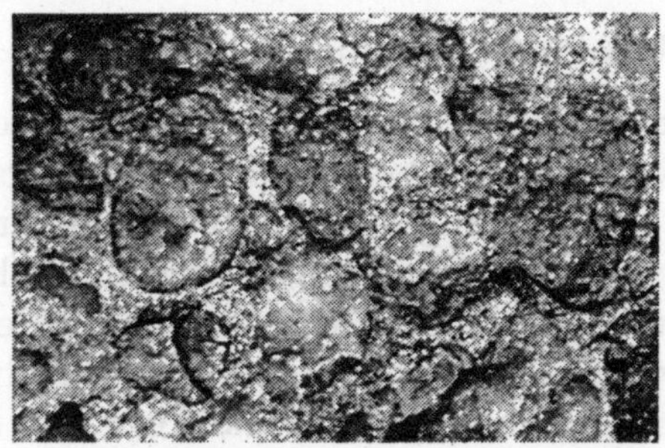

Fig. Metamorphosed Conglomerate with Tourmaline Needles Cutting the Other Constituents.

It is inferred that the rock was once a conglomerate and that the tourmaline could not have been present when the conglomerate was deposited but must have grown in the solid rock long after deposition.

Apparently the original round pebbles have been squeezed and stretched, and the sand and clay originally present in the conglomerate matrix have been altered to mica and quartz. Similarly, igneous features such as phenocrysts or flow banding may locally be recognized in their metamorphosed equivalents.

Still other rocks show no trace whatever of the textures of known sedimentary and igneous rocks. Yet, in places, they grade imperceptibly into rocks in which partially obliterated igneous or sedimentary structures still survive.

From such relations, the conclusion that this last great

group of rocks represent altered or *metamorphosed* rocks is well established. We therefore recognize this third group as *metamorphic rocks.* There can, from the nature of their origin, be no sharp boundary between metamorphic rocks on the one hand and the igneous or sedimentary rocks from which they were derived on the other.

The survival of relic features such as pebbles or phenocrysts in metamorphic rocks must mean that the original rock did not melt in attaining the metamorphic condition. Instead, the alteration appears to have been accomplished mainly by recrystallization while the rock remained essentially solid.

Foliated Metamorphic Rocks

Most metamorphic rocks have a banded or layered structure called *foliation.* The layers may be coarse bands 1/24 inch or more thick, as in gneiss or layers thinner than a sheet of paper, as in slate. The rock usually splits readily along these planes. The foliation appears to record widespread pervasive movement within the rock mass, during which most of the original minerals were broken, streaked out, and recrystallized into new minerals. Because movement is essential to their development, the foliated rocks are called *dynamic metamorphic* rocks.

THE ORIGIN OF SLATE

When geology was emerging as a science, the coarse-grained, faintly foliated rocks were usually called gneisses and classified with the granites, and the fine-grained, well-foliated ones with the sedimentary rocks. For example, the metamorphic foliation in common slate, a rock much used for roofing, flagstones, and other building purposes, was erroneously considered to be stratification.

Geologic study of European and American slate quarries disclosed, however, that most slates have two distinct layered structures. One can be seen to cross and displace the other.

The older of these is parallel with other interlayered rocks like limestone or quartzite, and with alternations in grain size, colour, and composition; it is the true sedimentary stratification. The rock, however, does not break or split along the stratification planes as in normal sedimentary rocks except when the younger structure, the true metamorphic foliation, happens to coincide with it. When foliation and stratification do coincide, recognizable but distorted fossils are found locally on the foliation surfaces.

Fig. The Fossil Collected from Unaltered Limestone in Idaho

Where these fossils can be compared with like specimens from otherwise little-changed sedimentary rocks, they are seen to have been greatly thinned in the direction at right angles to the foliation and are stretched out parallel to the foliation.

If the foliation cuts the stratification at a high angle, however, it is almost useless to look for fossils, because the rock will rarely split along the stratification planes where the fossils occur.

Furthermore, each foliation plane, where it crosses the stratification, commonly breaks and offsets the stratification planes by a small (usually microscopic) distance, showing that it is a structure younger than the bedding. Thus, a fossil may be broken and displaced by so many microscopic shears as to be unrecognizable.

In most slate quarries the foliation is nearly parallel throughout, although the strata themselves may be thrown into folds and become highly distorted. The foliation cuts through the folds, intersecting the strata at all angles.

From the relics of stratification and the occasional battered fossils, it is clear that most slates were derived from fine-grained sedimentary rocks, such as shale. This is confirmed by chemical analyses which show no essential difference between shale and most slates, except that the slates contain less chemically combined water than the shales.

Field and laboratory studies show, however, that not every slate is derived from shale; some were formed from other fine-grained rocks such as volcanic tuff. Only by the use of the petrographic microscope and by X-rays can we unravel the true nature of the foliation. The minerals of slate are exceedingly fine grained—indeed, most are ultramicroscopic.

Fig. Fragments of Slate

X-ray studies show, however, that slate is made up largely of tiny flakes of minerals akin to muscovite, the white mica. These flakes are nearly parallel. Since mica has only one perfect cleavage which is parallel to the platy form of each grain, the cleavage of all the grains in slate lies nearly parallel.

Slate splits so readily into thin parallel sheets because of this cleavage parallelism. Evidently slate is merely shale that has been sheared and heated sufficiently to cause the original clay minerals composing it to recrystallize into tiny micaceous flakes. Movement along the innumerable surfaces of foliation offsets of the bedding, has granulated the original minerals and has lined up the newly formed mica, which continued growing during and after movement.

In some areas, slate may be seen to pass by imperceptible gradation into more coarsely crystalline rocks such as mica schist. In these more highly metamorphic rocks, every trace of the original stratification and of fossils has been obliterated. Yet, in chemical composition they are identical to the slate into which they grade. Although they may contain minerals different from those in slate, the new minerals in the transitional rocks can be seen to have grown from the micaceous minerals of the slate, gradually replacing and obliterating them just as the micas in slate replaced the minerals of the shale.

From such mineral relations and transitions in rock type, we conclude that even these rocks, showing no trace of original sedimentary structures, were, in fact, derived from the sedimentary rock shale. They are merely more thoroughly metamorphosed.

Recrystallization at Igneous

One of the simplest kinds of metamorphism is the recrystallization of rocks by an igneous intrusion. In many places where, for example, pure quartz sandstone containing marine shells has been invaded by a mass of gabbro or some other plutonic rock, shells cannot be found in the rock close to the contact. Instead, the sandstone near the intrusion contains sheaves of a lightcolored mineral called wollastonite.

The wollastonite formed by the reaction has no resemblance to either the quartz grains or the shells from which it was derived. Most fossiliferous sandstones, however, contain many substances besides quartz—usually a little clay, some limonite or ferromagnesian minerals, and various other impurities. These generally join in the reaction, yielding, not a simple mineral like wollastonite, but a more complicated compound, usually a member of an isomorphous series such as amphibole, garnet, or pyroxene.

Where wollastonite actually does appear, however, we can make certain additional inferences about the conditions of metamorphism because the wollastonite reaction has been carefully investigated over a wide range of temperatures and pressures in the laboratory. The results of this investigation are plotted.

From this graph, we infer that if a rock containing wollastonite was formed at depths equivalent to a pressure of 3,000 atmospheres, the temperature attained by the metamorphosed rock was at least 850° C. Locally, *contact metamorphism*, as the kind of metamorphism caused by the heat of igneous masses is called, is greatly complicated by the addition of various substances from the igneous mass.

That such additions have taken place is demonstrated by the finding of rocks of different mineral and chemical composition along the contacts—for example, calcium-iron silicates taking the place of calcite in limestone near a granite contact. These different minerals are attributed to hot gases and aqueous solutions escaping from the magma and carrying large amounts of iron, silica, sulfur, and other substances into the wall rock. Here they combined with the minerals already present to produce new minerals.

Recrystallization of the Stassfurt Salts

Deep wells show that the temperature in the outer crust of the earth increases, on the average, about 1° Centigrade for every 100 feet of depth. Therefore, if the rocks are buried to great depths, some of their constituent minerals may become unstable at the higher temperatures.

An interesting example of such *geothermal metamorphism* is afforded by the well-known salt deposits at Stassfurt, Germany. These salt deposits were studied in 1900-1905 by the famous Dutch chemist van't Hoff. Because of the associated sedimentary rocks, he reasoned that the salts had been formed from a shallow body of sea water, cut off from the ocean by a bar or other obstruction, and slowly evaporated in a desert climate.

Van't Hoff was puzzled by one feature of the salts, however. Although their bulk chemical composition is consistent with their being the residue left from the evaporation of sea water, the actual minerals of the deposits are quite different from those obtained by evaporating sea water to dryness.

Patiently and thoroughly, van't Hoff worked out the physical chemistry of the salts by many laboratory experiments, and, ultimately, produced many of the minerals found in the Stassfurt salts by holding sea water at temperatures well above normal surface temperatures but below boiling during the period of evaporation. Van't Hoff then announced that the average annual temperature of the earth's crust at the time when the salts were laid down must have been much greater than at present—perhaps close to the temperature of boiling water!

Geologists, though accepting van't Hoff's experimental data, could not agree with the conclusion. Fossil fish and mollusks are found in strata associated with the Stassfurt salts. It is inconceivable that the marine animals of that ancient time had lived in hot seas. From the relations shown on a detailed geologic map of the Stassfurt area, it is obvious that, after the Stassfurt salts had been laid down in an arm of an ancient sea, they were buried under several thousand feet of younger sedimentary rocks.

The rise in temperature because of this deep burial could have caused the already precipitated salts to recombine into new minerals stable in the hotter environment. The textures of the salt, as seen in thin sections under the petrographic microscope, give abundant evidence of postdepositional recrystallization of the solid salt beds. Thus, van't Hoff's conclusion as to the temperature of mineral formation was correct; his inferences as to the climate were absurd because he neglected the field evidence.

Salt beds are particularly susceptible to metamorphic changes; they recrystallize completely at temperatures that are still too low to cause notable changes in the silicate minerals of which most rocks are composed. In sandstones associated with the Stassfurt salts fossil shells are completely unaltered. Therefore, the rise in temperature was clearly not enough to cause calcite to combine with quartz and form wollastonite. It would have required much deeper burial to metamorphose the sandstones and shales at Stassfurt; only the sensitive salt beds were affected.

METAMORPHIC PROCESSES

Even these brief descriptions—and scores of additional examples could be given—show that many minerals are chemically stable only within a limited range of pressure and temperature. If brought into a part of the earth's crust where the temperature is higher, or where crushing and shearing take place, or where hot fluids attack them, many minerals break down and form other minerals stable in the new environment. Metamorphic rocks result from such transformations.

In some, the alteration affects only a few minerals of the original sedimentary or igneous rock, and the metamorphic rock re tains in the surviving unchanged minerals or textures clear evidence of its origin; but more commonly, complete transformation has occurred. From relations revealed by geologic maps, we know that nearly all metamorphism takes place deep within the crust of the earth—far below the depths we can reach in mines and wells. Therefore, in general, we must infer the reactions that take place from the resultant products, although we can also get valuable clues from experiments made possible by laboratory equipment designed to imitate conditions deep within the earth.

Unfortunately for our purpose, knowledge of the chemistry of solids, and particularly of solids under shearing stress, is still meager as compared with the vast amount of information chemists have obtained on reactions in solutions.

SIGNIFICANCE OF FOLIATED METAMORPHIC ROCKS

Where large areas of metamorphic rocks-particularly the foliated rocks called crystalline schists—are found at the surface, deep erosion has taken place. Such a generalization is based on observed relations of metamorphic and nonmetamorphic rocks in many parts of the world. We infer from these relations that the textures and minerals characteristic of foliated metamorphic rocks were formed at great depths, under conditions of high pressure and temperature.

Thus, we also interpret other rocks of like mineral composition and texture as having been under like conditions—even where the local structural relations do not permit direct proof of deep burial. We reason that a thick covering must have been eroded from any area where crystalline schists are now seen at the surface.

CLASSIFICATION OF METAMORPHIC ROCKS

Since any rock may be recrystallized by metamorphism and there are several different ways in which metamorphism may take place, the number of different kinds of metamorphic rocks is large. It is possible, for example, to find at least five different metamorphic rocks all with the chemical composition of basalt—and indeed, all derived from basalt—yet each differing from the others in texture, mineral composition, and general appearance.

A greater variety of minerals appears in metamorphic rocks than in either sedimentary or igneous rocks. Therefore, metamorphism is best studied with the petrographic microscope, aided by the principles of physical chemistry and, of course, primarily by many field observations.

Metamorphic rocks, like sedimentary and igneous rocks, are classified on the basis of texture and composition. The principal textures to learn for determinations made without the aid of the microscope are:Gneissose Texture. Coarsely foliated; individual folia 1 mm. or more, even up to several centimeters thick.

The separate folia are commonly lenslike or wavy, and generally differ in composition; feldspars, for example, commonly alternate with dark minerals. Mineral grains are coarse, easily identified.Schistose Texture. Finely foliated, forming thin parallel bands or flat lenses. Individual mineral grains are distinctly visible. The minerals are mainly platy or rodlike—chiefly mica, chlorite, and amphibole. Equidimensional minerals like feldspar and pyroxene are not abundant.Slaty Texture.

Very fine foliation, producing almost rigidly parallel planes of splitting due to the parallelism of microscopic and ultramicroscopic grains of platy minerals such as mica.Granoblastic Texture. Unfoliated, or only faintly foliated. Mineral grains are large enough to be visible. Corresponds roughly to the granular texture of igneous rocks.Hornfelsic Texture. Unfoliated. Mineral grains commonly microscopic to ultramicroscopic, though a few may be visible. Corresponds to the aphanitic texture of lavas, from which it can be distinguished only with difficulty.

SECOND-RATE PLANET

AMONG the nine planets that speed endlessly around the sun is this terrestrial sphere. It is neither the largest nor the most massive, and astronomers are apt to remind us not only that it is a second-rate planet but that our sun itself is only a third-rater among the other stars. But because the earth is our abode, to us it is of supreme importance.

Fig. Terrestial Sphere

One of the first things we learn about the earth is that it is round. Some of us may even remember that a simple proof is that large objects gradually disappear below the horizon as one moves away from them. This is at best only a qualitative argument, but there are several precise proofs. As an example of the scientists' method of attacking such problems, let us follow through the steps involved in determining both the shape and size of the earth.

A little study of a circle will show that the property of circularity involves certain relations between the length of an arc of the circle and the angle made by the radii limiting this arc. This arc is so related to the angle a that if we choose another are equal in length to AB, such as CD, then the angle b must be of the same size as angle a. In the same manner, an arc twice the length of AB will be subtended by an angle twice the size of a.

We may now argue that if every section through the earth is a circle, then equal distances on its circumference will be subtended by equal angles made by the radii from the center of the earth to the ends of the arcs. In order to measure these internal angles, we must make use of astronomy.

Let us, therefore, imagine two observatories about 2,100 mi. apart along a northsouth line. In each observatory we shall suspend a plumb line, which is nothing more than a string with a weight on the end. This line will point to the center of the earth if the earth is round.

Now at the first observatory a telescope is pointed at a star directly overhead. At the same instant, the telescope in the second observatory is pointed at the same star; but because the second observatory is some 2,100 mi. from the first, the star obviously will no longer be directly overhead. Thus there will be an angle between the plumb line and the direction of the telescope.

This angle is measured, and it is found to be about 30°. We now make a diagram like the adjacent one. At A is the first observatory, with the plumb line and the telescope pointed directly overhead at the star.

At B is the second observatory, with its telescope also pointed at the same star, but with an angle between the telescope and the plumb line. Note, however, that the two telescopes themselves are parallel, because the distance to the star is so great that the rays of light reaching the earth from it are practically parallel at any point on the earth.

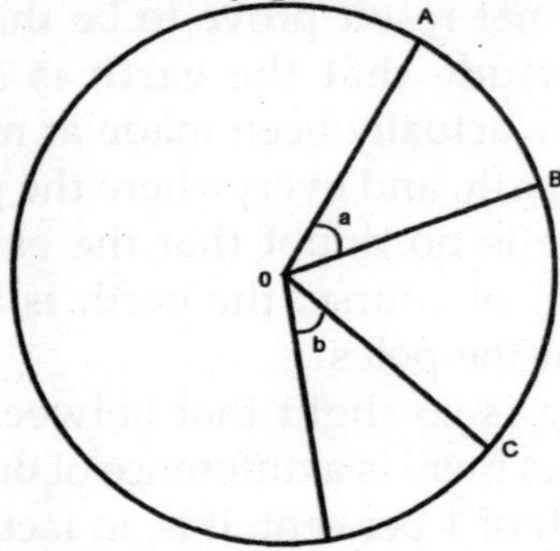

Fig. Angle a is the Same as Angle *b*, Provided Arc AB = Arc CD

Remembering that the angle between the telescope and the plumb line of the second observatory is 30°, we next project the plumb lines into the interior of the earth, where they meet. The angle made by their junction is also 30°, because the lines AC and BD are parallel, and the line OB cutting these parallel lines makes an equal angle with each.

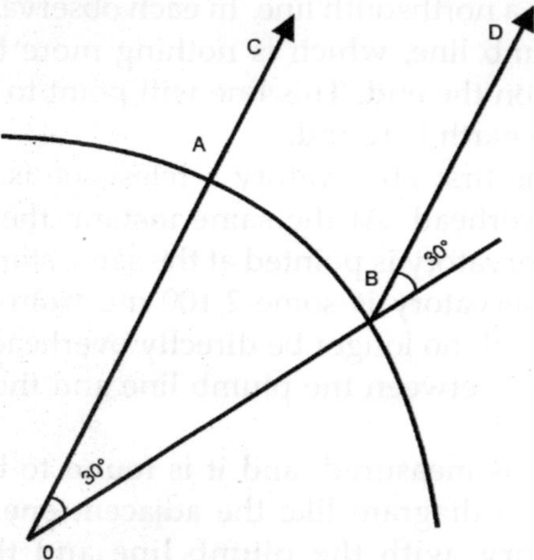

Fig. Two Observatories 2,100 mi. Apart Simultaneously Observe the Same Star

Thus, it follows that an angle of 30° at the earth's center subtends a distance of about 2,100 mi. along the surface. If we next divide through by 30, we see that an angle of 1° must subtend a distance of some 70 mi.

Should the same experiment be repeated elsewhere on the earth, and the net result prove to be the same, then we might safely conclude that the earth is spherical. Such measurements have actually been made at many places over the surface of the earth, and everywhere the general relations hold, so that there is no doubt that the earth is spherical. Precisely speaking, of course, the earth is a spheroid, and flattened slightly at the poles.

This flattening is so slight that between the polar and equatorial diameters there is a difference of only 27 mi., which is less than one-half of 1 per cent. It is, in fact, so insignificant that it amounts to only about ½ in. on a 10-ft. globe, and consequently is not shown even on the largest of such representations of the earth.

We Have by these experiments found not only that the earth is spherical but also that every degree of the interior angle subtends about 70 mi. on the surface. Since there are 360° in a complete circle, we need only multiply our 70 mi. by 360 to obtain the *circumference*, or distance around the earth.

When this is done, the result is approximately 25,000 mi. Once we have the value of the earth's circumference, we can easily find its *diameter*, or the distance through the earth. We simply divide the circumference by 3.1416 (the value of the old familiar *pi* of school days) and the answer is about 8,000 mi., which is the approximate diameter of the earth.

As long as we are on the subject of measurements, let us pursue it a bit farther. If we know the diameter of the earth, we can find its *volume*, because the volume of any ball is 3.1416/6 times the diameter cubed. If we put in our known values, we find that the volume of the earth is nearly 260,000,000,000 cu. mi.

Similarly, since the surface of a sphere is equal to *pi* times the diameter squared, we may compute the *surface area* of the earth. Such a computation shows us that there are about 197,000,000 sq. mi. on its broad surface. But large as this area may seem, it is only about one twelve-thousandth the area of the sun's great surface. In other words, the earth's surface area is as much smaller than the sun's as tiny Switzerland is smaller than the earth's entire extent of land and sea.

Another important item concerns the *mean density* of the earth, or its mass per unit of volume. To find this, it is necessary first to determine the mass of the entire earth and then to divide this by its volume. The mass of the earth can be determined by several methods, but perhaps the simplest is that known as the Cavendisk experiment. * By this procedure the attraction of two large lead balls for two small silver balls is measured, and then compared to the attraction between the earth and one of the smaller balls.

From the exceedingly delicate measurements made in this experiment it is found that the mass of the earth is 6×10^{21} tons, which is the number 6 followed by 21 ciphers. To obtain the mean density of the earth, then, we divide the mass by its volume; whereupon the density is found to be approximately 5.5. in other words, a cubic foot of average earth-substance weighs about 5.5 times as much as a cubic foot of water. But right here we stumble onto an interesting thing.

The average density of the familiar rocks of the earth's surface is only about 2.7. Thus the density of the earth's interior must be greater than 5.5, in order to counterbalance the lightness of the surface materials.

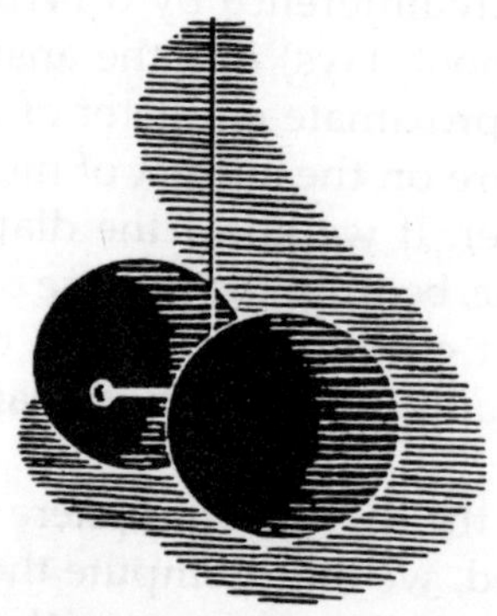

Fig. The Cavendish Experiment

For the purposes of later discussions we also ought to have before us a fairly complete picture of the earth's motions in space. Most proofs of the *earth's rotation* are difficult to grasp without some knowledge of physics.

The one usually cited depends on the fact that a heavy pendulum tends to swing in a fixed direction in space as long as no outside forces disturb it.

Evidently, then, if such a pendulum is set into motion on a rotating earth, the earth should rotate beneath the pendulum without affecting the direction of swing of the pendulum. Such, in fact, is the result of the pendulum experiment. When first set into motion, it is directed so that the bob swings in a north-south plane.

As time goes on, it is found that the pendulum continually departs from this direction, and that in the Northern Hemisphere the plane of swing tends to rotate in a northeast-southwest direction. It must be borne in mind, however, that an observer on the earth has the same motion as the earth, and therefore he will feel himself at rest and will notice that the direction of swing of the pendulum appears to change. This departure would increase indefinitely were it not for friction stopping the swing of the pendulum.

This experiment was first performed by Foucault in Paris, in 1851, and in his honor the device is called the Foucault pendulum. The theory of the Foucault pendulum is entirely beyond the scope but it is interesting to note that observation and theory agree so well that there is no doubt of the actual rotation of the earth upon its axis.

The motion of the earth about the sun, or the *revolution* of the earth, also can be demonstrated in several ways. One of the simplest is by the periodic shift in the position of the nearby stars with respect to the more remote stars. In the adjoining figure the earth's orbit around the sun, S, is shown diagrammatically.

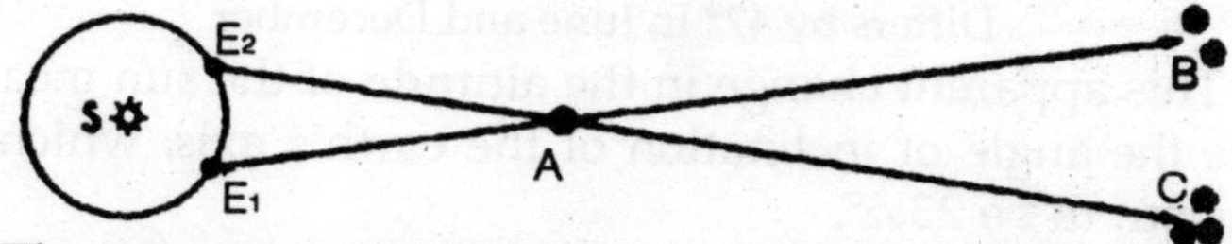

The apparent shift of the nearby star proves the earth's revolution. The sizes and distances are obviously not true to scale self appears in two successive positions, E_1 and E_2, several months apart in its journey around the sun. Star A is one of the nearer stars; and in the distance are two groups of stars, B and C, representing more remote constellations.

Notice that when the earth is at E_1, the star A, as viewed from the earth, appears to be in group B, but that later, when the earth is at E_2, star A appears to have moved to a position among group C.

Thus star A periodically appears to shift back and forth between groups B and C. Such shifts actually are noted among the nearer stars, and the phenomenon is cited as the parallax proof of the earth's revolution.

The *inclination of the earth's axis* to the plane of its orbit is another feature of this second-rate planet into which we should briefly inquire. This inclination is demonstrated by the fact that at any point on the earth's surface the altitude of the sun above the horizon at noon on June 21 and December 21 differs by 47°.

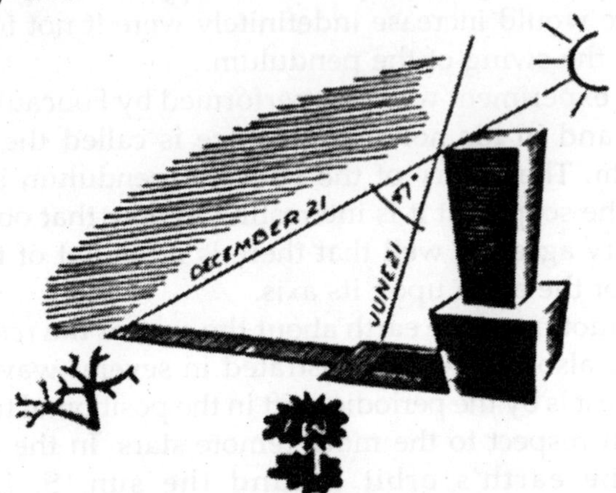

Fig. The angle of the Sun's Rays Differs by 47° in June and December

This apparent change in the altitude of the sun measures twice the angle of inclination of the earth's axis, which thus turns out to be 23½°.

In the adjacent figure the line OS, from the center of the earth to the center of the sun, lies in the plane of the orbit. Now at O let us erect a perpendicular OA, and then draw the earth's axis, OP at an angle of 23½° to OA. This angle, POA, measures the inclination of the earth's axis.

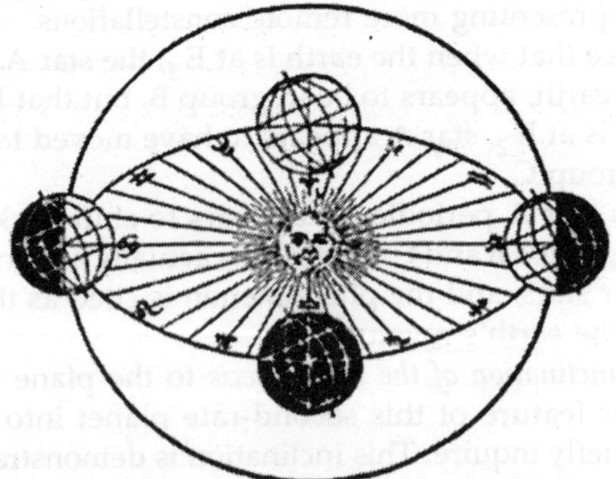

Fig. The Earth in its journey Around the Sun.

Inasmuch as the Equator of the earth is at right angles to its axis, it follows that the equator is inclined 28½° to the plane of the earth's orbit. This is line EOE2. It is the inclination of the earth's axis that causes the seasonal changes on the earth. The earth, like any rotating body, tends to keep its axis always oriented in the same direction in space, regardless of the earth's position in its orbit.

Notice, then, that when the earth is at the right in the diagram, the region receiving perpendicular rays from the sun lies 23½° *north* of the Equator; whereas when the earth is at the left, the perpendicular rays strike the surface 23½° *south* of the Equator.

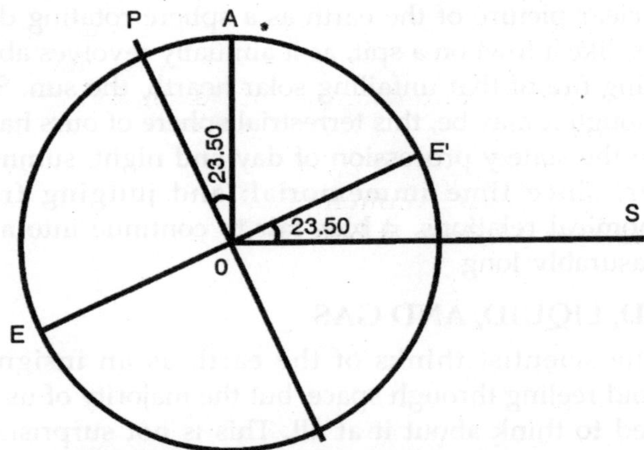

Fig. Diagram of the Inclination of the Earth's Axis

In the course of a year, then, the sun seems to swing through an angle of twice 23½°, or 47°, in its apparent motion above and below the Equator. When the earth is at the right, it is summer in the Northern Hemisphere, because at that time the northern half of the earth is receiving the more nearly perpendicular rays from the sun. In like manner, summer in the Southern Hemisphere and winter north of the Equator occur when the earth is on the left.

The several points which have been discussed furnish us with the modicum of astronomical data that we need as a background for topics to be presented later.

Item	*Value*
Shape	Very nearly spherical
Average circumference	24,800 mi.
Average diameter	7,910 mi.
Volume	2,599 × 108 cu. mi.
Surface area	197 × 106 sq. mi.
Mass (weight)	6,1 × 1021 tons
Average density	5.5 times that of water
Period of rotation	24 hr.
Period of revolution	3651/4 days
Inclination of axis	231/2°

By this time the reader should have in his mind's eye a fairly clear picture of the earth as a sphere rotating daily on its axis, like a fowl on a spit, as it annually revolves about the warming fire of that unfailing solar hearth, the sun. Second-rate though it may be, this terrestrial sphere of ours has taken part in the stately procession of day and night, summer and winter, since time immemorial; and judging from its astronomical relations, it bids fair to continue into a future immeasurably long.

SOLID, LIQUID, AND GAS

The scientist thinks of the earth as an insignificant spheroid reeling through space, but the majority of us are not inclined to think about it at all. This is not surprising. The earth usually is taken for granted as the stable part of our environment. Only during cataclysmic disturbances such as earthquakes, volcanic eruptions, or catastrophic floods are the unthinking ever reminded that this terrestrial sphere is indeed a mobile earth; is, in fact, an unimportant cosmic speck swayed by a myriad of external forces, wracked by a legion of internal stresses.

We all suffer from a provincial attitude, partly because of the limited part of the earth that we inhabit. Perhaps 90 per cent of the earth's population lives in a narrow zone between sea level and.... thinks of the earth as an insignificant spheroid reeling through space.... about 1,000 ft. above the sea. This is an extremely thin zone, when contrasted with the 4,000 mi. of the earth's radius. Thus it may be well to pause here a while and attempt to acquire a more cosmopolitan view of the world as a whole.

IT IS difficult for us to visualize the earth in its entirety. Even extensive travel on its surface does not help. It is only through the study of maps or globes that its larger features may be understood at a glance. Let us, therefore, take an ordinary geographic globe—say 12 in. in diameter—and look at it from a distance of about 2½ ft. We will then be seeing the earth, on a proper scale, as if from the porthole of some superstratospheric rocket ship distant about 20,000 mi. from this terrestrial sphere.

Let's not be disappointed at the smoothness of the globe; the world's highest mountains themselves are not very impressive from such a distance. In fact, on the scale involved in a 12-in. globe, the highest points above sea level, some 6 mi., would be about 1/100 of an inch high. Thus, Everest itself is scarcely a bleb in the thin film of varnish that covers our little man-made geographic globe.

As we watch the earth for 25 hr., it turns completely around, and we readily observe that most of the surface is covered with water. The continents occupy only about one-fourth of the earth's surface, or only 54 million of the globe's 197 million square miles. One of the features of the earth that we cannot see from our vantage point is its envelope of air. Shifting clouds, however, would suggest the presence of an atmosphere and blot out some of the surface features. Hence these armchair observations have some advantages over those made from a rocket ship.

But coming down to earth, and landing safely, we may recognize three states of matter comprising the globe as a whole. The main solid body of the earth is composed of rocks and soil, and is called the lithosphere. Partly surrounding this solid body is the hydrosphere, composed largely of the oceans, which keep some three-fourths of the solid earth from contact with the atmosphere, the layer of gas that surrounds this entire terrestrial sphere.

The lithosphere itself is known only to a very shallow depth. No wells or mines penetrate much more than 2 mi. below the surface, nor were any of the rocks now exposed on the earth ever more than a relatively few miles below the surface. In fact, on the scale of our geographic globe we have direct knowledge of scarcely more than the layer of varnish on the surface.

Although we are inclined to think of the oceans as being very deep, and the atmosphere around us as extending for considerable distances upward, they are mere films wrapped around the solid body of the earth. The average depth of the oceans is about 2 mi.; but this is an insignificant part of 4,000 mi., the earth's radius.

In similar fashion, even if the true atmosphere extends out for as much as 150 mi. from the earth's surface, nevertheless this value is only about 4 per cent of the earth's radius. Thus we may safely speak of the air and water *films* about the earth, because on a world-scale they are little more.

But insignificant as these features might be to an astral being, they are quite impressive to us. Viewed from our limited zone on the earth, we justly admire mountain-climbers or stratosphereflyers who reach dizzy heights, or those intrepid explorers of the oceans who descend a half mile or more into the deeps. We shall accordingly return to our customary restricted environment and look at some of the relief features of the earth in more detail. We should, however, at least keep the enlarged viewpoint in the back of our minds for later reference.

The major features of the lithosphere are the continents and ocean basins. If we include the shallowly submerged continental shelves with the continents, then we may say that about one-third of the earth's surface is occupied by the continental elevations and the remainder by the ocean basins. An idea of the differences in altitude between the ocean bottoms and the land surfaces may be had from the fact that the greatest oceanic depths are over 30,000 ft., or about 6 mi., and the highest mountains are nearly that high above sea level.

Thus there is a maximum relief of over 60,000 ft., or

roughly 12 mi., on the lithosphere. Viewed provincially, these are large figures; but in the light of our new cosmopolitan knowledge, we readily understand that the highest mountain and the deepest abyss are truly insignificant irregularities on the relatively smooth surface of this terrestrial sphere.

To help us visualize the distribution of irregularities on the earth's surface, let us consider the adjoining chart, called a *hypso- graph*. The vertical scale expresses elevations above or below sea level, and the horizontal line represents the total area of the earth.

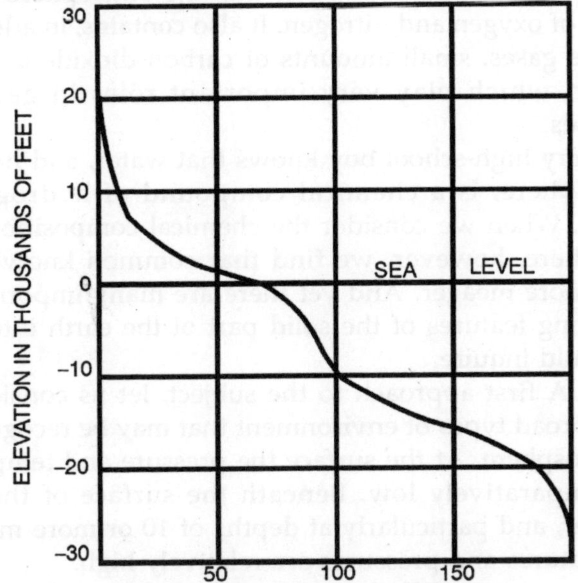

Fig. Hysograph Showing the Distribution of Relief on the Earth's Surface

The curve itself traces the distribution of relief on the globe. Notice how relatively little area is occupied by either the high mountains or the greatest oceanic depths. By far the greatest portion of the continents is within a few thousand feet of sea level; and most of the wide expanse of ocean bottom is between 10,000 and 20,000 ft. below sea level.

Land features on the continents themselves may be divided into mountains, plateaus, and plains. Mountains are high lands of small summit area, prominent above their surroundings. Plateaus are high lands with considerable summit areas, rising abruptly above their surroundings on at least one side. Plains are low lands with little relief. In addition there are the smaller relief features, such as hills and valleys, familiar to all.

By and large, the major physical features of the earth are pretty well known to everyone. Not so, however, the chemical nature of the several paris of the earth. Most people have learned at one time or another that the atmosphere consists mainly of oxygen and nitrogen. It also contains, in addition to the rare gases, small amounts of carbon dioxide and water vapour, which play very important rôles in geological processes.

Every high-school boy knows that water, and hence the hydrosphere, is a chemical compound of hydrogen and oxygen. When we consider the chemical composition of the lithosphere, however, we find that common knowledge is much more meager. And yet there are many important and interesting features of the solid part of the earth into which we should inquire.

As A first approach to the subject, let us consider two rather broad types of environment that may be recognized in the lithosphere. At the surface the pressure and temperature are comparatively low. Beneath the surface of the earth, however, and particularly at depths of 10 or more miles, the temperatures and pressures are relatively high.

The pressure effect can be understood easily when we remember that every cubic foot of rock must exert a pressure on the rock beneath it. A cubic foot of common surface rock weighs about 150 lb., and 10 mi. down the 52,800 cu. ft. of rock exert a pressure at the bottom of the column equivalent to thousands of pounds to the square inch, in contrast to the mere 14.7 lb. exerted by the atmosphere at the earth's surface. Likewise the temperature increases with increasing depth, and at a depth of 10 mi. it certainly is several hundred degrees higher than at the surface.

The point to this discussion of environments is that the

kinds of chemical compounds that tend to occur in the lithosphere depend in large measure on the environment.

We usually speak of the surface environment on the one hand, and deep-seated environments on the other, meaning by the latter an environment beneath the surface at a depth great enough to involve a considerable increase in temperature and pressure.

Now it happens that many rocks are formed deep beneath the surface of the earth, and as long as they are in their original environment of high temperatures and pressures they remain unchanged. Thus we speak of such rocks as being "in balance," or in *equilibrium,* with their environment. Broadly speaking, the deepseated environments tend toward the production of complicated chemical compounds, whereas the surface environment is more suitable to simpler compounds. The rocks which are formed deep beneath the surface may be brought nearer to the surface through the operation of geological changes, and thus they pass from a deepseated environment to the surface environment.

Obviously, of course, our direct observation is confined pretty well to surface conditions, but in deep canyons we may find types of rock from far below. Likewise there are large areas, where rocks have been brought to the surface by geological changes and may still be found largely unaltered by surface conditions.

It is from such rocks that we reconstruct and interpret the conditions of the deep-seated environments, those great laboratories in which the rocks of the earth are largely made. Furthermore, direct experiments in our chemical and physical laboratories, made at high temperatures and pressures, supplement and confirm this evidence from the rocks.

A great many samples of all kinds of rock have been collected and chemically analyzed, so that we may confidently say something about the average composition of the outer part of the lithosphere. Tables have been prepared of the ultimate composition of the lithosphere in terms of the percentages of the elements present.

Element	percentage
Oxygen.	46.46
Silicon.	27.61
Aluminum.	8.07
Iron.	5.06
Calcium.	3.64
Sodium.	2.75
Potassium.	2.58
Magnesium.	2.07
	98.24

When we look at such a table, we find to our surprise that out of the ninety-two elements, only eight are present to the extent of 1 per cent or more by weight.

Let us look here at the part of such a table that includes the eight most abundant elements. An outstanding feature of this table is that oxygen is by far the most abundant element of all. We do not mean to say that the oxygen exists in its free state. Obviously, it is chemically bound up in the rocks. Similarly with the seven other elements: not one of them occurs in its elemental state in the rocks; all of them are combined in one way or another into chemical compounds.

At this point some of us shall probably have to take a deep breath and hang on for a moment, because it may be easy to fall by the wayside. But if you are at all familiar with these ele- ments, you know that, with the exception of oxygen and silicon, the other six elements are true metals; silicon has some metallic attributes, but in its properties it lies between the true metals and the non-metals.

All seven of the solid elements, however, do form definite oxides, and the suggestion is strong that the fundamental chemical unit with which we have to deal is the oxide. At any rate, if we talk about oxides we shall find that the discussion is much simplified.

For example, metallic oxides tend to form *bases,* whereas the oxides of the non-metals tend to form *acids*. Silicon oxide shares this non-metallic attribute, at least in the presence of metallic oxides, and the result is that silicon oxide has a strong tendency to combine chemically with metallic oxides to form a silicate. For example, if magnesium oxide and silicon oxide combine chemically, a compound called *magnesium silicate* results. In chemical shorthand the reaction is expressed thus:

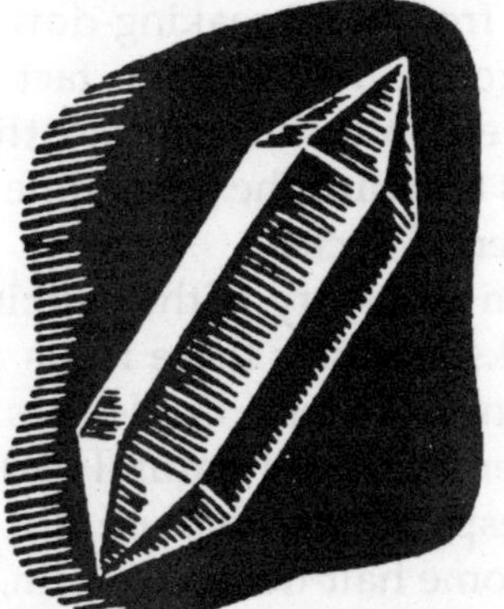

$$MgO + SiO_2 \rightarrow MgSiO_3$$

The resulting compound is one of the kinds usually formed under deep-seated conditions. Commonly, more than one of the metallic oxides combines with the silicon oxide to form double or triple silicates, like *calcium aluminum silicate,* $CaAl_2(SiO_3)_4$, which may be written $CaO.Al_2O_3.(SiO_2)_4$. It is thus a combination of calcium oxide, aluminum oxide, and silicon Oxide. Now these very silicates are the naturally occurring minerals that we find among the rocks of the lithosphere.

A mineral is a naturally occurring chemical compound; and although we may not be aware of it, we already are familiar with several minerals. They commonly occur as crystals, that is, in definite geometric forms bounded by plane surfaces. The simpler types of crystals may be cubes or prisms; but complex forms occur, as the accompanying sketches show. Rocks are combinations of minerals in various proportions, or occasionally are made of a single mineral. Almost invariably the minerals making up a rock are present as a physical mixture of crystals or crystal fragments.

It is the crystalline, as opposed to the fragmental, nature of rocks that largely determines their classification. Rocks composed mainly of interlocking crystals are called crystalline rocks; whereas others, composed, of crystal fragments, commonly with broken or rounded edges, are called fragmental rocks.

The fragmental nature of this second type suggests that they were formed from the breaking-down or wearing-away of the crystalline rocks. This is so in fact but the story is no simple one. Indeed, we shall spend the next dozen the geological changes to which the crystalline rocks are subjected to form the fragmental rocks.

We shall begin our story with the minerals contained in the crystalline rocks, because these rocks are associated with the deep-seated environment. Geologists recognize literally thousands of minerals among natural earth-products.

But do not despair. Kind Nature has decreed that if we are familiar with some half-dozen of them, we shall be able to talk intelligently about most rocks.

Let's go back for a moment to the eight elements we mentioned. If we make a list of oxides that are formed from these elements, we have the following:

Silicon oxide, SiO_2	Calcium oxide, CaO
Iron oxide; FeO	Magnesium oxide, MgO
Iron oxide, Fe_2O_3	Potassium oxide, K_2O
Aluminum oxide, Al_2O_3	Sodium oxide, Na_2O

Iron forms two oxides (ferrous and ferric), while the other elements form one each. Some of the silicon oxide combines with other oxides to form silicates. Out of the welter of such silicates that are known, only two groups are just now important from our point of view. They are the ferromagnesian minerals, and the feldspars. In addition to these silicate minerals, silicon oxide itself commonly occurs as a mineral, in which case it is called quartz.

Look for a moment more closely at the mineral groups we have just listed. As the name of the first suggests, it is composed largely of iron and magnesium oxides. Actually, a number of individual mineral species are included in the ferromagnesians, but they have so much in common that we may well group them under this one term.

In general the ferromagnesians are chemical combinations of magnesium oxide and iron oxide with silicon oxide, and in addition they may contain calcium oxide and aluminum oxide. If you are interested in actual chemical formulas, we may list one of them which is fairly typical of the group: $Ca(Mg, Fe)(SiO_3)_2$. This is a calcium-bearing iron-magnesium silicate, commonly called *pyroxene.* The ferromagnesians are usually dark green to black in colour, owing to their high iron content.

The feldspars, in contrast to the ferromagnesians, are generally light in colour. They range from white through cream to pink and pale green. The feldspars are combinations of the oxides of aluminum, calcium, sodium, and potassium with silicon oxide, and they form rather complex silicates. The chemically minded may be interested in knowing that one common feldspar has the formula $(K,Na)AlSi_3O_8$, and another is $CaAl_2Si_2O_8$. The first of these is a sodium-potassium aluminosilicate called *orthoclase,* and the second is a calcium aluminosilicate called *anorthite.* Both are naturally occurring minerals.

We shall be quite chagrined if our readers feel it necessary to memorize any of these complicated chemical formulas; they are included merely as a shorthand method of indicating which of the eight most abundant elements they contain. Whenever we have occasion to mention these minerals in a chemical sense we shall point out the elements again, and thus refresh your mind.

However, if you are one of those hardy folk who insist on remembering formulas, then remember the one that we shall introduce next. The ancient Greeks believed that the mineral quartz was water frozen so firmly that it would never thaw again. This belief may have arisen from the finding of crystal-clear quartz in the Alps, among the perpetual snows that swathe those majestic peaks. We are more prosaic nowadays, and we recognize quartz as silicon oxide itself, the chemical formula for which is SiO_2.

Quartz, as a matter of fact, is one of the most abundant minerals on earth. The sand of beaches and streams is practically all quartz, broken and ground into fragments by the forces of nature. The abundance of quartz suggests that in nature there is an excess of this oxide above what is needed to form silicates with the metallic oxides.

The minerals we have been discussing occur most commonly in a particular group of the crystalline rocks, formed far beneath the surface of the earth, at high temperatures, and usually from liquid rock. Everyone is familiar with the phenomenon of volcanism, at least to the extent of knowing that there are vents in the earth's surface from which molten rocks pour.

This lava comes from beneath the surface, obviously; but much molten rock never sees the light of day. It is such lava, buried beneath the surface, protected from rapid cooling by a layer of other rocks above, that cools and crystallizes into rocks which contain the primary minerals we have cited. Before taking up in detail the changes that these crystalline rocks undergo to form the fragmental rocks, we must concern ourselves with one other important aspect of the physical set-up of the earth: the earth's sources of energy, and the passage of that energy through the economy of this terrestrial sphere.

ENERGY AUDIT

Business corporations periodically issue financial reports to indicate their solvency, and to suggest, perhaps, the desirability of investing in their stocks and bonds. A complete report includes not only a balance sheet but also a statement of profit and loss which summarizes the income and expenditures of the firm. A study of the sources of income and the channels of expenditure generally yields a clear picture of the nature of the business and the efficiency with which it is operated. Now this terrestrial sphere has sources of income and channels of expenditure of energy.

Suppose we constitute ourselves a planetary commission to audit the earth's income and outgo of that extremely important commodity. To do so, we must first consider the sources of energy inand contrast with them the come outgoing channels. We should also consider the manner in which the incoming supply of energy is distributed over the earth; and finally we should set up a statement of profit and loss, and note whether the balance is to be drawn in black ink or red.

The first and most obvious source of energy income is the sun. With the aid of suitable instruments and appropriate calculations, astronomers have been able to determine that the earth receives 1.94 cal. per min. per sq. cm. This is called the *solar constant* and, except for small variations, it appears to remain fairly uniform over long periods of time.

The stars, too, send us radiant energy; but when we attempt to evaluate it in the same terms as that of the sun, it is found to be an extremely small amount—so small, indeed, that it is to be compared with the augmented income that would accrue to the Standard Oil Company for selling an additional gallon of gasoline a year. For the purposes of our audit, then, we may quite safely ignore the energy income from the stars, as far as its recognizable effects on the earth are concerned. In Addition to its one main external energy source, the earth has a supply of internal energy.

When one descends into a mine, or lowers a thermometer into a deep well, he finds that the temperature rises at a fairly uniform rate toward the earth's interior. On the average, this increase of temperature is about 1°C. for every 100 ft. of descent. The name geothermal gradient is applied to this downward increase of temperature. Everybody knows that if we place a hot object in contact with a cold one, heat will flow from the hotter body to the colder.

This flow of heat expresses itself to our senses as temperature decreasing away from the hot object; and, indeed, we can easily tell which way the heat is flowing by moving in the direction of decreasing temperature. If we apply this simple test to the earth, we find that the temperature decreases toward the surface, so that the heat itself must be flowing outward from its interior.

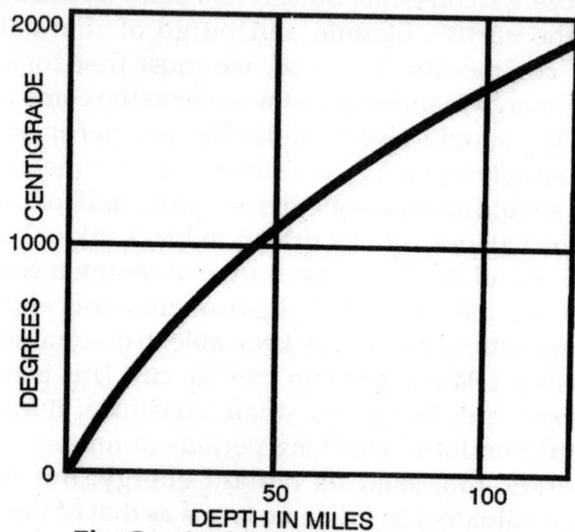

Fig. Graph of Probable Temperature Increase Beneath the Earth's Surface.

We on the earth's surface thus have two main sources of energy income; and of the two, that from the sun is by far the greater. But, although the heat radiated outward from the earth's interior is small in comparison with the energy received from the sun, it is nevertheless very much larger than the energy received from the stars; and it is of great importance in geological processes such as volcanism and the deposition of many of the ores.

Let us now consider how the energy income of the earth is distributed. If we confine our attention first to the external source of energy, we may see that the astronomical relations of the earth have much to do with it. The earth's spheroidal shape, the inclination of its axis, and its movements on its axis and in its orbit around the sun cause this external energy to be distributed over the earth's surface in varying quantities: a variation from place to place and from time to time.

Consider for a moment some point on the earth's surface,

such as Chicago, which is situated about 42° north latitude. During the day the sun warms the surface and raises the temperature. At night some of the heat is radiated off into space and the temperature drops. Thus does the rotation of the earth affect the energy relations.

In summer the sun is more nearly overhead at noon, and its rays pass almost vertically through the atmosphere. In winter the noon sun is lower down toward the southern horizon, and hence its rays must follow a longer route through the atmosphere, with a consequent greater absorption.

The result is that in summer more energy is poured directly on Chicago than in winter. The spherical shape of the earth is brought into the picture when we contrast Chicago with some point nearer the Equator.

There the rays are more nearly vertical the year round, so that annually a greater quantity of energy is poured down near the Equator than farther north at Chicago. It should be recalled, of course, that the number of hours per year that the sun is above the horizon is the same for all latitudes; but the actual distribution of heat depends partly on the angle at which the rays strike the earth.

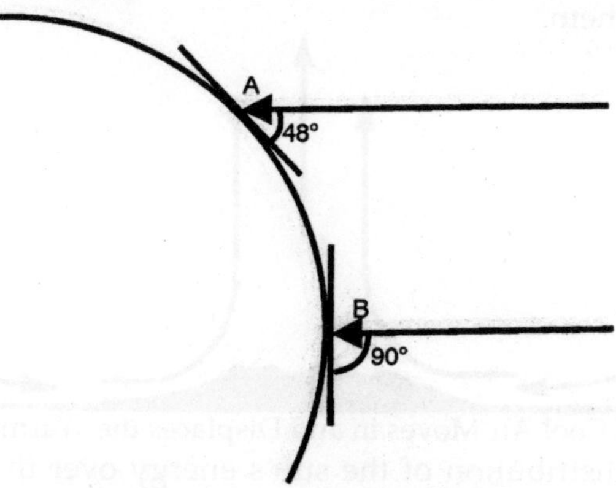

Fig. Effect of Latitude on the Angle of Inclination of the Sun's Rays on the Earth.

In addition to the astronomical factors, there are certain terrestrial features that affect the distribution of the sun's energy over the earth. The amount of absorption by the atmosphere increases with its content of water vapour, carbon dioxide, dust and several other minor constituents. Clouds reduce the amount of sun's rays directly striking the earth's surface, and differences in the nature of land and water determine largely the increase of temperature that follows the absorption of a given amount of sunlight on the surface.

The principal factor that controls temperature rise (assuming a given amount of absorbed heat and a given mass) is the *specific heat* of the substance. * Water requires about five times as much heat to raise its temperature a given amount than does an equal weight of rock.

Thus with a given amount of absorbed sunlight, land areas will rise in temperature rapidly, while water areas will not. Of the heat which strikes the surface, some is directly reflected, of course, and so does not contribute to the temperature rise.

Three are certain aspects of the sun's energy that we should consider. Much of the solar energy which penetrates the atmosphere is converted to heat, and it is almost entirely as heat that it is utilized on the earth. A number of important processes depend on this supply of heat and its distribution, so that it will be pertinent to pause here for a moment to examine them.

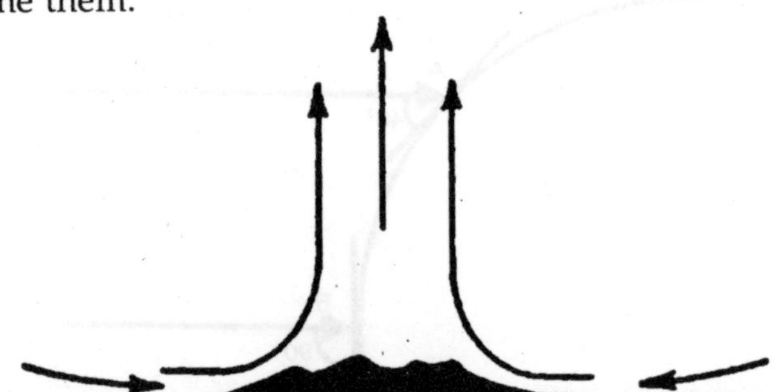

Fig. Cool Air Moves in and Displaces the Warmed Air

The distribution of the sun's energy over the earth, and the wide fluctuations in its amount at any given time or place, result in a differential heating in the earth's air and water films. Since heated air expands and becomes lighter, colder and heavier air may flow into the area of heating to displace the warmed air.

Thus are set up atmospheric circulations. In addition to the circulation of the winds, there is a constant passage from liquid water to gaseous water vapour, owing to evaporation from the surfaces of seas and lakes. This water vapour is swept along with the winds and eventually condenses back to water, falling as rain or snow.

The rain feeds and swells the streams, which ultimately carry the water back to the ocean basins. Thus there are continual circulations within the films, and as the winds and waters flow over the earth they modify its surface. Everyone has seen examples of these air and water movements in wind storms that sweep clouds of dust with them, or in cloudbursts that swell small streams into torrents and load them with mud and débris.

All of these circulations require energy. As winds blow, they dissipate some of their energy in friction; and from other causes as well, they gradually lose their energy and die down. Similarly, evaporation requires energy, which is stored in the vapour as latent heat, only to be released again when the vapour condenses. We begin to see here some of the devious paths followed by the energy that reaches the earth, but it is too early yet to summarize the process.

The circulations of the earth films cause a number of important phenomena known as geological changes. Air and water, or ice, acting on the solid part of the earth, modify its surface, by carrying away the transportable rock débris, in such manner that the irregularities of the surface, like hills and mountains, are slowly worn away and leveled down. The material carried off is eventually deposited in the seas, those great receiving-basins of the earth.

Thus the two films serve as active geological agents, and the surface of the lithosphere is the object acted upon. These interactions of the films and lithosphere are summarized in the term gradation, the great leveling-down process of geology. *This is the first of three important geological processes*.

In Additional to the inorganic world itself, we recognize a fourth sphere about the earth; this is the biosphere. It may also be likened to a thin film, because the life zone of the earth is confined within very narrow limits indeed. The biosphere extends from the ocean bottoms to a point several miles above the earth's surface, and that is all. Yet this limited zone is extremely important in the distribution and use of the earth's supply of energy.

Plants absorb sunlight directly and convert it into potential chemical energy, which is stored in such substances as cellulose and starch. Animals may then eat the plants and use this energy for their own life-processes.

Part of the plant material may, however, be buried beneath the surface of the earth as coal, and there the potential energy remains locked up until released by the magic wand of fire. In similar fashion organic matter may be buried and later give rise to valuable pools of petroleum, with their high stores of potential chemical energy.

Fig. Energy from Plant to Man

Eventually, however, most of the energy stored up by plants and animals is released again through the slow processes of decay, and thus contributes to the outgoing energy which no longer figures largely in the economy of the earth.We are now almost ready to prepare our statement of energy profit and loss, and we shall do so as soon as we have looked a bit more closely at the earth's interior sources of energy. Just as we do not know certainly what the ultimate sources of the sun's energy may be, so are we uncertain about the cause of the earth's interior heat. We may, however, reason intelligently about the matter.

There is a strong likelihood that a considerable part of the energy is released by *radioactive decay*. In fact, chemical analyses of common rocks indicate that their radioactive content is sufficient to account for a good portion, if not all, of the heat escaping from the earth's interior.

It seems likely, however, that other sources of heat are also present. For example, we know that within the earth there are tremendous pressures due to the weight of layers and layers of rock from the surface downward. The possible compaction of the earth's interior that may result from this could well generate heat in the struggle between the resistance of the earth body to shrinkage and the indomitable forces of compression.

Whatever the sources the heat is nevertheless conducted always to regions of lower temperature; and in the earth this means toward the surface.

We have ample evidence, however, that all the energy Within the earth is not merely dissipated as heat. As a matter of fact, it is by no means certain that the heat being conducted outward from the interior of the earth represents even the major part of the earth's internal energy. The great compaction within the earth may there set up stresses and strains; and when the limits of adaptability are reached, the earth yields by deformation.

Fig. Earthquakes are an Evidence of the Earth

We have ample evidences of earth deformation; *earthquakes* are an example. Such deformations or movements of the solid part of the earth, either rapid or slow, are called diastrophism. Usually justments, and when they occur, the local area is shaken by earthquakes. Serious as they are from a human viewpoint, earthquakes are thus merely shivers and spasms within the earth body.Movements are Very Very Slow, Internal Energy.

Diastrophic events may express themselves in many ways, and later on we shall devote a series For the present we may point out that in all its aspects diastrophism involves energy in some form or other, and thus belongs in our ledgers. So important is diastrophism in earth economy that it is called the *second great process of geology*.

The energy of the earth's interior tends to be radiated outward into space, but of course it must first be conducted to the surface. During this conduction, large bodies of rock may become heated above their melting-points, despite the great pressures, and so liquefy.

Movements of this liquid rock. or lava, within and upon the earth, give rise to volcanism, the *third great process of geology*. If the liquid rock finds access to the surface, we have the spectacular phenomena of volcanic eruptions.

If it solidifies again underground, the results are no less far reaching, but of course not so spectacular, for, although the outpourings of lava on the earth cause havoc and destruction, the more quiet intrusions of lava among the rocks below the surface commonly give rise to valuable and extensive ore deposits.Wenow have all the data we need to complete our audit. We saw that the incoming energy arose from two sources—the sun and the earth's interior.

In the transformation of the sun's radiation through the energy scheme of the earth, the short incoming light waves are largely converted to heat energy, and as such are used in the activities of wind, water, and animal life, or are stored as potential chemical energy in plants and various fuels. Simultaneously, some of this heat energy is dissipated outward into space as radiations of long wave-length. *

Likewise the energy from the earth's interior is partly used in diastrophic adjustments and in volcanic activity; but eventually it, too, escapes as heat radiation.

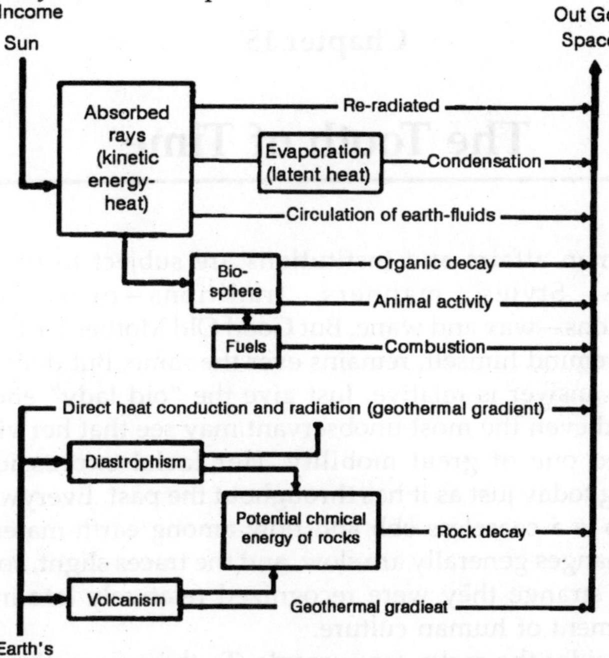

It thus seems that there is today essentially a balance between the incoming and outgoing energy, with perhaps a slight excess for the surplus account, stored up as potential energy in one form or another.All these various aspects of the energy flow are shown in the appended diagram, which represents our condensed statement of profit and loss. it may not resemble an accounting record as much as an electrical circuit, but that is a minor matter.

The essential thing is that it furnishes us with a picture of the ebb and flow of energy upon and through the earth, and demonstrates that, although the percentage of profit may not be very high, nevertheless this terrestrial sphere is still a going concern, and certainly is in a position to continue actively in business for many years to come.

Chapter 15

The Tooth of Time

Human affairs and institutions are subject to decided changes. Styles, manners, traditions—even entire civilizations—wax and wane. But Good Old Mother Earth, one likes to remind himself, remains ever the same. But does she?

The answer is relative. Just give the "old lady" enough time, and even the most unobservant may see that her visage is indeed one of great mobility. Her facial expression is changing today just as it has throughout the past. Everywhere about us is a ceaseless ebb and flow among earth-materials. These changes generally are slow, and the traces slight, so that it is not strange they were recognized relatively late in the development of human culture.

Consider the rocks, for example. To the average layman, if anything is firm and stable, it is rock; and this almost universal belief has given rise to innumerable sayings to the effect that something is "hard as rock." And yet there is abundant evidence that even rocks are subject to changes: that under surface environments they tend to break down into débris.

In fact, this gradual decay of rocks is common knowledge, when one really stops to think about it. Everyone knows that the writing on old tombstones is gradually obliterated by the tooth of time, and the blocks of most old stone buildings show signs of weakness and decay. Pitted surfaces develop, dark stains appear, or the blocks crack and crumble away.

Few of us, however, have considered the implications of such slow changes extending over long periods of time. If marked changes occur within a few score of years, how much more apparent will they be in thousands of years.

Geologists have accumulated abundant evidence that under surface conditions rocks are relatively easily altered. The process of alteration is called rock-weathering. The nature of this weathering process depends largely on the kind of rock and on the specific conditions of its environment. In some places the surface of the rock becomes dull and stained, and pitted and crumbly.

In others the surface of the rock remains reasonably fresh, but thin shells break from its surface and joints and cracks appear in the body of the rock. The first kinds of change are chemical; the second are mainly physical. Both are included under the general term "weathering."

The net effect of most weathering changes is that hard and firm rock ultimately becomes a mass of small fragments and grains, which in most cases accumulate around the base or on the surface of the parent rock. This loose débris on the earth's surface is called mantle rock. Because it consists mainly of small particles, it is subject to the mechanical forces of wind and running water. These agents may sweep the particles away from their parent location and deposit them elsewhere to form a second generation of rocks. In this sense weathering is the first stage in the ever continuing processes that wear down the land surfaces.

During the discussion of shallow and deep-seated environments it was pointed out that quartz and the silicate minerals (ferromagnesians and feldspar) were in equilibrium with their environment as long as that environment remained one deeply buried beneath the earth's surface.

But if the rocks containing these minerals are later exposed to the atmosphere by some geological change, the minerals which were formed under high temperatures and pressures are no longer permanent. In fact, the silicate minerals are relatively unstable and tend to break down to simpler compounds when exposed to the chemically active gases of the atmosphere, which are *oxygen, carbon dioxide,* and *water vapour*.

A complete analysis of rock-weathering deserves an entire volume, and so we must confine ourselves to certain fundamental notions. The principal points that we shall emphasize here are that

- Weathering is an approach toward a state of equilibrium under surface conditions and that
- This approach toward a stable state usually involves the breaking-down of the rock into finer débris.

ROCK is a poor conductor of heat; and when the sun's rays pour down on exposed rocks, the surface layer is heated and expanded more than the interior. This expanding shell sets up strains within the rock and weakens it. At night the reverse process takes place, and the outermost layer contracts more than the inside.

If these expansions and contractions continue for a long enough time, the rock may finally yield by shelling off its outer layer. This loosened shell then fails from the parent block and lies at its base. Thus is exposed a fresh surface, ready for a repetition of the process.

In detail this phenomenon is probably much more complicated than we have indicated, and involves some chemical changes as well as a yielding to physical forces. Time is undoubtedly an important factor.

Fig. Rocks Crack and Crumble
Under the Attack of the Weather.

Another aspect of temperature change is the disrupting effect of ice formed in rock crevices. All rocks, however dense in appearance, are penetrated by crevices and cracks, and into them rain water percolates. If the temperature drops sufficiently, the water freezes, and on freezing expands to about eleven-tenths its former volume. If there is no direction in which the expansion may take place freely, the ice exerts a pressure against the rock that may be great enough to disrupt it.

These are examples of physical weathering. Such physical changes take place most rapidly when high altitude is combined with steep slopes, as on mountainsides. Here the extremes of temperature are more pronounced, and the broken fragments rapidly fall away from the parent rock. The blocks and fragments accumulate at the base of the mountains and build up talus slope, which reach up the sides of the mountains and may ultimately completely bury them.

Fig. Talus Slope at Base of the Twin Sisters

Simultaneously with the mechanical or physical weathering of rocks, chemical agents attack them and produce various decomposition products. When rocks are weakened and broken by physical forces, the resulting fragments have a total surface area much greater than that of the original rock exposure. Inasmuch as the speed of chemical reaction depends in part on the intimacy of contact between the reacting bodies, the greater surface area becomes an important factor in decomposition.

The active gases of the atmosphere are dissolved in rain water which percolates among the fragments of broken rock. Thus these agents are able more effectively to perform their work of chemical decomposition.

An important distinction between *physical* and *chemical* weathering must be made clear. Although physical disintegration changes the size and shape of the parent rock masses, the minerals originally present remain intact. With chemical reaction, on the other hand, comes a change in the very composition of the minerals themselves, as well as a change in the physical state due to the formation of new compounds having their own physical properties. Chemical weathering, therefore, is a much more complicated process than physical weathering.

The chemical decay of minerals depends in large part on the fact that the surface environment contains the active chemical agents oxygen, carbon dioxide, and water vapour, which are rarely present in deep-seated environments. These three chemical agents tend to produce *oxides, carbonates,* and *hydrated compounds,* respectively, by combining with the primary minerals of the rocks that originated far below the earth's surface.

The low pressures and low temperatures of the surface environment, then, demand oxidized, carbonated; and hydrated minerals for stability. Among the primary minerals thus far mentioned, quartz is already an oxide, and it is indifferent to chemical combination with either carbon dioxide or water. Consequently it remains practically unchanged under surface conditions. Not so the minerals of the silicate groups.

The ferromagnesians, for example, may contain only partially oxidized iron, with the result that decomposition of the molecules sets in during the further oxidation of the iron by atmospheric oxygen. During this decomposition the rest of the silicate molecule may combine chemically with water and thus become hydrated.

Just as the ferromagnesians decompose in the presence of atmospheric agents, so the feldspars may suffer chemical decay. The calcium oxide may be released from the silicate molecule and may unite with carbon dioxide to form a very common carbonate called calcite. Similarly the sodium and potassium may be released to form various soluble salts. Meanwhile, the silicate part of the feldspar molecule Becomes hydrated by combination with water, much as the ferromagnesians do.

The net result of this hydration and decomposition of the silicate molecules is a mixture of débris that we call clay. Since clay may consist of several minerals, depending on the composition of the original material, we apply the term clay minerals to the group as a whole.

We have introduced two new mineral names here. Let us pause a moment and look at them. We said the clay minerals are hydrated silicates. That means they still are silicates, but now water has become an essential part of the molecule. Here is the formula of one of them: $H_4 Al_2 Si_2 O_9$, or $Al_2O_3 \cdot 2SiO_2 \cdot 2H_2O$, commonly called *kaolin*. Other clay minerals may contain magnesium.

Clays are composed of very small particles; and, as everyone knows, they are adhesive and slippery when wet. In the formation of the clay minerals, then, the hard and firm silicate minerals break down to finer débris.

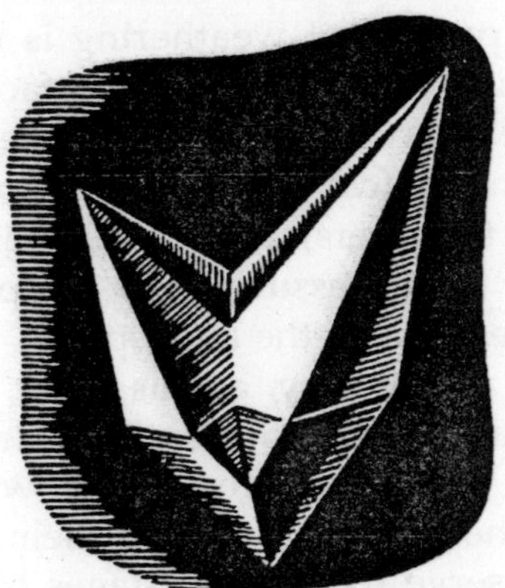

Fig. Dogtooth Spar. A Fanciful Name

This is the important point to understand. The other mineral, calcite, is calcium carbonate, $CaCO_3$. It almost invariably results from the decomposition of calciumbearing primary minerals. In some cases calcite is found in crystals that resemble somewhat the canine teeth of dogs, from which the fanciful name *dogtooth spar* arose. In comparison with most other minerals, calcite is relatively soluble in water, and especially so in the presence of carbon dioxide.

Hence, under many conditions of weathering the calcite may be carried away in solution by waters percolating through the ground. Quite similarly the soluble salts that are formed from the sodium and potassium in the silicates are carried away in solution and ultimately contribute to the salt supply of the oceans.

We may now summarize the process of chemical weathering something like this:

Primary Minerals	*Weathering Environment*	*Secondary Minerals*
Quartz	⟶	Quartz
		Clay minerals
Feldspars →	(By oxidation, carbona- { tion, and hydration) }	Calcite
Ferromagnesians		Salts of the sea

The whole process of weathering is thus an approach toward equilibrium or balance under surface conditions. With the exception of quartz, the primary minerals largely break down to new minerals (called secondary minerals because they are derived from the others) which are in adjustment with their environment of low pressures, low temperatures, and the chemically active gases of the atmosphere.

The careful reader may, at this point, recall that began with some emphasis on the slowness of the weathering process. If it is so slow, how do geologists know the exact stages through which the rocks pass during their decay? Simply by being able to observe rocks in all stages of weathering, and from this evidence building up a connected sequence of events.

At the risk of becoming a bit technical, we shall let the reader look behind the curtains and see the methods by which the stages of weathering are studied.Suppose we collect a sample of a partially weathered deepseated rock from near the surface of a quarry, and then take a series of samples farther and farther down into the fresh, unaltered rock.

We grind down chips of the rock with an abrasive until they are only a fraction of a hundredth of an inch in thickness. These *thin sections* are then examined under a microscope. Let us start at the bottom, and see what the fresh, original rock looks like. For simplicity's sake we shall concern ourselves only with the feldspar minerals, which quite adequately the process.

Here we shall have to turn to Plate and look at the photomicrographs toward the bottom. In the circle to the left, notice the large central crystal with its alternate white and dark bands, which fresh, unaltered feldspar. Notice that it is crystal clear. Now look at the middle picture.

See how the upper crystal of feldspar is blotched and mottled, although the light and dark bands (narrower in this case) may still be seen. The blotches are incipient changes, in which the feldspar is beginning to show the effects of surface conditions. When we ex- amine the photograph on the right, we see a much later stage of weathering in a sample taken from nearer the surface. In this thin section only a "ghost" outline of the feldspar crystal remains.

The original composition has almost wholly changed to clay minerals, although a few suggestions of the bands may still be seen. The fourth and final stage, it would be a handful of crumbly soil from the surface, where the originally crystalline rock has broken down to heterogeneous débris.

Not alone are there physical and chemical agents of weathering, but many biological agencies as well. Lichens, for example, cling to the surface of rocks and gradually dissolve some of their constituents for plant food. These attacks commonly produce dull and pitted surfaces where the plant acids have leached the rock.

In addition to this solvent action, plants may also exert mechanical forces by their growing roots. It is not uncommon to find trees growing from cracks and joints in rocks, where their enlarging roots have forced apart the rock by a type of wedge action. Animals also contribute to the ultimate decay of rocks in several ways. Burrowing animals dig up buried material from the mantle rock and expose it to the more concentrated action of the atmosphere.

In this connection most persons are familiar with Darwin's classical researches on earthworms, wherein he showed that in many places annually the earthworms on each acre of land ingest and cast aside 10 tons of soil, constantly exposing fresh material at the surface.

All of these factors contribute in some degree to the weathering process and afford one of the many examples of the close integration of geology with the biological as well as the physical sciences.One of the most important factors in rock-weathering is the climate. In arid regions physical changes may predominate and chemical reactions be relatively inconsequential.

Oxidation of the disintegrated rock, however, is fairly common, although hydration is relatively rare. Hence, in the formation of iron oxides the red ferric oxide may predominate. The absence of reducing agents such as organic matter (humus) tends to keep the iron fully oxidized, so that red and brown colors among ancient rocks (such rocks are called red beds) are regarded by many as an important feature in identifying desert conditions of the geological past.

In humid climates chemical weathering predominates, because the active atmospheric gases are dissolved in the rain water that enters and seeps through the rocks. Here hydration is pronounced, and the net result is a thoroughly decayed rock. Temperature also plays a rôle in this case, because the speed of chemical reactions increases rapidly with slight rises of temperature.

Hence in humid tropical regions weathering proceeds most rapidly, produces the most highly weathered débris, and consequently very thick soils.

Once the surface becomes covered with a mantle of decayed rock, however, the percolating rain water has to circulate farther to reach fresh surfaces, and hence its dissolved gases will be less concentrated. Furthermore, the cover of mantle rock protects the fresh rock below from sudden or extreme temperature changes, so that the physical forces become less effective. Thus arises the commonly observed situation of a zone of mantle rock, with its surficial covering of *soil*, blanketing the fresh rock below, and with the degree of weathering becoming less pronounced downward. Some of the factors involved in the formations of soils and mantle rock.

Fig. Bedrock Grades Upward Rock and Soil

If the weathered products accumulate indefinitely, weathering of the underlying rock virtually ceases. into mantle Other forces, however, are continually at work preventing this situation from developing in all but a few favored localities. Usually there is a fairly constant removal of the surface materials by gravitative pull down slopes, by winds, and by running water.

The surface of the mantle rock is thus gradually removed; and as the mantle thins, weathering processes are again able to penetrate more deeply. In this way there tends to develop a balance between the rate of weathering and the rate at which surface agents carry away the débris.

In the final analysis the removal of the weathered rock débris depends on the force of gravity. It is the component of gravity along the slope that determines the flow of water in streams, and winds are generated by relative gravitative effects between warm and cool bodies of air. There is one mode of transportation, however, that depends directly on gravitative pull and which involves no moving medium with its frictional grip on the débris.

We have already touched upon it: it is the downward tumble of loose blocks of rock from steep slopes. When rocks crack and break during weathering, fragments result. If this disruption occurs on cliffs or steep slopes, the loosened débris immediately falls or rolls down, and collects as talus at the bottom.

The slope of the talus is adjusted to the *angle of rest* of the blocks; it is the steepest angle at which the blocks may lie. Further breaking of the blocks in the talus pile causes shifting of the material, so that over long periods of time there is a gradual movement downslope.

A somewhat similar phenomenon takes place along river valleys, where the mantle rock on the valley walls gradually moves downward under gravitative pull. This slow movement is called *slope creep*. It is aided by the lubricating effect of water which percolates among the soil particles.

In some situations, such as in steep mountainous regions, the entire mountainside may give way and relatively rapidly slide downslope. Such landslides are often destructive. In these movements underground water also may loosen or lubricate the material, so that the force of gravity may more readily set it in motion.

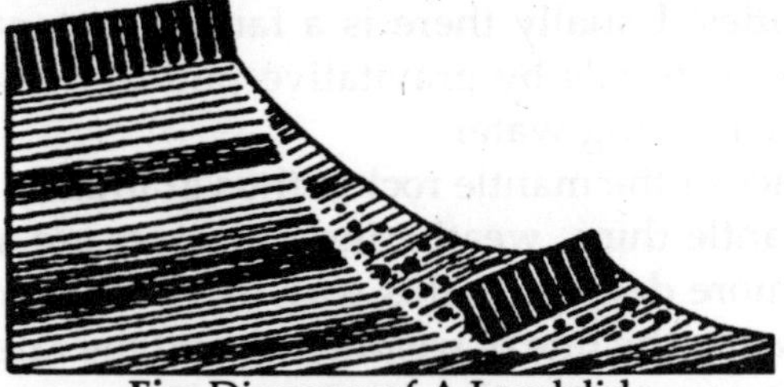

Fig. Diagram of A Landslide

These last two cases of movement are somewhat transitional between gravity alone as the operating force and the movements of surficial agents, like wind and water, which transport the débris directly. In fact, most movement of rock débris is carried on by the geological agents, which are wind, running water (rivers and streams), glacial ice, shore agents (waves and currents along shores), and ground water, which seeps and percolates through the body of the rocks.

The current weathering breaks down the massive rocks into smaller débris, which is thus subject to movement by the geological agents. We may accordingly turn our attention to these agents, considering first some of the underlying principles of transportation that are common to several of them.

WINDS AND TURBULENT MOTION

A modern, open-type freight car is about 50 ft. long and has a capacity of 50 tons. To transport 400,000,000 tons of gravel, sand, and mud requires 8,000,000 such cars, enough to make a train about 80,000 mi. long. Such a train would girdle the earth more than three times. And think of the freight bill!

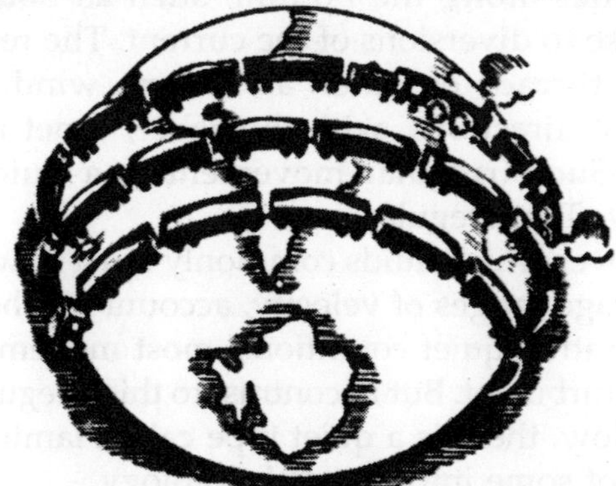

Fig. A Train of Cars 80,000 mi. long

It is estimated on good grounds that annually the Mississippi River carries enough sediment to fill such a train of cars. Yet the Mississippi is only one of multitudinous rivers on the earth; how much more, then, must be their total load Add to that total the dust carried annually by the winds, and the great quantities of sediment shifted along the seashores, and the imagination reels under the resulting figures. There is no escaping the fact that this old globe is in the transportation business in a big way.

All of this transported rock débris must come from somewhere on the earth's surface. Unfortunately, too much of it is derived from our plowed fields, where wind and water are stripping off good topsoil at an alarming rate. The widespread interest in the prevention of destructive erosion of our farm lands attests the importance of the subject in national economy.

It is pertinent, therefore, to consider this geologic process somewhat closely. Accordingly, we propose to develop in some detail the principles underlying the movement of rock débris, in order to see just how it enters into the complete geological picture.Winds usually are not smooth flows of air over the surface, but rather they act as alternations of gusts, eddies, and whirls. The velocity of the wind also varies considerably from place to place and from time to time.

In similar fashion, when water flows over a stream bed, the irregularities along the bottom, such as boulders and stones, give rise to diversions of the current. The result is that intermingling threads of water, as it were, wind and whirl about in many directions, with, however, a net movement downstream. Such irregular movement in a fluid is called turbulent flow. The irregular.

Surfaces over which fluids commonly must flow in nature, and their average ranges of velocity, account for the fact that, except under rather quiet conditions, most movements of air and water are turbulent. But in contrast to this irregular motion or turbulent flow, there is a quiet type called laminar motion which is also of some importance in geology.

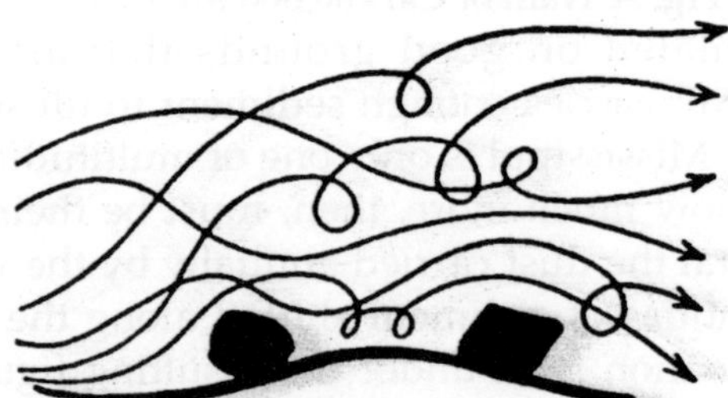

Fig. Eddies and Intermingling Threads of Liquid in Turbulent Flow

Turbulent motion in air and water, then, involves certain upward and sideward currents, as well as the forward mass movement. The irregularities of movement, especially those with upward components, may lift objects of small size up into the main stream, and so carry them along. The greater the turbulence, broadly speaking, the greater is the lifting effect on the particles.

The moment the particle is lifted off the ground into the

liquid, however, it tends to settle again. The settling rate, or terminal velocity, * of the particle depends on its size, shape, density, and the nature of the fluid.. Accordingly, we shall content ourselves with those whose effects may readily be seen; the more complicated ones involve technicalities beyond the scope of this volume.

Most everyone would predict that if a sphere and a cube of the same material and volume are placed on a smooth level table, the sphere could more easily be set in motion by blowing on it than the cube.

Furthermore, it takes no physicist to see that this result depends partly on the fact that the cube rests on the table with a greater part of its area than the sphere, Applying the same principle to a mixture of rock particles *of a given volume* but of different shapes, we should expeet that a gentle breeze will roll away the more spherical grains and leave the others, unless, indeed, the wind is strong enough to move them all.

This action of choosing certain grains on the basis of size or shape is called selective transportation, and it is an important factor in the work of wind and water. In fact, when a gentle wind acts on the dried sand of the seashore, it actually does roll the more spherical grains along with it, and leaves the flatter ones behind,

This accounts for a phenomenon that you yourself may observe with a low-power magnifying glass, namely, that in general the sand grains in sand dunes are more rounded than those on the beach.For a still better understanding of this situation, let us distribute a layer of sand grains of various sizes and shapes on a table-top and turn an electric fan on them. Most commonly the immediate effect will be a whirling cloud and a layer of débris over all the furniture; but if the fan is adjusted to move all but the largest grains, the selective action of the air current may be observed.

In the somewhat complex shifting of grains that results, it will be possible at least to distinguish that the very small grains of all shapes are whisked up into the air, the larger spherical grains are rolled along the surface of the table, and the largest particles remain behind. The grains that are rolled along constitute the traction load, the particles that are carried into the air constitute the suspension load, and the material left behind is the lag sediment. The traction load, being confined to the surface of the ground, may be stopped by any obstacle in its path; whereas the suspension load, carried higher up, may move over the obstruction.

When later we analyse the factors causing deposition, we shall see that the deposits resulting from the action of wind or water are pretty intimately bound up with this difference in the traction and suspension loads.

But for the moment let us look more closely at the suspension load. We have seen that it is made up of a mixture of very small particles of all shapes, ranging from flat to spherical. As soon as these rise above the surface, they are acted upon by gravity and tend to settle. Now there is a property of spheres that is unique among solids, whatever their shape, namely, a sphere has the minimum surface area per unit of volume.

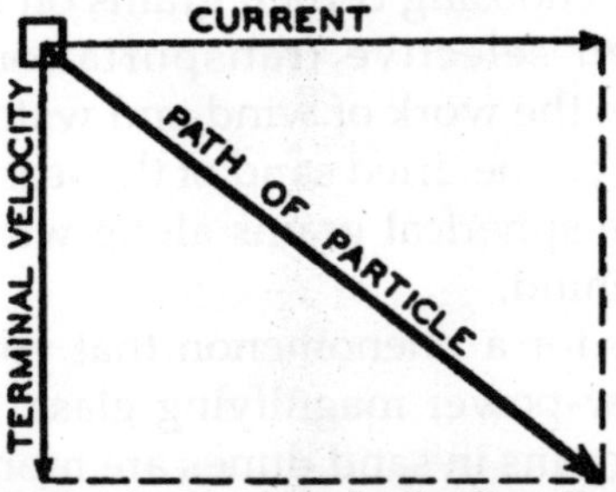

Fig. Path of a Uniformly Settling Particle Acted on by Forward Motion of a Fluid

Of particles having the same volume, then, the flat ones will encounter more resistance from the air because of their relatively greater areas, and so they will tend to settle more slowly than the spherical grains. As long as the lifting effects of the air currents are greater than the settling velocities of the grains, the grains will rise. But there are so many intermingling currents in turbulent air that, as the grains rise, they are shifted about in a most complex manner. If they are shifted from an upward current to one having a horizontal movement, the grains will immediately start to fall. In general they will not fall back to the starting-point, however, because as they settle the wind carries them forward.

The path they follow, therefore, is a line that trends forward and downward simultaneously.IF WE now consider the entire picture, we see that the action of the wind has been to sort out and classify the miscellaneous mixture of débris that results from rock-weathering. The various loads that are carried away, as well as the material left behind, have certain characteristics in terms of size, shape, and density.

The density, in turn, depends on the chemical or mineralogical composition of the grains; so it is apparent that many factors are involved. Because of the ways in which rock débris is sorted during transportation, we may expect that, if the resulting deposits are studied in detail, we should be able to learn something about the transporting agent. What we have been saying about wind applies equally well to running water, for in general the principles of transportation are the same for the two agents.

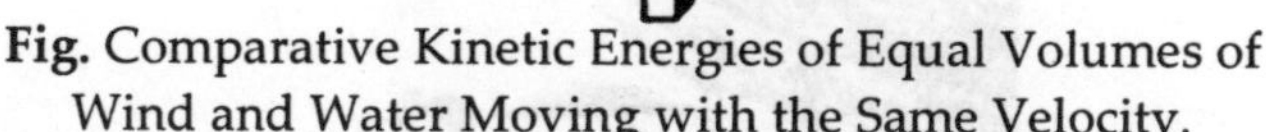

Fig. Comparative Kinetic Energies of Equal Volumes of Wind and Water Moving with the Same Velocity.

We may note, however, that the differences in the densities of the two fluids partly determine their carrying-capacities. That water can move larger objects than wind is not surprising when we remember that the density of air is about one onethousandth that of water, or, in other words, it is only one one-thousandth as heavy, volume for volume.

Consequently, if equal volumes of air and water are moving with the same velocity, the kinetic energy, on which the effective work depends, is a thousand times as great in the water as in the air. In fact, the average range of wind velocities, combined with its low density, pretty well confines wind action to the smaller particles like dust and sand. On the other hand, water suffers no such restrictions; and indeed it is capable of moving very large boulders.Winds and running water are the most obvious transporting agents that display turbulence, but they are not the only ones. Ocean waves, as a matter of fact, also are turbulent as they dash against the shore within the line of white-capped breakers.

Waves are peculiar in one respect, however, in that they display a *pusating effect*, in a degree much more pronounced than in the general forward movement of wind and running water. We shall accordingly defer our treatment of ocean waves to a later point, especially since certain aspects of wave work also involve laminar motion.

Let us now look somewhat more closely at the movements of wind and see what sort of work it performs on weathered débris. It requires no great knowledge of geology to understand that wind work is most effective in arid regions. There the decayed rock at the surface is usually quite dry and powdery, and it has no protective cover of vegetation. In desert regions, then, wind may be one of the most effective agents of transportation, and during windstorms may carry loads of sand and dust far and wide.

Fig. Dust Storm on the Western Plains

Not only deserts, but even our western plains, have witnessed the havoc and destruction of wind action. The plowed fields of those western states, where rainfall is scarcely 10 in. a year, are peculiarly susceptible to wind work.

The years 1995 and 1995 saw dust storms of such magnitude that practically complete destruction of large areas of farm land followed in their wake. The great clouds of dust were borne upward into the atmosphere and carried as far as the Atlantic seaboard, where they constituted dust storms of no mean magnitude.There is more to wind work than we have mentioned. Not only is the sand and dust swirled about by the wind, but as it moves it may be blown against other rocks and act on them like a sand blast.

In fact, wind abrasion, as this scouring effect is called, is of considerable importance in the arid Southwest; many of the prominent cliffs of varicolored sandstones there have been shaped and polished largely by wind abrasion. Commonly the traction load of sand, blown along near the ground, may also undermine the rocks and develop curious pedestal effects.

Among the various rocks that may readily be acted on by wind abrasion are sandstones. Composed, as they are, of sand grains loosely cemented together, individual grains are easily torn off by the blast of wind-borne débris, and they themselves join the load carried on by the wind. Their erosional effectiveness is found, of course, in the fact that they lie within the range of sizes that can be most easily swept along by winds. As wind abrasion goes on, the rocks affected are worn away, so that the landscape is gradually changed.

Naturally enough, the material carried by wind tends ultimately to be deposited somewhere as the winds die down or as obstacles are met in the path of the traction load. The deposits thus built up often assume characteristic shapes, and are composed of typical sizes of débris. Ordinary sand dunes, familiar to everyone, are wind deposits.

In their formation, once again the selective effect of turbulent agents enters the picture, and we shall have to devote some space to these depositional effects after we have considered a few other transportational agents.The similarities between wind work and the work of running water are due mainly to similarities in their movement. In detail there are wide differences between them, because, whereas winds sweep over large areas simultaneously, rivers are confined to their valleys and concentrate their work along narrow sinuous paths. We may with profit, then, turn our attention to running water as a geological agent.

Chapter 16

Valleys and Streams

Except in the most arid parts of the country, it is impossible to drive a car many miles without crossing bridges. The observant thus tend to recognize that there are several kinds of streams, some rushing through narrow rock gorges, others winding across wide flats with gentle valley walls in the distance; all of them, however, are confined to some kind of valley or trough cut below the general level of the land.

Furthermore, everyone knows that main trunk rivers have tributaries, and these in turn have smaller branches, so that the whole drainage pattern, as spread out on a map, bears some resemblance to the branches of a tree. Finally, we all know that major rivers eventually flow into the oceans.

If we start at any point along a river and follow it upstream, we note as we pass successive tributary-mouths that the main river becomes a little smaller. Finally, near the headwaters the river itself has become a small stream, in some cases beginning in a swampy tract, in others tumbling down a mountain side as a brook, or merely heading in a dry ravine among the hills.

There are other features to be noted too. The valley flat, or flood plain, becomes narrower as it is followed upstream, until eventually it disappears and the stream occupies all of the valley bottom. The valley walls then become steeper, and the slope of the valley bottom itself increases toward the head.

Now let us see what happens at the very head of the valley during a rainstorm. Let s choose a simple case, where the terminating ravine is cut into mantle rock. When rain falls on land, it first soaks down through the spaces between the soil particles until the surface is thoroughly wetted. Then, as more rain falls, the water tends to collect in irregularities and to form tiny streamlets which flow downslope.

As the volume of water increases, it gains energy and sweeps along the finer soil particles in its path. By this process it carves a little trough, which may assume the proportions of a fair-sized gully if the land slopes away abruptly. Nearly everyone has seen such gullies along the sides of hills, especially where the soil is reasonably free from a protective covering of vegetation.

The rain water and its toad of débris are led by the slope into the ravine at the head of our valley. Here, during rainy weather, the concentration of water may become considerable; and as it tumbles along it carries mud, sand, and even pebbles along with it. When the rain has ceased, the ravine again becomes dry; but in the meantime several important things have happened.

In the first place, the rain water has washed rock débris down the slopes into the ravine, and the water there has carried much of this débris along, to supply it to the permanent stream farther down the valley. As a consequence of the removal of this material, the ravine gnaws a little farther into the hillside.

Thus the valley is lengthened, and this lengthening always takes place in a headward direction. Now, during any one rainstorm the amount of material swept from a square yard of land surface may not be great; but, given enough time, it becomes obvious that the valley will increase both in length and depth. Furthermore, the gully that we saw develop will itself increase by headward growth, and the rainfall that enters it will flow into the main ravine. Thus the gully becomes a lengthened tributary of the main stream and will in time develop branches of its own.

The reader at this point may well ask where this process will end. In brief, we may reply that the typical stream pattern is gradually enlarged by this headward growth until it comes into direct competition with another drainage system working in the opposite direction. Each of these headward-cutting stream systems will then control a series of slopes leading to itself, with the result that a divide will lie between the opposing ravines. At this point a state of equilibrium is reached between the two drainage basins, and successive rainstorms will lower the divide, but, except under unusual conditions, the two drainage basins will remain separate.

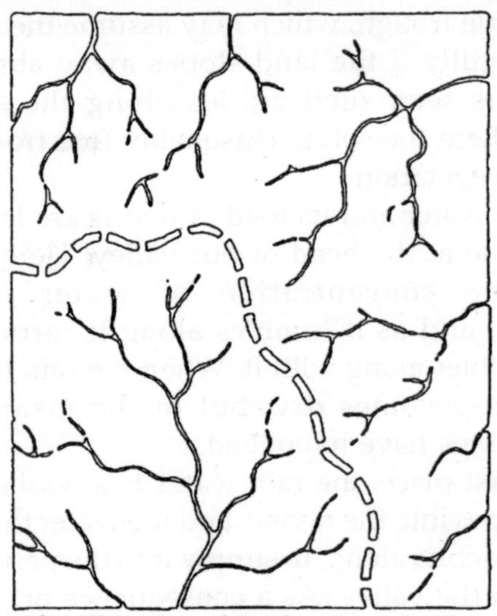

Fig. The Line of Dashes Locates the Divide between Opposing Drainage basins

In the foregoing discussion we assumed a simple case in which a supply of weathered rock was always available for the stream.

This situation, however, is by no means universal. When rain falls on a bare slope of solid rock the clear rain water merely flows over the surface and does hot develop a gully. For water must have tools with which to work before it can attack solid rocks. In the first situation there was a layer of soil or decayed rock débris which was fine enough to be picked up by the running water.

Then carryingaway of this débris must, of necessity, leave a hollow in the soil; and since water in rivulets flows as a continuous stream, the hollows must be shaped like the path of the water, which is sinuous and winds down the slope. If water which carries sand and pebbles with if now flows over solid rock, the particles strike against the rock surface and abrade it like sandpaper. Thus are other grains torn out; and these in turn act on more rock, gradually carving from even the most resistant rock a path for the water to follow.

This wearing-away of rock masses by running water, or by air or ice, for that matter, equipped with tools for abrasion, is an important part of erosion, or the wearing-away of land masses by transporting agents.

The carrying-along of weathered rock débris (and, in fact, the process of weathering itself) is also a part of erosion, since it tends to Wear away the land surface; and the tools thus furnished to the stream are in turn able to cut away the solid, and as yet unweathered, rocks that they encounter.

Observations of streams and their activities have been carried on in many parts of the world, and there is no doubt that the network of rivers everywhere is the result of just such headward encroachments of streams, spread out over great lengths of time.

We are able, then, to state with certainty that streams serve the function of delivering the rain water back to the oceans, and at the same time they transfer tremendous loads of débris from place to place over the earth. Eventually, all débris carried by the streams is dumped into the sea; and, given enough time under stable conditions, land areas may be worn down nearly to sea-level by this process of erosion.

These almost boringly simple principles of stream development have not always been recognized. Indeed, not much more than a century ago a controversy raged over even the manner in which valleys originate. It seemed unthinkable to many that streams could themselves have carved their valleys from the land.

A series of catastrophic events was called upon, during which great clefts opened in the earth, which later became occupied by their present streams. Even the source of water from rainfall was not recognized. Writers spoke of great subterranean chambers which fed the streams and which were tapped by the very catastrophes that formed the valleys.

There were, however, some observers who noted that in

practically every case the size of a stream bore a definite relation to the size of the valley it occupied, that tributaries entered the main stream with an acute upstream angle, and that at the juncture of two streams the water generally met at the same level.

To them it seemed that, if valleys were purely accidental features due to earthquakes, for instance, there should be no reason why this systematic relation should hold in practically every case. Hence in opposition to the catastrophic notion there arose the present, almost self-evident, idea that streams themselves accomplish the work of cutting their own valleys.

One of the first geologists who recognized that valleys and other land features are due to the slow work of everyday forces, was James Hutton of Edinburgh. In 1795 he supported his views in a two-volume work on a Theory of the Earth, but unfortunately his style of writing was involved and his ideas did not gain the audience they deserved.

Consequently, his friend, John Playfair, became his disciple and set out to explain Hutton's ideas "in a manner more popular and perspicuous than is done in his own writings." But Playfair was no mere Boswell to his Dr. Johnson; rather he was a Huxley to his Darwin.

He himself gave numerous examples of the principles announced by Hutton and attracted a respectable group of followers. Opposition remained strong, but with the accumulation of more and more data the fundamental truth of the new doctrines became clear. Hutton's observations were admirably summed up by himself in his now classic quotation, that "in the economy of the world I find no traces of a beginning,—no prospect of an end."

We are at present much farther along than Hutton or Playfair in out knowledge of the details of stream action; and among the several factors that can be evaluated, there are some that deserve our present attention.

All the factors that control the velocity of a stream along its channel are not easy to evaluate, because of the complexity of some of them. But we may at least discuss several of them that will help us to understand how a stream transports its load. Certainly, the *slope* of the bed is involved; and if is easy to see that the greater the slope the larger the component of gravity that operates on the water. The *discharge* of the stream, or the volume of water flowing through a given cross section per unit of time, is also involved: the greater the discharge, the higher the velocity.

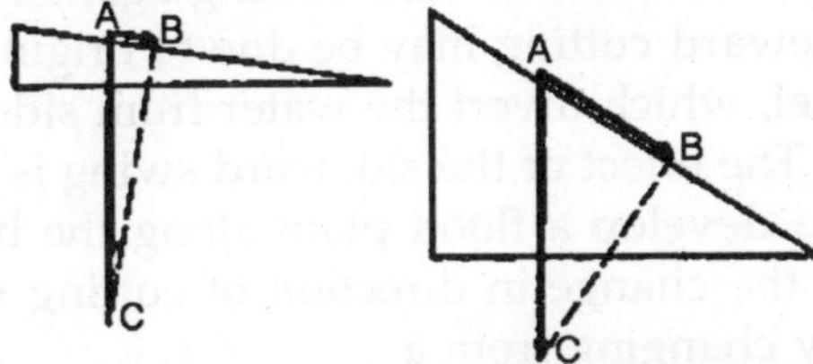

Fig. Effect of Gradient on Stream Velocity.

Likewise the *form* of the channel, whether it is broad and shallow or narrow and deep, is an important factor because of the frictional effects of a large area of stream bed in contact with the water. The frictional effect is also dependent partly on *channel roughness*, for a smooth bed offers less friction to the flow of water than a rough one. We may for out purposes, however, summarize these complexities as three main factors—the slope, the volume or discharge, and the frictional effects.

Whatever factors may predominate in a given case, the velocity of the water determines its kinetic energy, which in turn controls its carrying-power; and consequently the carrying-power is related to ail the factors we touched upon. As soon as material is picked up by the stream, some of the stream's energy is used in the transportation; and as a result the velocity of the stream is diminished.

The gradient of a stream is determined by the difference in altitude of the head and the mouth and by the distance between these two points. If a valley is incised into a plateau, where the relief is great, the gradient may be steeper than where the valley is cut into the walls of a rounded hill. The lowest level to which running water commonly may flow is sea level; and since gravitative forces seek to drain water to the lowest possible point, the waters of most streams ultimately find their way into the oceans.

The depth to which a stream may cut is determined in part

by the original configuration of the land and by the difference in elevation between the head and mouth of the stream. There must inevitably come a time, then, when the stream has excavated its valley about as deeply as local conditions will permit. When this point is reached, the excess energy of the stream is expended in sideward cutting against its banks.

This sideward cutting may be due to original sinuosities of the channel, which divert the water from side to side as it flows along. The effect of the sideward swing is to widen the valley and to develop a flood plain along the bottom of the valley. Thus the change in direction of cutting results in the stream valley changing from a

V-shaped notch, with relatively steep walls and a stoop gradient, to a flat U-shaped valley, with more gentle valley walls and a gentle gradient. With continued widening of the valley the stream is unable to cover the entire width of the flood plain, and hence it winds about the valley bottom in a series of loops or *meanders*.

Now, if there is any critical time in the life-history of a stream, it is this one, where the downward cutting becomes less pronounced and the sideward cutting predominates.

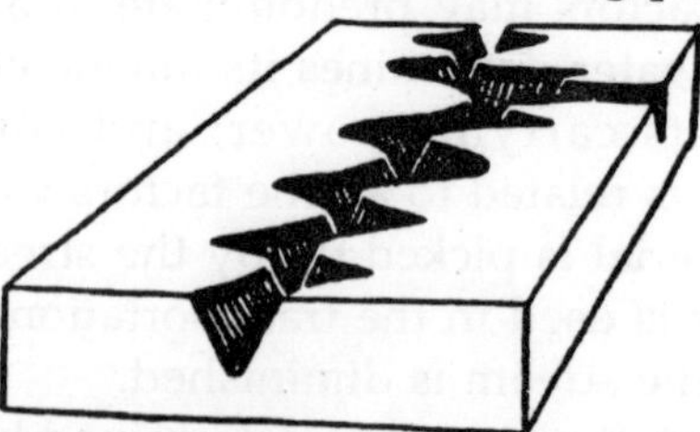

Fig. Block Diagram of A Youthful Valley, Showing its Typical V-Shaped Cross Section

The change in direction of cutting at this stage results in the stream valley changing its cross-sectional shape, as we have seen. The point at which the flood plain begins to develop is easily recognized. The shift in the direction of dominant cutting is expressed by saying that the valley has passed from youth to maturity.

The downward-cutting stage, with its steep gradients, steep walls, rapid movement, and small volume, marks the youthful stage of the stream's life-cycle. The flood-plain stage, characteristic of its maturity, is marked by sideward cutting, a lessened gradient, and a greater volume of water.

Finally, with a still greater width of the flood plain, an even lesser gradient, and slow movement, the old-age period of the stream is entered. The distinction between maturity and old age is one of degree rather than of kind, and it is not as sharply marked off as the change from youth to maturity. Refer, in this connection, to the photographs of streams and valleys in Plate 8.

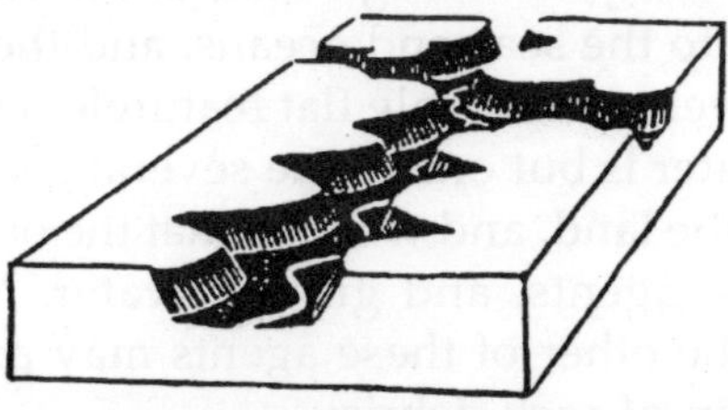

Fig. Block Diagram of A Mature Valley, with its Broad U-Shaped Cross Section, Due to the Presence of a Flood Plain

We must remember that we have here a dynamic process- that, while the stream is continually cutting headward in its youthful, lengthening ravines, the older parts of the valley are approaching the tower level to which they may cut, and that near the mouth the river may even be in the old-age stage.

This accounts for the fact that, as a river is followed downstream, the configuration of its valley changes markedly. IT IS important to recognize that this age classification of valleys is not based on time in years, but rather on the characteristics.

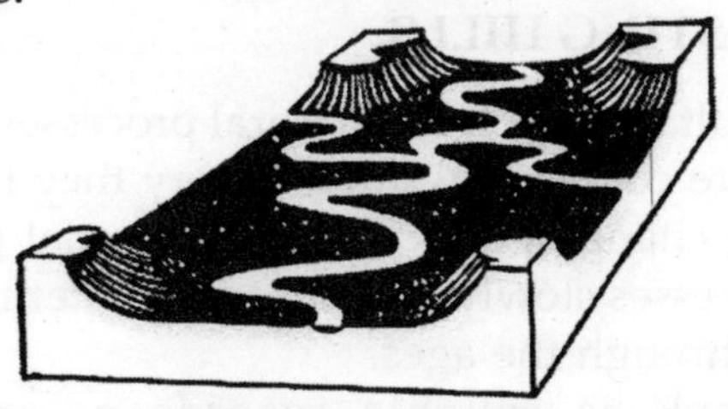

Fig. Block Diagram of a Valley in Old Age.

The stream meanders at will over the broad good plain

of the valley itself in terms of the stages through which it passes. In fact, the length of time required for such a set of events depends upon the nature of the rocks, on the climate, the relief of the region, and so on, so that some streams may pass through their entire cycles while others are slowly passing through maturity.

The great importance of streams from an earth-science point of view is that they act as agents in wearing down the land and in carrying rock débris from higher to lower elevations. Ultimately, if enough time is afforded them, this débris is carried to the seas and oceans, and the general level of the land is lowered until only flat featureless plains remain.

Running water is but one of the several geological agents that wear away the land, and we may that the others are wind, glacial ice, shore agents, and ground water. Now, in some regions one or the other of these agents may predominate in the transportation of rock débris:

Thus, in arid regions the work of the wind may be of great importance, and along the coasts ocean waves and currents may be most effective. By and large, however, it is the streams and rivers that are the most important erosive agents.

Given sufficient time, whole regions—nay, entire continents even—may be worn down essentially to sea level by the relentless gradational activities of streams. Before we consider some of the other agents, therefore, we may complete out picture of running water by turning now to the *regional,* as opposed to the *single-valley,* aspects of running water and its geological work.

THE EVERLASTING HILLS

IT IS difficult to conceive of natural processes so slow that during the entire course of written history they have effected no outstanding changes; and yet our terrestrial globe knows many such processes slowly but inexorably altering its surface configuration through the ages.

As far back as reliable records go, the general configuration of land and sea bas been about the same, and the ruins of the most ancient civilizations lie in geographic settings not markedly different from those when pomp and ceremony graced their age-old halls.

True, everyone realizes that some changes have taken place. Earthquakes, volcanic outbursts, tidal waves—all have left their marks here and there on the earth's surface. By and large, however, the same valleys are still present, the same mountain ranges, the same seas and oceans.

We today use the same passes over the Alps that were traversed by Hannibal, and the same majestic glaciers wind their frozen way along the same rock-ribbed clefts. Little wonder, then, that as far as average human experience goes, the hills are considered everlasting.

Notwithstanding this apparent permanence of hills and mountains, Tennyson was right. The hills *are* shadows. Every trickling rivulet, every gust of wind, carries with it some rock débris. Therefore, given time enough, whole land areas will be carried away by wind and water, each agent constantly at work reducing the continents toward sea level. True, it may require millions of years; but after all, a million years is not so very long in comparison with the awe-inspiring length of geological time.

For definite evidence, which we shall examine later, indicates that the age of the earth is to be reckoned in billions of years. Even the most case-hardened skeptic will grant that lots of things can happen in a billion years, though it is doubtful if any but the most imaginative can grasp the real meaning of such a great lapse of time. If the process of wearing away the land is so slow from a human viewpoint, how do we know that any vital changes do take place?

Observations by geologists in many lands have shown that, from an erosional point of view, there are areas in all stages of development; and by classifying these areas we are able to discern the underlying pattern of regional erosion.

Thus it is possible to work out a sequence of events that not only explains the land forms themselves but actually permits the development of a logical theory of the origin of land forms, relating each feature to the geological processes that developed it.

You may remember that we anticipated the present discussion when we described how streams develop first a small valley in which downcutting is dominant, followed later by a dominance of sideward cutting which widens out the valley and forms a flood plain. Furthermore, as the valley increases in length, it develops tributaries, until an entire area is covered by a network of streams.

We saw how all these features fitted hand in glove with the idea of a gradual development of stream valleys through several stages. In our first consideration of stream valleys we confined our attention to a single valley system; but of course there are many valley systems in any given area of a continent, and all of them are going through similar stages of development. Consequently, there should also be cycles of development through which entire areas go as a result of normal stream development.

The study of numerous regions does, in fact, permit us to pick out a sequence of regional forms. For example, there are large tracts of land in Utah and Arizona, called the Colorado Plateau, where thė elevated land surface is nearly fiat between the gorges and canyons chat here and there deeply incise it.

In other regions, as in much of West Virginia, the land surface is mainly in slope, with very little level land, and with streams cutting into every hill and mountain. Finally, we have examples of regiȯns, such as the lower Mississippi Valley, in which the land surface is nearly flat. In such areas a winding stream, which is not confined within a deep gorge as in the first example, flows on a wide, shallow plain with very gentle valley walls.

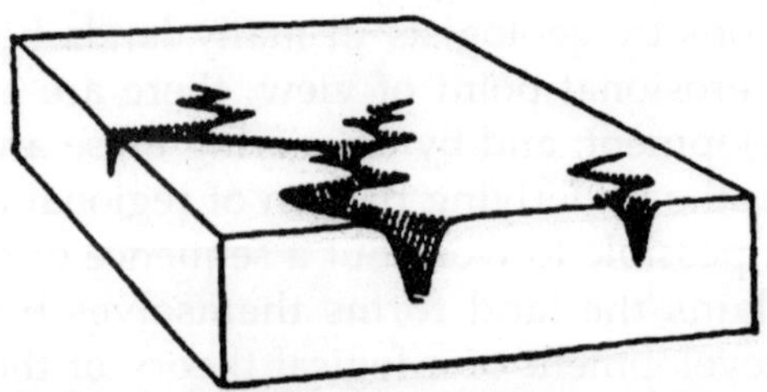

Fig. Block Diagram of A Youthful Region.

Betweent these several types of surface are all gradations,

which a detailed analysis shows result from a continuous process. We may sketch the broad outlines of regional development by considering some large plateau area, high above sea level, ignor

Note the dominance of uplanding for the present the question of how the plateau originated. The elevation of the plateau gives steep gradients to streams; and as time goes on, we may visualize the slow but unceasing excavation that produces valleys in the plateau, starting at the edges and gradually lengthening toward the interior.

The valleys will be deep and steepsided because of the relative height of the plateau; and the tributaries will be similar to the main streams, but smaller. In the course of events, them we may picture to ourselves an area in which the flat top of the plateau still stretches out in unbroken monotony for many miles, but with here and there a localized drainage pattern made up of a main gorge and its tributaries, the gorges perhaps several thousand feet deep.

If we allow time enough for the main gorge to be cut down as far as the base of the plateau, it may even have a narrow strip of floodplain adjacent to the stream. Between the flat top of the plateau and the flat stretch of the flood plain there is considerable relief; and to distinguish the several levels we call the plateau top the upland, the valley walls the slope, and the floodplains the bottom land.

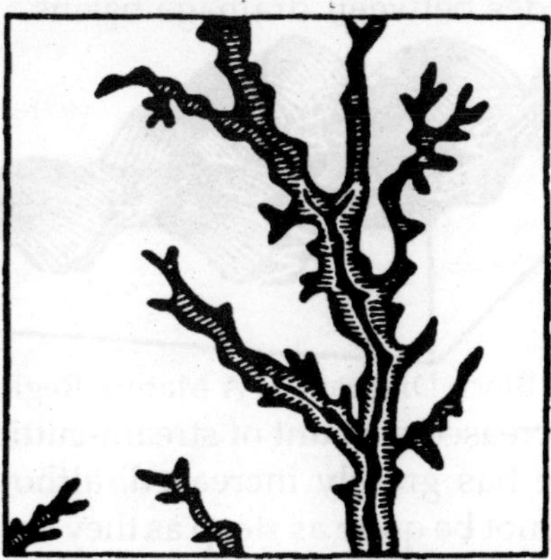

Fig. Map of Youthful Region.

Let's stop a moment and see whether we can describe the area in this stage of development a little more precisely. We see before us a region in which most of the surface as seen on a map.

The white area is upland. Note the narrow flood plain at the of the valley is still an unbroken upland. Here and there, where the streams have penetrated the plateau, there are steep slopes leading down to the valley bottoms. The total amount of slope, again considered as an area on a map, may be only a few per cent of the total.

Finally, along the bottoms of the larger canyons there are narrow flood plains here and there, making up the very small remainder of the total map area. Since we started out with 100 per cent upland, it is obvious that the present stage of development cannot be very far along, because there still is a great preponderance of upland left. Consequently, we call this a youthful stage, on the basis of the ratios of upland to slope and upland to bottom land.

While we have been considering this first stage, the process of erosion has continued; and, coming back now a million or so years later, we find that the streams have grown headward farther and farther into the plateau and have spread a network of tributaries over most of the region. We may still find a few small areas, however, where the original flat top of the plateau remains; but, if so, these are now merely narrow, flat-topped divides between drainage basins.

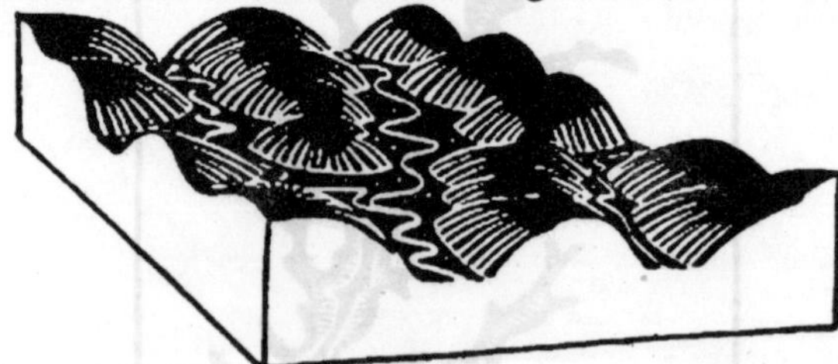

Fig. Block Diagram of A Mature Region.

With the increased amount of stream-cutting, the amount of land in slope has greatly increased, although the slopes themselves may not be quite as steep as they were in the earlier stages. The major streams, too, have developed wider flood plains, and as a consequence we find that the ratios of upland, slope, and bottom land have changed considerably. The upland has been decreased to a small percentage of the total map area or has been obliterated entirely.

Meanwhile the area in slope has increased to dominance in the region; and the bottom land itself has grown to some extent in area.

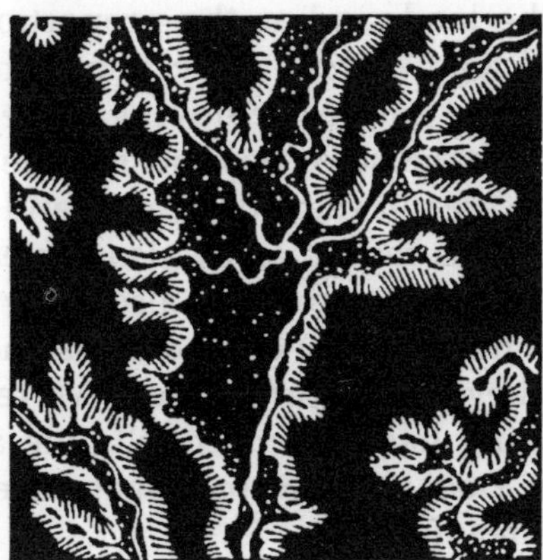

Fig. Map of Mature Region.

Here the slopes clearly predominate, and the characteristic youthful aspects are gone. We accordingly describe the region as mature on the basis of the high ratio of slope to bottom land and upland. Somewhere in the transition from youth.

The upland has entirely disappeared, and the area in flood plain (dotted) has increased maturity the last remnants of the original upland entirely disappear, so that slopes occupy all of the region except for the still relatively small amount of bottom land.

When the land is mainly in slopes like this, we naturally expect that the number of streams and tributaries will be at a maximum, because it is the streams that are responsible for the slopes. This is actually the case, and during the stage called *middle maturity* the streams are as closely crowded as they can be, and the drainage pattern has reached its greatest development. In this stage also, the amount of débris that is being carried away is at a maximum.

We may with advantage pause here to remind the reader that all during this erosion of the land, the streams are carrying away rock débris. The process we are describing is thus *unidirectional*: all the factors involved tend in one direction, the direction of *degradation* or downcutting.

During regional youth, when streams are few, the amount of débris carried off is relatively small; but as more streams incise the land, and as the proportion of slope increases, the total load carried away also increases, until in middle maturity the rate of removing material reaches a maximum, as we have said. Beyond middle maturity, we shall find that the area in slope decreases; and as it does, the amount of rock débris carried away also decreases.

While we are pausing, we may also draw an interesting contrast between the maturity of land areas and the maturity of individual valleys. A mature valley, we recall, is one in which downcutting has become negligible and sideward cutting dominant. Does it follow that a mature region will have mainly mature streams?

Not at all, because if the number of streams has been increasing all the time, it is obvious that there must have been considerable high land available so that new tributaries could continually cut headward. These new tributaries will be in their dominantly downcutting stage, so that, as a matter of fact, most of the smaller streams in a mature region are themselves youthful.

This fact ought not to cause confusion, however, as long as we remember that we are talking about two different things when we contrast stream age and regional age. Furthermore, as emphasized earlier, the ages are not measured in years, but in stages of development. According to the natural laws that control the development of both streams and of regions, it follows that during maturity in a regional sense the conditions are most suited to youthfulness in the majority of tributary streams.

Now, as we follow our regional development still further, the reader will be able to anticipate that, as the numerous streams of middle maturity continue their work, they will gradually deepen their valleys, and finally more and more will reach the lower level of downcutting. From this point on, sideward cutting will increase relatively, and the total amount of bottom land in the area will correspondingly increase.

Consequently, a few large streams will take the place of many small ones. In a stage somewhat later than middle maturity, then, we should expect to find that many of these streams have relatively wide flood plains and that the slopes are no longer nearly as steep as they were.

In fact, the divides between adjacent streams tend to become rounded off into gentle hills, and much of the ruggedness of the earlier stages is lost. The upland, we saw, disappeared long ago; and with the widening of the flood plains the ratio between slope and bottom land is increasing in favor of the latter.

The stage will inevitably be reached, given enough time, when the total area occupied by bottom land will be greater than that occupied by slopes. When this situation develops, we say that the region has reached old age. These three principle stages of regional age are shown in the photographs of Plate 9.

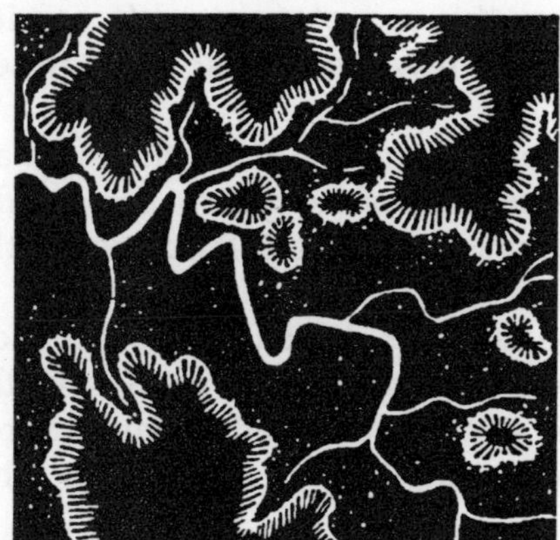

Fig. Map of Region in Old Age. The Bottom Land Predominates

Here, then, are three distinct stages through which entire regions may pass; and as we should expect, there are all possible gradations between the stages, because this is a continuous process and moves along as smoothly as time itself, it will be interesting to venture a bit farther and inquire what the very final product of this whole process is.

Fig. Block Diagram of a Peneplain; The Ultimate Stage of Stream Erosion

Well, as the streams continue carrying away the débris, the slopes become still more gentle, the divides are lowered still more in height, and the main streams become more and more sluggish as they reach a state of complete equilibrium between their valley bottoms and sea level.

No stream can cut much below sea level, of course; and in addition, there must be a slight gradient from the ocean upward into the continent, to allow the water in the streams to flow into the sea.

A flat, or nearly featureless, plain, with sluggish streams that slowly wind toward the seas? All about your point of vantage stretches this monotonous surface, with the horizon broken only by a few low hills; it is, indeed, a nearly horizontal line.

If you can picture such a plain, then you are visualizing the final peneplain stage of the cycle of erosion, the lowest level to which streams can reduce a land area if they are given unlimited time in which to work.

We Realise that it is difficult to obtain such a long-time point of view on short notice, but we believe that the logical sequence of the events described should impress the reader with the essential simplicity of the process when it is shorn of details.

There are complexities galore in actual practice. The rocks of the area may not be all of the same kinds, so that some will yield more readily than others to weathering and erosion; the original laud may not have been a plateau at all; and finally, various other types of events may interrupt the process so that not enough time is afforded for a completion of all the stages. Changes in the levels of land and sea may, and usually do, interrupt the procedure, so that one cycle of erosion becomes superimposed on another until such a degree of complexity is reached that even a full-fledged geologist finds great difficulty in deciphering the record.

Fortunately, there are many regions available for study in which the complexities are not insurmountable. It is from these that the underlying principles have been drawn; and, armed with the laws that control the situation, geologists are pretty well able to attack even the more special and complex situations.

We may see, then, why we can state with reasonable assurance that land areas undergo systematic changes with time, because in nature we find regions in all stages of erosional development. It is a monument to human thinking that, from the apparently confusing and chaotic abundance of land forms, it has been possible to set up a logical system of events that both explains and systematizes the many details of land sculpture.

Furthermore, the possibility of relating these details to the underlying erosional processes places a finishing touch, as it were, on our ability to understand the devious ways in which nature does her work.

The study of land forms, and the classification of them in accordance with the processes that formed them, is called physiography, an important branch of earth science. In the discussion of this subject we have been concerned with the land forms that result from the erosive work of streams, but of course streams are only one of several geological agents.

Each of the agents—and here are ineluded wind, glacial ice, the waves and currents of the oceans, and ground water—have their own sequences of land forms, due both to erosion and deposition.

For we must not forget that while erosion takes place at one point, deposition takes place somewhere else. Thus far we have not said much about these depositional types of topography, but their time and place will come.

Finally, in addition to the particular work that each agent performs, usually more than one act simultaneously on a region, thus giving rise to a whole series of related cycles moving along side by side.

As an example of such a situation we may think of most any high mountain region where wind, streams, and glaciers are all contributing to the downcutting process.

The past history of the earth is marked by numerous cases in which whole masses of land have been cut down, from lofty mountains to actual peneplains.

The process we have been sketching is thus not purely hypothetical, or one that is just starting for the first time; but rather it is a process which has seen fruition on many occasions in the past. So important are these ancient peneplains in marking off stages of the earth's history that later we shall have occasion to mention them time and again.

Chapter 17

Underground Water and Laminar Flow

Athanasius Kircher, one of the greatest writers on natural science in the seventeenth century, thought that all rivers and streams arose from underground lakes or reservoirs, which in turn were supplied by conduits leading to them from the oceans. When we remember that this was back in 1665, before many scientific observations had been made, we see that Kircher's idea was rather ingenious.

He at least recognized that the lithosphere contained abundant water, and that this water somehow escaped to the surface, where it contributed to the supply in rivers and streams. In short, he sensed what we know quite well today, namely, that water is one of the most ubiquitous things on earth. The hydrosphere, which covers some threefourths of the globe's surface, is named after it; and the atmosphere contains on an average about I per cent of water vapour. The lithosphere also has its own important share of water.

Whenever rain falls on the ground, some of the water inevitably sinks down among the interstices of the rock and thus enters the body of the lithosphere. This is called ground water, and locally it may completely saturate the pores of the rocks. Ground water circulates beneath the surface, and during its travels it performs work which is geologically important. Suppose we accompany it for a part of its journey and see what happens.

We Mentioned the pores in rocks. To the average person the very phrase porous rock" seems a contradiction, and it may be well to explain further. We already know that rocks are not the impregnable, everlasting things we may have thought they were; so it is not difficult to understand also that most rocks are not perfectly dense.

They are made up of solid particles like crystals and grains, among which are open spaces, so that a varying amount of pore space is present in almost every rock, at least near the surface of the earth. To make the notion concrete, imagine a rock made up of small spheres, packed together as closely as possible. Then there would still be about 26 per cent of open spaces among the spheres.

Here is an interesting thing, however: no matter what size the spheres are, any given volume of them packed in the same manner has exactly the same amount of pore space.

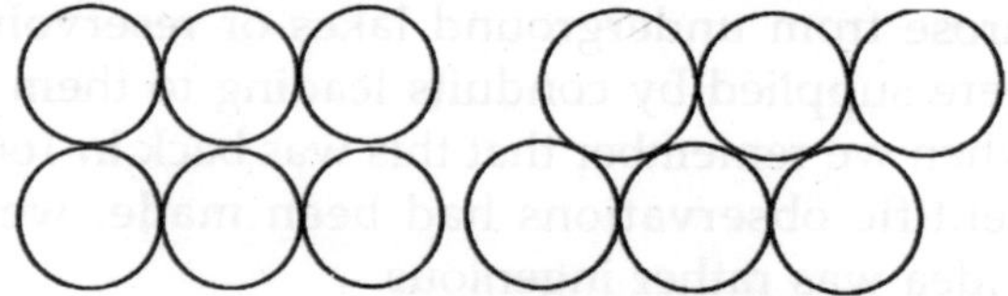

Fig. Spheres Packed in two Manners.

That means that a cubic foot of rock made up of spheres one-hundredth of an inch in diameter would have as much water-holding capacity as a cubic foot of rock made up of spheres 1 in. or more in diameter. With more irregular grains, and also with mixtures of various sizes, this porosity of course varies considerably.

Rock porosity is obviously important in the general subject of ground water, but even more significant is the rate at which the water is able to circulate among the rock pores. Consider our 1-in. spheres, for example, in which the open spaces will be rather large. We can easily see that water would have little difficulty in moving about between such spheres. Permeability is a measure of the ease with which fluids move among the pores of rocks.

Hence a rock made up of these 1-in. spheres would be quite permeable. Now if the spheres are gradually decreased in size, the spaces between them also become smaller, and consequently the water must pass through smaller and smaller openings in the rock. Eventually a stage is reached in which the total surface area becomes so great that the spheres begin to attract an appreciable part of the water by molecular forces.

When this happens, it requires a greater pressure or a longer time, or both, for the water to move through the pore spaces. Permeability, then, plays an important rôle in the movement of ground water.

Near the surface, of course, the rocks may be shot through with cracks and crevices, so that the water has many lines of descent; but with increasing depth, such openings are less common, and the water in large part must pass through the pores of the rocks. The latter may be likened to a sort of continuous sieve, with openings of various sizes.

As a general rule, ground water moves very slowly, because of the variations in the permeability of the rocks it encounters, and also because of the character of the forces which set it in motion. We shall discuss these forces later, but for the present perhaps we had better get back to the surface and see what is going on there.

Three things may happen to rain that falls on the land. It may evaporate and return to the atmosphere, it may run off down the slope, or it may sink into the ground. If it goes back into the atmosphere, it ultimately contributes to the rainfall somewhere else; if it runs downslope, it feeds into streams and so contributes to the work of running water.

If it sinks down into the earth, it becomes ground water, and so enters our present bailiwick, Now the relative proportions of rain water that follow one or another of these three paths clearly depend on the conditions of the locality.

In a region of steep slopes, most of the rain flows directly to streams; but in a region dominantly fiat, a fairly large amount of water sinks downward into the ground. Atmospheric conditions, climate, and a number of other factors control the rate of evaporation.

Whatever proportion of rain water enters the soil and rocks beneath the surface tends to sinks; straight down under the influence of gravity until it reaches a zone where the rocks are already completely saturated. Everybody knows that the soil is not usually sopping wet to the surface, but that, when wells are dug, it is necessary to penetrate some distance downward before a supply of water is found. The level below which the rocks are saturated is called the ground-water table, and in any given region it bears a certain relation to the configuration of the land surface above it.

A considerable amount of fairly exact data has been collected on the subject of the ground-water table, particularly by observing the height of water in wells. From such data it is easy to construct a map or surface showing the relationship of the ground-water table to the land surface above.

The adjacent diagram shows the general relation of these two surfaces. Note that the ground-water table is at the same level as the stream in the valley, and that it rises beneath the divides, so that literally the hills are mounds of water as well as of land.

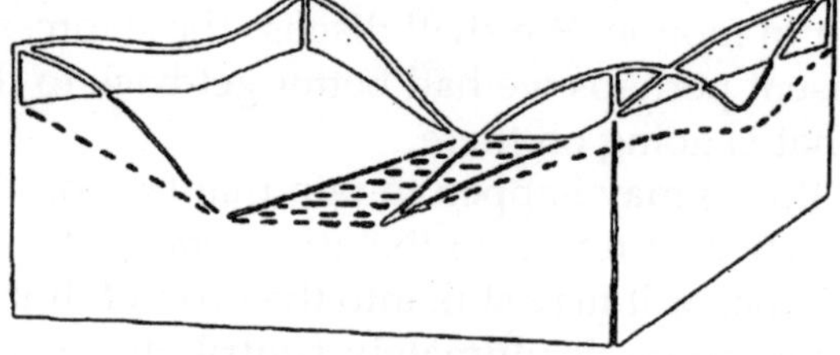

Fig. The Surface of the Ground-Water Table

The fact that the ground-water level coincides with the stream level is important, because it explains why streams continue to flow ecen after the rain has ceased. The water that entered the ground during rains is slowly fed to the streams, and so maintains them in the dry weather. This is not universally true, of course, because ravines and gullies usually are quite dry between rains.

The reason, obviously, is that these little fellows have not yet cut down far enough to intersect the permanent ground-water table. As the ravine is followed downslope, however, a place is reached where it carries a small permanent stream. The head of this stream marks the level of the ground water.

We see from the foregoing discussion that where the groundwater table intersects the surface, water issues from the earth. Thus *springs* are nothing but seepages of ground water where the land surface is cut by the water table. Naturally enough, the water that seeps out usually comes from a higher level.

There are many complexities among individual springs, and we shall not pursue them all. In some cases the water issues from more permeable beds; at others there may be hydrostatic pressures involved; and so on. In the main, however, it is reasonably accurate to say that wherever the land surface dips down beneath the ground-water table, there water will issue from the earth. Lakes and swamps, then, are merely depressions in the land where the ground-water table comes out of concealment into the open.

The upper level of ground water can be clearly established. Let us see now whether there is also a lower level, or whether, conceivably, the ground water fills all rocks far down into the interior of the earth. Observation again comes to our aid, for we know that in many deep mines and deep wells the rocks are quite dry.

Surprisingly enough, this is only a matter of some thousands of feet. In other words, it is not unusual to find that deep wells, such as are drilled for petroleum, penetrate through a zone of wet rocks, where water is abundant, down to a zone of dry rocks where little or no water is encountered. It seems to be pretty generally true, then, that a completely saturated zone of rocks occurs only relatively near the surface—say within a mile or two at most.

And yet there is some water at greater depths, because steam is given off abundantly by most volcanoes, and it is generally believed that this steam may come from considerable depths. Whether it is water that originally arose as ground water percolating farther and farther down, or whether it is some original source of water deep within the earth, we are not certain.

It is likely that the increasing pressures on the rocks far beneath the surface may have an effect on the depth to which the ground water penetrates. Eventually, if one considers a point deep enough beneath the surface, the pressures of the overlying rocks are so great that none of the rocks buried there has open cavities or pores; but such a depth is a matter of miles rather than of feet.

One reason why the problem is not completely solved is that the number of deep wells that go below the ground water is limited. After all, most wells are drilled for water itself, and so they stop when they penetrate some short distance into the ground water.

Occasionally, a rather curious situation is met when wells are drilled. The water, instead of remaining at about the level of the ground-water table, rises up into the well, and in some cases flows out on the surface. Such wells are called artesian. The explanation of these wells is not difficult in the light of what we already know. Consider a layer of highly permeable rock between two that are relatively non-permeable or impervious.

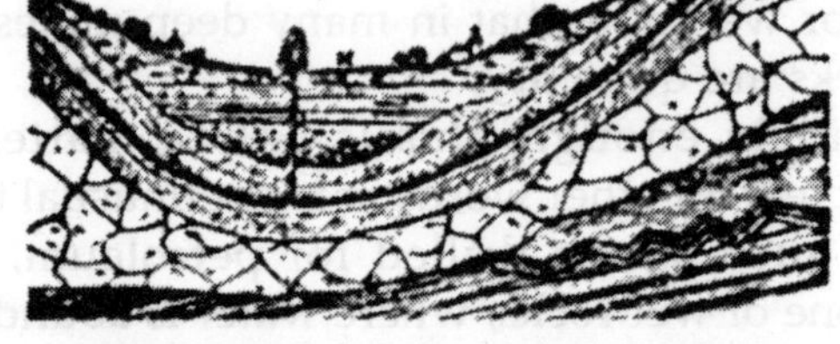

Fig. Conditions Necessary for An Artesian well.

If these rock layers are inclined downward from the surface, the ground water within the permeable bed is confined by the impervious layers and thus exerts a pressure against its surroundings. If man drills a well through the overlying impervious layer, he affords the confined water an opportunity of relieving its pressure, which is promptly done by some of the water flowing up through the well.

Artesian wells are very important, economically speaking. Many cities and towns depend entirely on artesian waters; and even in Chicago, with a huge take at its doorstep, many industrial firms have used artesian water. One of the important sources of this water is a sandstone which is exposed at the surface in Wisconsin at altitudes higher than Chicago.

These beds dip downward beneath the surface toward the

south, affording a means of circulation of rain water from the more elevated exposed catchment area to the metropolis distant several hundred miles.

But after all, artesian conditions demand special relations among the rocks. More usually there are alternate layers of rock of varying degrees of porosity and permeability through which the ground water as a whole tends to circulate. When we come to consider this general circulation of ground water, we find ourselves faced with some problems, for it is difficult to make direct observations on the movement of water beneath the surface of the earth.

Wedo know definitely, however, that ground water moves very slowly. The water percolates among the rocks at a rate that may be expressed as a few hundred feet per year.

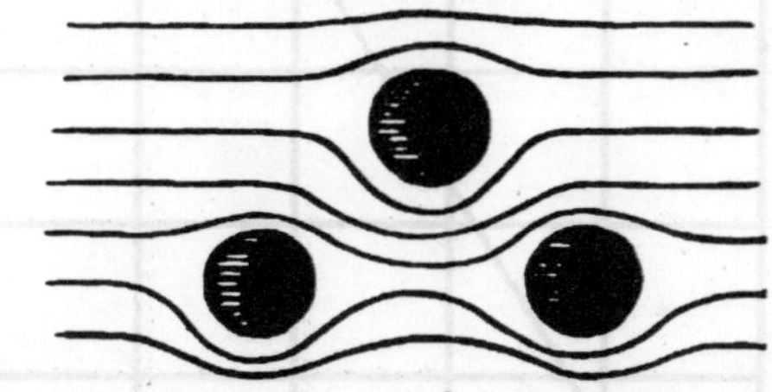

Fig. Streamlines in Laminar Flow

This sluggishness is a result of the great frictional resistance the water encounters as it migrates among the small interstices in the rocks. Ground water displays no turbulence in its slow movement among the pores. Here is a case where water flows with another type of movement, which is called laminar flow.

In laminar flow, any given thread of the liquid retains its identity and flows smoothly alongside its neighbors. It is a sort of streamline flow which adjusts itself to irregularities in its path by curving around them, rather than by setting up whirls and eddies in the current.

Laminar flow involves the concept of *viscosity*, which is an internal friction in the liquid, so that there is a drag of the liquid layers one on the other. Now, water actually is a viscous fluid; but the running water of surface streams, because of its relatively high velocity and the irregularities in its path, seldom displays laminar motion. Beneath the surface, however, laboratory experiments have indicated that the slow movement of water through rock pores due to gravity is definitely laminar.

Some readers may be interested in the experiments that study the flow of water and other fluids through rocks. In the field of petroleum geology it is important that the nature of the movement of oil and its associated water be known. Consequently, cores of actual rock are mounted in suitable apparatus and water

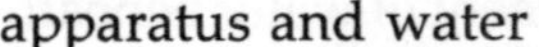

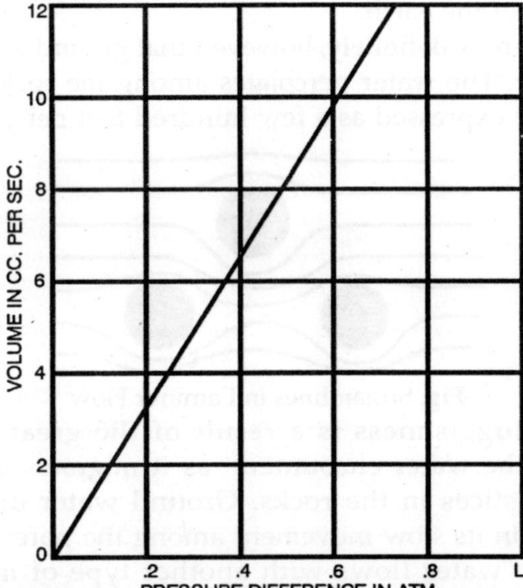

Fig. Graph Showing Flow of Water through Cisco Sandstone. or other liquids are forced through under pressure.

The amount of water flowing through per unit time and the pressures that cause this flow are measured. These observations constitute the factual data. From theoretical grounds it is known that if the volume of water flowing through the rock is directly proportional to the difference in pressure at the two ends of the rock core, then the flow is laminar.

In other words, the graph of the experiment is a straight line in laminar flow; whereas, if the curve departs from a straight line, then turbulence is present, for the energy dissipaled in eddies and whirls reduces the volume below its expected theoretical amount. Graphs, such as the accompanying one, are based on experiments of this sort. Covering a wide range of pressures and velocities, they prove that the ordinary gravity movement of water in rocks is laminar.

To get some idea of the causes of ground-water circulation, we shall have to assume a rather simple situation in nature. imagine a valley between two divides, in which it has been raining "for forty days and forty nights," as it were, until all the pore space in the rocks of the divides is filled with water. Now imagine the rain to cease. The excess water on the surface will flow down to the stream and be carried away. Meanwhile the surface of the ground water in the rock coincides with the surface of the land.

This underground water also has a gradient, then, down which it may flow. Gravity is able to operate on the mass of water, and a circulation results. Some of the water, perhaps a large proportion of it, tends to move down parallel to the land surface toward the stream.

The slope of the ground-water surface will tend to remain about parallel to the land surface, but with an ever increasing departure in its higher portions due to the greater circulation from those higher points, Meanwhile the stream itself is exerting an effect on the circulation of the ground water. But here we must pause for a bit of theory.

If two ships pass each other at sea, their captains have to be careful so that their ships do not approach each other too closely. If they do, then the streamlines of water movement between the boats are constricted, and a powerful suction pressure is generated between the ships. As a result they may be drawn together and collide.

Now a somewhat analogous thing is going on in our picture. As the water in the stream flows along its valley, it generates a suction pressure in its immediate vicinity which tends to draw the ground water into the stream itself.

Thus are set up slow and almost imperceptible circulations

which may theoretically extend to considerable depths below the gradient that is established by gravity alone. For practical purposes, however, effective circulation probably does not extend much below the level of the main stream in the region. Below some such level the slight mass movement that may be present is probably negligible.

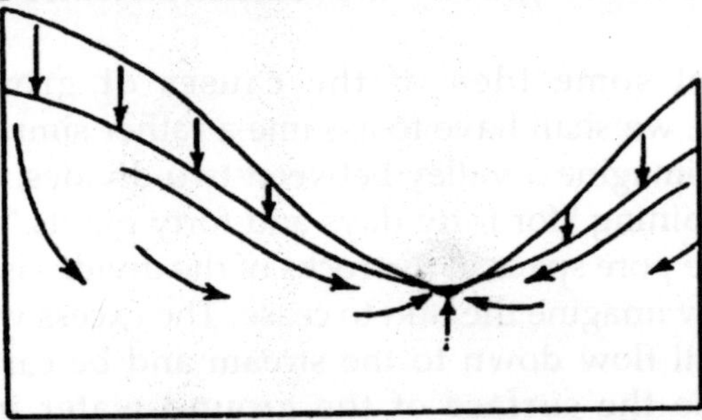

Fig. Ground Water Descends Vertically.

The very slow circulation of ground water, combined with the fact that it is passing through what may be considered a continuous sieve, means that most ground water does not transport solid matter.

By far the greatest transportation work of ground water is done by solution, and consequently we should expect its action to be limited to the more soluble minerals. In fact, *limestone,* among all the common rocks, is most susceptible to ground-water attack, because the calcite of which it is composed is the most soluble of the commoner minerals.

When we talk about the solubility of calcite, we do not want our readers to think that it compares with the solubility of sugar, for example. As far as ordinary things go, calcite is relatively insoluble; but in contrast with most of the rocks that occur in nature, it is readily subject to solvent action by percolating waters.

To make the thing more specific, we may say that only about 0.001 gm. of calcite can be dissolved in 100 gm. of carbon-dioxide-free water. That isn't much in comparison with sugar, say, of which about 200 gm. will dissolve in 100 gm. of water. But if we compare calcite with feldspar, we find that calcite is many times as soluble. After all, even a slightly soluble material, acted on by water for a long enough time and in great enough volume, will be completely dissolved. This is the case in nature.

There is, however, an important factor that tends to make calcite even more subject to attack by solution than the figures just given would indicate. When there is any carbon dioxide in the water, then the solubility of the calcite goes up by leaps and bounds. That is because the calcite is converted into *calcium bicarbonate,* $Ca(HCO_3)^2$, as expressed in the reaction:

$$CaCO_3 + H_2O + CO_2 \rightarrow Ca(HCO_3)_2$$

Calcium bicarbonate has a solubility of about 0.04 gm. per 100 gm. of water, which makes it about forty times as soluble as calcite itself. It takes no expert to see where the carbon dioxide comes from. As the rain falls through the atmosphere, it dissolves some of that gas from the air; and as it percolates through the rocks, any calcite in its path soon goes into solution.

Suppose rain falls on an area in which limestone lies just beneath a thin layer of soil. Near the surface, where the loose soil affords an easy path for the water, there will be a percolation of carbon-dioxide-charged water coming in contact with the limestone.

As the water runs down the cracks in the rock, some of the limestone is dissolved, the cracks are widened, and there is an even easier path for later rain water.

The limestone also varies in permeability here and there, as all rocks do, so that in the course of time the percolating waters will dissolve out a regular series of passages. When they do, caves and caverns result. Surprisingly enough, these scenic underground features, which annually attract so many tourists, are due chiefly to the unobtrusive work of ground water operating on soluble rocks.

As a general thing, caves are in those zones where the circulation of the ground water is most pronounced. That is, the lower levels of a series of caverns tend to be adjusted to the level of the major stream in the area. If the caverns become large, and are dissolved out near enough to the surface, their roofs may lack support and consequently fall in.

Fig. The Unobtrusive Work of Ground Water.

The steep-walled circular or oval openings that result are called sinkholes. They are relatively common in some cave areas, as we should expect. Sinkholes may also be formed by the enlarging of crevices leading down into the limestone; but whatever their origin, they act as funnels to lead water down into the rock layers, and thus help to carry on the work of solution. In regions having numerous sinkholes most of the rain that falls on the surface may run down into the sinks, with the result that surface streams may be largely or entirely absent.

Almost everyone who has visited a cave or who has seen pictures of them seems to be more impressed by the curious for, mations in them than he is by the cave itself. From the roof of the cavern may depend long slender "icicles" of stone; and upward from the floor spring similar, but generally more squatty, grotesque deposits.

These formations are only indirectly due to the solvent action of ground water; they are the direct results of its depositional activity. The "stonecicles" that hang from the ceiling are called stalactites, and those that rise from the floor are stalagmites. In some eases they join; others are sheetlike instead of cylindrical: but one and all are due to a redeposition of calcite dissolved elsewhere by the ground water.

Most caves have entrances, and so there usually is air circulating through them. Thus there is a chance for evaporation. If rain water, circulating downward through the rocks above a cavern, dissolves some calcite along its journey, it may drip from the roof of the cave. As it reaches the air it evaporates in part; and when it does, calcite is deposited. Thus by a slow process stalactites are formed and elongated.

Meanwhile the water that drips to the floor may evaporate there, and thus build up a stalagmite. As usual, this simple picture is complicated in nature, and we find that relative concentrations of carbon dioxide, pressures, and especially temperature changes enter into the interplay of factors that decide whether the ground water is to be in a dissolving or depositing mood.

Again we must forego technicalities, but we can explain that if the environment entered by the water is one of relatively high carbon dioxide concentration, solution will be the tendency, whereas a lower concentration of carbon dioxide will cause deposition. There really is a nice balance here in any given cast and it affords one more example of the approaches toward equilibrium that Nature seems so anxious to achieve.

Those among our readers who like to study chemical equations will be interested to note that the following reaction really expresses the relationship between ground-water solution and deposition: For example, if conditions tend toward an increase of CO_2, the lower reaction is driven in an upward direction, and this drives the upper reaction toward the right.

Ground-water work is not confined solely to the solution and deposition of calcite, for it plays an important rôle in many geological phenomena which can scarce be mentioned here. For example, ground-water deposition of minerals about some nucleus may form *concretions* which range in size from microscopic objects to symmetrical structures a dozen feet across.

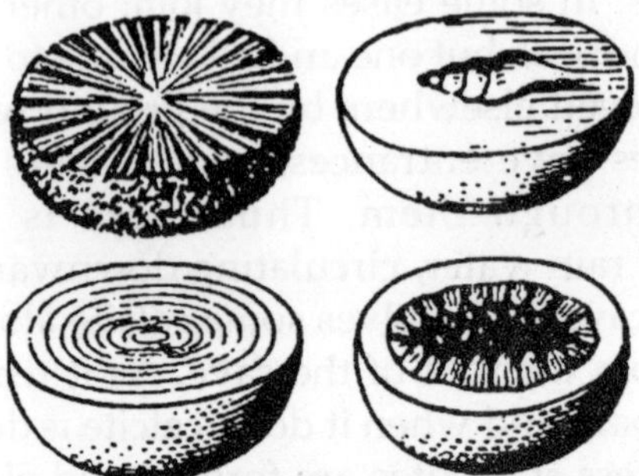

Fig. Concretions are Formed when Ground Water Deposits Material about A Nucleus.

The deposition of minerals in rock pores is responsible for cementing together sediments deposited by other agents. Ground-water deposits in larger cavities or fissures may result in the formation of *mineral veins.*

Furthermore, deep beneath the zones of ground-water circulation, there may Still be a diffusion of dissolved materials; and when such diffusing solutions meet, chemical reactions may take place.

Thus are formed some types of ore deposits. In fact, very commonly ground water acts to concentrate otherwise thinly distributed ores, and so contributes definitely to human economy.

Geysers, those openings in the earth from which steaming hot water is ejected at intervals, are more spectacular evidences of ground-water activity. Yellowstone Park is the region par excel. lence for these phenomena, and annually thousands of tourists be come acquainted with this relatively rare aspect of ground-water work.

But ground water alone is not sufficient for geyser action: there must be a source of heat, as well as peculiar conditions among the fissures in the rocks. Since the source of heat is volcanic, the subject of geysers is obviously transitional between the current topic and volcanism.

With our treatment of ground water we have finished two important aspects of the geological work of water. Our consideration thus far has been largely with regard to *rain* as the uitimate source of the water. But what if it *snows* instead?

The snow may melt or evaporate, as most of it does; but here and there may be an excess of accumulation over melting.

Fig. Beehive Geyser, Yellowstone.

RIVERS OF ICE

Nature is infinite in her moods. Consider water, for example. Under moderate temperatures it tumbles along the surface, constantly adjusting itself to irregularities in its path, seeking ever the lowest level it can find. Raise the temperature somewhat and the water vaporizes, disappearing into a tenuous gas which finds itself at the mercy of even the most capricious winds.

But now reverse the conditions, allow winter to hold the water in its frozen grip, and the turbulent spirit of liquid or gas is stilled. But is all movement necessarily lost? By no means. Almost everyone has read about the movement of those great rivers of ice called glaciers. Yes, given suitable conditions, even the apparently solid ice may flow; and on this curious planet of ours it is not so unusual to find just such favorable conditions for glacia movement.

Fig. Rivers of Ice

When precipitation takes the form of snow instead of rain, the blanket of snow does not run off down the slope, but accumulates until it evaporates or melts; in the latter case the *meltwater* at once runs off.

Now under certain favorable situations, such as high latitudes and high altitudes, the heat of the summer sun is not great enough to melt or evaporate all the snow of the preceding winter.

Hence there is an accumulation of snow from year to year, which finally assumes considerable proportions. Large snow fields are built up in this manner.

Their lower limits form the permanent snow line of the mountains. Near the Equator this snow line oc urs only on the tops of the loftiest mountains; in the middle latitudes it is found at much lower elevations; and in frigid zones it may descend to sea level. Thus, from Equator to poles there is a descending surface above which snow may remain all the year around.

Although all the snow does not melt, some of it does, and part of the meltwater, percolating down into the snowbank, recongeals to form granular particles of ice. Similarly, the pressure of the overlying snow aids in converting the deeper layers into ice granules, which pass downward into porous ice; and toward the base, where the pressure is greatest, the ice may be densely compacted.

The pressures developed by the overlying snow and ice tend to seek relief, and this relief is found by the lowermost marginal parts of the ice mass moving down the slope of the mountain. Thus are glaciers born, and where movement begins is commonly regarded as the beginning of the glacier.

Glaciers do not move with a velocity anything like that of water. The ice moves slowly and ponderously: it is a matter of feet per year rather than feet per second as in streams. The movement of glacial ice, indeed, is no simple matter, as we shall shortly see; at present we are concerned mainly with the fact that it does move downslope.

Fig. The Snow Line Descends from the Equator Toward the Poles

As the glacier moves downward, it tends to follow any irregularities along the surface. Commonly, glaciers follow pre-existing stream valleys down the mountain side. These we call valley glaciers. The distance downslope that the glacier can extend depends in part on the increase of temperature in lower altitudes and in part on the rate at which snow is accumulating in the snow field above.

There tends to be a balance reached in any given situation; and consequently glaciers may be large or small, and long or short. In high latitudes the glaciers may persist right down to the sea, and there break up into *icebergs*. Elsewhere the snout of the glacier reclines along the valley, and equilibrium is reached when the rate of arrival of new ice to the margin is just balanced by the rate of melting. From the terminus of the glacier issues a river which carries off the meltwater.

VALLEY GLACIERS

Valley glaciers are fairly common in the mountainous

areas of the world, especially in middle and high latitudes. The Alps are famous for their glaciers; it was there that they were first studied, and for that reason valley glaciers are often called *Alpine glaciers*.

In the United States valley glaciers are rather rare, being confined largely to the higher peaks of the Cascade and Wind River Mountains. Farther north, in British Columbia and Alaska, there are thousands of them, many of which persist down to sea level.

The similarities of valley glaciers to rivers may be seen by considering the snow fields above, which constitute their catchment basins; their confinement to valleys; and their movement downslope. In addition, valley glaciers may have tributaries, so that two or more glaciers may merge to follow along a single valley from their point of juncture. In contrast to these similarities, of course, there are striking differences.

In Addition to valley glaciers there are much larger masses of glacial ice called continental glaciers or *icecaps,* which cover all elevations of the region. Antarctica is covered with such an icecap, as is the greater part of Greenland.

In fact, an area about twice the size of the United States, or nearly 6,000,000 sq. mi. of the earth's surface, is today covered by such ice. These icecaps are much thicker at the center than at the edges, so that there is a surface gradient from the center outward

Fig. Greenland with its Cap of Ice

This surface gradient causes a slow movement in such ice

masses also, and the position of the outermost edge again represents an equilibrium between the rate of movement and the rate of melting, or the rate of breaking into bergs at the seaward margin of the ice field. Present-day icecaps are largely confined to the high latitudes, but in the geological past they covered much larger areas of the globe.

Glaciers, like rivers and wind, wear away the surface over which they move. The débris they pick up is largely carried to the terminus of the glacier, and there some of it is delivered to the streams of meltwater that flow on down the valley. Much of the rest of it is literally dumped near the margin of the ice as a heterogeneous pile of débris.

In many respects, however, the carrying of material by glacial ice strikingly differs from transportation by turbulent agents. This difference depends largely on the motion and physical properties of glacial ice as compared to those of other agents. In the first place, glaciers move ponderously downslope at a slow and snail-like pace.

Furthermore, glaciers are composed of solid ice granules, and thus are not liquid. As a matter of fact, as far as direct observation goes, ice behaves as a solid; its outer portions yield suddenly and discontinuously, like any other brittle substance. Great slicing or shearing movements transect the ice at an angle, so that masses of ice are literally shoved over the lower portions, as the glacier yields to the forces operating on it.

In detail the structure of the ice often shows thin laminae arranged essentially parallel to the bed of the glacier. Relative movement takes place along these laminae, which may be zones of weakness in the ice. There is a suggestion that these laminae may also reflect a more continuous plastic yielding within the body of the ice, similar to laminar motion.

The net effect of all factors involved in glacial movement is a complex yielding to gravitative forces; and whatever else may be involved, turbulence, in the sense that we have used it, plays no part in the story. Furthermore, there is no sorting effect in glaciers.

In essence the ice simply freezes over, and carries with it, every movable thing, including rock masses literally torn from the glacier bed. Obviously, the shapes, sizes, or densities of the débris are not the determining factors in such transportation.

The absence of selectivity means that there is no traction or suspension load in the ice, in the sense that we earlier discussed. Ice, in fact, carries a heterogeneous load of débris of all sizes and shapes, intermingled haphazardly. The load ranges from immense blocks of rock down to the finest clay particles; and although much of it is concentrated at the base, yet in valley glaciers part of the load results from blocks falling on the surface.

Similarly, the slicing or shearing movements within the ice carry some of the basal portions of the load up into the main body of the glacier, and even to its surface. In contrast to the heavy loading in some zones, there may be layers of ice almost entirely free from débris, but this is rarely or never true of the basal portions of the ice.

If the frictional resistance of the basal load against the glacier bed becomes very great, the ice may be unable to move the load along. Then that basal portion of the ice becomes stagnant; and the overlying ice, less heavily loaded, may slide over it, using the basal layer as its new bed. Such complications naturally render an exact evaluation of all the factors involved in ice movement a bit difficult.

Since glacial ice carries its load in a frozen grip, the fragments at the base of the ice are held rigidly; and as they are shoved along the glacier bed, they literally gouge out great scratches and grooves in the underlying rock. It is easy to see that a really good-sized glacier may account for a tremendous amount of erosion under these conditions.

Fig. Glaciated Pebbles are Marked by Parallel Scratches

A glacier is thus like a huge plane scraping off the bedrock over which it moves. Even large rock hills are profoundly modified or entirely worn away by the overriding ice. If a part of the hill remains, the side which faced the ice movement is, in most cases, decidedly rounded off.

It Might be predicted that the erosive power of ice should give rise to typical land forms, and detailed investigations have shown that it does. In the first place, valley glaciers scour out the valleys they occupy, and change the cross sections of water-carved valleys considerably. Instead of being confined in their erosive action to the bottom of the valley, as water is, the glacier may occupy large parts of the valley.

Thus the ice has an opportunity of cutting the bottoms and sides simultaneously, and in this manner it produces a typical valley form of its own. We say that glaciated valleys are like broad round U's in cross section.

At the heads of valleys occupied by glaciers are huge amphitheaters called cirques, They are due to erosion, occasioned apparently by a plucking action at the head of the glacier.

Alternate thawing and freezing, combined with an accompanying movement of compacted snow and ice, result in the excavation of a notch in the mountain side. As this increases in size, it develops steep walls and ultimately becomes a conspicuous feature in the landscape.

Fig. Bridal Veil Falls Plunges from A Hanging Valley.

Since some glaciers are larger than others, cirques vary considerably in size. More than that, glaciated valleys also differ in the degree of glacial scour. Thus it happens that where two glaciers merge, or where a glaciated valley passes an unglaciated tributary, the larger valley may be carved more deeply into the rock, so that the bottoms of the valleys do not coincide in level. When the ice later disappears (as even glaciers do when the climate becomes warmer), a hanging valley results.

Present-day streams, flowing through formerly glaciated valleys, commonly develop waterfalls at the juncture of the main valley and the hanging tributary. Bridal Veil Falls in Yosemite is a striking example. Plate includes several of the more common erosive features of glacial ice. It is to be expected that when ice, charged with rock fragments, moves over the glacier bed, the fragments, as well as the floor, should be scratched and ground.

The exact results depend on the relative hardness of the rock masses; but in general we do find within ice-transported débris a large number of pebbles with somewhat flat or soled sides, bearing sets of parallel scratches on them. These *soled pebbles* are evidence that the ice held them rigidly while it moved along, otherwise the sides would not be flat, nor would the scratches be parallel.

In some cases there are two or more intersecting sets of parallel scratches on the same face, indicating that as the pebble was moved along it struck one or more snags and was slightly rotated.

We may see now that glacial ice is individualistic in its movement, in its mode of transporting rock débris, and in the land forms that result from its erosive action. In fact, each geological agent has associated with it certain land forms and deposits; and from the geological point of view, the important thing is to recognize and understand how these features arise.

Thus far we have considered four of the five geological agents, largely in terms of their transporting work and the erosive land forms that accompany the removal of rock débris. We have yet to consider the opposite phase of the subject, namely, the *depositional* features of these geological agents. In the long run, erosion is balanced by deposition in nature; and it will be convenient for us now to turn our attention to this opposite aspect of our subject.

END OF THE LINE

What goes up must come down" is a homely old saying, but it applies quite well to the transportation of rock débris. The material that is picked up at any point by wind, water, or ice must be deposited somewhere else. This is inevitable because the energy of the transporting agent is limited, and eventually this energy is dissipated in the very act of transporting or by the many sources of friction that the agent encounters.

The transporting-capacity of a stream depends in large part on its velocity. As long as the velocity is increasing, the load may also be increased. If the velocity remains constant, the stream may continue to pick up material until it is utilizing its maximum carrying-capacity, when it is said to be *loaded*. If the velocity decreases after this loaded state is reached, some of the débris must be dropped. Consequently deposition also depends on the velocity, but in quite an opposite sense from transportation.

In the upper reaches of a stream valley the gradient is steeper than it is farther along its course. Now, in a perfectly frictionless fluid the velocity would continue to increase as long as any slope remained, because there always would be some component of the force of gravity operating on the fluid. In nature, however, the sources of friction are legion, and a stream's velocity tends to decrease as the slope becomes more gentle.

There is an inner friction within the water itself, turbulence consumes some energy, and there is the friction of the moving mass of water against the banks and bottom of the stream. The tendency toward decreased velocity due to energy losses through friction is in part counteracted by an increase in the volume or *discharge* of the stream as it flows along its course. An increase in discharge tends toward increased velocity.

It may also act in the opposite sense, however, because a

decrease in the amount of water will decrease the velocity of the stream. Evaporation is constantly taking place along rivers, and in some cases they may lose part of their water because of seepage through the bed. There is thus a balance of forces in any given case which tends to determine the conditions of transportation or deposition that will apply.

Now, although the velocity determines whether transportation or deposition will take place, the velocity itself depends largely on changes in the gradient, changes in the volume or discharge of the water, and in the friction it encounters from its channel.

We should accordingly expect sites of deposition to be those very places where the stream gradient decreases appreciably, where decreases in volume take place, or where the frictional effect increases. The best examples of gradient decrease are to be found at the bases of mountains.

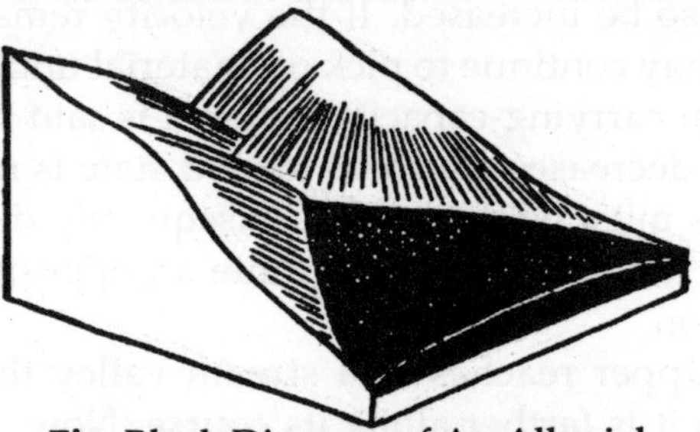

Fig. Block Diagram of An Alluvial Fan at the mouth of A Mountain Valley

A mountain stream, tumbling down a rather steep gradient with a high velocity, carries a heavy load of débris, as it actively erodes its channel.

At the base of the mountain the land surface flattens out rather abruptly into a plain. This relatively abrupt change in the gradient results in quite as sudden a decrease in the stream velocity, and hence there is deposition at this critical point. In fact, large aprons of deposited material, which spread out in fanlike forms, are found at the foot of many steep slopes where streams issue from the mountains. These deposits are called alluvial fans and are due to conditions precisely like those outlined above.

Another example of deposition, involving a change in volume (and hence in discharge) of the stream, may be mentioned. Some streams arise in regions of appreciable rainfall, and during their course flow across lands that are more arid. Such a stream is the Humboldt River, which rises in the northeastern part of Nevada and flows over the arid plains of the Great Basin.

Evaporation and seepage losses are high in the arid part of its course, and the result is that the amount of water continually diminishes as the river flows along. Ultimately the river dries up completely, about 200 mi. from its source. With a decrease of volume comes a decrease in the kinetic energy, and hence in the carrying-capacity of the stream. Finally, of course, when all the water has been dissipated there can be no more transportation at all.

In the lower reaches of valleys the flood plains are usually very wide and flat, so that during spring floods the streams commonly overflow their banks and overrun their flood plains. Since the velocity of the stream in its channel is greatly increased during such highwater stages, the floodwaters may be carrying a considerable load of débris. Now the moment the flooding begins, the sheet of water

Fig. Block Diagram of A Natural Levee

That spreads sideward out from the main channel meets with considerable friction along the bottom, and thus is greatly retarded. This is due to the large area of contact which such a shallow sheet of water has with the flood plain. What happens is simply explained:

As soon as the velocity is checked, some of the load is dropped; and since the greatest checking of the velocity takes place right at the main channel edge, it is there that the bulk of sediment accumulates.

Thus are built up natural levees, or low, broad ridges of débris along the channels of these old streams. When the volume of water in the channel again subsides, these levees stand above the new level of the stream and serve to confine the water within their bounds.

People living along many old valleys build up these natural levees to greater heights, in an attempt to prevent the floodwaters from flowing over onto the flood plains. The lower reaches of the Mississippi Valley afford good examples of such levees, and everyone is familiar with the havoc raised when destructive floods do tear away or overflow these ridges of earth.

The building of natural levees results from a velocity decrease due to frictional forces. Similar frictional situations also occur at the mouths of streams, where rivers empty themselves into lakes or seas. What happens is that the velocity of the river is checked by the inertia of the relatively quiet waters into which it flows. Thus there is a dissipation of energy due to friction. The Mississippi River will again serve here as a good example.

The river empties into the Gulf of Mexico; and as it does so, the velocity of the water is checked to such an extent that most of its load is dropped. In the absence of strong currents along the shore at that point, the deposited material accumulates near the mouth of the river.

In the course of time a delta has been formed, which grew out into the bay as a fanlike deposit spread out widely from the river's mouth. Such deposits usually assume the form of a rough triangle, which is the shape of the Greek letter delta; and from this symbol they derive their name.

In the formation of deltas an equilibrium is maintained between the deposition of the load that is continually arriving and the material already deposited. As the delta grows in length, its upper surface, being essentially at sea level, must be adjusted to allow the débris to be carried out to its edge. Thus it is necessary that a slope develop in the direction of the river's flow.

This slope is formed by the deposition of débris toward

the landward edge of the delta, and in this manner the upstream portions of the deposit gradually emerge above sea level. Plate 14 includes some of the more common features formed by stream deposition

When the differences between traction and suspension loads are taken into consideration, we find further interesting details arising concerning sedimentary deposits. The suspension load is generally finer than the traction load, which we saw.

Consisted of particles too large to be picked up and yet small enough to be rolled along the bottom. In any given stream—or air current, as a matter of fact—there is usually a gradation between the two types of loads.

The traction load, however, is the marginal portion, and it will tend to be dropped at the slightest decrease in velocity. This effect is brought out most strikingly in the case of wind. A current of air carrying dust in suspension and sand as a traction load may move over an area where vegetation tends to slow down the velocity of the wind,

As soon as the velocity is decreased, the traction load is deposited, but the suspension load may be carried right over the obstacles and so escape deposition. The fact that the traction load of sand is stopped by obstacles means that the sand will be deposited in mounds or hills; and as soon as some sand is dropped, it acts as an additional barrier.

In this manner sand dunes are built; and such deposits are definitely localized as mounds, hills, or ridges, depending on the supply of sand, the direction of the winds, the obstacles, and other factors.

Sand dunes are fairly common land formes, and they are generally found near sources of loose sand. Hence we find them along coasts and also in arid regions where windwork has an opportunity to manifest itself.

Dunes have typical forms which shed light on their mode of formation and their slow migration over the surface. For dunes do move; and cases are not unknown where even towns have been engulfed by the sand, only to appear again when the sand moves beyond.

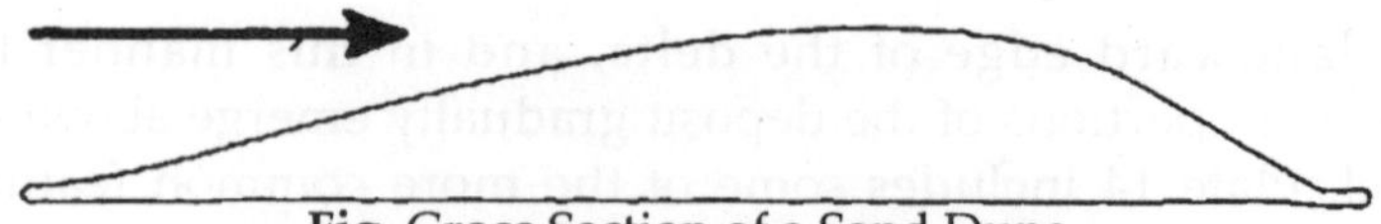

Fig. Cross Section of a Sand Dune,
Showing the Direction of the Prevailing Wind

In profile the dunes present a gentle slope to the wind, and a steeper slope to the lee. The grains of sand are blown up the gentle slope and tumble over the lee edge, where they lie at the angle of rest of loose sand, about 30°. The continual movement of sand from the front to the rear slope of the dunes results in a slow movement of the dune as a whole. Many dunes are rather low features, but hills up to 200 ft. high are not uncommon.

On a much smaller scale are the tiny *ripples* that commonly form on dry sand; and if they are watched, it is seen that they move with the wind. Here also is a migration of the sand over a gentle windward slope and down a steeper lee slope. The development of the ripples appears to be a response to conditions demanding minimum friction between the moving air and the semi-mobile grains of sand.

Between the ripples are tiny horizontal eddies, which allow the wind to blow over the crests and over these horizontal eddies with less friction than would be encountered over a smooth field of sand.

Fig. Loess Stands in Steep Walls During its Subsequent Erosion.

The suspension load of dust in the air is carried beyond

the localized obstacles and remains in suspension as long as the wind has enough motion to keep the particles off the ground. When the wind dies down, as winds inevitably do, owing to energy dissipation, the suspension load also tends to settle, not locally in mounds, but over a wide area. The net effect is that the suspension load forms a mantle of fine material over the landscape, which may be fairly uniform in thickness over large regions.

There is a common wind deposit called loess, with exactly these characteristics. It is formed from the suspension load of winds, and it is disseminated widely as a blanket which, unlike other deposits, clothes hills and valley slopes indiscriminately. Loess has the interesting characteristic that it is capable of standing in very steep walls when it is itself subjected to later erosion.

Such simple analyses as the foregoing indicate the importance of distinctions between traction and suspension loads, because in the interpretation of deposits they afford a means of determining the conditions that prevailed during the formation of ancient rocks on the earth.

We must not gain the impression from our earlier discussion of traction and suspension loads that the materials deposited from the traction load are always quite spherical or that those from suspension loads are always flat. During transport even the partitles in suspension may become more rounded from mutual impacts and other causes, and new loads are often picked up in the course of travel The net result is that the final deposit may have quite an assortment of various shapes and sizes.

There is an extremely important characteristic of deposits laid by turbulent agents, and that is the sorting of the deposits. From the miscellaneous débris made available for transport by weathering and erosion, the turbulent transporting agents sort out, or select, particular sizes, shapes, or densities by virtue of the differences between the traction and suspension loads, as well as by the carrying-power of the agent itself.

When later these loads are dropped, the deposition is also somewhat selective, as we have already seen. The net result is that there is a fairly definite sequence of deposits sorted according to size and composition in any depositional environment.

Thus arise the materials that we call *gravel, sand,* and *mud*. It is important to see that only through sorting action is it possible to separate the miscellaneous débris of weathering into a series of sediments having restricted ranges of sizes, shapes, or densities. Further, the particular agent that does the sorting often leaves other marks of its action on the material, and these furnish us with additional clues regarding ancient sediments.

Another striking feature of deposits left by turbulent agents is bedding. We saw that decreases in velocity cause selective deposition, so that a layer of débris is laid down. Later a slight shift in conditions or a difference in the load may bring about the deposition of another layer. These layers of material may then be separated by a distinct juncture plane, known as a bedding plane.

The layers between these planes are called *beds,* or strata; and consolidated or cemented deposits showing such features are called stratified rocks. These rocks, composed, as they largely are, of fragments of weathered and broken deepseated rocks, are part of the group of fragmental rocks.

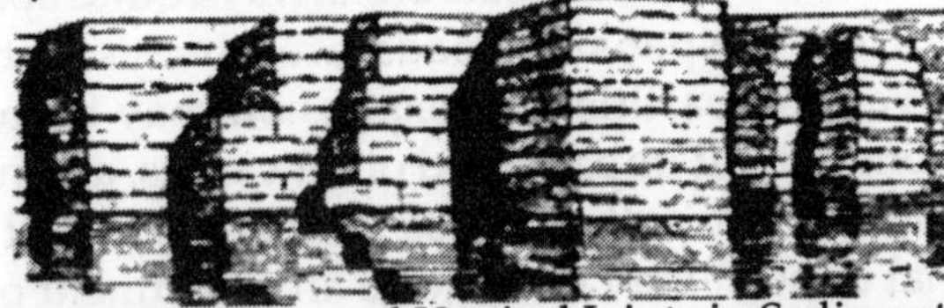

Fig. Horizontal Bedding and Vertical Joints in Sedimentary Rock.

Stratification of the sort just mentioned is due to selective transportation and deposition. But all stratified rocks are not necessarily fragmental rocks. Successive lava flows, poured out of volcanoes, may harden into layers of rock which also have a stratified appearance, although they lack the other characteristics of sediments.

We should understand from this, however, that stratification is not confined to the sediments; indeed, not even all sediments are stratified.

In this connection the deposits left by glacial ice are in striking contrast to the bedded deposits of wind and water. The fact that ice is incapable of sorting out the load that it picks up means.

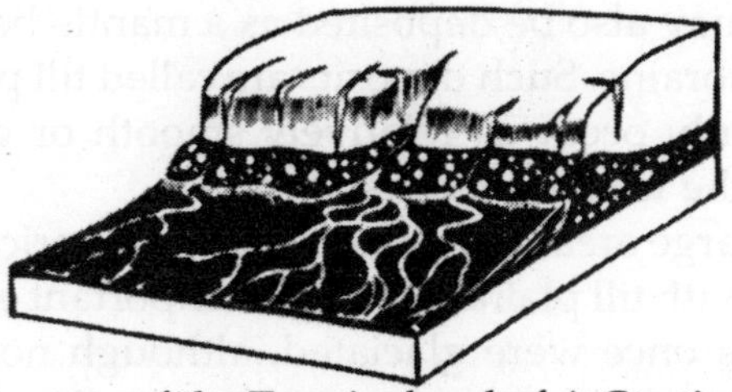

Fig. Block Diagram of the Terminal end of A Continental Glacier.

The morainic ridge lies beneath the edge of the ice, and farther back under the ice is the till plain. The outwash plain, with its streams of meltwater, is in the foreground that the deposits left by the ice must necessarily reflect the same heterogeneity as the original débris.

This situation is, in fact, met in nature, because the most outstanding feature of ice deposits is, that they are unsorted. In the deposits we find a heterogeneous mingling of all sizes and shapes, just as it was left when the glacier melted and dumped its load at the point of melting. This heterogeneous sediment laid by glacial ice is called glacial till.

Suppose a glacier moves down slope with its load until it reaches a level where the ice is completely wasted away. Here the bulk of the load is dropped, because with no more ice there can obviously be no more carrying. Consequently, at the ice edge there will be a great accumulation of débris, deposited in the same heterogeneous manner that it was carried in the ice.

Of course, the meltwater will pick up some of this material and carry it along, but in this case the load is distributed in accordance with the laws of stream transportation and deposition. What we find then is a large ridge of unbedded and unsorted débris at the edge of the glacier, and stretching out from this ridge a series of bedded sands and gravels that have been carried away by the meltwater.

Such ridges built at the edge of the ice are called moraines, from their wall-like aspects; and the gravels and sands beyond them are called outwash plains.

Not Only is glacial till piled up in moraines along the ice edge, but it may also be deposited as a mantle beneath the ice behind the moraine. Such deposits are called till plains because they commonly occur as relatively smooth or gently rolling plains after the ice has disappeared.

In fact, large areas of northern North America and Europe are covered with till plains, which are important evidences that these regions once were glaciated, although no remnants of the great icecaps now remain on either continent.

Even in the case of the débris that is fed to the meltwater by the ice, there are other deposits in addition to outwash plains. For example, streams of meltwater flowing from the glacier's edge may carry material with them and, owing to changed gradients or pressures near the point of emergence, may deposit part of their load as hills of gravel or sand.

These features, commonly found on moraines, we call kames. Plate 15 inełudes several of the more common features of ice deposition.

In fact, the work of glacial ice affords an interesting example of the great complexity of nature when examined in detail. The intermingling of ice-laid and water-laid deposits is so typical of glacial phenomena that we usually think of the associated outwash deposits as a part of glacial action itself. In a more advanced text we would call the outwash deposits *glaciofluvial* deposits, to indicate that both ice and water had a hand in their formation.

Just as there is a close relation between the geological agents and their deposits in nature, so there may be a close relationship between erosion and deposition itself. Thus far we have kept transportation and deposition somewhat separate, but that has been largely to keep unity in our discussion. We shall combine the two, however, in turning to the next geological agent, and endeavor to show how both erosion and deposition may take place side by side in the development of certain land forms.

Chapter 18

Folds and Faults

There is a curious kind of sandstone that can easily be bent by hand when it is in thin slabs. We say it is "curious" because everyday experience tells us that rocks are quite brittle. They may be chipped with a hammer, and one would be surprised to find a rock that merely bends under the blows. Our piece of sandstone, however, is a sort of trick rock. It is composed of quartz grains rather loosely held together with flakes of another mineral (mica); and as the rock is bent, the sand grains readily slide over the mica flakes.

Despite the brittleness of common rocks, we do find them in nature contorted into all kinds of intricate folds and plications. This is especially true in mountainous regions, where diastrophic forces have acted on the rocks. As a matter of fact, the conditions of pressure, temperature, composition, and the time element itself, are the factors that largely determine whether rocks will bend or break.

Our direct experiences with rocks occur in the surface environment with its low pressures and relatively low temperatures. It is under such circumstances that rocks break rather than fold. Further, the time at our disposal is limited. Even so, a century or two is sufficient for some changes. For example, covers of old tombs occasionally are found to sag downward a bit. The rock (commonly limestone) adjusts itself to its unsupported condition by a slow bending, even under surface conditions.

The pressure and temperature effects may be brought out by experiments that have actually been performed in laboratories. Certain substances, like salt, ice, and calcite—even some types of rock—can be made to flow when they are confined by great pressures. By *flow* we mean a fractureless change of form, for under great confining pressures the material cannot break or shatter, but must adjust itself to its environment by an actual plastic yielding.

The flow of rocks within deep-seated environments is likewise a plastic yielding, and it takes place partly through a recrystallization of the constituent minerals. The present we shall examine their larger features, the folds and other structures shown by rocks that have been subjected to diastrophic stresses in the deep-seated environment.

Suppose we consider a cubic foot of rock somewhere down beneath the earth's surface say 10 or 15 mi., in what we have called the "deep-seated environment." Our cube is thus part of a column of rock 1 ft. square that extends down from the surface to the depths far below.

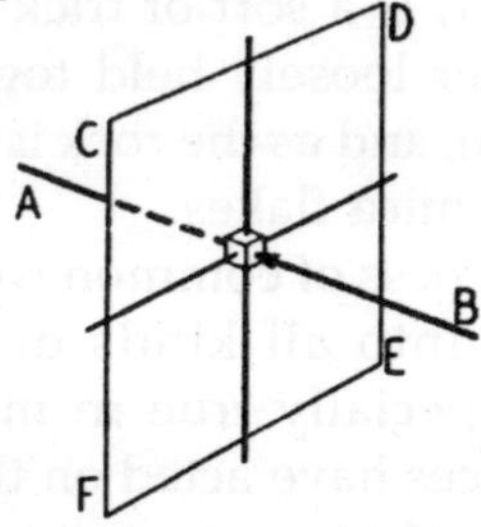

Fig. Deeply Buried Cube of Rock Acted on by Diastrophic Forces

Hence it will be under considerable pressure from the overlying rocks, and at a respectable temperature. The pressure under normal conditions will be static, and the cube will remain at rest because all the forces operating on it are equal.

But now suppose that, owing to adjustments in the earth, the cube is acted upon by diastrophic forces from two opposite sides. What will happen? The rock cannot very well shrink into a smaller volume, because it is already quite compact. There may be a slight adjustment of volume due to a recrystallization of its minerals; but on the whole, the tendency will be somehow to ease the pressures that operate on it by escaping from them.

The directions in which this relief will find expression lie in a plane at right angles to the diastrophic forces. In the adjacent figure the cube is acted on by diastrophic forces along the line *AB*, and the plane *CDEF* contains the directions of maximum relief.

But will the relief be upward, downward, or sideward away from the compressive force? In actual mountain-making, of course, there is a *zone* subjected to these forces, rather than some fixed cube. Hence, whatever yielding takes place affects the entire zone. Our cube, however, serves to fix attention on a particular part of the zone affected.

The question of the direction of relief cannot be answered positively, because we cannot get down to examine the situation. All we can do is to examine the eroded belts of folded rocks along mountain chains and see what they tell us. This much we know:

The belts of folded rocks are usually linear, and in many cases they are symmetrical on both sides of the belt. Further, they show an elevation relative to their surroundings, so that in the folding the whole series of rocks has been uplifted. The accompanying block diagram sketches the usual relation of a mountain belt to the diastrophic forces that formed it.

We are reasonably certain that some relief has been upward and sideward along the plane of maximum relief. As to possible downfolding simultaneous with the other adjustments, we are not certain.

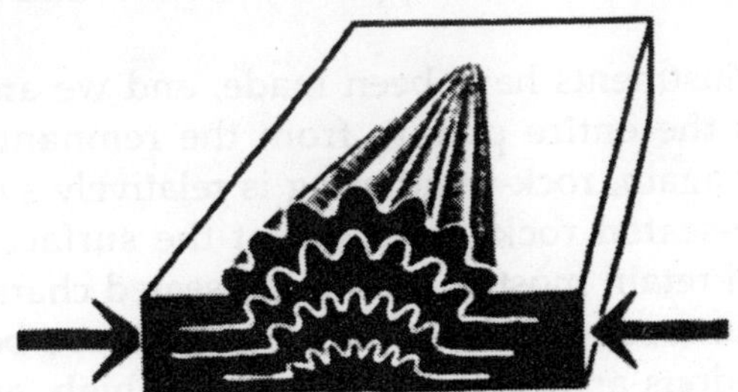

Fig. Portion of an Elevated Belt of Folds

We have to pick our way slowly and painfully, distinguishing between what we know and what we infer, for, in dealing with this subject we are approaching one of the outstanding problems of present-day geology. The nature and operations of diastrophic forces are not fully known, partly because the scale on which earth operations take place is too great for us to encompass in experiments; nor can we make direct observations at the localities where the deep-seated disturbances take place.

Confined, as we are, to surface environments, our observations include mainly the surficial aspects of these deep-seated diastrophic adjustments.

The upward relief of rocks under differential stresses results in those great elevated belts of rock that we call mountains; and usually within them the rocks are found to be highly folded, or greatly mashed, and quite commonly altered from their original appearance. Fortunately, gradational agents attack these elevated tracts and incise deep valleys among the mountain folds, so that the very roots of the mountains may be laid bare.

Then, by observing how many layers of rocks are involved in the folding, and their thicknesses, we are able to tell how deeply such rocks were buried before the folding occurred. In fact, it is by examining eroded rock structures like this that we learn most of what we know about deep-seated environments.

Although we are able thus to reconstruct many of the environmental factors involved in deep-seated diastrophic adjustments, we are handicapped by the fact that the show is over.

The adjustınents have been made, and we are forced to reconstruct the entire picture from the remnants we find. Fortunately again, rock-weathering is relatively slow, so that many deep-seated rocks, exposed at the surface by recent erosion, still retain most of their deep-seated characteristics.

In Some cases we find that the rocks have been folded into tremendous arches, thousands of feet high, and equally broad. Such upfolds we call anticlines, and they may be rounded, domelike affairs or elongate bulges of rock. Corresponding downfolds are called synclines. In some cases we find an alternation of upfolds and downfolds occupying a wide belt. The Appalachian Mountains are a case in point.

Fig. Anticlines and Synclines in the Jura Mountains.

Now these folds of rock need not be symmetrical or even upright. Nature is complicated in her ways, and we find folds with all degrees of complexity.

That is, if we imagine a plane through the center and along the length of a fold (the *axial plane*) the plane will be vertical if the fold is upright; but it may be inclined at any angle, so that the folds may be inclined or even recumbent if they lie over on their side.

Similarly the *axis* of the fold, which is the line along which the axial plane cuts the top or bottom of a folded stratum, may not be horizontal but may dip away at an angle. In that case the fold is *plunging*; and by the time several of these complexities are put into the picture, the reader may be left far behind.

Fig. Complexities of Rock Folds.

While an uninterested reader may dismiss the topic as abstract, the geologist dares not. On just such complexities may depend the location of valuable ore bodies or extensive pools of petroleum.

For the geologist, in order to understand and interpret nature, must devise ways and means of untangling her complexities and making them available for economic exploitation or for scientific study.

But these complexities of the folds are only part of the story. We said that erosion promptly attacks the uplifted and folded belts, so that the picture is further obscured by the absence of much of the structure and the burial of other parts of it beneath mantle rock and soil.

Fortunately, the remnants of folds afford ample evidence of their former extent, and usually the bedrock is sufficiently exposed at the surface so that the essential features of the picture may be reconstructed.

When a geologist studies an area, he hunts around for outcrops, or patches of the bedrock exposed at the surface. A common place to find them is along valleys where streams have cut away the superincumbent layer of mantle rock. Ditches, railroad cuts, and other man-made excavations may also expose the bedrock. Hence you may find geologists in the most out-of-the-way places.

Several observations are usually made on the rock outcrops. For our immediate purposes we are interested in the rock struc tures; and to work them out, two rather precise measurements are made on each outcrop. For simplicity let us assume that we are dealing with sedimentary rocks, because then we may speak about the *attitude* of the beds. Sedimentary rocks are best adapted to such studies because they are originally laid almost horizontally, and any subsequent tilting may easily be seen.

The first measurement is made along the plane of the bed, and the angular inclination of this plane from the horizontal is called the dip. At right angles to the direction of dip, where the horizontal plane intersects the bedding, is the strike of the bed. The strike is expressed as the compass direction of this horizontal line. The adjacent diagram will give you the complete picture.

Thus, dip and strike afford the necessary data to tell us what the direction and amount of inclination of the bed is at that point. In the meantime the exact location of the outcrop is also noted on a map.

Fig. Sedimentary Rocks Showing Direction of Dip and Strike.

After the dips and strikes of all the outcrops have been measured and plotted, some conspicuous rock layer is chosen as a standard.

Then, by connecting the indicated strikes of this rock layer, the structure of the region is determined. It may be a single syncline or anticline, miles in extent. On the other hand, there may be a succession of small folds or any combination of large and small.

Furthermore, the surface expression of the land may not reflect the structure at all, because erosional agents, cutting away in terms of the relative resistances, attitudes, and elevations of the rocks, may so disguise the original landscape that hills exist over the synclines, and valleys over the anticlines.

While plastic deformation, and its attendant rock-folding, is taking place within the deep-seated environments, diastrophic adjustments may also be taking place near the surface. Here the confining pressures are less, and the relative nearness to the surface may permit sudden crushing effects to operate on the rocks.

Hence they may yield by fracturing instead of by flow, so that joints and fractures may be formed in the rocks. If there is any slippage along the fractures, as a result of the rock on one side moving relative to that on the other, the result is a fault. Faultplanes, then, are surfaces along which relative movements take place.

When the compressive forces are unusually great, or when the rocks are unusually resistant to folding, nearly horizontal fractures may develop; and along these fractures whole mountain masses may be shoved over the underlying rocks. We call these features thrust faults, and in our own Northern Rockies such faults occurred, in one of which a great slice of rock, 10,000 ft. thick, was shoved eastward for at least 18 mi. over the rocks beneath.

This thrust fault resulted in a shortening of the earth's crust by some 18 mi. Thrust faults, then, like mountain-folding, move hand in hand with the concept of a shrinking earth.

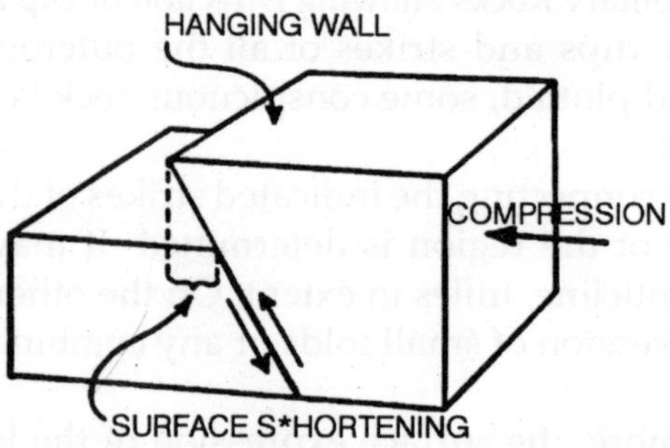

Fig. Block Diagram of a Thrust Fault, Showing Surface Shortening

In addition to thrust faults, in which the main result is a shortening of the surface, there are normal faults, which result in surface elongation.

In order that you may clearly see why shortening or elongation of the surface accompanies faults, we have included the adjacent figures.

Note that in both cases the fault plane dips down at an angle, so that the two surfaces which bound it may be called the hanging wall and the foot wall.

Notice that 'n the thrust fault the hanging wall has gone *up* relative to the foot wall, while in the normal fault the hanging wall has gone *down*. The resultant shortening or elongation is indicated in the figures.

In actual practice it is this relation between the walls that is observed by the geologist; and from it, when he finds the evidence that shows the relation clearly, he is able to tell which kind of fault is involved and in which direction to continue the search for a faulted ore body or coal seam. Examples of faults are included in Plate 23.

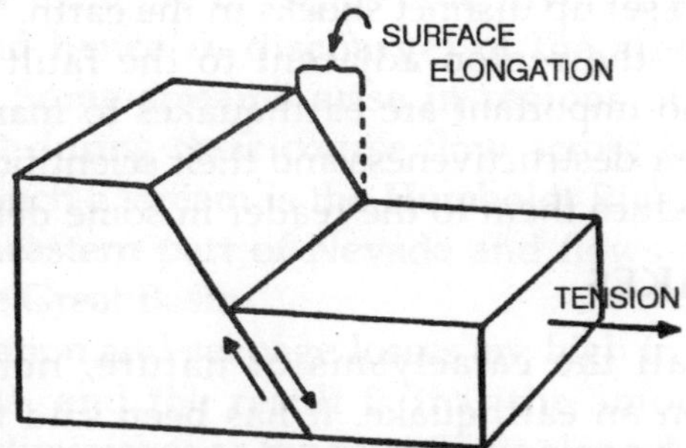

Fig. A Normal Fault, Showing Surface Elongation

Faults may take place horizontally as well as vertically. They, in common with folds, may display any degree of complexity. There are some regions so intricately faulted that they defy full understanding.

There also are regions in which folds and faults are so intimately intermingled that the task of untangling them is wellnigh hopeless. Futhermore, folded and faulted regions may later be subjected to more folding or faulting or both, and the final picture may be one that gives nightmares even to a specialist.

Now while folds and faults are the two main effects of diastrophism in a broad sense, there are other effects in detail. Thus chemical rearrangements that we mentioned in passing have profound effects on the nature of the rocks. New minerals are often formed, and the rocks take on new appearances to such an extent that we consider them a separate group of rocks. They are called the metamorphic rocks, either of the two broad classes of rocks, igneous and sedimentary, may be altered to metamorphic rocks.

He must not gain the impression that mountains are folded up over night. No, it is an extremely slow process, as most geological phenomena are. We have pretty good evidence for believing that some mountain ranges are today in process of growing, but of course that growth is not revealed to a casual glance.

While folding may be extremely slow in its development, that need not necessarily be true of faulting. In fact, we know it isn't true, because faults often take place with enough suddenness to set up distinct shocks in the earth. When these shocks occur, the region adjacent to the fault suffers an earthquake. So important are earthquakes to mankind, both because of their destructiveness and their scientific value, that we shall introduce them to the reader in some detail.

EARTHQUAKES

Among all the cataclysms of nature, none is more terrifying than an earthquake. It has been said that fear of earth tremors increases with increasing experience, whereas in almost any other peril familiarity breeds a certain degree of indifference.

Whatever the reason for this, assuming it to be true, there doubtlessly is involved our unconscious acceptance of the earth as a firm and unyielding foundation. When even this moves beneath our feet, it is not surprising that one's consternation should exceed his judgment.

Combine with this the jumble of shattered houses, of gaps and fissures that open in the earth, of springs and streams gone suddenly dry, the low rumble of earth noises, and the dust, confusion, and conflagration that are summed in the word *earthquake,* and the cumulative effect is hardly surprising.

So Common are earthquakes that the instruments used to record them show several thousand a year. Most of these are imperceptible tremors, of course; but it must be true that somewhere on the earth every day some people are subjected to greater or less disturbances.

How is it, then, that so few of us have ever experienced an earthquake? Simply because earthquakes, despite their abundance, occur in certain restricted areas on the earth and because, beyond a relatively small radius, the effects of the shocks are not perceptible to our senses. Records covering many years show that most earthquakes are confined to certain belts about the earth. Many of these belts are close to continental margins.

The accompanying map indicates the extent of these belts; and if you will examine it in detail you will realise that practically every earthquake you have ever read about has occurred at some locality within these zones. We cannot say that no earthquake will ever occur outside of these zones; but beyond the limits of the heavy belts, earthquake insurance is pretty much of an idle gesture.

A great deal of evidence collected from many parts of the world shows quite conclusively that most major earthquakes are due to faulting. Quite commonly fissures are formed in the ground during major earthquakes, and these fissures are found to be the surface expressions of faults.

The movement along fault planes may be relatively sudden, and this movement sets up powerful vibrations, which give rise to waves that radiate outward from the center of the disturbance. It is the sudden shock of these waves that does some of the damage during earthquakes.

Fig. Earthquake Fissures Often Mark the Position of Faults.

At present we are concerned with the causes of earthquakes and their surface effects.

In many cases the movements during earthquakes take place along already existing fault zones, which may bear evidences of numerous displacements. One of the most noteworthy earthquakes of recent times was that of San Francisco in April, 1906.

This earthquake resulted from a displacement of the rocks along the nearly vertical *San Andreas fault*, a zone of weakness that can be traced for many miles through California. The maximum movement along the fault was 21 ft. in a horizontal direction by displaced fences, roads, and the like.

Faults, we may recall, can take place in any direction,

depending on local conditions, although it is true that vertical movements are more common than horizontal ones.

Rocks, contrary to popular fancy, are quite elastic. When an external force is applied, the rocks suffer an elastic deformation, or strain. When the strain due to these forces becomes too great, the rock yields. Near the surface the yield is quite sudden and may result in fractures or actual movement along fault planes.

The conditions leading up to an earthquake may then be pictured as the gradual accumulation of stresses within the rocks until they are so great that the rock must yield. Imagine the sides of a fault tightly pressed together, so that movement can only take place through the overcoming of tremendous frictional resistance. If forces tending to produce movement along the fault plane are set up, the rock resists the movement until its elasticity is exceeded. Then a sudden movement takes place, and the vibrations set up during these movements may be quite intense.

This interpretation of earthquakes has direct evidence to support it. Certain slow displacements of landmarks, demanded by the theory, have been reported along the San Andreas fault; and such slow displacements represent the accumulation of stresses within the rocks adjacent to the fault, so that, when the release occurs, the rocks "snap" into their new positions by an *elastic rebound*.

The accompanying three block diagrams represent the picture as a whole. In the first block a pre-existing fault plane is shown. As forces accumulate in the zone adjacent to the fault, we may visualize a tendency on the part of the rock to adjust itself by a slight deformation, as the second block fancifully suggests.

No slip has yet occurred along the fault plane, however, but as the elastic limit is reached, the rocks yield by relatively sudden or jerky movements, and the earthquake damage is done. The third block pictures this final stage, after the faulting has occurred. Once relief is afforded, the stage is set for a repetition of the process.

These examples refer to an earthquake in which the

movement is horizontal, but obviously a similar set of conditions could apply to movement in a vertical direction.

The theory just presented is relatively recent; the whole subject of earthquake study is still rather new, but the mass of data now available fairly well establishes the broad outlines of the subject. In addition to direct field observations, scientists nowadays have instruments of such sensitivity and precision that even the feeblest earth tremors are faithfully recorded, and cases are not unusual where the center of the disturbance is determined from these records before even the news reports tell of the damage.

It was not always thus, of course; until shortly before the turn of the present century, the study of earthquake phenomena was confined largely to qualitative observations. Some of the data obtained from these earlier investigations are interesting and important. The exact geological significance of earthquakes was but dimly realized, also, and even today there is much that remains to be explained.

One thing is certain, however: the advent of modern seismology, or the scientific study of earthquakes, has opened up a field which has shed important light on the question of the nature of the earth's interior. More of this anon; at present suppose we journey backward in time about a century and pick the story up there.

In 1846, Robert Mallet, an Englishman, published a memoir on the dynamics of earthquakes. Here, for the first time, were applied the laws of wave motion in solids to the phenomena of earthquakes. In England, where Mallet lived, however, earthquakes are rarities, and so his experience was more theoretical than actual.

It is reported that when he was awakened one night by the tremor that stirred Britain in 1852, he did not recognize it as an earthquake. Despite this lack of actual experience, Mallet believed that he held the key to many earthquake secrets. His opportunity came late in 1857, when the great Neapolitan earthquake occurred.

So great was the destruction in this Italian earthquake that news of the catastrophe did not reach the outer world until about a week later. Mallet at once proceeded to the stricken area and began his researches. His principal aims were to determine the exact center of the earthquake disturbance and to find out how deep within the earth the disturbance occurred, for Mallet argued that both of these ends could be gained by a study of the surface effects of the quake.

You undoubtedly know from newspaper accounts that an earthquake may first present itself as a rather sudden shock, accompanied by billowy movements, as though waves were passing along the surface of the ground. The very suddenness of the shock may contribute in part to the damage, although the amplitude of the earthquake waves, the duration of the quake, and resonance in building structures all fit into the complete picture.

The general result is that after an earthquake of destructive violence, there usually is a center of maximum damage, surrounded by areas in which the destruction diminishes. The sequence ranges from towns that witness the dramatic events sketched in our opening paragraph to more distant towns and villages in which the greatest damage is perhaps the stopping of clocks or the rattling of a few. dishes.

Naturally enough, any disturbance originating at some center dies down as it moves outward, much as the ripples in a pond diminish in size as they move away from the point where the stone is dropped in.

Fig. Isoseismal Lines of the Neopolitan Earthquake of 1857, After Mallet

Now, what Mallet did was to set up a scale of damage suffered, a comparative scale of the intensity of the quake at anypoint. A tour of inspection of the towns in the stricken area then afforded him data for drawing on a map a series of curves which enclosed areas of equal damage.

These lines are called isoseismals, and they enclosed elliptical areas increasing in size as he moved away from the center of greatest damage. The innermost isoseismal included the zone in which the towns were prostrated, the next larger one those where major parts of the towns were destroyed, and so on with decreasing effect outward.

For locating the center of the disturbance and its depth, below the surface, Mallet devised an ingenious method. Suppose a disturbance occurs beneath the surface at 0 (called the focus of the earthquake), situated a distance OA beneath the surface.

Waves are sent out in all directions from O, and we shall consider one of them, OB. This strikes the surface at B, at an angle b with the surface. This angle Mallet called the *angle of emergence*. Also, if B is connected with A, the line AB has some given Compass direction; that is, it has an azimuth a with respect to the line AN, supposing N to be the north point.

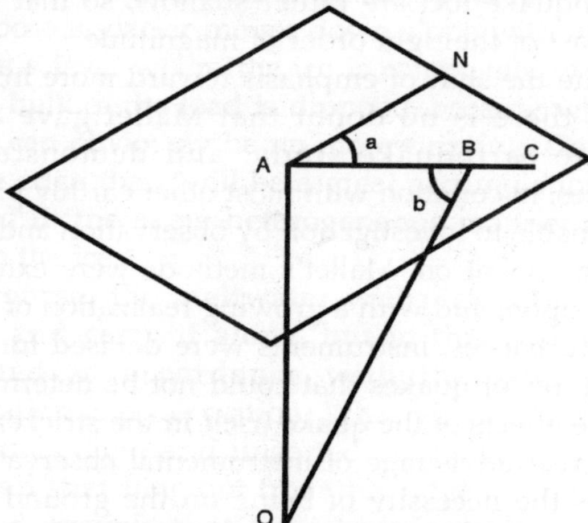

Fig. Mallet's Method for Locating the Epicenter and Focus of an Earthquake

Now Mallet argued that the direction of travel of the earth wave OB, its angle of emergence b, and the direction AB should be determinable from the effects of the shock at B. The cracks in damaged buildings or walls should be at right angles to the wave direction OB; and overthrown monuments and the like should be found lying in the direction BC, with their bases toward the point A, which is called the epicenter of the earthquake.

As a result of this study he found that the epicenter of the Neapolitan disturbance was not a single point, but a somewhat linear zone (as we would expect if movement along a fault plane caused the quake). Neither was the depth to the focus the same for all measurements, but the average depth was found to be about 6½ mi.

A Numer R of technical objections can be found to Mallet's methods; but the point remains that his was the first fairly successful quantitative study of earthquake phenomena, and Mallet's work laid the cornerstone for many investigations since his time.

Nowadays, instruments are used much more extensively, and reliance is placed on mathematics to determine the depth of focus. It is an interesting verification of Mallet's work that most earthquake foci are rather shallow, so that his results were at least of the right order of magnitude.

Despite the shift of emphasis toward more instrumental methods, there is no doubt that Mallet gave a decided impetus to earthquake study, and demonstrated that earthquakes, in common with most other earthly phenomena, were amenable to investigation by observation and inference.

As time went on, Mallet's methods were extended and improved upon; and with a growing realization of the nature of the disturbances, instruments were devised for recording many features of quakes that could not be determined from the surface effects of the quake itself in the stricken area.

The great advantage of instrumental observation is that it obviates the necessity of being on the ground during or after the earthquake; the instrument may be located far from the disturbance, and yet from the record it traces the distance to the center of disturbance and the violence of the shock can be determined.

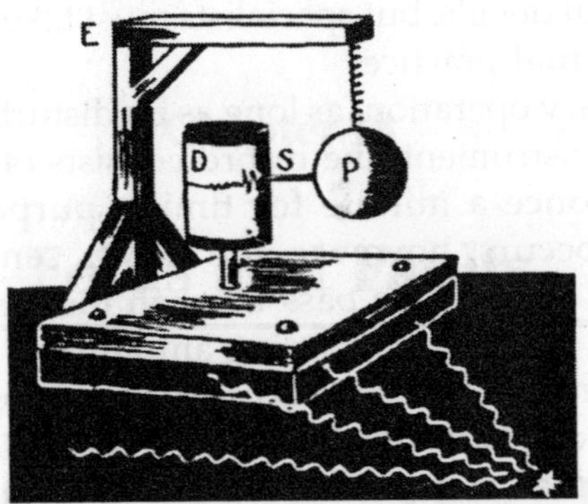

Fig. Simplified Diagram of A Vertical Component Seismograph

The instruments used to detect earthquake waves are called seismographs, and in principle they are quite simple. A seismograph is essentially a pendulum with a long period. A much simplified diagram of one is shown herewith. A heavy weight, P, is suspended by a spring from a rigid frame, E, firmly fastened to the earth, often by means of a pillar of concrete resting directly on bedrock.

On a platform of this frame is a revolving drum, D. A stylus, S, attached to the weight, rests against a sheet of paper wound around the drum. Now imagine an earthquake suddenly vibrating the frame E, The frame, including the drum, will then oscillate a bit, while the mass P, because of its great inertia, will momentarily remain at rest.

Thus a wavy line will be traced on the paper, recording the passage of the earth wave. This is called a *vertical-component seismograph*; and in practice two instruments are used, one to record the vertical components of the wave and the other to detect the horizontal components.

Actually, of course, modern seismographs are much more refined and complicated than this. Timing devices are put on, so that the exact moment of the wave's passing is recorded, and usually the record is made by a beam of light on photographic paper instead of with a stylus. Further, the natural vibrations of the pendulum must be severely damped, so that it does not unduly distort the earthquake waves with its own periodic motion. We shall not bother the reader with these technical details, but specialists must take careful account of them in actual practice.

In ordinary operation, as long as no disturbance is passing through the instrument, the record consists of a straight line, interrupted once a minute for timing purposes. When an earthquake occurs, however, the waves sent out from the center of the disturbance pass through the machine, and the line becomes wavy, with varying amplitude.

The record thus obtained is called a seismogram, and from it is read much more than one might suspect at first glance. Here is shown a sample of such a record from an earthquake a thousand or so miles from the instrument. Notice the letters p, s, and l that are indicated below the curved line. They mark the appearance of the three wave trains that make up the a seismogram, or earthquake record.

The letters p, s, and l mark the arrival-time of the three sets of waves usual record. The first is called the primary wave, because it travels fastest and arrives first; the second is called the secondary wave, and the third is called the long wave. The exact nature of these waves will be developed in our sequel because they afford such an interesting example of agreement between theory and practice.

Aside from any theory, however, it was noted that these waves are separated from each other on the seismogram by time intervals which increase as the distance from the disturbance to the observatory increases. Thus, at an Observatory a few hundred miles from the earthquake the times of arrival of the waves may differ by only a few minutes, whereas from more distant earthquakes they are separated by greater and greater intervals.

An Important practical application is made of this timelag between the earthquake waves. It was found by plotting the data of many seismograms that there is a definite relation between the time of arrival of the p waves and s waves, for example, and the distance from the Observatory to the quake.

This relation holds regardless of the geographic location

of the earthquake or of the observatory; in short, the time-lag is a function of distance only. Hence, to determine the distance between the observatory and the disturbance, it is only necessary to observe the time in minutes, say, that elapses between the arrival of the first two waves.

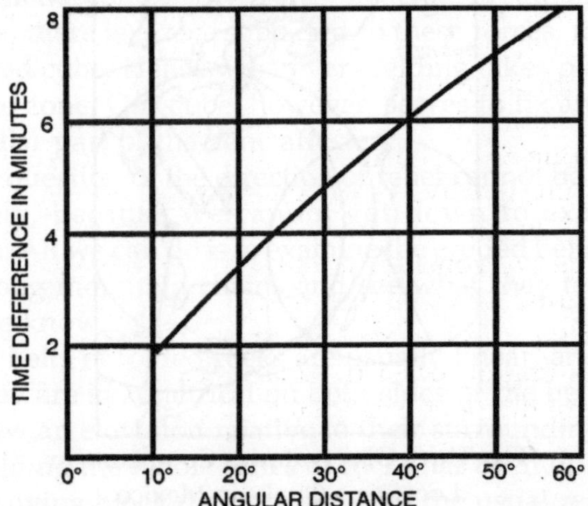

Fig. The Relation between Distance, and the Difference in Time of Arrival of the *p* and *s* Waves

Graphs have been prepared to show this relation between timelag and distance, and the adjacent diagram is one somewhat simplified. The time difference in minutes between the arrival of the p and s waves is shown along the vertical axis, and the angular distance to the epicenter is given along the horizontal axis.

Suppose, for example, that the p wave is recorded on the seismogram at 7:02 A.M. and the s wave at 7:07 A.M. Here is a difference of 5 min.; and by finding this value on the vertical axis, the corresponding distance is seen to be 30°, or about 2,100 mi.

Now, for any one station to determine the *direction* to the epicenter is somewhat tedious. It requires quite a bit of calculation, so that several observatories usually co-operate in locating the stricken area, for, notice that if each station can determine the distance, then three stations working together may locate the center of the shock.

This is how it is done. A circle is drawn about each observatory as a center, with a radius equal to the computed distance to the quake. The intersection of the three circles is the desired location.

Fig. Three Observatories Co-operate in Locating a Quake in Mexico

The example just given serves as a striking contrast to the much less refined methods that were available in Mallet's time. During the earthquake of 1857, the area was so stricken that news of the quake was delayed nearly a week; nowadays the recording and location of quakes are a matter of hours only. Earthquake waves travel at the rate of several miles per second, so that almost immediately after the shock seismographs for many miles around record the disturbance.

Not only have methods of studying earthquakes improved tremendously in the last third- or half-century, but much more is now known about the immediate causes of the destructive effects. It has been amply demonstrated, for example, that structures on mantle rock and on loose sedimentary deposits suffer much more than those on solid bedrock.

In the great San Francisco quake of 1906 it was the buildings erected on filled-in land that suffered most; the same was true of the Messina earthquake of 1908 and the Tokio quake of 1923. From an engineering viewpoint, then, earthquake research has done much to avoid the extreme destruction of earlier shocks, both by new types of shock-resisting construction and by recognizing the probable centers of greatest damage in regions subject to earthquakes.

The reason why earthquake damage is greatest on loose or unconsolidated rock is not hard to understand. As long as the earthquake waves pass through dense rock, they merely vibrate the particles; but when they enter loose materials, their energy may be transformed to an actual shifting of the particles among themselves.

If the floor is tapped smartly with a hammer several feet away, no movement of the floor is perceptible, but the marble is made to jump several inches into the air. The energy here has been partly applied to movement of the loose marble itself.

Inasmuch as earthquake waves are partly transmitted beneath the earth's surface during their propagation from the focus, we may expect that a theoretical study of the waves themselves could shed light on the nature of the earth's interior.

This is exactly the case: earthquake waves are our most powerful weapon for studying the deep interior of the earth.

Chapter 19

The Earth Model

THE INTERIOR OF THE EARTH AS REVEALED BY EARTHQUAKES

The deep interior of the earth is inaccessible, and no rays of light penetrate to let us see what is below the surface. But rays of another kind penetrate and carry with them their messages from the interior. The Earth has been found to have elastic properties that allow movement set up at the source (focus) of an earthquake to radiate into the interior and to spread over the surface.

In a strong earthquake the whole of the Earth is set vibrating. At some distance from the focus, depending on the strength of the shock, the movement is no longer perceptible, but sensitive seismographs can record the waves that emerge at the surface. The records provide data from which knowledge of the Earth's interior may be gained. From seismic studies we have learned that the Earth consists of a core surrounded by a mantle on which there is a crust; inside the core there is a small inner core.

Towards the end of the nineteenth century it was realized that earthquake movements extend to great distances from the focus. Systematic recording of earthquakes began, and the very important results gained at an early stage greatly stimulated interest in this new science.

The elastic waves that radiate into the Earth are of two kinds, having different speeds of travel: P waves (*undae primae*), in which the particle motion is longitudinal, and S waves with transverse particle motion. The speed of S waves is roughly

60 per cent of that of P waves. The arrivals of the waves are marked by groups of oscillations in the seismograms, and at moderate distances from the focus they usually stand out clearly. The seismogram obtained at a distance of 18-61 from the focus. P and S appear on it. The large oscillations succeeding S are due to surface waves.

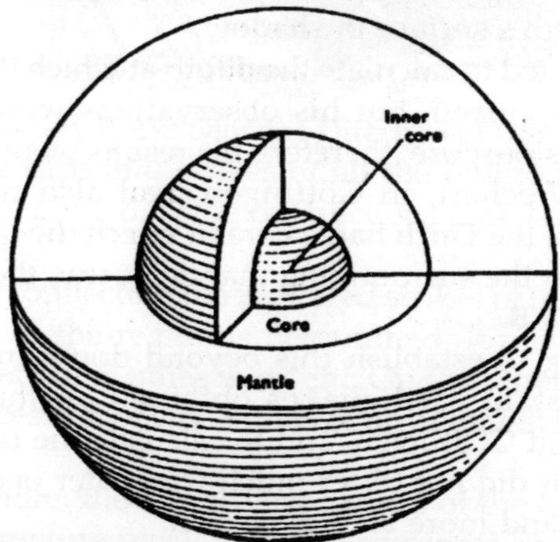

Fig. Mantle, Core, and Inner Core of the Earth.

We measure the arrival times of the waves today usually with an accuracy of not less than 1 second, and deduce the travel times of the waves from focus to recording station, if the location of the focus and the time of occurrence of the earthquake can be found.

When a fair number of travel times to points at different distances from foci at approximately the same depth (the normal depth is about 10-20 km) are available, timedistance tables can be set up, giving the travel times of each type of wave over the various distances. It is customary to present them as time-curves.

The first attempts to construct time-curves for the P and S phases revealed that the average speed, as determined along the chords, increased with focal distance, indicating the wave velocity increased with depth. As a consequence of this the rays are not straight lines but have an upward curvature.

At about 100° focal distance P and S became small and at somewhat greater distances could not be detected. Clear P phases reappeared at about 140°, but they had been delayed, so that their travel times did not fit on to the continuation of the time-curve for shorter distances.

This observation was made in 1906 by R. D. Oldham, who drew the conclusion that deep in the Earth there was a decrease of velocity, causing the rays to bend downwards so as to leave part of the Earth's surface in shadow.

He attempted to calculate the depth at which the decrease of velocity oc- curred, but his observations were not good enough for this purpose; therefore his results were very much in error. E. Wiechert, in Göttingen, had also come to the conclusion that the Earth had a core in which the velocity was smaller than in the surrounding mantle. It was the first great achievement of B.

Gutenberg to establish this beyond doubt by means of records of distant earthquakes obtained on the Wiechert seismographs at GÓttingen, and to calculate the radius of the core. His result did not differ much from later ones obtained from modern and more abundant data.

Gutenberg made use of the time-curves for P and S up to about 103° established by Wiechert and Zöppritz, and of the wave velocities derived from them.

At that time formulae had been developed by means of which transmission times to different distances could be calculated when the velocity as a function of the distance from the Earth's centre was known, and also formulae by means of which the wave velocities as a function of depth could be obtained when the time-curve was known. The ray emerging at the distance where the P curve broke off would graze the core.

The belated P waves observed from a little beyond 140° onwards had passed through the core, but the velocity of these waves in the core could not be derived from the formulae used for the mantle, for these formulae break down when there is a discontinuous decrease of velocity, as at the core boundary.

But when a velocity distribution in the core was assumed the travel times could be calculated and Gutenberg varied his assumptions until the calculated travel times agreed with those observed. The velocity as a function of depth was then known for the whole of the Earth to a first approximation.

On theoretical grounds it was to be expected that the rays would be reflected on reaching the surface of the Earth and would be reflected and refracted at discontinuity surfaces in the interior, such as the boundary of the core, partly as rays of the same kind and partly transformed into rays of the other kind.

Thus at great distances from the epicentre (the point on the Earth's surface directly above the focus) waves would be arriving along many different parths, producing oscillations in a seismogram and marking phases more or less prominent according to the energy, carried. When the velocity distribution within the Earth is known, it is possible to calculate the transmission times along all the different paths.

When earthquake records are examined a great many of the anticipated phases can be identified but some of those originally expected to be present are not found, namely all those that would have come as S waves through the core. Since phases may be present without being very clear, many years passed before it was definitely concluded that transverse waves were not transmitted through the core.

At the surface of the Earth a fluid does not transmit transverse waves, and therefore we say that the core is fluid, although in other respects it may not resemble a fluid as we know it.

The shadow zone for the P phase extends from about 105° to 143É epicentral distance. With modern highly sensitive instruments the P phase is found not to be completely absent in this range: in strong earthquakes it is usually faintly recorded. The appearance of P waves in the shadow zone may be due either to diffraction around the core boundary or to a spreading of the rays caused by a small gradual decrease of velocity just outside the core.

In addition to this faint P phase there is, in the shadow zone, another later P phase that is faint at the smaller distances. On Gutenberg's original Earth model its presence could not be explained and it was vaguely ascribed to diffraction.

However, as seismographs improved it was more and more clearly recorded, and an explanation was required.

In 1936 the writer pointed out that the presence of a small inner core in which the P velocity was greater than just outside it would fully account for the occurrence of the phase, for it would cause enough incident rays to bend upwards strongly enough for part of them to emerge in the shadow zone. Gutenberg and Richter accepted the existence of this inner core and calculated its radius and the velocity distribution in it.

Later H. Jeffreys proved that diffraction could not account for the phase in question flected once and twice at the surface of the Earth. PS starts as P and is reflected as S is part of a seismogram recorded at an epicentral distance of 108°. P is here the P wave reflected at the boundary of the inner core. SKS starts as an S wave, traverses the core as a P wave, and is again an S wave after leaving the core. SKKS is also an S wave outside the core and a P wave inside, but it is reflected when, from inside, it meets the core boundary.

The wave velocity in the crust is smaller than in the mantle underneath, and therefore the waves coming through the crust are refracted and bent upwards when they meet the mantle.

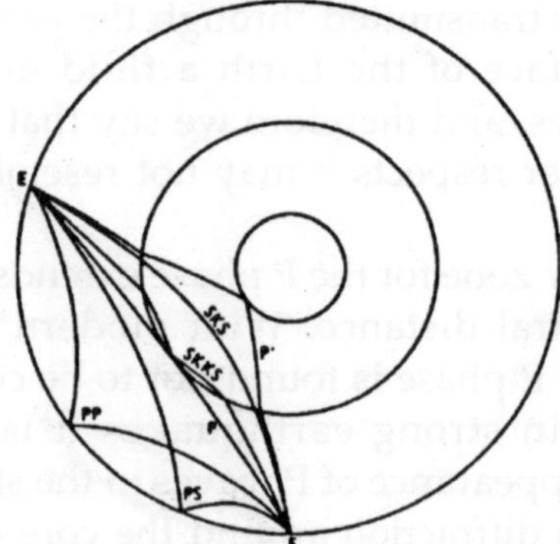

Fig. Rays from Focus *F* to Epicentre *E* at Distance 108°.

There will be a range of distance within which both the refracted and the direct waves emerge at a certain distance the refracted wave overtakes the direct wave because it travels faster in the lower layer. But its path is longer, and energy is lost on refraction, and therefore the corresponding phase in a seismogram will be smaller than that due to the direct wave. There therefore appears a small P phase succeeded by another much larger P phase.

This was observed by Mohorovi, who gave the correct interpretation. He was interested in finding the depth of the discontinuity but did not have the means of determining it with any accuracy. Actually it turned out to be extremely difficult to arrive at reliable values for the depth, though many different methods were employed. The best results have been obtained from explosions, which can be looked upon as artificial earthquakes.

They can be timed with great precision, and the focus is exactly known; when they are well recorded at suitable distances more useful data are obtained than can be derived from earthquakes.

It now seems to be established that the discontinuity is at a depth of between 30 and 40 km under most continental areas. Under the deep oceans it is at a much smaller depth, only about 10-15 km under the water surface.

While the evidence for the existence of the Mohorovi?i? discontinuity and of the other subdivisions of the Earth mentioned above is very clear, precise determination of their depths is difficult. It depends on very precise determination of the velocity variation throughout the Earth, and this in turn depends on precise determination of the travel times of the direct P and S waves and of some of the reflected and refracted waves.

Much important work has been done along these lines, a great deal of it in the 1930s. In the course of this work it appeared that subdivisions of the mantle have also to be considered, but the evidence for them is not of a very precise nature, and their location is uncertain.

It is a difficult and lengthy process to construct good time-distance tables or time-curves. To obtain accurate travel times we require to have the focus and the time of occurrence of the earthquake accurately determined. As a rule this cannot be done directly, because there are not enough observations close to and around the epicentre.

It is therefore necessary to make use of time-curves already in existence, and there is then the risk of transferring errors from these curves to the new travel times. When time-curves have been determined from them it may therefore be desirable to have the elements of the earthquakes redetermined and the whole process repeated.

For the construction of the time-curves it has been customary to use a graphical method, plotting travel times against distance and drawing a smooth curve through the cluster of points thus obtained. It is a somewhat arbitrary process, and, in common with other smoothing procedures, it is apt to smooth away or faultily to introduce changes of slope or curvature.

This is serious, because it is such changes that indicate the existence of more or less abrupt changes of velocity or velocity gradient in the Earth. Many time-curves have been constructed in the course of time, and the existence of various so-called discontinuities has been derived from them.

Two sets of time-distance tables or time-curves are now available which are far better than any previous ones. They are due to Gutenberg and Richter and to Jeffreys and Bullen. In 1928 Gutenberg published his Frankfurter Laufzeitkurven based on a large number of observations. In collaboration with C. F. Richter he greatly extended the work.

In 1936 they jointly published the first part of "On seismic Waves", containing time-curves for a great many phases. Amplitude variation was considered for the fixing of the distances at which the curvature of the time-curve was either greater or smaller than usual.

The amplitudes of the recorded waves should be relatively large at the distances where the time-curve bends strongly, and amplitudes should be small where the curve is straight. Amplitudes were measured and used, and although very precise information is not derivable in this way, some useful indications were obtained. "The International Seismological Summary"for which foci and times of origin of individual earthquakes are determined.

For the reduction of the data down to 1928 inclusive the

Zoppritz tables were used, but it became more and more apparent that the times given by these tables departed seriously from actual travel times.

In 1928 Jeffreys began his very important work on travel times by a preliminary revision of the Zöppritz tables, using I.S.S. data. Later he undertook a thorough revision in collaboration with K. E. Bullen. A great quantity of data was used, and for the first time statistical methods were applied and the accuracy of the results obtained was evaluated.

This implied difficult and extensive work because of the complicated processes involved. In part, new methods had to be developed. After the main work many special investigations followed, improving somewhat the first results. In 1940 the Jeffreys- Bullen "Seismological Tables"were published, and these are now being used for the I.S.S.

From these tables Jeffreys calculated the variation of velocity with depth. On the whole, the velocities of P and S waves increase down to the core boundary. Jeffreys found the velocity of P waves to be 5.6 km/sec in the upper crust, below the sediments, and to be 13.6 km/sec at the core boundary, where it decreases to 8.1 km/sec.

With increasing depth the velocity increases steadily in the outer core to about 10.4 km/sec. In the inner core it is nearly constant at 11.2 km/sec.

The velocity does not increase uniformly with depth all through the mantle. In the uppermost mantle the velocity increases slowly, but at a depth of a few hundred kilometres a strong velocity increase sets in, as indicated by a bending of the time-curve around 20° epicentral distance. Below a depth of about 1000 km the velocity again increases more slowly. On the whole, the velocities as derived from the Jeffreys- Bullen tables are probably not far wrong, but there are serious uncertainties.

This is chiefly because it is so difficult to determine the slope of the time-curve accurately. The boundaries of the mantle regions in which the velocity gradients differ are indicated by changes of slope and curvature of the time-curve, but our data are not accurate enough for us to say exactly where these occur. It has been found particularly difficult to determine the velocity variation near the boundary of the inner core.

A vast amount of data, in part more reliable, has accumulated since Jeffreys and Bullen constructed their tables. New seismographs have been developed, from the-records of which the arrival times of the phases can be read with greater precision. In addition, explosion work has entered the picture.

It has for a great many years been used for the exploration of the crust, especially for the finding of oil; since the second World War more effective explosives have been available, and many of the explosions have yielded results also for the upper mantle. Some large accidental explosions have also provided seismologists with new data.

It was then found that the velocity just below the crust was greater than that derived from the Jeffreys-Bullen tables. Jeffreys drew attention to this, and in later work he provided corrected P tables for distances up to 30° for Europe. For, to complicate matters, it turned out that there were regional differences not only in the crust but also in the uppermost mantle. From the new tables it appears that the strong increase of velocity gradient in the mantle at first placed near 400 km depth is likely to occur at a much smaller depth, probably between 200 and 250 km.

However, attempts to fix the depth accurately have as vet met with unsurmountable difficulties.

The study of the Earth's interior is approached from many different directions. The upper mantle plays an important part in many investigations, and geophysicists in various fields look to seismologists for precise and detailed information. Have we any means of supplying such information?

K. E. Bullen's presidential address to the International Association of Seismology and the Physics of the Interior of the Earth at Toronto in 1957 had the title "Seismology in our Atomic Age". He pointed out that atomic explosions had far greater energy than the chemical explosions it is possible to use, and that they send waves deep into the interior of the Earth. Some atomic explosions have been recorded by seismographs, but there has been a reluctance by the authorities concerned to give prior information about the exact location and time of the explosions.

Though this has reduced the value of these explosions for seismological purposes, a few important results have been obtained, results that seemingly could not be derived from earthquake observations.

There is no doubt that if an opportunity arose to record atomic explosions, by many seismographs placed at suitable distances from the source, information would be obtained that would help us to solve the problems that now confront us.

We are here, as Bullen said, in a tantalizing position, for the tools we so much need exist, but they are not very likely to be placed in our hands. With ever-increasing fear of the perilous effect of atomic explosions, seismology can scarcely hope for any to be organized for its special purposes.

However, if the intention to explode bombs for other purposes is made known beforehand and the location and time are communicated, as has been the case in some recent instances, it should be possible for seismologists to make some use of them.

It is also well to remember that great masses of earthquake data are as yet unreduced and that, skilfully handled, they may yield fruitful results. New methods are also forthcoming in earthquake studies. The surface waves, not dealt with here, spread over the surface of the Earth, but they penetrate to some depth below it; in large earthquakes they may penetrate to considerable depth.

The intense study of surface waves carried out in recent years, especially at the Lamont Geological Observatory, has provided a new approach to the exploration of the crust and the upper mantle.

While seismology teaches us a great deal about the interior of the Earth there is certainly very much more we should like to know. If we ask what are the materials inside the Earth, in what state they are, what is their density, and so on, seismology alone does not supply the answer. We have to look to other branches of geophysics and to other sciences such as geology, the physics and chemistry of the Earth's interior, and also to astronomy for additional information. Much attention has been given to pertinent questions in recent years, but the results are, in part, highly controversial.

Important results on the density variation throughout the Earth were obtained by K. E. Bullen. The velocities of the seismic P and S waves depend on the density of the transmitting material and on the elasticity as characterized by the rigidity and the incompressibility.

These three quantities cannot be derived from the two velocities, but estimates can be arrived at when information from various other sources is taken into account. Bullen finds that the density increases in the mantle from about 3.3g/cm3 just below the Mohorovi?i? discontinuity to about 5.5 g/cm3 at the bottom of the mantle It then jumps to about 9.5 g/cm3 and increases to 11.5 g/cm3 at the bottom of the outer core. In the inner core there is a strong increase.

The rigidity that represents the resistance to shearing stress increases in the mantle, until at the bottom it is nearly four times that of ordinary, steel. In the outer core the rigidity is quite small, and this is what we mean when we say that the core is fluid. On the other hand, the incompressibility or the resistance to pressure does not change materially at the boundary of the core.

The pressure when evaluated was found to have reached about 11/3 million atmospheres at the bottom of the mantle and about 4 million atmospheres at the centre of the Earth. Following a long line of argument. Bullen arrives at the conclusion that the inner core is likely to be solid.The composition of the continental crust varies a great deal from one region to another. Granite is one of its main constituents, whereas this material is absent under the deep oceans, where the crust is much thinner.

The upper mantle is believed to consist of ultrabasic rock rich in olivine. A transition is likely to take place, perhaps from one form of olivine to another, in the region where the seismic velocities increase more strongly than elsewhere, this region beginning at a depth of a few hundred kilometres. It has for a long time been believed that the core consists of iron and nickel, but recent investigations have led to the conclusion that the outer core possibly consists of material not much different from that of the mantle but transformed under the prevailing high pressure. The inner core is still believed to consist chiefly of iron and nickel.

THE RADIOACTIVE EARTH

Without the heat from Radioactivity it is probable that we would have had no atmosphere or oceans. Even if the ocean had existed, no land would have risen above it. Indeed, it is probable that the earth would have had a bare, rocky surface like the moon's, scorched by the sun in daytime and bitter cold at night.

But the story of the earth is a story of heat. Throughout earth history large amounts of energy have been continuously expended in mountain building, volcanism, and other activities which have formed the continents, oceans, and atmosphere.

Except for the actions of the surface agencies, driven by heat from the sun, the energy comes from the interior of the earth and must have been at one time in the form of heat. To try to explain the occurrence of this thermal energy, we must consider two principal sources: the heat inherited from the formation of the earth and the heat generated in the breakdown of radioactive elements.

There are many arguments in favor of believing that the earth formed at a relatively low temperature. If this is true, a uniform distribution of the radioactive elements that we estimate were contained within the earth would have heated it sufficiently to have caused it to melt or partly melt. It is purely by chance that the sequence of events which we believe followed was such that the bulk of the earth stopped heating up again and remained fairly stable.

These purely chance events are as follows. First, if the mantle of the earth is solid and there is no convection (transfer of heat by movement of fluid material) in it, it must lose heat by the slow process of conduction in the upper regions. If there is convection, from melting or otherwise, the heat can be rapidly transported to the surface.

Therefore, the temperature can never get very, much above that necessary for melting. In the lower regions heat may be carried out by radiative transfer.

Second, as soon as melting begins, there probably would be a migration of the molten radioactive substances upward because they crystallize at lower temperatures than compounds of magnesium, silicon, and iron. They would be forced upward in the liquid as the solid material settled downward.

If in any region the heat-producing elements have not moved close enough to the surface, the temperature will rise locally to the melting point and a further upward migration will occur. Eventually they will have come close enough to the surface so that there will be no further melting. Enough of the heat generated will be lost by conduction to the surface so that a stable solid mantle remains.

Gradually, following that time, the radioactive elements will decay slowly, and their heat production will diminish. This would tend to stabilize the mantle so that it is at some temperature below the melting point of its most fusible components. But if the chemistry of the radioactive elements had been otherwise and they had settled into the core, the earth would be continuously melting, losing heat by convection and solidifying again.

THE MAJOR HEAT-PRODUCING ELEMENTS

Let us now examine the amount of heat given off by radioactive elements and estimate what abundance of these elements would cause melting in the mantle. The element uranium breaks down through several stages to form a stable end product, lead. As it undergoes successive transformations toward this stable end product, the isotope uranium 238 gives off 8 alpha particles as well as numerous gamma rays and beta particles.

Summing up the energies of all these emitted particles and rays, we find that a total of 47.4 Mev (million electron volts) of energy has been expended for each atom of uranium 238 that breaks down to form an atom of lead 206. Since 1 Mev is equivalent to 3.83 X 10-14 calories, it can be calculated that one gram of uranium in equilibrium with its daughter products is continuously giving off 0.71 calories per year.

Similarly it can be calculated that the isotope uranium 235, of which atom bombs are made, is giving off 4.3 calories per gram per year when in equilibrium with its daughter products. Thorium and its series give off 0.20 calories per gram of thorium per year.

The only other important heatproducing element is the isotope of potassium, K40. This gives off beta particles and gamma rays at a rate that yields 27 X 10-6 calories per gram of total potassium per year.

Average granite and volcanic rock contain approximately the following amounts of these radioactive elements:

Rock Type	*Uranium parts per million*	*Thorium parts per million*	*Potassium %*
Granitic rocks	4	14	3.5
Dark-colored volcanic rocks	0.6	2	1.0

Thus the radioactive components of the average granite can produce 7 microcalories of heat per gram per year. Other rocks that make up the bulk of the crust produce somewhat less heat than granites, and it is estimated that the average rock in the crust above the Mohorovi?i? discontinuity probably produces about 2 microcalories of heat per gram per year.

There is a measurable amount of heat continuously flowing to the surface of the earth. Measurements over continental areas have indicated that this amount averages about 1.2 microcalories per square centimeter per second. The amount of heat flowing in oceanic areas has been difficult to measure, but several measurements have been made.

This is done by dropping a probe from a ship so it penetrates the mud on the bottom of the ocean for some distance. Refined temperature devices are then used to record the difference in temperature between two points on the probe. By determining the thermal conductivity of the mud, it is possible to calculate the amount of heat flowing upward from the earth into the ocean water.

Surprisingly, it turns out that approximately the same amount of heat is flowing from the interior of the earth in the oceanic areas as in the continental areas; namely, about 1.2 microcalories per square centimeter per second.

From all these figures it can be calculated that the average continental crust down to a depth of 35 kilometers produces about 1/2 microcalories per square centimeter per second from the radioactive breakdown of uranium, thorium, and potassium. This is about one half of the observed heat flow to the surface. It means that only about one half of the heat flowing to the surface comes from a depth of greater than 35 kilometers.

If you measure the amount of heat flow and estimate the thermal conductivity of the materials in the crust and below the crust, it is possible to estimate the increase in temperature with depth. The temperature-depth relationship. Note that the production of the heat in the crust greatly reduces the thermal gradient (rate of heat flow) at depth.

It follows that the temperature at depth is very much less than would be expected if one simply measured the temperature in deep mines or other openings in the earth near the surface and extrapolated this information to great depth.

In fact, if there were no radioactivity in the crustal rocks, the observed temperature gradients at the surface would require that the mantle be molten at a shallow depth; this is not in agree ment with the known geological stability of the region. The facts, therefore, support the hypothesis that the radioactive components of the earth are largely concentrated in the near-surface layers.

By calculating the temperature at which materials would be molten at depths of 100 or 200 kilometers, it is possible to estimate how much radioactivity must be in the near-surface rocks in order to keep the temperature gradient within known bounds. Attempts to do this have indicated that at least 0.2

part per million of uranium and 0.7 part per million of thorium on the average must be in the rocks down to 100 kilometers depth under oceanic regions. Since these amounts of radioactive elements would supply much of the observed heat flow to the surface, there must be little heat left flowing from the interior.

Thus we arrive at two important conclusions. The first is that very little of the original heat stored in the earth at the time of its formation is being lost, and, therefore, the earth is not cooling down at an appreciable rate, if at all. Secondly, we conclude that the major part of the earth's heat is coming from the breakdown of radioactive elements.

Since almost all the breakdowns occur within the near-surface regions of the earth, it is reasonable to infer that some process has moved the radioactive elements to this location from a presumably homogeneous distribution at the time of the earth's origin.

MIGRATIONS THROUGH THE MANTLE

Again we see the need for some process to have brought up from within the earth the materials that make up the oceans and atmosphere and the radioactive elements. In support of this requirement, we see that uranium, thorium, and potassium do in fact belong to the group of elements that form compounds of rather low stability and therefore would be most likely to move to the outer part of the mantle in any process in which partial melting was involved.

It is interesting that these conclusions do not violate the concept of an earth that is composed of materials similar to the iron and stone meteorites. The proportion of radioactive components in iron meteorites is very small indeed and would contribute a negligible amount to the heat of the earth if it were made of similar material.

The amount of radioactive components in stony meteorites has been measured carefully, and it is a striking coincidence that the amount corresponds very closely with that needed to give rise to the observed heat flow in the earth if it were uniformly of stony meteoritic composition. The fact that the radioactive components have migrated to the outer part of the mantle does not alter this interesting and supporting evidence.

Thus we see an earth in which the central part is rather slowly changing, if at all, in temperature and losing heat to the outside very slowly. Near the surface, however, there is an important balance in which the heat produced by radioactive elements can flow to the surface without causing melting unless the system is disturbed in some way.

If at the margin of a continent the accumulation of sediments formed a low-conductivity blanket which also contained added amounts of heat-producing elements, this might be enough to cause melting at a depth of 100 or 200 kilometers and give rise to volcanic activity and other effects related to mountain building.

There has been much discussion and difference in opinion about the possibility of major convective overturns in the mantle down to the core boundary as a result of inhomogeneous distribution of heat sources. This process of convection could give rise to surface activity also and could be the cause of mountain-building events. In either case it is the heat from radioactivity that provides most of the energy for the dynamic events that have occurred at the earth's surface throughout geologic time.

THE AMSOC HOLE TO THE EARTH'S MANTLE

The earth sciences panel of the National Science Foundation (NSF) had just completed two days of hard work evaluating some 60 odd projects requesting research grants. The difficult job completed we relaxed to reflect on our work. Many of the suggested projects were excellent and most were of high caliber. Walter Munk remarked, however, that not a single one of them was of such a nature that a really major advance in Earth Science would result.

He suggested the panel invent a project which might strike directly at the roots of a major problem and forthwith suggested a hole to the Moho. If the writer deserves any credit for past or future association with this project, it is that he took

Munk seriously and prevented momentarily adjournment of the panel. In the few minutes remaining he proposed referring the project to the American Miscellaneous Society for action. This was not a joke. It was done for several very good reasons.

Amso has no constitution, no officers, no members, only founders, many of whom are distinguished scientists. Here was a society which could act immediately -- no need to wait for the next council meeting to have the proposal referred to a committee which would report to the council a year later.

Gordon Lill was appointed chairman of the committee to proceed with the project and there followed the breakfast meeting in California mentioned by Bascom at which the specific plans were laid out. NSF being unable to grant funds to a society without officers or a membership list made it necessary that amsoc get a reputable sponsor.

The National Academy of Sciences-National Research Council very kindly, and I might say, courageously, gave the Amsoc committee a respectable home. NSF funds for a feasibility study were then forthcoming and, once in hand, Bascom was appointed Executive Secretary of the committee. The committee's history from this point on is one of record and can be left to historians.

THE METEORITE ANALOGY AND THE MANTLE

In 1850 Boisse proposed that the Earth's interior might be analogous in composition to meteorites. It was a brilliant proposal for its day and in a general way it is probably correct. In detail, perhaps, it is today being taken too seriously. It depends on how good the meteorite sample is, considering only those seen to fall, and how reliable a sample this is of the average composition of the body or bodies from which the meteorites were derived.

The great majority of stony meteorites are analogous to volcanic rocks such as crystal tuffs, lithic tuffs, and breccias. Are we only getting a sample of the outer surface of the meteorite parent? Olivine nodules from the Earth's interior where ejected from a volcano as bombs are found to be friable, the crystals loosely held together. This is probably due in part to sudden decompression and in part to the lack of a cementing matrix. The mantle of the meteorite parent body might well behave in a similar manner upon being suddenly extracted from its environment. Were this the case the small particles resulting would be swept up by the Sun leaving only the somewhat more lithified stony meteorites and irons in orbits which might collide with the Earth.

The decompressed densities of the inner planets (and the Moon) are not the same. This must mean that their compositions are not exactly the same. It may merely indicate a difference in the degree of oxidation of the iron present It could mean that the initial bodies forming the solar system varied somewhat in composition or that their initial compositions have been changed by some process which caused losses to surrounding space.

The uncertainties in the meteorite analogy cannot be resolved by further speculation. A hole or holes are required if we are to know something more definite about the chemistry of that 84 per cent of the volume of the Earth called the mantle. (The crust is less than one per cent of the Earth and we know quite a lot about it from observation.

It seems quite reasonable to accept the meteorite analogy to the degree of postulating a Ni-Fe core.) A half-ton sample of mantle would probably give more specific information than the 1400 odd meteorites now in collections. The meteorites could then be properly evaluated to give much additional information. It seems foolhardy to put an enormous effort into attempts to sample the moon or even the planets without finding out what is 5 km below the sea floor. The information from the hole is necessary to attack

CHOICE OF SITE

Drilling a hole in the Moho in a continent is not at present possible because it involves depths near 100,000 ft and temperatures too high for modern drilling equipment. The high temperatures also present difficulties in that many desirable measurements could not be made because the electronic devices and electrical cables could not withstand such conditions. Oceanic islands were considered during the early stages of the feasibility study. Aside from the fact that these are of volcanic origin, and hence one would be'drilling into the underpinnings of a volcano, the depth to be drilled would be too great to reach the mantle.

A small volcanic island rising from the ocean floor in depths of 15,000 to 18,000 ft would be about 40 miles wide at its base. Volcanic islands are in isostatic equilibrium. They represent loads in excess of the strength of the crust. If one assumes a density of 2.8 g/cc for the volcanic material this load would depress the Moho to 78,000 ft. Assuming the lowest reasonable density rather than the most probable one would give a depth in excess of 60,000 ft.

This leaves only one alternative, drilling a hole from a barge in the deep sea. The Moho can in many places be reached at depths of 30,000 to 35,000 ft below sea level and only about half of this need be drilled, but the overlying water presents some unique problems. The ship must either be anchored or maintained in position by some sort of automatic positionkeeping system.

In choosing a site the weather conditions must be considered. In general this led us to look for one at a latitute less than 25°. It must, if possible, be within 500 miles of a port where supplies and repairs can be obtained and exchange of personnel becomes feasible.

The heat flow on the ocean floor must be less than two microcalories per cm^2 per sec. Assuming a conductivity of 0.005 cgs units, two microcalories gives a gradient of 40°C. per km or 200°C. approximately at the Moho which is somewhat too high for existing well logging instruments.Two areas seem to fulfill the above conditions.

One is about 120 miles north of San Juan, Puerto Rico, and the other south of Los Angeles from Guadalupe Island toward Clipperton Island. Seismic work is now under way to find places in these areas where a depth to the Moho is favorable and where the heat flow is comparatively low

Chapter 20

The Mountain Ranges and Ocean Basins

One of the unexpected discoveries in earth science in the previous century was that of a fundamental difference between continents and ocean basins. Ocean basins are not merely the low lying parts of the Earth's surface flooded by salt water but are great, relatively steep-sided, structural depressions.

In fact, there is, too much water for the size of the ocean basins, and parts of the continents are now flooded and probably have been flooded through a great deal of the geologic past.

Precise measurements of gravitation attraction in major mountain ranges, continental areas, and over the ocean basins showed an even more unexpected feature. The continents and mountain ranges do not represent extra loads of rock superimposed upon the Earth's crust, but are masses of lighter rock floating in a denser substrate.

An iceberg floats above the water much in the same fashion, buoyed up by deep submerged roots. The great mountain ranges of the world and the continental masses similarly have deep roots of light rock penetrating down into the denser crust, and thus the mountain ranges and continents float at elevations appropriate to the depth and size of these submerged light roots.

Thus, all the major features of relief of the surface of the Earth show mirror image features within the crust. The major mountain ranges float at high elevations because they are buoyed up by light rocks. The continents float at intermediate elevations with roots of intermediate depth, and the deep oceans are underlain by thin layers of light rock.

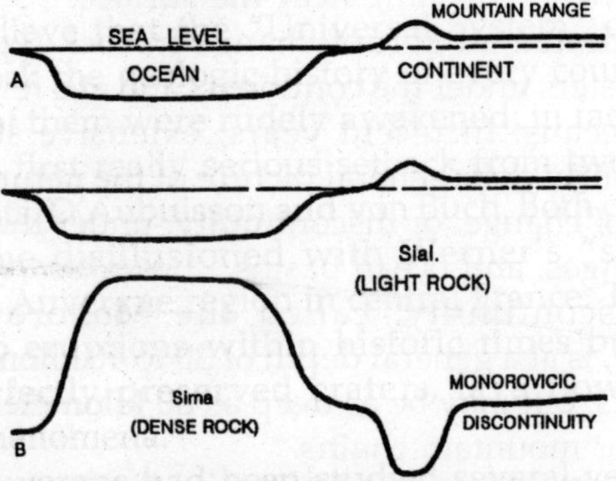

Fig. Profiles through the Earth's Crust.

Seismologists, from the study of earthquake waves, have shown that the Earth's mantle is solid to the depth of the outer core, some 2900 kilometers. The observation that large mountain ranges and continental masses float on the crust of the Earth at elevations appropriate to the size and density of their roots implies that rocks at shallow depths in the Earth's mantle, although solid, have little strength and can flow in respons e to small stresses given sufficient time.

This deduction is strengthened by the observation that rocks in deep, eroded, old mountain chains are intensely contorted and folded, plain evidence that, at high pressures, solid rocks can flow readily and do not have great strength. -

Recognition that continental rocks are lighter and more buoyant than oceanic rocks gave rise to the concept that the crust of the Earth is made of two contrasting materials: sialic material, rich in silicon, the alkalies and aluminum, making up the continents; and simatic material, richer in iron and magnesium, making up the denser rocks below the floor of the ocean and lying under the sial of the continents. The sial is assumed to be granite or granodiorite in composition, and the sima is assumed to be basaltic in composition.

Early in this century, the Yugoslav seismolog Mohorovicic, obtained evidence from seismogra earthquake waves, traveling a few tens of kilom the surface of the Earth, gave records showing sharply higher speeds for both shear and compressional waves than earthquake waves traveling near the surface.

This indicated an abrupt change in rock types at a few tens of kilometers under the continents and at a few kilometers under the oceans. In recent years, extensive studies have produced a fairly clear general picture of the nature and depth of this level of change, or discontinuity, under the continents, mountain ranges, and ocean basins.

This discontinuity, called the Mohorovicic or M discontinuity, is at a general depth of 30 to 40 kilometers under the continents, but may be as deep as 60 kilometers under the roots of major mountain chains.

It is as shallow as 4 to 5 kilometers below the floor of the deeper parts of the ocean. The discovery of the M discontinuity seemed to confirm the notion that the crust is fundamentally made up of two different kinds of rock material. The discontinuity itself appears to be the boundary between these, the sialic rocks above and the simatic rocks below.

The rocks below the M discontinuity have seismic velocities and densities which suggest that they may be even richer in magnesium and iron than normal sima of basaltic composition. Consequently, they are, by some, called ultrasima. However, throughout this paper the word sial will be applied to the lower velocity rocks above the M discontinuity, including the range of basalts to granites, and the word sima will be applied to the denser rocks below the M discontinuity.Prior to and along this general picture, the concept of isostasy developed.

This is the notion, previously discussed, that the lighter continental rocks float at an appropriate depth, depending on their mass and mean density, in a denser substratum. As rock is eroded from the tops of continents and mountain ranges they tend to float up higher and higher, renewing their relief, permitting erosion to continue.

Four facts, however, sharply contradict this picture of a sialic crust of varying thickness floating on a simatic substratum of different chemical composition and different density.

- Large areas of continents, long near sea level, have been uplifted many thousands of feet in the air. Further, this uplift seems to have taken place rather rapidly in terms of geologic time.
- Sediments of low density, filling troughs along the margins of continents, apparently are able to subside into this higher density substratum.
- Inasmuch as radioactive, heat-producing elements are associated with sialic rocks, one might expect heat flow through the thicker parts of the Earth's crust to be much greater than through the thinner parts of the Earth's crust. However, as a first approximation, heat flow through the crust of the Earth is approximately the same through continents, mountain ranges, and ocean basins.
- The lifetime of continents and mountain ranges is vastly greater than rates of erosion would suggest.

Let us examine each of these apparent facts and their consequences on the hypothesis of sialic continents floating on a simatic basin. The problem of the uplift of large plateau areas is one which has puzzled students of the Earth's crust for a very long time. Regions which are at sea level, or near sea level, may, over a relatively short geologic time span, such as a few million years, be uplifted several thousands of feet.

The Colorado plateau and adjacent highlands is an example. Here, in an area of approximately 250,000 square miles that apparently stood at sea level for several hundreds of millions of years was uplifted approximately one mile vertically some 40,000,000 years ago in early mid-Tertiary time and is still a high plateau. The Grand Canyon of the Colorado has been carved through this great uplifted plateau.

Given an Earth with sialic continents floating in a denser simatic substratum, what mechanism would cause a large volume of low standing continents to rise rapidly a mile in the air? Furthermore, evidence from gravity surveys suggest that the rocks underlying the Colorado plateau are in isostatic balance, that is, this large area is floating at its correct elevation in view of its mass and density.

Recent seismic evidence confirms this, in that the depth to the M discontinuity under the Colorado plateau is approximately 10 kilometers greater than over most of continental North America.

Thus, appropriate roots of light rock extend into the dense substratum to account for the higher elevation of the Colorado plateau. We have then a double-ended mystery, for the Colorado plateau seems to have grown downward at the same time that its emerged part rose upward.

This is just as startling as it would be to see a floating cork suddenly rise and float a half inch higher in a pan of water. To date, the only hypothesis to explain the upward motion of large regions like the Colorado plateau is that of convection currents. Slowly moving convection currents in the solid rock, some 40 to 50 kilometers below the surface of the Earth, are presumed to have swept a great volume of light rock from some unidentified place and to have deposited it underneath the Colorado plateau.

A total volume of approximately 2,500,000 cubic miles of sialic rock is necessary to account for the uplift of the Colorado plateau. While it is not hard to visualize rocks as having no great strength at the high pressures and temperatures existing at depths of 40 to 50 kilometers, it is quite another matter to visualize currents in solid rock of sufficient magnitude to bring in and deposit this quantity of light material in a relatively uniform layer underneath the entire Colorado plateau region.

The Tibetan plateaus present a similar problem, but on a vastly larger scale. There, an area of 750,000 square miles has been uplifted from approximately sea level to a mean elevation of roughly three miles, and the Himalayan mountain chain bordering this region has floated upward some five miles, and rather late in geologic time, probably within the last 20,000,000 years.

The quantity of light rock which would need to be swept underneath these plateaus by convection currents to produce the effects noted would be an order of magnitude greater than that needed to uplift the Colorado plateau, that is approximately 25,000,000 cubic miles. Even more troublesome than the method of transporting all this light rock at shallow depths below the surface of the Earth is the problem of its source.

The region from which the light rock was moved should have experienced spectacular subsidence, but no giant neighboring depressions are known. A lesser but large problem is how such enormous quantities of light rock can be dispersed so uniformly over so large an area.

This evidence of uplift and downsinking of various crustal blocks, with the blocks always remaining in approximate isostatic balance, does not seem to harmonize with the view of a floating sialic continent on a denser substratum where one might expect to find little variation in elevation with time.

The second problem, that of the subsidence of troughs, is of equal difficulty. The rivers of the world carry enormous quantities of sediments seaward. Most of this sedimentary burden is deposited within a few tens or hundreds of kilometers of the shore line and little is transported to the deep ocean basins. Thus, elongate prisms of sediments are built up parallel to the shores of certain regions where great quantities of sediments are transported to the sea.

The crust, in response to this added load of sediments, begins to buckle downward. Troughs filled with sediments appear, paralleling the coast line. The chicken and the egg argument enters here, for it is not entirely clear whether deposition of sediments generates the troughs or whether the troughs are formed first and are later filled with sediments.

However this may be, one such trough now in the making is along the coast of the Gulf of Mexico on both sides of the mouth of the Mississippi River. Surprisingly enough, this trough deepens at about the rate new sediments are added to it. Thus, the sediments are always deposited in relatively shallow water.

Fundamental laws of physics are violated and on a large scale if this downwarping is produced directly by continued loading of sediments. These deep troughs filled with sediments may contain 50,000 to 100,000 feet of sediments and may be 1000 or more miles long and 100 miles in width. The mean density of the sediments, even compacted under a load of

10,000 feet of other sediments, is approximately 2.4 to 2.5. The rocks displaced in the downwarping trough are known to be denser, with a mean density of 2.8 to 2.9.

By what mechanism do light sediments displace denser, crystalline rock? These troughs of sediments, like the plateaus considered earlier, always appear to be in isostatic balance. If the conventional is to be sustained, dense rock must automatically be removed from below the bottoms of these sedimentary troughs at approximately.

The same rate that they receive sediments from the rivers which feed them so that the troughs balance and float with their upper layers of sediments under a few tens or hundreds of feet of water most of the time.

The problem of the mechanics of the formation of deep troughs of low density sediments is heightened when their full history is considered. Many are known in the geologic record. In most, sediments accumulate for perhaps a hundred million years and reach a total thickness of as much as 100,000 feet.

These thick, highly elongate lenses of sediments may then be slowly folded and uplifted to form mountain ranges which may initially stand as much as 20,000 feet high. Surprisingly, the geologic record shows that a large fraction of the mountain ranges of the world have been formed from rocks of these thick, geosynclinal troughs.

Extensive volcanic activity may accompany and continue beyond the time of the formation of the mountain ranges. The mystery, then, of the downsinking of the sedimentary troughs, in which low density sediments apparently displace higher density rocks, is heightened when we note that these narrow elongate zones in the. Earth's crust, downwarped the most, with the greatest accumulation of rock debris, shed by the higher portions of the continents, become in turn the mountain ranges and the highest portions of the continents.

The third of the major problems connected with the postulated sialic continental area and simatic oceanic region is that pointed out by recent measurements of flow of heat through the crust of the Earth.

A considerable number of recent measurements have been

made of temperature gradients and rock conductivities within the outer part of the Earth's crust. Careful temperature profiles have been made within many of the accessible deep mines and in numerous wells and tunnels.

From these data, a fairly reliable picture has developed of heat flow within the Earth's outer crust, although measurements are not nearly so detailed or as numerous as is to be desired. The rate of escape of heat through most continental areas appears to be approximately 1.2 microcalories per centimeter per second. It has been known for many years that most of the heat escaping from the Earth is radiogenic heat, generated in the Earth by decay of radioactive isotopes of uranium, thorium, and potassium.

Little or none of the heat escaping from the Earth is primary heat, inherited from an initially hot Earth. In fact, there is no compelling evidence that the Earth was molten in its youth or even formed from hot material. We know that the rocks near surface today appear to be in fairly reasonable thermal balance. The rate of heat escaping from them to the surface of the Earth is very close to the rate at which heat is generated in them by radioactive decay of certain elements.

Over the last twenty years, extensive data have been accumulated concerning the distribution of the radioactive elements. Uranium, thorium, and potassium are 10 to 100-fold as abundant in the light silica-rich rocks as they are in denser simatic material, rich in magnesium and iron, and low in silica.

Consequently, we might expect that radiogenic heat in the thick sialic continents should be vastly greater than the heat generated in the presumably radioactive-poor simatic material underlying the ocean floors.

Further, we would expect heat flow to be greatest in the thickest parts of the continents, that is, in mountainous regions buoyed up by thick roots of sial rich in radioactive elements. A number of studies of heat flow through the continents have been made over the last two decades.

These studies have been made by examining the distribution of temperatures and rock conductivities down deep wells and along tunnels. Surprisingly, these studies show almost no correlation between mean elevation of land mass and heat flow through the Earth's crust.

This was most unexpected because all the broader regions of higher elevation are presumably underlain by thick zones of light rock which, from all determinations, should be richer in radioactive elements.

Nonetheless, it was confidently expected that heat flow through the floor of the ocean would be a fraction of that observed in the continental land masses. The first measurements of heat flow through the floors of the ocean were reported in 1952 by Sir Edward Bullard. These determinations of heat flow were ingeniously made by inserting probes containing thermisters into the muds on the floors of the oceans.

Startlingly, the heat flow determined by these measurements through the floor of the ocean was almost identical with that measured in continental and mountainous areas. Later results by Revelle and Maxwell (1952 and unpublished), although indicating wide ranges of heat flow from place to place in the oceans, have only affirmed the earlier observation that heat flow through the ocean floor is essentially the same as that on the continents.

There seem to be only two possible explanations for this most unexpected discovery: either the concentration of radioactive elements in the rocks below the floor of the ocean is the same as that in rocks which make up the continents or else heat is transferred by some special mechanism from deeper down in the Earth to near-surface sites underneath the oceans.

If the concentration of radioactive materials in the few tens of miles below the floor of the ocean is the same as in a few tens of miles below the continents, then our previous view that the floors of the oceans are underlined by radioactive-poor sima and the continents were underlain by radioactive sial certainly cannot be right.

The alternative explanation, equally difficult, is that high temperature rocks from deeper down in the Earth are convectively carried up to near-surface environments below the oceans. Thus, heat escape through the floor of radioactive-poor oceans is fortuitously approximately the same as heat escape through the radioactive-rich continents.

The fourth problem, that of the long lifetime of continents and mountain ranges, is perhaps the most difficult of all. The rivers of the world strip tremendous quantities of rock debris off the continents each year and deposit it in the oceans. The Mississippi, for example, contains about onehalf weight per cent of solids as it flows into the Gulf of Mexico. Each

year, it brings to the Gulf of Mexico approximately 750 million tons of dissolved and solid material. The great rivers are steadily wearing down their basins. Calculations show that the Missouri River lowers its drainage basin about one foot in each eight thousand years, and that the rate of erosion for the entire United States approximates one foot in 10,000 years.

At this rate, all the land masses of the world would be eroded to sea level in something of the order of 10-25 million years. This is particularly surprising in view of the fossil record. Land animals and plants have been known on the surface of the Earth for well over 300 million years, and the sedimentary record indicates high land masses extending back at least two billion years.

Much geological evidence indicates that the ancient continents were in approximately the same place as the present continents and that continents have existed more or less as they are today and for a period of at least two billion years. How do we reconcile an erosional lifetime for continents of something like 25 million years with a known lifetime of something of the order of two billion? Why has not all the continental sial been uniformly distributed through the ocean basins?

The mountain ranges bordering the continents and interior to the continents present an even more difficult problem. The rates of erosion along the slopes of steep mountains are many times those of lower lying continental land masses. The lifetime of mountains, therefore, must be far less than the 25 million years estimated for continents.

In contrast to this reasoning, however, is the geologic

record which strongly suggests that the Appalachian Mountain Range has existed more or less where it is today and, as far as we know, with reasonably similar relief for the last 200 million years, shedding sediments both to interior valleys and coastward.

Thus, we see orders of magnitude discrepancy between lifetimes of mountain ranges and continents, estimated on the basis of known rates of erosion, and the lifetimes of the mountains and continents as indicated by the geologic record.

Even though we assume that mountain ranges and continents are somewhat analogous to icebergs that float up as their exposed portions are melted away, the presumed depth of roots of the mountain ranges and thicknesses of the light continental rocks permit extension of the estimated lifetime of continents by no more than tenfold that based on present erosion rates and mean elevations.

Thus again, the notion that the rocks which make up the continents are grossly different in composition from those underlying the ocean basin does not seem to hold up, for we would expect that the rain waters washing over the continents would have long ago dispersed the continental rocks into the oceans.

These four major observations then — persistence of continents and mountain ranges in spite of high erosion rates, the relatively uniform values for heat flow in continents and ocean basins, subsidence of marginal troughs in response to loading by low density sediments, and uplift of plateaus once worn to sea level — suggest the inadequacy of the traditional view that continents represent masses of low density silica and aluminarich rock floating in the denser media of sima, iron, and magnesium-rich rock.

Recent theoretical studies by Gordon J. F. MacDonald and experimental work by Robertson, Birch, and MacDonald and by the writer, as well as interpretation by J. F. Lovering, suggest a different structure of continents, a structure which simultaneously explains most of the observed phenomena associated with continents, mountain ranges, and ocean basins and accounts for the four major stumbling blocks in existing theory. This new model of the Earth's crust stems from theoretical considerations largely confirmed by recent experimental work in the field of high pressures.

Very many crystalline solids undergo polymorphic inversions to denser phases when subjected to high pressures. The behaviour of matter at high pressures has been extensively investigated by P.W. Bridgman who has demonstrated literally hundreds of polymorphic inversions among common substances in the pressure range 0-100,000 atmospheres. Graphite and diamond form, for example, a familiar polymorphic pair.

At sufficiently high pressures and temperatures graphite may be converted to diamond. A temperature of 1500 K and a pressure of 100,000 atmospheres is sufficient for the conversion, and, indeed, many thousands of carats of diamonds are now being made annually by General Electric Company by subjecting carbonaceous material to high temperatures and pressures. It has long been noted that basalts and eclogites, rocks with sharply contrasting mineralogy, have essentially identical chemical composition.

Table I

	Eclogite	*Plateau Basalt*
SiO_2	48.12	48.80
TiO_2	.85	2.19
Al_2O_3	10.42	13.98
CaO	9.99	9.38
MgO	14.22	6.70
FeO	13.92	13.60
Na_2O	1.45	2.59
K_2O	.58	.69

Eclogite, however, contains no feldspar; instead, it is made up of jadeitic pyroxene and garnet. The mean density of eclogite is 3.3 gm/cc, that of gabbro is 2.95 gm/cc. As eclogite is the denser of the two phase assemblages, it is the rock which must exist at the higher pressures.

The density contrast, about 10%, between gabbro and eclogite is almost the same density contrast believed from seismic evidence to exist at the M discontinuity, although the contrast at the discontinuity has usually been assumed to be a chemical contrast rather than a phase contrast.

Indeed, Fermor, Holmes, and Goldschmidt suggested that M discontinuity might be a phase contrast and that the rocks below it are eclogite. Their suggestion received little discussion or acceptance but has been recently revised by G. J. F. MacDonald on the basis of calculations of the pressure-temperature conditions controlling the phase change of nepheline plus albite to jade and of albite to jade plus quartz.

These thermochemical calculations have been confirmed by experimental work of Robertson, Birch, and MacDonald and by the writer. These two experimental studies, though in disagreement in detail, confirm the calculations based on thermochemical data that, at pressures of 15,000 to 25,000 atmospheres, depending on temperature, the nepheline plus albite undergoes a polymorphic change to jade, and albite undergoes an inversion to jade plus quartz at slightly higher pressures. Further, an experiment made by me on basalt glass showed that, at 500° and pressures below 10,000 bars, basalt glass crystallized as gabbro.

The major mineral component is feldspar. At pressures above 10 kilobars and at a temperature of 500°, the amount of feldspar decreases and, finally, basalt glass crystallizes directly to a rock made up dominantly of jadeitic pyroxene. Identification of phases were by X-ray. Significantly, 500° and 10 kilobars are approximately the temperatures and pressures estimated at the M discontinuity underneath the continents. It thus appears that the M discontinuity may reflect a phase change from gabbro to eclogite rather than a change in chemical composition.

This phase change will account for the observed change in seismic velocity from approximately 6.5 kilometers per second to 8.1 kilometers per second and a change in density from 2.9 to approximately 3.23. Thus, the older suggestions of Fermor, Holmes, and Goldschmidt are supported by field measurements, theoretical calculations, and recent experimental work.

If the discontinuity caused by a phase change takes place at a depth of 30 kilometers, a depth equivalent to a pressure of approximately 10 kilobars under the continent, how do we account for the much greater depth to the discontinuity under mountain ranges and the much shallower depth to the discontinuity under the oceans?

The answer lies in the fact that the change takes place at a different pressure for a different temperature. As near as can be told from the computations and from the experimental data, the slope of this phase change is approximately the same as the Earth's pressure-temperature gradient.

Consequently, if it is assumed that the Earth's temperature increases a little more rapidly per foot of depth under mountain ranges than under continents generally, the transition will take place at a vastly greater depth. If it is assumed that the Earth's temperature increases with depth a little more slowly under the oceans than under the mountains and continents, the transition is at shallow depths, the single transition explains the varying depths to the M discontinuity under the oceans, mountain ranges, and continents.

We assume that there are variations in temperature from continents to ocean basins to mountain ranges, and, consequently, we would expect variations in heat flow. However, the necessary variations in heat flow to account for these different depths of intersection are exceedingly small, well within the range of observations and are certainly not the threefold variations in heat flow that we would expect if the continents and mountain ranges were thick zones of radioactive-rich sial and the ocean was underlain by radioactive-poor sima.

The Earth's pressure-temperature gradient is almost the same under the oceans as is the slope of the phase change. The intersection here is assumed to be at low pressures and temperatures. Because the temperature is very low, reaction rate of the phase change might be expected to be very slow and the response of the discontinuity position under the oceans might be extremely sluggish to small changes in temperature and pressure. Thus, we may not always have thermodynamic equilibrium under the oceans.

Early in this discussion it was noted that the relief of the Earth's crust is a direct function of the thickness of the zone of light rock. If the thickness of the zone of light rock reflects the depth to the M discontinuity, which it almost certainly does, the relief of the Earth's crust can be interpreted as mirroring the various temperature gradients in the upper part of the mantle.

The four major problems of the surface of the Earth, discussed earlier, seem satisfactorily explained by phase transition. A chemical contrast at the discontinuity is unnecessary. The rocks on both sides of the M discontinuity may thus be of the same composition and the depth to the discontinuity may be a function of very slight temperature variations from place to place in the Earth's crust.

The uplift of continents, once at sea level, to high plateaus would be a consequence of warming the rocks near the M discontinuity a few tens of degrees. When this happened, the phase change would migrate downward to much greater depths. The dense rock below the discontinuity would become light rock and the volume increment would float the continents to higher levels.

Thus, convection currents are no longer needed to transport millions of cubic miles of light material underneath the continents in order to float them higher into the air.

Similarly, the long lives of mountain ranges are explained. As the tops of mountains are eroded away, pressure at the discontinuity deep below the mountains decreases. Dense rock at the discontinuity would be converted to light rock, so light roots underneath the mountains would be recreated to keep them floated to high elevations.

The downsinking of sediments in troughs is also explained by the phase transition. If sediments from a mountain range were rapidly removed and deposited in troughs, the first effect of loading would be to increase the pressure at the base of the trough with very little change in the temperature.

Consequently, the discontinuity would migrate toward the surface. The trough would sink, not only because of the added load of rock at the surface, but because light rock would be converted into dense rock at the discontinuity below the trough with a consequent decrement in volume of material below. Thus, the short-time effect of rapid sedimentation is one of sinking.

A most interesting long-time effect appears. The added new sediments filling the trough are of low thermoconductivity and possibly richer in radioactive material than the surrounding rock. Consequently, given sufficient time, the temperature would slowly rise at the bottom of the trough, and, although the discontinuity would first migrate surfaceward under response to loading, it would ultimately migrate downward under response to the rise in temperature owing to the blanket of poorly conducting sediments rich in radioactive elements deposited in the trough.

Thus, troughs might sink for considerable time and then be uplifted to form mountain ranges as the roots of the trough deepen with warming of the base.

This implies that mountains are generated largely because of vertical motion and not lateral thrust. A good deal of the faulting and folding of rocks in mountain ranges is assumed to be the result of load. By this thesis, the major folds and faults associated with mountain chains are gravitational in origin, though concomitant lateral thrust of other origin is not excluded.

The problem of the relatively uniform heat flow to the surface of the Earth is readily explained by the phase transition concept. The earlier crustal models assumed continents were made up of silica-rich and radioactive element-rich rocks. Thus, continents should, but do not, show heat flows several times that of oceanic areas.

If the bulk composition of continental rocks were not vastly different from the bulk composition of oceanic rocks, we would expect relatively uniform heat flew from place to place in the Earth's crust. This is indeed what we do find. The precision, however, of measuring heat flow is not sufficiently great to exclude. The possibility that minor variations in temperature do exist from place to place in the Earth's crust. In fact, it is necessary to appeal to these minor variations to account for the existence of ocean basins, mountain ranges and continents on the basis of a phase change as discussed here.

If we assume the M discontinuity to be a phase change, many questions are answered, but other questions are also raised. The phase change cannot be a simple solid-solid phase change inasmuch as the major minerals involved are of variable composition.

Consequently, the change must take place over a considerable depth interval and should not be a sharp change taking place at a fixed depth. The data of seismology bear on this problem. They permit the interpretation that the discontinuity may take place, instead of at a given depth, over an interval of as much as 10 kilometers under the continents.

This is within the requirements of the change. However, more difficult problems emerge when oceanic areas are considered. The discontinuity under the oceans is very shallow and apparently takes place over a very narrow depth interval. In fact, the pressure interval seems much too narrow for it to represent the gabbro-eclogite change.

However, further experimental work needs to be done to measure precisely the required pressure interval and more refined seismic work will be necessary before we know exactly the distribution of seismic velocities below both the oceans and the continents.

DEVELOPMENT OF THE HYDROSPHERE AND ATMOSPHERE

Abstract. A satisfactory hypothesis of the development of the hydrosphere and atmosphere depends upon evidence from many sciences and the solution of many other fundamental problems of earth history. But because it is so closely related to many other problems, any progress toward' unravelling the history of the hydrosphere and atmosphere limits the range of permissible speculation about such distantly related questions as the origin of the solar system, continents, mountains, and living organisms.

Several hypotheses of the source of the earth's air and waters are examined for their consistency with established principles and observed geologic evidence, and special attention is given to the probable composition of the early atmosphere.

Hypotheses of the origin of the atmosphere and hydrosphere fall into two chief categories:

- That all air and water of the earth are residual from a dense primitive atmosphere that once enveloped a molten globe;
- That they have accumulated at the earth's surface by leakage from the interior.

The quantities of water, carbon dioxide, organic carbon, nitrogen, sulfur, etc., that have been or are now part of the earth's atmosphere and hydrosphere may be estimated within reasonable limits of uncertainty and these "excess" volatiles afford a basis for testing chemical consequences of the alternative hypotheses.

Several writers have suggested that the primitive atmosphere may have been composed largely of CH_4 and NH_3. However, the equilibrium constants for reactions of these and other gases, combined with the evidence of the "excess" volatiles, indicate that CO_2 and N_2 are much more likely.

The stabilities of methane and ammonia depend upon the presence of free hydrogen; and the escape rate of hydrogen from the earth is such that methane probably could have persisted in significant amounts in the early atmosphere no more than 10^3 to 10^8 years. For all but a relatively brief period at the very beginning of earth history, the atmosphere probably contained CO_2 and N_2 rather than CH_4 and NH_3.

When the consequences of a dense atmosphere of CO_2 and N_2 (but with almost no free O_2 or H_2) are examined, it is found that several chemical effects (such as the quantity of rocks that would have to be weathered, of sodium dissolved in sea water, and of $CaCO_3$ deposited on the sea floor very early in early history) are not borne out by the observed geologic record.

From this and other lines of evidence it seems extremely improbable that the present atmosphere and hydrosphere are residual from any such dense primitive atmosphere.

Instead, it seems likely that the atmosphere and hydrosphere have accumulated gradually during geologic time by the escape of water vapour, CO_2, CO, N_2, and other volatiles from intrusive and extrusive rocks that have risen more or less continuously from the deep interior of the earth.

The amount of free oxygen in the early atmosphere is a separate problem that cannot be solved until the evidence of the earliest rocks has been appraised more fully.

Current hypotheses of the origin of life appear to require a reducing atmosphere, yet it seems likely that oxygen has been accumulating from the photodissociation of water vapour ever since the earth was formed. The oxidation of ferrous iron and sulfides in the earliest sediments may have kept the oxygen content very low, and life may have begun in local reducing environments.

In a symposium on the crust of the earth no apologies are necessary for considering the probable history of air and water, even though they are not hard rocks. If we could be sure that the earth was somehow born fully equipped with ready-made continental masses and a crustal layer, then the hydrosphere and atmosphere would probably have influenced the rocks of the crust only superficially, by way of such processes as the erosion of uplifts, the deposition of sediments, and the development of living organisms.

But if, as many geologists and others think likely, the crust, the continents, and the ocean basins have evolved from changes and dynamic processes that have operated through the geologic past, then the hydrosphere and atmosphere are almost certainly inter-related in origin and subsequent history with the deeper-seated processes of petrogenesis, epeirogeny, and mountain making.

Just what the mechanism of this inter-relationship may be is to the writer a fascinating problem that touches many aspects of earth history but its discussion will not be attempted in this paper. The origin and subsequent history of the hydrosphere and atmosphere are still unsolved problems and probably will long remain so.

The questions involved in these problems depend for their answers upon evidence from a surprising range of scientific disciplines — not only the fields of geology, geochemistry, meteorology, oceanography, and hydrology, which come readily to mind, but also such seemingly unrelated subjects as astronomy, seismology, nuclear physics, biochemistry, and others.

A truly satisfactory hypothesis of the development of the hydrosphere and atmosphere must wait until nearly all other broader problems of earth his- tory and constitution have been solved. But precisely because it is so closely related to many other problems, any progress that can be made toward unravelling the history of the earth's air and waters touches also these many other problems and thereby limits, at least to some degree, the range of permissible speculation regarding them.

It would be difficult if not impossible to make specific acknowledgments to the many colleagues who in informal discussions have contributed at least indirectly many of the ideas presented in this paper. The writer is especially indebted to his friends G. P. Kuiper and H. C. Urey of the University of Chicago and G.C.

Kennedy and George Tunell of the University of California at Los Angeles for reading the manuscript and offering many helpful suggestions and criticism. None of the anonymous colleagues or four reviewers, however, are to be held responsible for any of the views expressed, as one of them has kindly been at some pains to make clear.

ALTERNATIVE THEORIES OF ORIGIN OF THE HYDROSPHERE AND ATMOSPHERE

Several years ago the writer reviewed the available evidence that bears on the origin of sea water. This review sought to assemble critical evidence, to find or formulate alternative hypotheses that appear consistent with the facts, and then to examine these hypotheses for consequences that can be tested by the actual geologic record. The present paper outlines those portions of this earlier review that bear upon the probable chemical composition of the earth's early atmosphere and seem to merit more detailed consideration at this time.

Source of "Excess" Volatiles

When the probable quantities of rocks weathered and sediments deposited during geologic time are compared, it seems evident that most of the major rock-forming elements (Si, Al, Fe, Ca, Mg, Na, K, etc.) in sedimentary rocks and all the dissolved bases in sea water have been derived from the weathering of earlier rocks throughout the past.

This is not, however, an adequate source for a group of other materials (H_2, CO_2, C1, N, S, and several others), all of which are much too abundant in the present atmosphere and hydrosphere and in ancient sedimentary rocks to be accounted for solely by rock weathering.

For these materials (which for convenience may be called the "excess" volatiles) some other source is required. In seeking this other source, we encounter head-on the central problem of the origin of the hydrosphere and atmosphere.

Only two possible sources of these "excess" materials appear to have been suggested. Either the waters of the present ocean and all the other "excess" volatiles have been inherited from a primitive ocean and atmos- phere, or they have risen to the surface from the earth's interior during the course of geologic time.

The first possibility is sometimes stated about as follows: If the earth were once molten throughout, then much of the water and many materials were at one time volatilized in a hot primitive atmosphere. Later, as the earth cooled, the water vapour condensed and formed a primitive ocean. According to this hypothesis, the present atmosphere and ocean are simply residual from a hot primitive atmosphere.

The alternative hypothesis lacks the simplicity and vividness of the first one. This explanation depends upon some complex and relatively unfamiliar process or processes of "degassing" of the rocks of the earth's interior, and these complex processes get the hypothesis deep into problems of physical chemistry and petrogenesis. Two distinct variants of this second hypothesis may be recognized:

- That the water and other volatiles came from the earth's interior within some brief period very early in earth history;
- That these volatile substances have been escaping from the earth's interior gradually and at about the same rate throughout much of geologic time.

After correction for quantities released by rock weathering (in units of 1020 grams)	
H_2O	16,600
Total C as CO_2	910
C1	300
N	42
S	22
H	10
B, Br, A, F, etc	4

Chemical Consequences of Dense Primitive Atmosphere

In the review paper previously mentioned, an effort was made to follow through, in a semiquantitative way, some of the consequences of these alternative hypotheses, and to find corollaries of one or the other that would permit critical tests by field or laboratory evidence. One of these several lines of consequences from alternative hypotheses was an effort to estimate the chemical effects that might be expected.

- If all the "excess" volatiles were present at one time in an early ocean and atmosphere (regardless of whether inherited from a primitive atmosphere or erupted suddenly from the interior early in earth history)
- If, on the other hand, the volatiles accumulated gradually to form the present ocean and atmosphere.

With a primitive ocean and atmosphere made up of all the "excess" volatiles indicated the different elements and compounds would dispose themselves between solution in the ocean water and as gases in the atmosphere, depending on their individual chemical properties and their quantities.

Under the postulated conditions and with the given gross composition (which would permit essentially no free oxygen or hydrogen), it was assumed that, at the outset before rock weathering became important, the predominant gases would be CO_2, N_2, and H_2S, rather than CH_4, NH_3, and other possibilities. With an assumed temperature of 30°C as a basis for estimating the distribution of materials between ocean and atmosphere, it becomes possible to work out the sequence of chemical changes that would be expected.

On this hypothesis the ocean and rain water would at first be intensely acid. This acid water would decompose bare rock with which it came in contact and in so doing would dissolve some of the bases from these rocks. As a result, the acidity of the water would decrease, and the chemical effects of dissolved gases would come into play.

As decomposition of rocks proceeded, more and more dissolved bases, chiefly Ca, Mg, Na, K, and Fe, would be carried to the sea; as soon as the appropriate solubility products were exceeded, $CaCO_3$, FeS_2, and other compounds would be precipitated on the sea floor.

Under the postulated conditions the atmosphere would have been made up largely of CO_2 at the outset, but as carbonates continued to precipitate from the ocean more and more of the original CO_2 would be subtracted from the atmosphere. Eventually the concentration of this gas would be decreased to some point at which the most primitive forms of life might conceivably have existed.

The highest partial pressure of CO_2 that any living organism is known to tolerate may be selected, rather arbitrarily, as something that might have approximated the CO_2, content of an early Precambrian atmosphere.

From this train of argument it becomes possible to deduce, as chemical consequences of a dense residual atmosphere, the quantities of rocks that would have to be decomposed, of bases that would have to be dissolved in sea water, and of $CaCO_3$ that would have to be deposited in order to reduce the original CO_2 of the atmosphere to a concentration at which primitive life might have begun.

The results of this calculation appear distinctly unfavorable to the hypothesis from which we started. The amount of rock that would have to be decomposed under this hypothesis before life began would be very large—considerably larger in fact than seems, from independent estimates, to have been decomposed in all of geologic time, both before and after the advent of living organisms.

Similarly the total amount of Na dissolved in sea water would have been much greater, even before life began, than the amount which from other evidence appears to have been dissolved in all earth history — that is to say, the hypothesis leaves no place for solution of any additional Na after early Precambrian time.

Furthermore, the quantity of carbonate deposits required means that the early Precambrian should have been a time of abnormally great carbonate deposition throughout the world; more limestone and dolomite should have been deposited then than now remains in rocks of all ages.

The known geologic record indicates relatively few thick limestones of early Precambrian age; the record suggests instead that the bulk of the earth's limestone was deposited much later, during Palcozoic time.

Chemical Consequences of Gradual Accumulation

If the same type of reasoning is now applied to the alternative hypothesis (*i.e.*, that the "excess" volatiles have accumulated gradually to form the present ocean and atmosphere), the results are significantly different. The same assumptions may be made regarding which gases would be stable, the decomposition of rocks, and the solution of bases by acid waters.

But where for the other hypothesis an upper tolerance limit of CO_2, in the atmosphere was assumed to measure the quantity of rocks decomposed, Na dissolved, and carbonates deposited before life became possible, for this alternative hypothesis we may assume that the CO_2 content of the atmosphere never rose above this same limit.

From these assumptions calculations show that carbonates, for example, would begin to precipitate from the sea water when less than one-tenth of the total water and other "excess" volatiles had accumulated in the atmosphere-ocean system.

Thereafter, if CO_2 continued to be released into the system at a relatively constant rate, rock decomposition would proceed, and limestones would be deposited at a more or less constant rate down to the present time. That is to say, with this alternative hypothesis we are not led to expect embarrassingly large quantities of rocks decomposed, Na dissolved, and carbonates deposited in the early Precambrian, as we were with the first hypothesis.

The train of chemical consequences just outlined is but one of several independent corollaries that may be deduced from the opposed hypotheses. These will not be discussed here, except to mention that all such corollaries that the writer has considered and has tried to follow through semiquantitatively seem to lead rather compellingly to the same conclusion — namely,

That the geologic evidence appears distinctly to favor the hypothesis that the atmosphere and ocean have accumulated gradually from some process of "degassing" of the earth's interior and that this process must have operated at a more or less steady rate through geologic time.

GASES IN THE EARLY ATMOSPHERE

In trying to estimate the initial chemical character of the postulated early atmosphere and ocean, the writer had followed traditional theory in taking carbon dioxide, free nitrogen, and hydrogen sulfide as the principal gases likely to be stable under the assumed conditions.

However, a number of investigators in various fields have recently considered somewhat the same general problem and have taken other gases, particularly methane or ammonia, or both, as more likely to have been the dominant constituents of the earth's early atmosphere.

Poole also at one time advocated the prevalence of methane in the early atmosphere but more recently he has rejected this view and concluded instead that a primitive atmosphere of carbon dioxide, free nitrogen, and water vapour seems more probable.

The reasons that have led these writers to consider methane or ammonia, or both, as major constituents of the early atmosphere are probably several, but they may include one or more of the following considerations:

First, we know that hydrogen and helium greatly exceed in abundance all other chemical elements in the spectra of the sun and stars. For significant downward revision of previous estimates of the relative abundance of helium, *see* Neven Underhill. If hydrogen were at one time very abundant in the atmosphere of the earth, then methane and ammonia, rather than carbon dioxide and nitrogen, should have been the dominant gases.

A second consideration is the fact that methane and ammonia are the most abundant gases in the atmospheres of the major planets, Jupiter, Saturn, Uranus, and Neptune. These planets are all more distant from the sun and more massive than the earth, and it does not necessarily follow that these gases were once present in the same abundance on earth.

Third, the hypothesis of Oparin and Horowitz is widely attractive to scientists in many fields. This postulates that before ozone became a significant constituent of the earth's atmosphere, complex organic compounds were synthesized by photochemical processes; that the most primitive forms of life originated in this way; and that these first self-duplicating molecules evolved into more specialized organisms as they consumed the supply of previously formed organic compounds.

This hypothesis seems to require a reducing atmosphere, and some of its advocates have held that it also requires the presence of either methane or ammonia, or both. Others, however, seem to think that the same general principle could operate quite as well in an atmosphere of carbon dioxide, provided no free oxygen were present.

Finally, Miller has succeeded in synthesizing two amino acids, the essential substances of proteins, by passing an electric discharge (the effects of which would be comparable to lightning) through a mixture of water vapour, methane, ammonia, and hydrogen.

Insofar as known, this experiment has not yet been tried with other combinations of gases.

The chemical character of the early ocean would of course have depended greatly upon what gases were dissolved in it. If the calculations in the writer's "Sea water" paper were based on the wrong gases in the atmosphere, then the specific conclusions reached regarding the probable chemical effects of a dense residual atmosphere and ocean would require modification.

It should be mentioned, however, that the over-all conclusion reached there — that the earth's atmosphere and hydrosphere are probably secondary and have accumulated gradually from "degassing" of the interior — is not affected by this question of which gases were dominant in the early atmosphere,

Because it rests upon several independent lines of evidence, such as the mineralogical and paleontological evidence that CO_2 must have been supplied to the atmosphere at a fairly constant rate throughout the past.

As indicated by the list of writers who have discussed various aspects of the subject recently, the question of the probable composition of the earth's early atmosphere is of considerable interest in a number of fields of science. The evidence afforded by the "excess" volatiles appears to bear directly on this question and to place some restrictions on the range of permissible speculation regarding it.

Inasmuch as this evidence of the "excess" volatiles seems not to have been considered as fully as its importance warrants, the writer has presumed to wander far outside his own field of science in order to present at least the high lights of it.

EQUILIBRIUM CONCENTRATIONS OF DIFFERENT GASES

Gross Composition of Dense Primitive Atmosphere from "Excess" Volatiles

The significance of the "excess" volatiles for this problem may be seen most readily if one tries to estimate the probable composition of an early atmosphere under the two alternative hypotheses previously described:

- Residual from a dense primitive atmosphere, or
- By gradual accumulation. The basic assumption is that the "excess" volatiles (that is, those materials present in today's atmosphere, hydrosphere, and biosphere and entombed in ancient sediments that cannot be accounted for by rock weathering) afford the best, and in fact almost the only, direct evidence of the probable composition of the early atmosphere, whatever may have been its mode of origin.

In order to learn which gases would be stable under various conditions, the writer's earlier estimates of "excess" volatiles may be taken as a starting point. Perhaps these earlier estimates could now be improved, but they are at best only estimates, and these readily available figures are used for the present calculations in the belief that any conclusions based upon them will still be valid qualitatively even when these estimates are improved by new data.

Table. Chemical Reactions Considered

(g) indicates the gas phase				
H_2O (g)			=	H_2(g) + ½O_2(g)
CH_4(g)	+	2H_2O (g)	=	CO_2(g) + 4H_2(g)
CH_4(g)	+	H_2O (g)	=	CO (g) + 3H_2(g)
CH_4(g)			=	C (solid) + 2H_2(g)
NH_3(g)			=	½N_2(g) + 3/2H_2(g)
2HC1 (g)			=	H_2(9) + C12 (g)
H_2S (g)			=	S (solid) + H_2(g)

Equations were then assembled of a number of gas reactions that seem likely to be important for this gross composition and for moderate temperatures and pressures. To keep the total number of components small, the possible oxidized gases of sulfur are neglected. This neglect probably introduces no significant errors: of the four principal gas-forming elements to be considered, sulfur is the least abundant among the "excess" volatiles; furthermore, with the gross composition given, little if any free oxygen is possible, and the sulfur oxides would therefore be negligible in amount.

The mean temperature of the present atmosphere, considering the distribution of its mass and temperature with height is approximately −20°C. The conditions of the present inquiry require an average surface temperature consistent with a liquid ocean but are not otherwise more narrowly specified. Experimental data for oxidation potentials,

Free energies of formation, and equilibrium constants are conveniently tabulated for the standard laboratory temperature of 25°C. If this be taken as the mean temperature of the postulated early atmosphere, it means, by comparison with the present earth, a climate tropical even at the poles. In view of numerous other sources of uncertainty in the calculations and estimates that follow,

It seems scarcely worthwhile to attempt the thermal corrections necessary to convert the equilibrium constants to a mean temperature near that of the earth's atmosphere at the present time. For these reasons 25°C has been used as the basis for calculation. The several degrees of dependence of these equilibrium mol fractions upon pressure, here measured by the exponents of the symbol P, are based on the ideal gas laws.

These pressure relationships would not hold strictly true, of course, for actual gases, but thy serve as a useful basis for approximation.

$(H_2).(O_2)^{1/2}$ $= 9.0 \text{ X } P1^{1/2}$

$(CO_2)\cdot(H_2)^4$ $= 1.3 \times 10^{-20} \times P^{-2}$

$(CH_4)\cdot(H_2O)^2$

$(CO)\cdot(H_2)^3$ $= 1.3 \times 10^{-25} \times P^{-2}$

$(CH_4)\cdot(H_2O)^2$

$(Csol)\cdot(H_2)^2$ $= 1.3 \times 10^{-9} \times P^{-2}$

(CH_4)

$(N_2)^{1/2}\cdot(H^2)^2 \qquad = 1.2 \times 10^{-3} \times P^{-1}$

(NH_3)

$(H_2)\cdot(Cl_2) \qquad = 2.0 \times 10^{-17} \times P^0$

$(HCl)^2$

$(Ssol)\cdot(H_2) \qquad = 1.6 \times 10^{-6} X P^0$

(H_2S)

P = sum of partial of gaseous components participatng in given reaction.

With the seven equilibrium constants and the six totals of C, C1, N, S, O, and H from the estimates of "excess" volatiles we have 13 equations and 13 components of unknown concentration participating in the seven reactions. Theoretically at least, this problem can be solved. In fact, however, the writer found it intractable.

The number of unknowns was therefore reduced to 12 by assuming that an early atmosphere in contact with an extensive ocean would be saturated with water vapour. For the assumed temperature of 25°C and the total pressure of all gases (which is to be found later in the calculation), this gives the number of mols of water vapour in the atmosphere. The scarcely perceptible effects of this arbitrary assumption upon the numerical values

In preparing the list of "excess" volatiles one finds that some of the chemical elements are more readily tallied in such an inventory than others. Carbon and chlorine, for example, are readily enough recognized as such wherever they occur, but the "excess" oxygen and "excess" hydrogen are relatively subtle concepts, and these materials are thus more difficult to identify with reasonable assurance. Strictly speaking, it is not quite logical to accept the estimates of both "excess" hydrogen and "excess" oxygen, for it would be simpler and possibly more accurate to assume that all the "excess" hydrogen was originally combined as H_2O, thus leaving no "excess" hydrogen.

But to make this assumption would prejudice the outcome of the present inquiry as it would mean almost no possibility of any CH_4 or NH_3 in the early atmosphere. It was therefore decided to use for these calculations all the 10 X 10^{20} g of estimated "excess" hydrogen.

As the number of unknowns had, for reasons mentioned above, been reduced from 13 to 12, the number of equations required for solution of the problem was also reduced to 12; because of the uncertainties involved in it, the separate estimate of 506 X 10^{20} g of "excess" oxygen was not used in these calculations.

One still needs to assemble the experimentally determined solubilities of each of these gases in water. When these solubility data and the equations are all duly tabulated, only the work of solving the problem remains. This proved to be a formidable chore. The amount of a gas dissolved depends upon the quantity of water, the solubility of the gas, and its partial pressure in the atmosphere.

But the partial pressure of the gas depends in turn upon how much of the total amount is dissolved in water and how the amount that still remains in the atmosphere compares with the total of all other gases present. And so on, almost *ad infinitum*. However, with sufficient patience, the problem can be solved numerically by a series of successive approximations.

When all the paper work is done, we have an estimate of the amounts of the different gases under equilibrium conditions. It is, of course, only an estimate. Perfect equilibrium might never be attained, even with geologic time.

Ammonia would react with CO_2 and HC1 to form carbonates, chlorides, and other compounds; these would dissolve in water but scarcely affect the distribution of the gases between atmosphere and ocean. Furthermore, no account has been taken of photochemical processes that would have operated on the gases.

Such photochemical reactions would doubtless form other and more complex compounds. But from what is known regarding the photochemical destruction of ammonia and methane, the net effect of such processes would be to decrease rather than to increase the proportion of these two gases.

Insofar as the results have significance then, they mean that, under the assumed conditions, free nitrogen in the atmosphere would at the outset tend to be about 490 times more abundant than ammonia, and carbon dioxide about 3800 times more abundant than methane.

The arbitrary assumption of a water-saturated atmosphere made to facilitate the computations scarcely affects the numerical value of the final results. If fully saturated with water vapour, the total atmosphere required by this hypothesis is so large that water vapour can make up only a very small part of it. For this reason the effects of this arbitrary assumption upon the final results are negligible.

Table Composition of Early Atmosphere and Ocean by Residual Hypothesis

	Atmosphere (per cent by volume)	*Dissolved in ocean (units of 1020 gm)*
CO_2	91	240
N_2	6.4	0.2
H_2S	2.0	12
HC1	0.2	310
H_2O (vapor)	0.2	
NH_3	0.01	14
H_3BO_3		7.4
HBr		1.0
HF		0.3
CH_4	0.02	0.004
Cl_2	0.01	0.2
CO	*	†
H_2	*	†
	Total atmospheric pres- sure -- 14.1 kg/cm^2	Mass of ocean water -- 16.6 x 10^{20} kg pH of ocean water -- 0.3
* Less than 1 ppm.		
† Less than 1 ppm of total dissolved substances.		

To appraise these effects, one may assume an atmosphere only one-tenth (instead of fully) saturated with water vapour and then carry through the same calculations as before. When this is done it is found that, of the volume percentages, only that of water vapour itself is modified by this change: it becomes 0.02 instead of 02 per cent; the others are changed by amounts too small. It is important to note that the composition is that of the postulated early atmosphere while still uncontaminated by the effects of rock weathering. As mentioned above, this intensely acid water of the ocean (and also of rain from an atmosphere of this composition) would promptly begin to decompose bare rock with which it came in contact.

The bases thus dissolved from silicate minerals would accumulate in the sea until they reached concentrations sufficient to cause precipitation of $CaCO_3$, FeS_2, and other compounds. Such precipitation would subtract acid radicals from the sea water, and consequently the composition of the atmosphere would in time be modified significantly. With an atmosphere of this original composition the most notable effect would be the subtraction of large quantities of CO_2.

The results of these calculations thus seem to confirm what was simply assumed in the previous paper — namely, that, with a gross composition equivalent to that of all the "excess" volatiles, the postulated dense primitive atmosphere would at the outset have been made up dominantly of carbon dioxide and nitrogen.

Gross Composition from Gradual Accumulation of "Excess" Volatiles

It is scarcely necessary to carry through the laborious calculations still another time in order to see which gases would be the dominant ones under the alternative hypothesis of gradual escape of the "excess" volatiles from the earth's interior. Under this hypothesis the total atmospheric pressure would never have been great at any time.

Neglecting for the moment the effects of rock weathering, the principal differences in relative abundances of the several gases in the atmosphere would be those due to the smaller volume of ocean water and to the effects of lower partial pressures of the different gases. Under these conditions smaller quantities of all gases would be dissolved in ocean water, and relatively more would therefore remain in the atmosphere.

For example, carbon dioxide, which is more soluble than

methane, would be relatively more abundant in the atmosphere under this gradual-accumulation hypothesis than under the alternative one of higher partial pressures of the gases and greater volume of ocean water.

A second and quantitatively more important difference depends upon the fact that the equilibria of those gas reactions that are pressure dependent would, under the gradual-accumulation hypothesis, be shifted toward the components that occupy more volume.

If, for example, the total atmospheric pressure were 1 kg $/cm^2$ instead of 14.1 kg/cm^2, the equilibrium ratio of CO_2 to CH_4 would be $(14.1)^2 = 200$ times greater than under the first hypothesis, other things being equal. Thus it seems clear that, under the hypothesis of gradual escape of volatiles from the interior, even more than under the other hypothesis of a dense primitive atmosphere, CO_2 and N_2 would be the dominant gases to be expected initially.

As volatiles continued to leak upward gradually from the "degassing" of the earth's interior, they would become mixed with an atmosphere and ocean already somewhat modified in composition by the effects of earlier rock weathering and deposition of carbonates.

With gradual addition of more volatiles and especially if free oxygen began to accumulate from the photodissociation of water vapour in the upper atmosphere and from the photosynthesis of carbon from CO_2 by plants, the compositions of both atmosphere and ocean water would approach ever more closely those of today.

OTHER LINES OF EVIDENCE

The present inquiry might be terminated at this point with the conclusion that the hypothesis of gradual "degassing" of the earth's interior leads to chemical consequences at the surface that appear entirely consistent with the observed geologic record.

But to close the discussion here would be to neglect other considerations and lines of evidence that deserve attention. It is obvious that some gross composition of the atmosphere and hydrosphere very different from that of the "excess" volatiles would lead by this method of calculation to a very different composition of the early atmosphere.

For example, one might assume, entirely *ad hoc*, an original source of methane and ammonia and thereby reach entirely different conclusions about the atmospheric composition. However, without independent evidence to support it, such an assumption would seem unwarranted. A brief survey is therefore attempted to see if lines of evidence neglected in the above discussion may point to some gross composition very different from that of the "excess" volatiles.

If, as seems probable, the atmosphere has grown by "degassing" of the earth's interior, one might look for evidence of some different gross composition of the early atmosphere in the gases now escaping from volcanoes and hot springs or in the gases occluded in igneous rocks and meteorites.

Reliable data on the composition of such volcanic and occluded gases are scanty. In most gases from volcanoes and hot springs and in gases occluded in igneous rocks that have been analyzed, CO_2, CO, and N_2 greatly predominate over CH_4 and NH_3, and the average composition of such gases agrees reasonably well with the gross composition of the "excess" volatiles.

This evidence is not so conclusive as it sounds, however, because it is uncertain to what extent such gases may have been contaminated by previous contact with the atmosphere or with sedimentary rocks and then recycled through a later stage of molten rock. Because of this possibility of contamination and the common view that meteorites probably afford the best available clues about the composition of the earth's interior,

The occluded gas content of meteorites may be particularly significant, for meteorites presumably have not been subjected to previous contact with the earth's atmosphere or to subsequent recycling.

The scanty data available indicate that the occluded gases in stony and iron meteorites are likewise relatively high in, CO_2, CO, and N_2 and low in CH_4 and NH_3.

Admittedly the analysis of occluded gases is difficult and the results uncertain, because the conbination of chemical elements can be modified significantly by the analytic extraction procedures used. But even with these uncertainties in mind, no important sources of methane and ammonia are indicated.

In the gases of stony meteorites, the type presumably most nearly representative of the silicate rocks of the deep interior, the atomic volumes of oxygen exceed the combined totals of carbon, hydrogen, and nitrogen, and oxygen exceeds hydrogen by an average factor of about four.

The available evidence therefore suggests that the gases now escaping from the earth's interior and also the occluded gases as an index of those most likely to escape, are of approximately the composition of the "excess" volatiles and that CH_4 and NH_3 are relatively uncommon.

If no reason seems evident for assuming an important source of CH_4 and NH_3 in the gases from the earth's interior, one may look for some process that might have produced these gases at the earth's surface or within the atmosphere. Ammonia has not been found, but small quantities of methane and hydrogen have been detected in the present atmosphere. Both these gases are subject to known dissipative processes.

Methane undergoes decomposition by normal oxidation and by photochemical, biological, and other processes and hydrogen is so light an element that it is escaping from the earth's gravitational field. It seems reasonable to assume that both gases are in a steady state and are being continually produced by processes that operate today and thus may possibly have operated more actively in the past.

The small amount of methane in the present atmosphere seems fully accounted for by strictly biological processes and it therefore affords no evidence of an important generative process that might have operated before the advent of life.

The hydrogen in today's atmosphere is of interest in the present inquiry because, if the amount were much greater sometime in the past, the chemical equilibria are such that CH_4 and NH_3 would also have been relatively abundant.

According to Paneth the atmosphere now contains about 5×10^{-7} parts by volume of H_2 or approximately 1.8 X 10^{14} g in the total atmosphere. This amount may be compared with the yield of the only natural process that is thought to produce hydrogen in significant quantities today. In the upper limits of the atmosphere, H_2O vapour is probably undergoing continual decomposition by ultraviolet radiation.

After reviewing the pertinent data, Kuiper and greatly lowered previous estimates of the rate at which this process operates. He concluded that approximately 7.7×1.0^{-16} mol H_2 O/cm.[2] see are being decomposed, which would mean the production of 2.5×10^{11} gH_2 /yr.

From Spitzer's modification of Jeans' fundamental equation for the escape rates of lighter atoms and gases from a planet, one may estimate that this quantity of hydrogen would escape from the 1.8 X 10^{14} g of the present atmosphere each year if the kinetic temperature of the upper atmosphere were approximately 3500°K.

However, the temperatures thought to prevail in the upper atmosphere today, 500 km or so above the earth's surface, are around 1500°K, with occasional bursts of radiation producing a maximum of something like 2500°K.

Taken at face value, therefore, this calculation suggests that hydrogen is now being produced more rapidly than it is escaping — in other words that it is accumulating toward some concentration at which the escape rate can eventually equal the production.

It is much more probable, however, that the discrepancy between the calculated and actual temperatures of the upper atmosphere is attributable solely to uncertainties in the data used for this calculation.

If the rate of photodissociation is less than Kuiper estimates or if the hydrogen content of the uppermost atmosphere is greater than Paneth's estimate, the discrepancy disappears.

The results of this calculation are merely suggestive of course. Spitzer points out that his equation is not reliable for very rapid escape rates, because with such rates diffusion of the gas becomes the controlling factor. However, the available information indicates that the present rate of photodissociation of water vapour in the upper atmosphere is probably quite adequate to explain the small amount of H_2 observed in today's atmosphere.

If for any reason the rate of photodissociation of water vapour were much greater sometime in the past while the escape rate of hydrogen remained the same, or if the upper atmospheric temperature which controls escape rates were much lower while the photodissociation rate remained the same, then the concentration of hydrogen in the atmosphere might have been greater than it is today.

Presumably, however, the rates of the two processes vary together with any changes in upper atmospheric temperature, and, without convincing reasons why some difference in the balance between the two processes should be expected in the past, it is unwarranted to assume that photodissociation of water vapour has ever caused a significantly higher concentration of hydrogen in the atmosphere than it does today.

The gases escaping from the earth's interior and those being produced at the surface afford no strong evidence of important sources of CH_4 and NH_3 in the past. But when we turn to consider possible sources from out in interplanetary or interstellar space, a gross composition very different from that of the "excess" volatiles is clearly suggested. The question then becomes whether such a possible atmosphere from outer space could have persisted long enough after the earth was first formed to leave chemical evidence of itself in the earliest rocks and in the subsequent history of the earth.

The hypothesis that the present atmosphere and ocean are largely residual from a dense primitive atmosphere fits particularly well with the pos- sibility that the earth may have brought with its solid substances from outer space an initial atmosphere very different in composition from that calculated above. This supposed initial atmosphere might have consisted not only of the "excess" volatiles as postulated above but also of all other chemical elements in the proportions observed in the solar and stellar atmospheres today. This would mean a very high concentration of hydrogen, which in turn would greatly favor the occurrence of CH_4 and NH_3 rather than CO_2 and N_2.

To make this assumption explicit, one might start with a primeval earth containing the present percentages of Si, Fe, and other heavier elements and calculate from cosmic abundances the quantities of lighter elements it should have contained originally.

This would give a protoearth approximately 190 times as heavy as the present earth and consisting of 80 to 85 per cent by weight of hydrogen. After allowing for oxygen to balance Si, Mg, etc., in the rockforming oxides, that remaining affords a measure of the huge volume of water to be expected on such a planet. To this set of assumed conditions the various gas-reaction constants and solubilities in sea water may then be applied as in the calculation described above for the "excess" volatiles.

This yields an initial atmosphere of about two-thirds helium and one-third hydrogen by volume, and minor amounts of other gases, among which CH_4 and NH_3 would be much more abundant than CO_2 and N_2

Whether or not this exceedingly hypothetical set of conditions could have persisted long enough to have chemical effects still perceptible in the earliest rocks of the earth depends largely on the rate at which the lighter gases and especially hydrogen would escape from the planet's gravitational field. If the kinetic temperature at the upper limits of this initial atmosphere could be estimated, the escape rates might be calculated from Spitzer's equation.

Assuming a temperature of 1500°K, as today, the initially tremendous quantities of hydrogen would decrease to present atmospheric concentration in only 3×10^5 years. However, the upper atmospheric temperature today depends largely on the properties of the two dominant gases, N_2 and O_2, and the temperature in an He-H_2 atmosphere would be different and difficult to estimate. It would probably be as high as or even higher than that of the present N_2–O2 atmosphere, and if higher the escape rate would be more rapid.

By an argument quite independent of the temperature or even of the exact mechanism of escape — *i.e.,* by comparing the energy needed to drive off the lighter elements with that received from the sun — Kuiper estimates that a period of about 108 years would be required in the early stages of earth formation for this escape. Or if this initial atmosphere escaped before the earth con- densed to its present size.

The details of the above calculations based on a protoearth composed largely of hydrogen and helium are not given here because, when one looks at this model critically, it is quite incredible. Modern views of the origin of the solar system and earth are that the great bulk of the lighter elements, and perhaps much of the heavier elements as well, escaped long before the earth condensed to anything approaching its present dimensions.

The results of these calculations are mentioned here merely as something in the nature of a limiting case. The admittedly incredible model, being very high in hydrogen at the outset, calls for an extremely high escape rate at the beginning; nevertheless, it is more favorable to the presence of large amounts of CH_4 and NH_3 in an initial atmosphere than most other assumptions one might make.

With lapse of time and depletion of hydrogen, the escape rate would become slower; and this extreme assumption then becomes approximately equivalent to other far more moderate assumptions that might be made about the original atmospheric composition.

Among recent discussions of the composition of the early atmosphere, two contributions by H. C. Urey are especially prominent. He concludes that methane and ammonia have probably been important constituents of the atmosphere for a large part of geologic time. Many of the arguments he presents in support of this conclusion have been discussed in the paragraphs above. However some of his points seem less than convincing to the writer, and these deserve separate consideration.

In discussing the high temperatures to be expected as stony and iron meteorites accumulated to form the earth, he lists several chemical reactions that should take place under these conditions and gives the equilibrium constants of these reactions. He then remarks: "From these equilibrium constants one sees that hydrogen was a prominent constituent of the primitive atmosphere and hence that methane was as well".

It is perfectly valid procedure to deduce quantitative corollaries from an assumption and then to see whether or not these corollaries are confirmed by observed facts; but here, unless the writer misinterprets the passage, Urey seems to accept the deduced corollaries as in themselves facts.

Urey merits a salutation from geologists for the equation,

$$CaSiO_3 + CO_2 = CaCO_3 + SiO_2$$

which he uses to show how limestone is formed from rock silicates and why the concentration of CO_2 in the atmosphere must have remained very low at all times. This equation, literally wollastonite + carbon dioxide = calcite + quartz, is so condensed and dramatic a metaphor that it is scarcely recognizable to geologists as a shorthand statement of the familiar over-all process by which rocks are decomposed and limestone deposits formed.

It is nonetheless a useful thumb-nail summary of the long and complex series of genetic relationships usually spelled out, laboriously but in a still oversimplified version, somewhat as follows:

- Carbon dioxide, sunlight, and water favor the growth of plants on the land and in the sea;
- Plants support the growth of all other organisms;
- Decaying organic matter yields CO_2 which, dissolved in ground water, forms carbonic acid;
- Calcic feldspars, pyroxenes, and amphiboles in crystalline rocks and carbonates in sedimentary rocks are decomposed by weak solutions of carbonic acid, with the formation of clay minerals and the release of Ca and other bases into aqueous solution;
- Ground waters carrying Ca and other dissolved materials discharge into streams and so into the ocean where solar evaporation causes gradual concentration;

- CO_2 in the atmosphere produces a small but significant concentration of CO_3 — ions in sea water;
- When the solubility product of CA++ ions and CO_3 - ions exceeds a certain value, $CaCO_3$ is deposited.

It so happens that an *increase* in dissolved CO_2 *decreases* the concentration of dissolved CO_3 — ions rather than the reverse so that an increase of atmospheric CO_2 would cause solution rather than precipitation of $CaCO_2$, thereby allowing Ca++ ions to accumulate in the sulfate-bearing sea water until gypsum as well as calcite would become insoluble. Consequently, at first glance,

Urey's chemical shorthand looks exactly backwards for the deposition of limestone. But looked at more carefully it is seen to be entirely valid, because in the long run the formation of carbonate deposits in the sea depends upon the supply of calcium from rock weathering on the lands; and rock weathering in turn depends, by way of decaying organic matter, upon the concentration of carbon dioxide in the atmosphere.

Properly interpreted, Urey's equation is a brilliant and useful generalization regarding the net effect of two basic and highly complex processes in the chemistry of the earth. Read literally, however, it certainly does not accurately represent the way limestone deposits are formed in nature. Even the broader relationship it expresses so succinctly between atmospheric CO_2 and rock weathering must be viewed with circumspection, because silicate rocks decompose very slowly in pure rain water that is uncontaminated by decaying organic matter.

In the writer's opinion, this equation cannot properly be used to estimate even approximately the maximum concentration of CO_2 in the early atmosphere before life had appeared on earth. This brief survey of the evidence from volcanic and occluded gases, from present-day processes by which CH_4 and H_2 are being formed, and from the cosmic abundances of elements has turned up no positive evidence of a gross composition of the atmosphere and hydrosphere that was significantly different any time in the past from that indicated by the "excess" volatiles. Calculations indicate that, with this gross composition, CO_2 and N_2 were probably the dominant gases in the earth's early atmosphere. No positive evidence is known, yet the observed facts do not contradict the possibility that the earth may have had, for a relatively brief period after it was first formed, a very different initial atmosphere which was soon lost or greatly modified by escape of the lighter elements.

The very fact that hydrogen is escaping from the upper atmosphere today precludes us from saying that hydrogen may not have been, even for a considerable time, much more abundant than it is today; and having escaped, it has left no record of itself in the "excess" volatiles that remain. If hydrogen were once abundant, then so also must have been methane and ammonia. This view is held by some writers, but from the available evidence the writer considers this possibility exceedingly remote.

PROBLEM OF FREE OXYGEN IN THE EARLY ATMOSPHERE

Little mention has been made thus far of the presence or absence of free oxygen in the earliest atmospheres. This is in itself a difficult problem distinct from those discussed above, but a few remarks regarding it seem appropriate here. The gases escaping from volcanoes, fumaroles, and hot springs and those occluded in igneous rocks and meteorites contain no free oxygen, except small amounts that seem clearly the result of contamination from other sources.

Likewise the estimates of "excess" volatile materials in the present atmosphere and hydrosphere and in ancient sedimentary rocks afford no evidence of any original free oxygen; in fact, they indicate less oxygen than enough to balance the "excess" carbon as CO_2. Thus whether the earliest atmosphere was residual from a dense primitive atmosphere or accumulated gradually by escape from the interior, no reason is known to expect free oxygen from such sources when the earth was first formed.

Just how soon thereafter oxygen began to accumulate is a

problem about which it seems advisable to withhold judgment until the evidence from the earliest rocks has been appraised more extensively.

The presence of somewhat higher ratios of ferric to ferrous iron in Precambrian slates than in average igneous rocks; of sedimentary iron oxides in the Precambrian iron formations of this and other continents; and of anhydrite and gypsum interbedded with limestone in the Grenville series suggests that at least some oxygen must have been available fairly early in geologic time.

Biologists have long been interested in whether free oxygen appeared before or after the advent of living organisms. The oxygen in today's atmosphere is being formed continuously by its release from CO_2 and H_2O during photosynthetic fixation of carbon by green plants.

Green plants require free oxygen in order to survive in the dark, but certain primitive types of bacteria are capable of fixing carbon in the presence of hydrogen sulfide, hydrogen, or other compounds. Thus living organisms may have come into existence long before free oxygen became available, a conclusion that is consistent with the apparent requirements of the Oparin-Horowitz theory of the origin of life.

On the other hand, free oxygen is apparently being produced continuously in the upper atmosphere by the photodissociation of water vapour and the resultant escape of hydrogen from the earth's field. From the best evidence available this process seems to be operating at a rate which, continued throughout geologic time, would yield about 5 times as much oxygen as that in today's atmosphere.

However, this quantity is only about one third to one sixth the total oxygen that has been removed from circulation by the oxidation of materials now buried in sedimentary rocks.

We are thus led to a paradox: Life probably could not have originated in the presence of free oxygen, yet free oxygen has probably been in process of formation in the upper atmosphere since the beginning of earth history.

Possibly this dilemma can be resolved by considering the quantities of ferrous iron and sulfides that evidently were oxidized to ferric iron and to sulfate during the earliest stages of rock weathering and also the quantities of CO that probably were oxidized to CO_2 following the earliest stages of vulcanism. Free oxygen could not have accumulated in the atmosphere in significant concentrations until this early oxidation was relatively complete. The appearance of even very small amounts of free oxygen in the early atmosphere would have had a tremendous effect on the equilibrium relations of some of the gases. For example, the reaction,

$$CH_4\,(g) + 2O_2\,(g) = CO_2\,(g) + 2H_2O(g)$$

at 25°C has the equilibrium constant,

$$\frac{(CO_2).\,(H_2O)^2}{(CH_4).\,(O_2)^2} = 1.9 \times 10^{140}$$

This in effect says that methane is thermodynamically unstable in the presence of even faint traces of free oxygen, and that, if any methane were present in such an early atmosphere, it would burn to carbon dioxide.

Ammonia is similarly unstable in the presence of free oxygen and oxidizes to free nitrogen. For these reasons it seems doubtful that the iron of the Precambrian iron formations could have been oxidized and the sulfates of the Grenville series could have been formed unless O_2 was sufficiently abundant in the atmosphere to oxidize any CH_4 and NH_3 that may have been present. This need not have required more than a fraction of a per cent of O_2 in the atmosphere.

It is questionable, however, that even such small amounts of O_2 would permit development of primitive life freely exposed to the open atmosphere, as apparently is required by the current version of the Oparin Horowitz theory. The problems here involved are exceedingly difficult, but they are truly basic to many sciences and for this reason call for much more attention than they have thus far received, either in the laboratory or in the field.

Specifically, the probable percentage concentration of free oxygen in the early atmosphere cannot be estimated even approximately until the evidence of the earliest Precambrian rocks has been examined much more extensively with this particular question in mind.

From such evidence as is presently available, however, it seems to the writer that the atmosphere, even very early in earth history, must have contained at least a small amount of free oxygen; that there were then, just as there are today, local environments in the ocean and in mud pools where free oxygen was absent; and that it was in such local reducing environments that the first organisms may have come into existence.

Encyclopaedia

of

GEOLOGICAL SCIENCE AND TECHNOLOGY

Encyclopaedia of Geological Science and Technology

Volume 3

Dr. Samayal Varadarajan

ANMOL PUBLICATIONS PVT. LTD.
NEW DELHI-110 002 (INDIA)

ANMOL PUBLICATIONS PVT. LTD.

Regd. Office: 4360/4, Ansari Road, Daryaganj,
New Delhi-110 002 (India)
Ph.: 23278000, 23261597

Branch Office: No. 1015, Ist Main Road, BSK IIIrd Stage
IIIrd Phase, IIIrd Block,
Bangalore-560 085 (India)
Tel.: 080-41723429
Visit us at: www.anmolpublications.com

Encyclopaedia of Geological Science and Technology

First Edition, 2009

ISBN 978-81-261-4125-8 (Set)

PRINTED IN INDIA

Printed at Mehra Offset Press, Delhi.

Contents

Volume 3

Preface

Geology is the science and study of the solid and liquid matter that constitute the Earth. The field of geology encompasses the study of the composition, structure, physical properties, dynamics, and history of Earth materials, and the processes by which they are formed, moved, and changed.

The field is important in academics, industry (due to mineral and hydrocarbon extraction), and for social issues such as geotechnical engineering, the mitigation of natural hazards, and knowledge about past climate and climate change.

The knowledge that geologists have known from the study of the earth a subject so large must be treated very briefly if it is to be presented between the covers of a single book; we have chosen to concentrate on the analysis of processes that are at work upon and within the earth, rather than to present a catalog of descriptive facts and terms.

We have felt, that the student is entitled to know something of the kind of evidence on which geologic conclusions are based, even though its presentation takes valuable pages that might be used to put forth more facts.

Author

other hand, he will be sure to point to the gaps in his knowledge if he finds them to be incomplete and realizes that he cannot arrive at the truth with absolute objectivity.

In such circumstances he may possibly revert to other methods in an attempt to bridge over the missing parts, and will make use of temporary hypothetical constructions for lack of solid facts. It often happens that the historian, who supplies from his own imagination the missing lines of his scientific prose, allows himself to be carried away by an unbridled poetical inspiration and soars to giddy heights. This is inevitable, yet he should never forget that hypotheses constitute a necessary evil and should be discarded as soon as contradictory facts come to light.

We may expect to find a similar "geopoetical" aspect in many a geological treatise, in addition to the normal geological prose. However, authors should always keep their theories strictly separated from descriptions and conclusions of a more rigorously documented kind.

We repeat it: geology is a historical science. The history of the earth is a most absorbing one and its unknown elements — many of which, will perhaps elude us forever — challenge us. The beginning of this history brings us into contact with Astronomy, particularly with Cosmogony, the science of the origin of the universe.

Immeasurable space and time encompass us. "As Rama looks out upon the Ocean, its limits mingling and uniting with heaven on the horizon, and as he ponders whether a path might not be built into the Immeasurable, so we look over the Ocean of time, but nowhere do we see signs of a shore". These lines appear towards the end of the second part of Suess' masterly work *The Face of the Earth.*

Science progresses with steady strides. New facts come to light and new ideas are born every day, and our conception of the structure of the universe changes accordingly. Our views have constantly to be revised and readjusted, but occasionally new aspects of far-reaching consequence shed such an unexpectedly different light on existing problems that its effects might be compared to those of a revolution.

Chapter 21

Space and Time

Geology, the science of the history of the Earth and Life, reaches back into the infinitely remote ages and depths of the Universe and extends its speculations to the origin and meaning of all organisms and inorganic matter. One generation after another has attempted to unravel the problems of the continents and oceans, or to decipher the origin and evolution of Life, or even the mystery of Man himself, who never rests in his unceasing quest for knowledge.

The historical succession of phenomena, their correlation and meaning, form the most attractive and interesting features of geology. This applies not only to the major outlines of development of our globe and the evolution of life, but also any other minor geological problem. Thus the geologist who is mapping a region keeps careful note of every detail of the rocks and strata.

Nevertheless, his object is not merely to determine whether granite, limestone, schists or other rocks occur within the area, nor will the presence of folds, overthrusts or other tectonic phenomena satisfy his curiosity. What he wants to know is the sequence of events through space and time, i.e. the geological history of that particular region up to the present day. The only reward for his painstaking efforts will perhaps be what Termier so enthusiastically described as *"la joie de connaître"*.

The geologist might be compared to the historian. The historian will make a careful study of any parchment that happens to fall into his hands and will reconstruct the past with the aid of its data if he finds them to be complete. On the

All that had hitherto been sacrosanct crumbles to the ground; hardly anything is left untouched.

The late American geologist Barrell, whose death, alas, came so prematurely, wrote in one of his brilliant articles: "The scheme of the Universe is more profound and the unknown is a little nearer than it was recently thought to be.

But such has been the progress of knowledge since man, in the days before the advent of science, naively regarded the earth, his home, as the center of the universe and the heavenly bodies as lights in a nearby firmament, created a few thousand years previously especially for his benefit".

In the light of the above, it would be advisable to begin with a brief outline of some of the features of the modern conception of space and time.

THE UNIVERSE, THE SOLAR-SYSTEM AND THE EARTH

We will begin with a short summary of cosmic dimensions. These may help us to obtain a better idea of the earth's humble place in the universe. We will then immediately pass on to a discussion of the origin of the earth. The earth's volume is more than a thousand times less that of Jupiter, and the volume of the latter is in turn a thousand times less than that of the sun.

The sun looks like a small star when compared to a giant of the type of Antares. Sixty million suns could fill Antares' space, but this giant is surpassed several times by the super-giant Epsilon Aurigae, which has a diameter 3,000 times that of the sun. The sun represents only a small element of a spiral nebula, the so-called galaxy, composed of about ten to a hundred thousand milliard stars. A great many examples of this type of "nebulae" are known and their diameters vary between 1,000 and 100,000 light-years).

The planets of our solar-system lie extremely far apart, but the galaxy appears to be even more thinly populated with stars than the solar-system with planets. A few comparisons from Jeans may some of these cosmic dimensions. Five apples, placed on our five continents — one on Europe, another on

Asia, etc., would provide us with a scale model of the dimensions of the stars and the intervening distances.

Supposing that — à la Jules Verne -we were to let ourselves be fired from the earth in a rocket, travelling at a rate of 5,000 miles an hour, it would take us two days to reach the moon. If we were to travel through the sun at the same speed, our journey would take us a week, and nine years would be required to pass through Antares.

Finally it would take us no less than five thousand million years to travel in this same rocket through a spiral galaxy such as the Andromeda nebula.

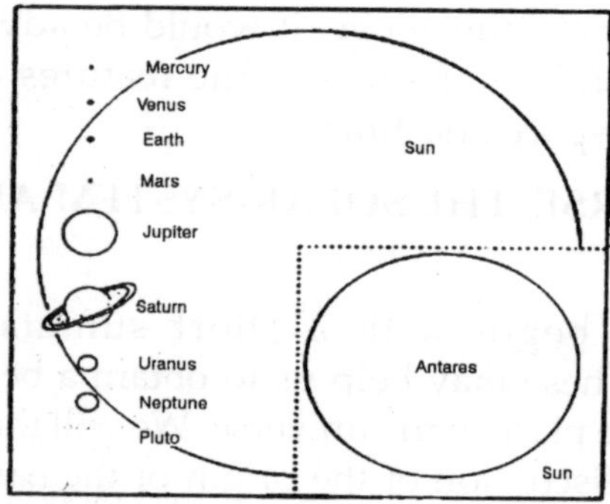

Fig. The Dimensions of the Planets as Compared to the Disc of the Sun, and the Sun as Compared to Antares.

Spiral galaxies are distributed less sparsely through space than the stars through a galaxy. If Amsterdam — to quote an example from de Sitter -were to represent the extent of our galaxy its next-door neighbors would be The Hague and Utrecht. Some 800,000 light-years separate us from the nearest spiral nebula. In other words, if viewed from this region today, the earth would present a somewhat unwonted aspect for Man would just be beginning to appear.

The spiral nebulae are distributed fairly evenly through space. The most distant ones—representing the very extremities of that part of the universe which is known to us at present — are separated from us by a thousand million light-years. This means that if it were at present possible to scan the earth through a super-telescope from one of these extremely distant parts, we should find that we were looking at the first and most primitive terrestrial organisms in our history, swimming around in Pre-Cambrian seas, and we should have to wait 700 million years to detect the first signs of life on the continents — that is, assuming that the distance remained unchanged during this lapse of time.

These cart-wheel shaped galaxies have a central hub. Thirty thousand light-years separate the sun from the center of its sidereal system and this galaxy rotates in approximately 200 million years according to observations by Oort, Lindblad and Plaskett.

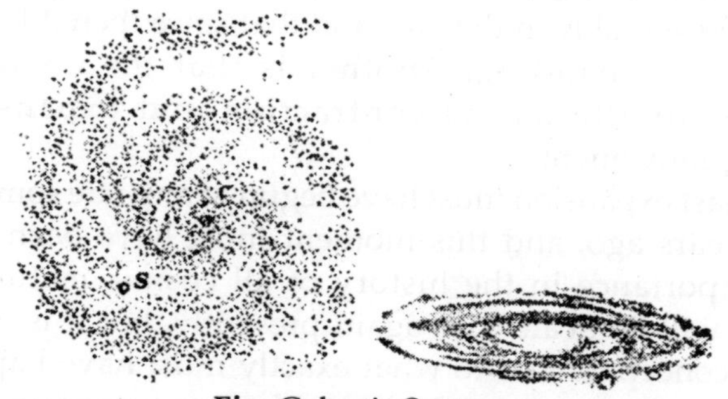

Fig. Galactic System.

One of the most sensational astronomical discoveries of the twentieth century was that all the star-systems appear to recede from our galactic system, as well as from one another. Everyone knows that the pitch of the whistle of a locomotive will drop as the engine rushes past and vanishes into the background.

The same phenomenon is observed when a source of light recedes fro us at high speed. It manifests itself to the astronomer as a red-shift of the spectral lines of a star. This enables him to calculate the velocity of the movement of the star relative to the sun from the amount of the red-shift. Observations have disclosed the remarkable fact that the velocities of the recession of the spiral systems from our own galaxy and from one another, increase proportionately with the distance between the receding systems.

By reversing the picture and imagining the galaxies to travel towards instead of away from one another, we are able to conclude that a tremendous quantity of matter was packed into a considerably smaller volume of space some 2,00 to 3,000 million years ago. The fact that all the galaxies move away from us, does not mean that we are remaining stationary. We should think in this respect of bits of straw which, while floating in a swiftly-flowing river, move away from one a nother as the river broadens.

The statement that the universe is expanding (which was first alluded to by W. de Sitter in 1917 on theoretical grounds) is one of several possible interpretations of a mathematical equation. One possible solution is that the universe had once shrunk considerably in the past and that since then it has been continually expanding. Another is that the universe is subjected to alternating contraction and expansion, a pulsating movement.

The last expansion must have begun at any rate some 2,000 million years ago, and this moment must have been one of major importance in the history of all cosmic matter. This question will be dealt with again presently, but it is obvious that our conceptions as to what exactly must have happened at that moment are, to say the least, of a very uncertain nature.

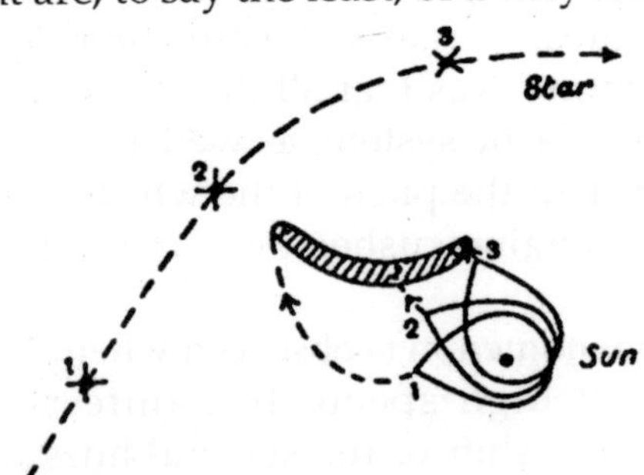

Fig. The Sun and the Formation of a Boomerang-Shaped Filament.

The following reflections appear in a systemation by de Sitter. It cannot be said whether the galaxies were already in existence before the catastrophe, or whether they originated from this turbulence, as it may be that the stars were distributed more evenly through space previously.

We know for certain, however, that all eventual deviations from absolute homogeneity existing at that time must have been strongly augmented by the terrific commotion". It would take us too far to mention the many interesting theories on the ultimate limit of the age of matter, which astronomers estimate to amount to between 5 and 10 billion years, or to deal with the absorbing results obtained as regards curved and finite space, and the correlation of matter and radiation, space and time.

The above quotation from de Sitter shows that the problem of the origin of the galaxies is far from being solved. This is equally true of the solar-system. There are still a great many conflicting views on this subject! Russell has reviewed the *embarras de choix* in a cleverly written book. One group of hypotheses assumes that there has been a close encounter between the sun and another star during some period in the past, and that both either approached one another very closely or collided.

The chances are that such an event did actually take place 2,000 million years ago. A much debated hypothesis of Jeans and Jeffreys surmises that the sun thus entered the danger zone of gravitational pull of a bigger star, and that the sun's surface rose toward it in the shape of a conical surface, from which a narrow filament would be produced.

The ejected material would then condense into separate cooling masses revolving around the sun, and would thus lead to the formation of the planets. Russell and Lyttleton are of the opinion that the sun might have been a binary star at that time, and that its smaller companion broke into fragments as the result of a collision with — or the near approach to — a passing star. These fragments would then have developed into the planets.

THE BIRTH OF THE MOON

The genesis of the solar-system presents many problems and is the subject of much conjecture. The same is true of the origin of the moon. The ratio of the volumes of the moon and the earth is 1 : 82. This is exceptionally high compared to that of Titan and Saturn (1:4700, the highest ratio found among the remaining satellites and planets).

Besides, the moon's orbital momentum — which, in the

case of other satellites, amounts to a mere fraction of the angular momentum of the accompanying planets -is five times that of the earth.

This means that the system of the moon and the earth is more like a binary star than a miniature planetary system. From this it might be assumed that the moon originated as a separate body during the catastrophe which resulted in the formation of the planets.

However, this hypothesis if true contains a few incomprehensible elements. It is difficult to see why two bodies, so unequal in size; should have originated during the catastrophe, and why these two bodies should have formed so close together. One thing is clear, however the moon has slowly been retreating from the earth ever since its formation.

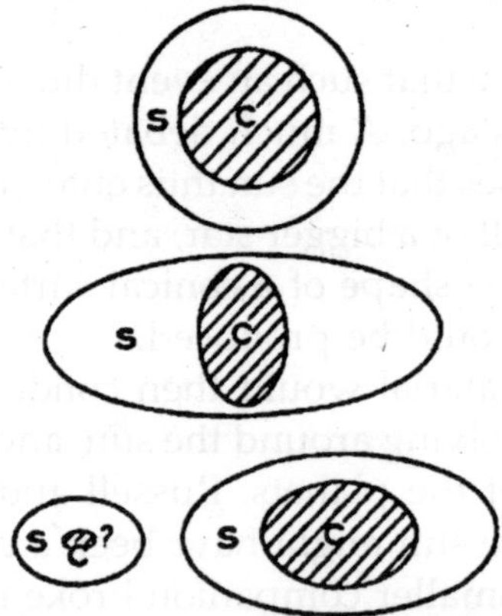

Fig.. Changes in the Form of the Earth During Resonance, and the Origin of the Moon; c, Core; s, Intermediate and Outer Shells

The friction of the oceanic tides has gradually slowed down the earth's rotation and diminished its angular momentum. The result is that the moon's recession from the earth amounts to about five feet per century. Higher tides are observed as we reach back into the past; the days are found to be shorter, and the moon is seen to revolve nearer the earth.

Of course, we sometimes wonder whether an entirely different hypothesis should not be preferred to the above, and whether the moon and the earth were not at one time united as a single body. This was the opinion of Sir George Darwin, who regarded the moon's disruption as a resonance effect, i.e. the result of a concurrence of the solar tide with the natural, tree period of vibration of the earth's cooling fluid mass. The initial small amplitude of the tidal movement would continue to increase steadily according to this theory, and the globe would ultimately grow unstable and disintegrate when a certain limit was exceeded.

This hypothesis is based on the assumption that the moon was severed from the earth during the earliest part of their joint evolution. The liquid earth-materials were probably already more or less differentiated by that time, the heavier minerals sinking, and the lighter ones rising and accumulating in outer layers of the molten planet.

According to the resonance-hypothesis of the moon's origin a great part of the earth's outer shells clearly entered the formation of the moon. And this would not only explain why the moon's density is less than of the earth, but also why its heavy core is comparatively smaller or non-existent, and why its external silicate shell is so much thicker).

Mohorovièiæ computed the thickness of the earth's outer silicate mantle at 60 kilometre, whereas at the moon it would amount about 400 km. It should be mentioned, however, at once that — according to Jeffreys — the moon's density could be accounted for as well under the hypothesis of its formation as a separate body near the earth. The premisses of Mohorovièiæ's calculation are of questionable nature.

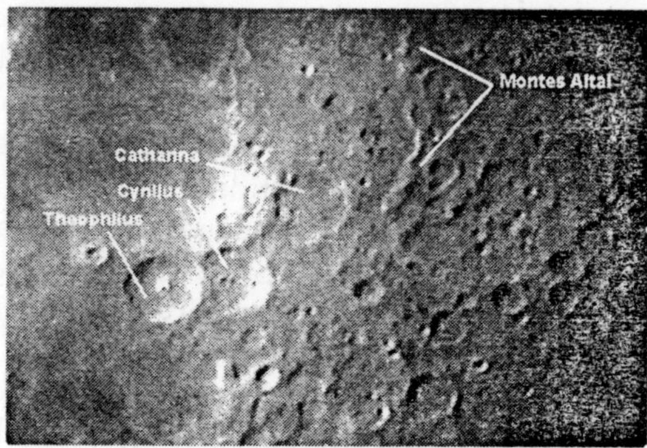

Fig. Theophilus

The publication of Mohorovièiæ dates from the year 1925. Fourteen years later Escher, not knowing of Mohorovièiã's results, once more calculated the thickness of the lunar sial shells. He started from the same premisses and câme to the same result. It was on this foundation that Escher built up his interesting interpretation of the moon's morphology.

A granitic magma abounding in gases is known to bring about a much more explosive kind of volcanism on earth than a basic magma, and it follows that the discharge .of gases from the moon's very thick sialic shell caused a huge amount of volcanism, resulting in types of volcanic explosive vents such as have never been equalled, either in abundance or magnitude, by similar terrestrial volcanoes).

THE EARTH'S INTERIOR

The supposed constitution of the earth's interior to which allusion was made in the foregoing pages, is largely based on the interpretation of seismic data.

When differences of stress in the rigid and elastic crust of the earth exceed the strength of the crust, the stress is relieved by a sudden slipping or rupture along a new or still existing fracture-plane. The spasmodic discharge of elastic stress which overcomes frictional resistance causes what is called a tectonic earthquake.

Two waves with unequal velocities, even in the same medium, are propagated through the earth's body. In the swifter waves the movement corresponds with the direction of propagation; they are longitudinal waves and the first to be recorded by the seismographs at some appreciable distance from the focus; they are called primary or P-waves.

In the slower waves the direction of the wave-movement is transverse to the, direction of propagation. They are marked in the seismograms as secondary S-waves. The velocities of P and S vary with the density and elasticity of the rocks through which they pass.

Their path is curved owing to the gradual change of density and elasticity of the rock masses through which they are propagated. Moreover, P and S are subjected to refraction and reflection at the boundary-surfaces of different kinds of rock-masses. Their arrival at the seismograph is marked by such notations as Pp, Ss.

Finally a third type of waves, travelling along the earth's surface is recorded by the seismograph; their arrival is indicated by L (Long waves) which shows some types of seismograms varying according to their distance from the focus or hypocentre and the spot situated immediately above it at the surface called the epicentre.

Entering into the secrets of seismology would lead us too far from the scope. This short introduction will serve only to furnish an idea of a few of its fundamental principles. For seismological science has led to far reaching conclusions regarding our conceptions of the earth's internal constitution. The time of arrival of the waves is automatically recorded by many stations.

Conclusions can be drawn as to the site of the epicentre and the probable depth of the focus, by studying the combined results of horizontal and vertical pendulums at these stations.

In addition it is by a study of seismograms that conclusions are reached regarding the situation of the major discontinuities in the earth's crust and its deeper interior. The first discontinuity is supposed to form the boundary between the rigid crystalline crust and the outer mantle.

According to the opinion of Goldschmidt and other geochemists gravitational differentiation caused the liquid rocks of the earth to form concentric shells around a central core. In their opinion the core would consist of nickeliron while a sulphide-oxide melt separated around it and an outer mantle was formed by the relatively light material of a silicate melt.

Then the discontinuity of 3000 km. was supposed to be identical with the supposed boundary between the core and the intermediate shell of sulphide-oxide. In a similar way the discontinuity of 1000 km was supposed to coincide with the boundary between the intermediate shell and the silicate melt of the outer shell.

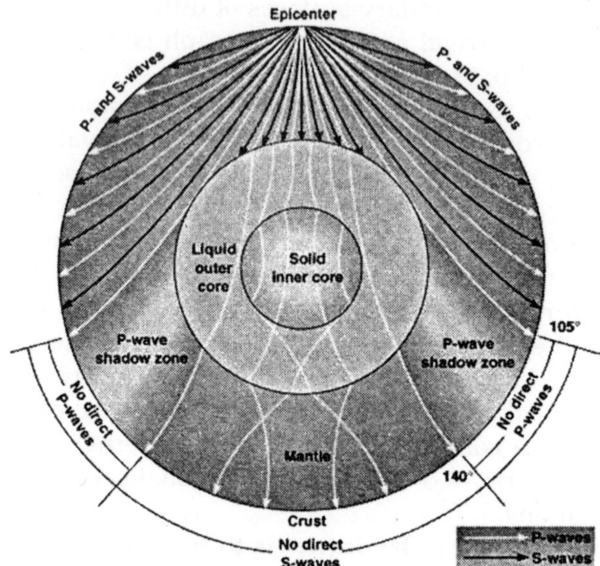

Fig. Seismic Waves through the Earth.

It is, however, doubtful whether the seismic discontinuities are due to sudden changes of the chemical composition of the earth's interior.

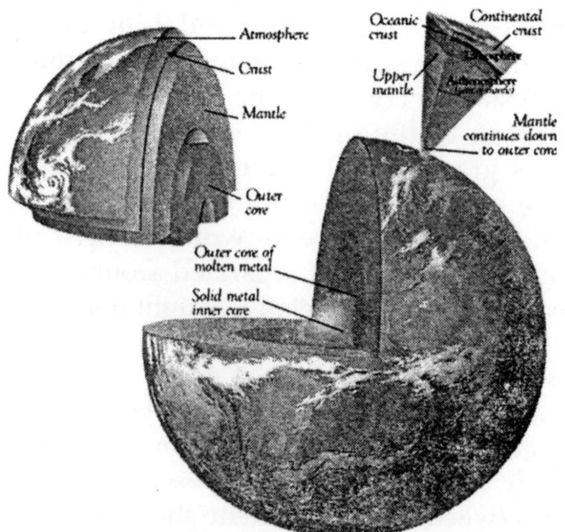

Fig. Internal Structure of the Earth.

According to Kühn it is much more probable that the major discontinuities represent changes of a purely physical nature. It is difficult to understand how complete differentiation of the molten planet could have been accomplished in a comparatively short period of its early infancy. And, in addition gravitational differentiation as far as the earth's centre seems improbable in as much as the value of gravity diminishes towards the interior becoming zero at the centre.

In a new hypothesis Kühn and Rittmann consider the interior of the earth to consist of undifferentiated solar material. Their opinion, in contrast to the older theory of a nickel-iron core.

Perhaps the truth lies somewhere between the theory of Kühn and the older conception. Certain arguments seem to favour the new theory of a core consisting of undifferentiated solar material. Possibly, however, the discontinuity at 3000 km separates the core from the outer layers which consist of more or less differentiated material. That convection currents possibly reached down to that level in a primordial stage of the earth.

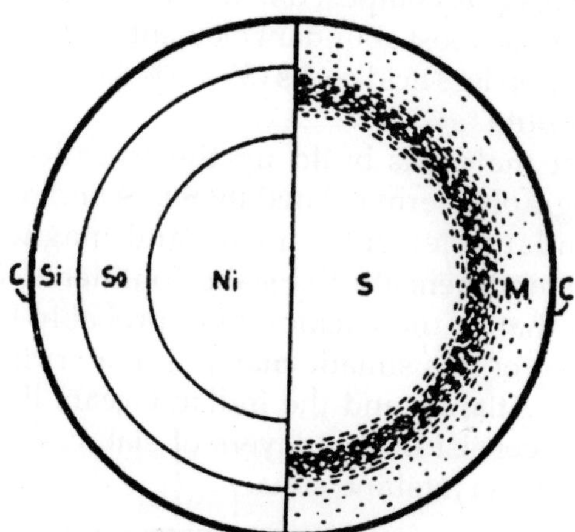

Fig. Two Different theories on the Constitution of the Earth's Interior.

Their action would explain some striking features in the shape and distribution of the continents and ocean floors according to a recent speculation ot Vening Meinesz. Convection was accompanied by a complicated process of differentiation of the outer layers. Loss of gases, especially of hydrogen at the surface, accumulation of sialic material in a comparatively thin outer layer and of heavier materials in the deeper shells are considered as the principal results of such a process of convection and differentiation.

The outer, dark-coloured zone the earth's crust, composed of some 40 to 80 km of crystalline rocks. This solid, elastic crust presumably rests on amorphous formations, i.e. rocks which, as a result of the high temperature at this depth, must be above their melting point. They are regarded as a fluid possessing a high viscosity.

Daly speaks of a "vitreous substratum". The schematic cross-sections of the continents have been drawn as white lenses in the dark zone. Petrographical, volcanological and seismic evidence has lead to the assumption that the continents are built up of light siliceous material (55–70% $SiO_{2)}$, viz. granitic "acid" rocks and sediments.

This constitutes the so-called sial, a term introduced by Suess (the word sial is composed of the first syllables of silicon and aluminium, the most abundant elements of the continents). The sial rests upon less acid rocks (35–55% $SiO_{2)}$, ranging from basalt to peridotite.

These last materials build up the so-called "sima" — another petrographic term created by Suess and mnemonic of its most abundant elements, silicon and magnesium. It is believed to extend beneath the sialic continents. One of the ocean-floors—that of the Pacific—may probably be assumed to be composed of this simatic material in a crystalline state. As regards the Atlantic and the Indian Ocean, their bottoms are thought to consist of thin layers of sial, covering — and resting upon -the crystalline sima.

THE AGE OF THE EARTH AND THE UNIVERSE

Similar results as regards the age of the earth are obtained by investigators in different branches of science, and show that 2,000 to 3,000 million years have elapsed since its creation.

These results are arrived at by petrographers and physicists in collaboration with geologists and also by astronomers.

These figures are specially impressive in as much as many of these scientists arrived at them independently and their importance is augmented by the fact that — apart from the earth — many other bodies, such as the moon, the — planets, the entire solar-system and the spiral galaxies, received the impulse to perform their present movements at the same moment. This moment seems to have formed the last critical date in the universe which since then has developed gradually into its present condition.

By means of no less than ten independent methods have scientists arrived at an estimate of the age of the universe since the last great catastrophe.

Chemists have shown that uranium disintegrates spontaneously into radium and helium at known systematic rates, and that radium will in turn disintegrate into lead and helium. Uranium is a mixture of two isotopes UI and AcU. The first, 139 times as abundant as AcU, disintegrates through the "uranium series" to lead which has a different atomic weight (Pb 206) from lead (Pb 207) that originates via the "actinium series", while lead deriving from thorium has again a different mass number (Pb 208).

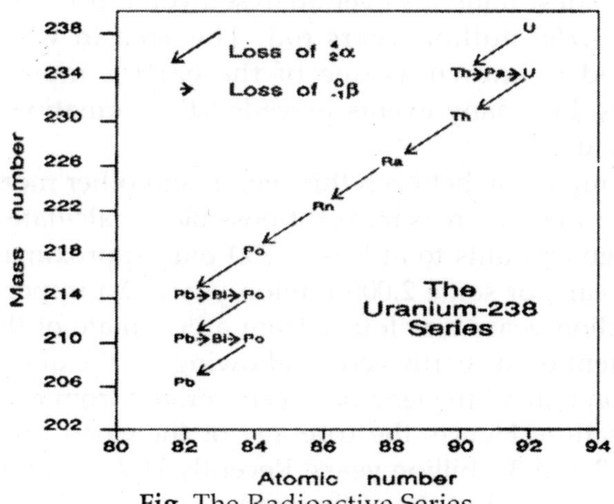

Fig. The Radioactive Series.

Common, that is non-radiogenic lead, has again a different atomic weight (Pb 207.21). An analysis tells us the contents of radiogenic lead (we will not examine the problems affecting the determination and analysis of the various isotopes of lead), and as the amount of radiogenic lead accumulating from a given quantity of uranium or thorium per unit of time is known, the age of any mineral containing radiogenic lead can be ascertained.

A second method is that based on the constant ratio observed between the amount of expelled helium atoms and the initial uranium. Helium is a gas, however, and can escape as such from a rock.

The values arrived at by this method will therefore probably be too low. In other words, the ages of rocks obtained by the helium method will also be too low. Many ingenious improvements have been recently introduced to remedy this deficiency, in an attempt to obtain useful results even from this method.

The ages of a great many minerals have been determined so far, and their analyses show that the ages of those rocks which could indeed be said to be older because of the geological position in which they were found, were invariably higher than others.

The oldest minerals ever analysed come from Manitoba, and are 1,750 million years old. The area in which they originated is known as one of the earth's most ancient structures, but many events preceded the formation even of this structure.

A comparison between this region and other more recent and better-known areas makes it possible to calculate that the earth's age amounts to at least 1,750 plus approximately 300 million years, or some 2,000 million years. An upper limit of 3,500 million years was found from an estimate of the entire lead content of the earth's crust, allowing for the questionable asumption that all the lead has been derived from radioactive disintegration. Hence, the true age of the earth probably is between 2 and 3,5 billion years. Recently Holmes found 3,000 m.y. as the probable age of the earth.

This figure has not only been arrived at by geologists. It was mentioned above that a similar result was deduced from the recession of the spiral galaxies.

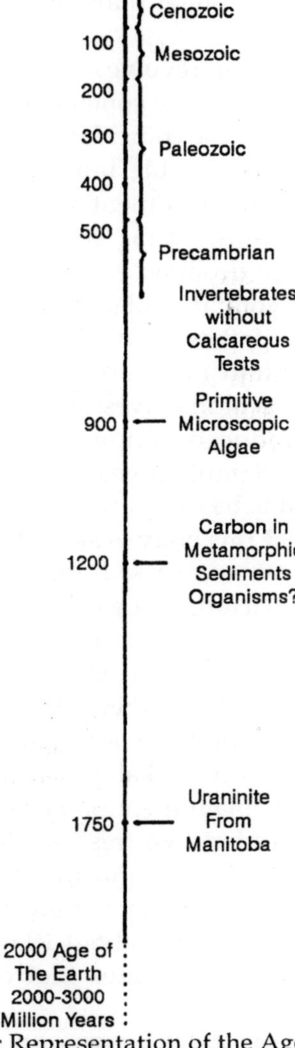

Fig. Graphic Representation of the Age of the Earth.

The spiral structure, too, may serve as a basis for further conclusions. Eddington showed that a galaxy could not possibly be preserved in a steady state for more than 109 years without considerable collapse or dispersal, and to this Bok added that "many spiral galaxies exhibit two well developed spiral arms and it is probable that such a structure cannot persist for much longer than ten to fifteen revolutions of the nucleus".

As the galactic system revolves once in every 200 million years, its age might likewise be concluded not to exceed 2,000 to 3,000 million years. These coincidences between astronomy and geology are truly remarkable, but we may add at the same time one more important argument.

We know for certain that some of the nickel-iron meteorites which have dropped on to our planet can be said to have originated beyond the solar-system. A minute amount of radioactive elements has been detected in some of these bodies, and their helium ratios have been determined. The eagerly awaited results of the investigation of these fragments of cosmic matter produced various figures, none of which exceeded 2,000 to 3,000 million years, however.

Another method is based on the movement of the moon. It was seen above that this body is steadily receding from the earth. If the process were reversed, the moon would be seen to be gradually approaching the earth.

Though admitting that it is impossible to tell at what moment the moon was severed from the earth, in as much as this process is itself purely hypothetical, Jeffreys estimated that the age of the moon was probably less than 4,000 million years.

This figure is again remarkably near that of the 2,000 to 3,000 million years estimate mentioned above several times.

Chandrasekhar added two fresh evidences by his study of the statistics of binary stars and the dynamics of galactic star clusters. The Pleiades star cluster includes some 200 stars in a spherical volume of radius about 10 light years which is a star density about twenty times that of the background "field" stars.

As a result of his calculations on the stability and dynamics of such aggregations of matter Chandrasekhar came to a rational means for an estimate of the probable rate at which a star cluster tends to disintegrate. His conclusion is that the average life of a cluster is about 3×10^9 years! The same author made another new contribution to the question of the time-scale of the Universe by his consideration of the binaries.

Binary stars are unstable because of their tendency towards disruption in as much as the distances of the nearby stars from the two components will be different, and consequently also their fluctuating tidal effects. According to a mathematical formula "binaries with separations between 1,000 and 10,000 astronomical units will be dissociated in times ranging from 7×10^{10} to 2×10^9 years."

This means that during an interval of time of about 1010 years a tendency must be expected of ever larger separations towards a final condition of statistical equilibrium. However, it has been found that "the larger separations occur with far less frequency than should be expected under conditions approximating those of equilibrium.

Accordingly, we should conclude that sufficient time has not elapsed for the tidal forces of the neighboring stars to appreciably modify the elements of binary orbits with separations in the range stated.

This implies that 10^{10} years represents a true upper limit to the time scale, and would suggest a time scale of the order of say 5×10^9 years." "The discussion of the mean lives of galactic clusters and the statistics of binary stars agree therefore in pointing to a time scale of the order of a few billion years."

Lastly, Jeffreys calculated from the eccentricity of the planet Mercury that the age of the solar-system was probably nearer 1,000 than 10,000 million years.

This concludes our summary of the various methods which make it possible to calculate a minimum value of the age of the earth and the universe. It is obvious that these methods provide converging evidence of great importance.

A few methods are dealt with more extensively by Holmes in *The Age of the Earth*. Data on the absolute age of terrestrial rocks are extremely important, for they give us an idea of the duration of the different geological eras and periods.

Therefore the question of absolute age determinations will be treated at greater length.

The Earth's Crust

We usually divide the history of the continents into two distinct parts. The first deals with the remotest ages and covers the extremely long PreCambrian era, of which very little is known. The second deals with the succeeding ages, from the Cambrian up to the present day. Much more is known of the earth's history since the Cambrian than of the preceding periods, for, generally speaking, its historical records have remained in a far better state of preservation than those of the earlier times.

Historical geology has often been compared to a book, a torn and tattered old manuscript the pages of which lie scattered far and wide.

The task of the geologist is to hunt up the missing pages and replace them in their original order. The sediment layers can be said to represent the pages of this geological history and the fossils the writing. It should be noted that fossils occur very rarely in the Pre-Cambrian and are almost entirely absent in the older parts of this era.

This explains why so little is known of Pre-Cambrian history. Its pages can only be replaced in the correct sequence with the greatest difficulty and the writing upon most of them has faded like that of an old palimpsest.

The principal events in the history of Life are listed in Table II, next to the time-scale. The Pleistocene and Holocene, encompassing the whole history of mankind from the earliest days up to our own, represent less than the thickness of the top line. Should our knowledge of the earth's history be communicated to a student in correct chronological proportion in the course of 50 to 60 lecture-hours, a bare two minutes would have to suffice for a discussion of the entire history of the Pleistocene ice-ages.

A comparison between the earth and various cosmic dimensions made us realise our planet's modest size. The preceding statement, however, shows that the history of the earth is at least 2,000 times as long as the whole history of *Homo sapiens.* One of the most remarkable features of that history is the ceaseless flow of alterations.

Mountains, seas, glaciers and deserts all seem immutable compared to the ephemeral span of human life. However, nothing is constant in geological time. Most visitors to the small though interesting town of Le Puy, in the south of France go solely in order to admire its beautiful old cathedral and to gaze upon the enormous bronze statue of the Virgin.

They stare in silent wonder at the steep rocks, each of which is crowned with a church or statue, but few realise that this strange scenery is merely due to the fact that the hard filling of a volcanic vent and some remnants of former lava flows are resisting erosion a little longer than the largely eroded tuffs and strata.

All those external forces, such as frost, solution, running water, the weather and wind, etc., which act unceasingly upon the terrestrial crust, are wearing down the mountains slowly but surely and levelling them with the sea. The peneplain constitutes the final stage of the largest mountains. In other areas the sea helps to abrade the coasts and this process, too, has a levelling effect.

Fig. Geomorphological Point of View.

Neptune invades the land to an ever-increasing extent, and the débris of the organisms which remain behind enable us to determine the relative ages of the various deposits. An illustration of one of these marine strata, which was deposited during the Cambrian in a sea invading the land over Torridon sandstone of late Pre-Cambrian origin.

Only a few of the remnants of this erosive action, or "monadnocks", as they are called, protrude from the peneplain in block 2 as evidence of the former relief.

These two blocks show more, however. Rivers transport detritus from the mountains to the sea, where it settles in large quantities. The land represents an area of erosion, but the sea is the most suitable place for the accumulation of great quantities of erosion products. One layer slowly accumulates on top of another, and this leads to a thick sequence of sedimentary strata.

Block 1 shows that sedimentation takes place in a trough-shaped depression of the earth's surface. The coarsest products remain near the coast, the finer material is carried further away and settles in thin layers at a considerable distance from the shore. The coastline extends slowly but surely into the sea, and the sequence of sediments steadily increases.

Block 2 shows that the floor of the basin of sedimentation also subsides. Without such subsidence the thickness of the strata would never be able to exceed the original depth of the sea at this point.

The accumulation of sediments, however, increases proportionately with the subsidence of the floor. A thickness of many thousands of meters (sometimes as much as 15,000 meters or more) is not an unusual occurrence.

Fig. Gedeh-Panggerango Volcano

Everyone probably knows that not only are the sediments which settle near the coast unlike the strata deposited in the deep-sea, but that these separate environments also contain different kinds of organisms. Various factors such as light, food, pressure, temperature, salinity, the movement of the water and a host of other influences all effect the zonal distribution of organisms in the sea.

An idea of the conditions under which these strata were originally formed can be obtained partly from their special lithological character but principally from the remains of former organisms. This total aspect of the sediments is known as the facies in geological terminology.

The facies of sediments deposited in a shallow sea in the vicinity of the coast (littoral), or in a few hundred meters of water (neritic sediments) are wholly unlike those which settled further away from the coast in approximately 1,000 meters of water (the so-called bathyal sediments) or at an even greater depth (abyssal sediments).

Littoral or neritic sediments cannot form in more than a few hundred meters of water, but we sometimes come across a sequence of neritic sediment totalling a few thousand meters in thickness, and it is clear that this must have been brought about by an accumulation of successive deposits during a slow subsidence of the sea floor.

The name geosyncline has been given to these subsiding furrows of the terrestrial crust, into which sediments accumulated to abnormally thick sequences.

Aschematic view of a geosynclinal cross-section is shown in block 2.The downward movement of these formations must be attributed to the influence of deep-seated forces, and the same forces will at a given moment put a stop to the subsidence.

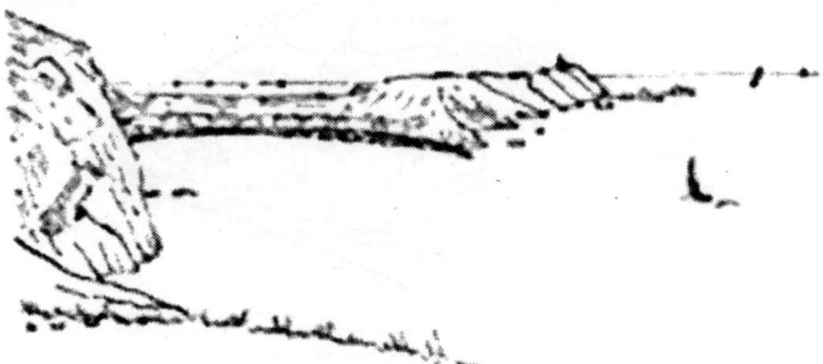

Fig. The Erosive Action of the Sea

A growing compression of the crust then causes the geosyncline to be thrown into spasms and its contents to be crumpled and folded. Block 3 depicts a more advanced stage. The whole zone will then tend to rise, and from the elongated belt emerges a mountain range which rises upwards very slowly, though with unremitting persistence, frequently to a height of several thousands of meters.

The landscape reproduced borders on the Caledonian geosyncline in the north-western part of the Scottish Highlands.

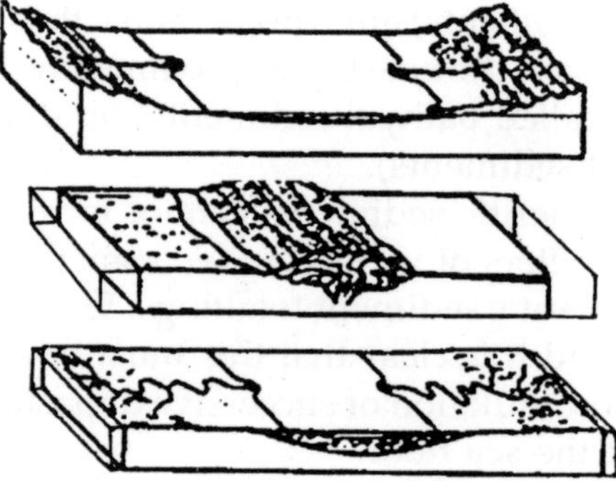

Fig. Diagrammatic Representation of a Geological Cycle.

To the left of it are seen Cambrian strata (these were deposited normally on an older Pre-Cambrian basement, consisting in this case of Lewisian gneiss). To the right, however, a large "nappe" of Lewisian gneiss is clearly seen to have been overthrust toward the left, i.e. over the edge of the geosyncline.

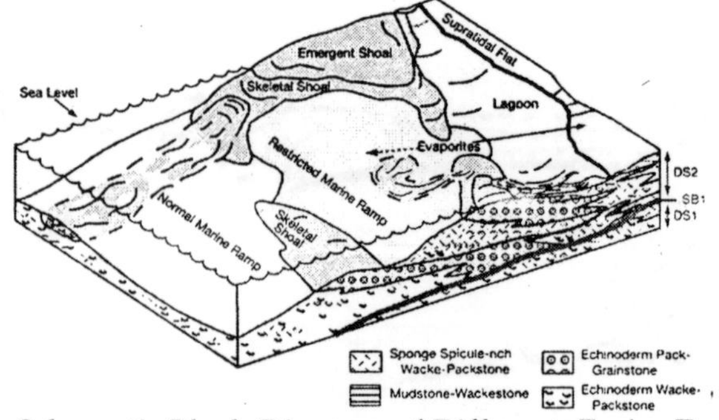

Fig. Schematic Block-Diagram of Different Facies Types.

The external forces embark upon their destructive action as soon as the folded chain emerges. A new area of subsidence and sedimentation is formed elsewhere a new geological cycle has opened.

Old Red sandstone resting unconformably on steeply dipping Silurian graywackes and shales. A plane of unconformity has a long history. The graywackes were deposited normally, i.e. just about horizontally, during the Silurian, and were subsequently steeply folded and elevated above sea-level.

Fig. Glencoul Overthrust.

Old Red Sandstone later settled on top of these eroded and peneplained marine strata. The period of folding of the latter corresponds with that in which a geosyncline was compressed over large areas in Scandinavia, Scotland and Ireland.

This last zone was later raised as a mountain-chain (Plate 1 indicates this belt with green lines). The actual Old Red Sandstone was supplied by detritus from these chains, and was later likewise worn away by erosion in extensive areas. Trough-shaped depressions, elongated in the trend of the Caledonian mountains, formed here and there, and the Old Red escaped erosion only in these depressions.

Fig. Silurian Graywackes and Shales.

Some of the troughs are bounded by faults in which we once more find the trend of the old Caledonian geosyncline. These faults are observed along the surface over a considerable distance.

Indeed, the relation between the scenery and structural history of an area is finely and instructively the whole of the morphology of Scotland of some of the most salient geological and "geomorphological" features of Scotland In a few cases geosynclinal subsidence and folding are known to have occurred several times in the same region.

In the Ardennes, namely — near Fépin, along the banks of the river Meuse — the Lower Paleozoic deposits (Revinian) of an ancient Caledonian geosyncline were strongly folded and covered unconformably by Lower Devonian (Gedinnian) strata, beginning with a basal conglomerate.

This younger sequence of the so-called Variscian geosyncline was later refolded, and the plane of unconformity was also strongly undulated.

Fig. The Highland Boundary Fault Near

The object of the preceding remarks, especially of the rough sketches, is to give an idea of the constant transformations that are altering the face of the earth. It should at once be added, however, that one of the most spectacular features of these transformations is periodicity.

The geologist comes across periodicity in many of the pages which he is so arduously deciphering, — in the sequence of the strata, for instance, and their contents of former organisms (stratigraphy).

He observes it elsewhere, in the pulse of the deep-seated forces that bring about subsidence first in one area and then in another, culminating in the folding of geosynclines; and again — in the intrusion of liquid melts or "magma" rising from some deeper part of the earth's interior; in the rhythmical invasion of the continents by epicontinental seas and the subsequent retreat of the latter (transgression and regression); he reconstructs the pulsation of the climates and the rhythmical evolution of Life.

"Nature vibrates with rhythms, climatic and diastrophic, those finding stratigraphic expression ranging in period from the rapid oscillation of surface waters, recorded in ripple-marks, to those long deferred stirrings of the deep-imprisoned titans which have divided the earth's history into periods and eras.

The flight of time is measured by the weaving of composite rhythms — day and night, calm and storm, summer and winter, birth and death — such as these are sensed in the brief life of man.

But the career of the earth recedes into a remoteness against which these lesser cycles are as unavailing for the measurement of that abyss of time as would be for human history the beating of an insect's wing. We must seek out, then, the nature of those longer rhythms whose very existence was unknown until man by the light of science sought to understand the earth".

Chapter 22

Mountain Chains

The first critical analysis of the earth's structural history was made by Ed. Suess, who published a monumental book on this subject: *The Face of the Earth*. Suess was not only the right man for such a task, but he also undertook it at the right moment. The amount of data had already become so voluminous by then that an attempt could be made, by grouping them systematically, to analyse the surface of the entire earth. It would be impossible for a single man to perform such a task to-day, since a judicial examination of the enormous amount of world-literature would require more than a life-time.

However, a series of regional treatises makes it much easier for us, at present, to find our bearings in this varied mass of publications and a further asset is, that the study of them requires far less of our time. Such synoptic works make it possible to form an idea of the whole region in question, without entering into a critical examination of all original data and fundamental details.

In this manner it is possible to obtain a survey of all that appears to be known at present of the structural history of the continents and to base an opinion on its general aspects.

This method had already been adopted previously by Born but he grouped his pictures regionally, i.e. continent-wise. We, on the other hand, have chosen a purely chronological sequence. Nevertheless, our two methods agree fully in one respect, i.e. the plain facts are stated separately, without being influenced by any pre-assumed hypothesis, and these are left to speak for themselves as they accumulate.

The data have lastly been combined into a synoptic picture serving at the same time as a chronological analysis of continental structural history. The following historical analysis will begin with the Cambrian, for the ages of the complicated structures of the Pre-Cambrian basement have only been determined in a few cases, and the study of their intricate patterns can therefore still be said to be in its infancy.

The intensive folding of the Ardennes had long been known to have occurred at a much earlier date than the tectonic activity of such mountains as the Alps, the Pyrenees and the Himalayas. The Upper Paleozoic chains were called the Hercynian Mountains, or — as this first name has now become obsolete — the Variscian (or Variscan). It had also been realized for some time that the final, intensive folding in Norway and Scotland occurred even earlier.

The mountains in these last two areas were called the Caledonides. Nevertheless, as more and more facts came to light it became clear that the Lower Paleozoic or "Caledonian" belts had not all originated during the same period.

Fig. Pentland Hills

Thus the Caledonides of North America, to mention but a few examples, were folded towards the close of the Ordovician, while those in Scotland and Ireland underwent a similar process some 30 million years later, towards the end of the Silurian (Gotlandian).

We speak of a Taconic epoch of compression in the first case, and of an Erian in the second. Several different periods of origin are also observed in Variscian and more recent mountains.

Another important point is that the epochs are distributed over the whole world. A comparison between said phases and their distribution will at once render clear that one movement was of particular importance in one group of mountains, while in another chain the influence of a different diastrophic epoch was especially strong.

It would not always be easy, however, to state concisely which epochs made themselves felt in Caledonian chains), and which other ones in Variscian), for Caledonian and Variscian epochs overlap. Moreover, the same applies to Variscian and younger convulsions.

Stille classified the post-Variscian Mountains as "Alpine" formations I have divided them into two large groups, for the final phase is now known to have occured during the Mesozoic — that is, before the major movements in the Alps — not only in the Sierra Nevada of North America, but also in many other areas, e.g. in the eastern and south-eastern part of Asia. These belts have consequently been grouped separately. The Cenozoic chains, beginning with the Laramide, will be found in Plate 4.

The different belts can thus be divided into four large groups: the Early Paleozoic (Caledonides), Late Paleozoic. The structure of the continents will be reviewed accordingly in Plate.

The accompanying table lists the diverse epochs of compression. A more diagrammatic view of their place on the time-scale will be found in Table II. We will now review the more general aspects of the data.

EPOCHS OF COMPRESSION

Our historical analysis of the continents begins some 500 million years ago, for too little is known of events prior to that time. The significance of such a statement will be realized if we pause to consider that the earth is thought to be approximately 2,000 to 3,000 million years old, for this means that we only begin to find sufficient evidence of the past at a time when 3/4 to 5/6 of the earth's structural history had already elapsed. In other words, only one fourth — or possible even less — of this history can be said to be legible. So little is known of the preceding era — the Pre-Cambrian -that we prefer to leave it alone altogether.

This Pre-Cambrian era, however, not only covers the greater part of time, but the Pre-Cambrian areas also occupy the largest part of the continents. It should furthermore be noted that the continents, including probably large areas beyond them, had already passed through an extremely lengthy and complicated evolutionary stage before the Cambrian, and that hardly anything is known of this part of their development.

Not only will a glance at the structural map in Plate 5 show that Pre-Cambrian formations occupy extensive areas, but the Pre-Cambrian basement is also known to crop out in almost all the larger Paleozoic, Mesozoic and Tertiary Mountains, proving clearly to what extent the younger part of the earth's history is engraved across the existing records of earlier ages.

We often read of the accretion of the continents, meaning that younger geosynclinal belts and folded chains are arranged consecutively around a Pre-Cambrian nucleus. In several cases these zones can be said to be younger the greater their distance from a nucleus. There are indeed areas which demonstrate the outward displacement of zones of folding.

It would not be right, however, to conclude from the above that a number of small blocks had existed in the Cambrian, which have grown larger and larger periodically, nor that the continents have consequently grown larger than at any previous date. On the contrary, it would be more plausible to assume that the continents had at one time been more extensive than they are at present.

Not only have parts of the continents foundered below sea-level since Pre-Cambrian time, but they have even done so until quite recently, and their subsidence occasionally attained great depths.

The present continents are mere fragments of onetime larger blocks. The same applies to the so-called "growth-zones". The Caledonides of Europe, North America and Asia originally covered wider areas than their actual fragments do to-day. The latter are bounded and cut off by the frontal chains of Variscian and later mountains.

The Mesozoic geosynclines of Asia, North and South America still contain the vestiges of the former Variscian basement upon which they were formed. Most of the Alpine chains of Europe and Asia are situated in a zone which had already been folded during Variscian time.

It cannot be denied, however, that there are cases in which the younger the age of the geosyncline, the greater the distance between the frontal chains and a Pre-Cambrian nucleus. Here, too, a glance at Pl. 5 will be enough to show that this applies to Asia, Australia, and partly to other continents. Certain Pre-Cambrian shields seem to form centres, around which originated a series of geosynclinal belts, spreading in ever-widening arcs.

From Australia the Caledonian, Variscian, Mesozoic and, finally, Tertiary zones migrate east-and north-eastwards in a centrifugal sequence. An identical centrifugal displacement is observed around the Angara Shield in Asia, though with some complications. For the chains were compelled to bend around the Tarim and Ordos Massifs, including other Pre-Cambrian nuclei, which had long since been rigid blocks, and analogous centrifugal waves spread from the Kara-Sea and Tsuktschen massifs.

The phenomenon of centrifugal migrations can be discerned to some extent in Europe, viz. in a southerly direction. These examples seem to imply that certain centrifugal impulses radiate from specific Pre-Cambrian areas. We might describe them as dynamic centers of tectonic activity.

Nevertheless, there can be no question of a "law" in this case, for it should at once be noted that such dynamic waves have never emanated from other Pre-Cambrian areas — neither from India, nor from Africa (with the possible exception of its extreme southern part), nor from the northern, western or southern part of Australia, nor from the eastern part of South America.

Leuchs therefore quite rightly described the block of India as a passive element in the structure of the Asiatic continent. It acted as an obstruction, against which Variscian, and subsequent Tertiary tectonic waves were shattered, and was responsible for the narrowing of the Tertiary zone of folding and the particularly intensive overthrusting and elevation of the Himalayas.

However, it would be a mistake to suppose that the formation and migration of various zones of folding were due exclusively to the mysterious activity of these "dynamic centers". For before passing on to an examination of their remarkable sequence through space and time, another equally important circumstance should be noted, viz. the world-wide occurrence of epochs of compression.

Data have frequently shown that simultaneous movements occurred on all continents. One phase was of course of greater importance in one, place than another, manifesting itself more intensely in one area than elsewhere.

But the growing amount of stratigraphic studies make it increasingly evident that the terrestrial crust was subjected to a periodically alternating increase and decrease of compression. This concurrence between epochs of folding is not only found in geosynclines, but also in other areas beyond them—in basins and regions where folding had already occurred at a previous date, and where less accentuated movements accompanied the folding of the geosyncline (shifting of blocks, faulting, sliding, and at times even folding). I feel there is overwelhming evidence that the movements are the expression of a common, world-wide active and deep-seated cause.

The term "world-wide" should not be interpreted to mean that all phases can be observed everywhere, or that these movements occur in all areas. On the contrary, the special constellation of the basement may cause two geosynclines to form side by side, one of which will be folded to an intense degree during a given epoch of compression, while the other will be hardly influenced by it. A striking illustration of this is found in the Cordilleras along the west-coast of North America. The following are found side by side in this area:

- The geosyncline of the Rocky Mountains, which was formed in the Pre-Cambrian and influenced by Laramide folding;
- The geosyncline of the Sierra Nevada (Triassic and Jurassic, Nevadian folding); and
- A geosyncline extending further west on a basement which was folded during the Nevadian revolution: the geosyncline of the Coast Ranges.

The same phenomenon can be observed in the Paleozoic Wichita geosyncline of Central-America. The Wichitas and Criner Hills, both of which are situated in this belt, were influenced during the Wichita and the Arbuckle epoch, but the Ardmore basin in the immediate neighbourhood of the geosyncline, including the Arbuckle chains, were only affected by the latter.

A certain similarity is apparent in the linear shift of specific epochs of folding in mountain-chains, e.g. in the Dobrutcha, Crimea, Caucasus; the Donetz and Ammodetic Mountains; and the Caledonides and Variscides on either side of the Atlantic Ocean.

These general aspects will help us with the further examination of the "dynamic centers". In which direction does that dynamic activity displace itself? The investigator would fain detect some regularity in its migration through space and time, some law, which at times we imagine can be formulated, though later it transpires that nature is more complicated than we supposed it to be.

The sequence of the folded systems seems quite regular in Asia and the eastern part of Australia, but a far less simple state of affairs prevails in Europe. The Caledonides have been preserved along the north-western margin of the Baltic Shield, and perhaps originally bent around in a south-easterly direction.

The Variscides, on the other hand, were not only partly

formed in a Caledonian basement but the front of these folded chains intersects the trend of the Caledonides in Ireland and England. This is also true of the younger folded chains in North America. The latter intersect the trend of the Variscian Marathon-Ouachita Mountains, and these in turn cross the Wichita-Arbuckle geosyncline.

Fig. Structural of Rocky Mountains

Besides, the idea of centrifugal migration breaks down entirely in America. For the younger Rocky Mountains extend on the eastern side of the older Sierra Nevada, along the Canadian Shield, and the even more recent formations of the Coast Ranges emerge on the other side of the Sierra Nevada. Such basins as Pugget Trough and California Valley formed again on the Canadian side of the Coast Ranges.

These examples make it quite clear that the centrifugal displacement of zones of folding confines itself to definite areas. Even in those regions where a centrifugal migration of zones occurs in the major systems, the following phenomenon can be observed a young trough of sedimentation originates on the continental side and is added as a last element to the folded chain. Moreover, intramontane troughs originate almost simultaneously.

Mountain-Building

Not all epochs of compression observed in the structure of a mountain are followed by mountain-building elevation of any importance. The Eocene "flysch" of the Alps furnishes ample proof that some folded chains were raised and subjected to the influence of intensive denudation after an Upper Cretaceous phase. The Miocene "molasse" was the product of the sub-aerial denudation of ranges which had been elevated following an Oligocene epoch.

However, intensive compression still persisted subsequently, during the Pliocene, and 'led to the severe overthrusting of the Helvetian nappes. Nevertheless, it was not until the close of the Pliocene that the Alps finally rose as a huge and lofty range of mountains. Similar conditions prevail in other Alpine mountains and older folded chains.

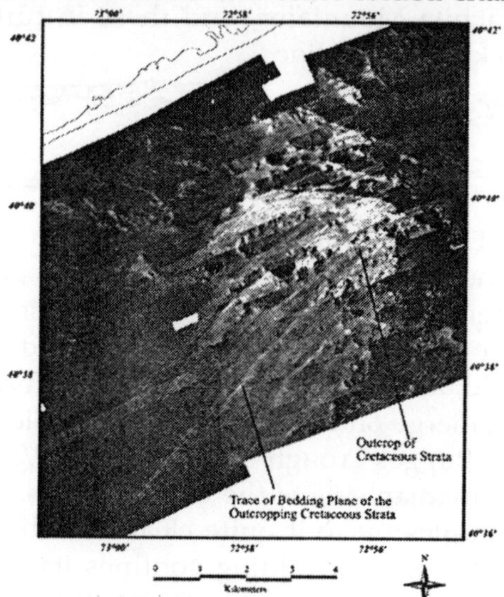

Fig. Cretaceous Strata

Besides, many examples show that the process of elevation may subsequently be repeated in a folded and peneplained system.

This second elevation, or "rejuvenating" movement, of the already considerably peneplained older chains, which continues from the late Tertiary up to our own day, has brought about the present aspect of the Rocky Mountains, including that of the Sierra Nevada of North.

America and the South American Andes, and must be held responsible for the disparity between the deeply eroded structures of the latter, with their many intrusions of batholiths, and those of other mountains, in which the last principal epoch of folding and subsequent elevation was of a more recent date.

All available data show that the largest expanse of mountain-chains since Cambrian time was occupied by the Variscides especially towards the Upper Carboniferous (post-Sudetic and pre-Mid Permian). The second largest area has been occupied by the Alpine system since the Pleistocene. The Caledonides (Taconic and Silurian), extending over large tracts, especially in the continents of the northern hemisphere were less extensive than the Alpine belts.

The Mesozoic Mountains covered the smallest areas, and surround particularly the Pacific. The Variscides and Alpine chains also show a circum-Pacific arrangement, but these belts likewise occupied large areas in an E. W. direction, along the southern side of the northern continents. Mountains have a very strong influence on atmospheric circulation, and this influence must therefore have been exceptionally severe in the Upper Carboniferous and Pleistocene.

Small arrows in Plates 1, 2 and 4 indicate the possible trend of the Caledonides, Variscides, and Alpine systems, whose chains are intersected by the coasts of the Atlanti.

Subsided Blocks

The presence of detritus in some folded chains can only be explained if the previous existence is assumed of areas of denudation which have since foundered deeply below sea-level. The most important submerged areas in maps 1-4 appear east of the Appalachians ("Appalachia"), west of the Sierra Nevada ("Cascadia"), west of Spitsbergen and Scotland (the "North Atlantic Continent" or "Scandia"), and, finally, south-east of Asia and east of Australia ("Melanesia"). These will now be treated accordingly.

Boesch is of the opinion that the coastal area of Nova Scotia represents the only visible portion of the hinterland of the Appalachian geosyncline which is commonly known as Appalachia or Nova Scotica. It is impossible to estimate the size of this hinterland, but there certainly must have existed a mountainous area large enough to supply the Appalachian geosyncline with sediments.

For not only was the foreland (consisting of the Laurentian Plateau, an extension of the Canadian Shield) covered by the sea, but Barrell was able to reconstruct an enormous Devonian delta in the Appalachians, and its structure shows that all the material came from the east. The very valuable seismic observations of Ewing and his collaborators reveal that the eastern part of Appalachia now lies buried from 3,000 to 4,000 meters below the shelf. This area would seem to have foundered more as a result of warping than of faulting.

An important part of the shelf-sediments probably consist of Triassic and Jurassic strata. The region of the Appalachian Mountains can therefore probably be said to have remained above sea-level during these periods, and to have supplied the eastern area, which was then subsiding, with sediments.

These deposits, together with Cretaceous and Tertiary strata, built up a considerable part of the shelf. If "Appalachia" can be assumed to have extended some 200 miles beyond the present coast (a conclusion arrived at both by Barrell and Schuchert), thick Paleozoic sediments — the erosion products of Appalachia — ought at any rate to lie burried in the Atlantic east of the shelf.

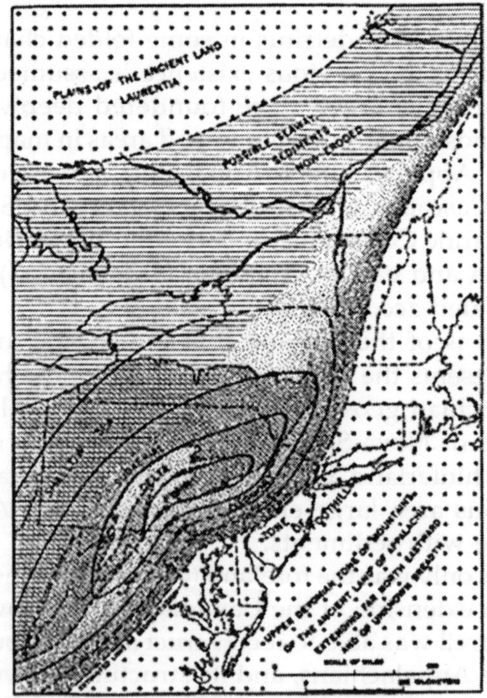

Fig. Reconstruction of a Devonian Delta.

Some geologists believe that a land-area south of the Ouachitas supplied the homonymous geosyncline with detritus in the same way as Appalachia had formerly supplied the Appalachian geosyncline with erosion products. This borderland was called Llanoria.

Van Waterschoot van der Gracht regards the Ouachitas as the mere frontal ranges of a far more important and larger system, most of which would at present lie beneath the Gulf Coast basin.

In the opinion of Stille a „borderland" generally has to be considered as part of a geosyncline that was folded during a previous epoch of compression and subsequently became a geanticlinal ridge.

The geosyncline and folded chains of the Sierra Nevada should not be regarded as extending along the edge of the Canadian Shield, as these for- mations verge upon a borderland known as Cascadia. The existence of a former area in the west was also deduced by Brock from the facies of Paleozoic and Mesozoic sediments in Canada. That such a mountainous area

The contours show original thickness of sediments. The diagonal lines indicate the area where the deposits still occur. The small dots represent fresh-water and brackish-water deposits.

The horizontal lines marine strata; and the large dots areas of denudation had existed in the west during the Mesozoic was a logical consequence, according to Born, of observations in this area, for the Mesozoic sediments of the geosyncline from which the Coast Ranges and the.

Sierra Nevada subsequently emerged, could not possibly have come from the east, since a wide zone, extending as far as — and including — the Rocky Mountains lay beneath the sea at that time. It seems highly improbable that the geosyncline of the Sierra Nevada would have been supplied with sediments from an area as far distant as the Canadian Shield, especially since the sediments had to be transported over a submarine ridge between the Nevadian and the Rockies' geosynclines.

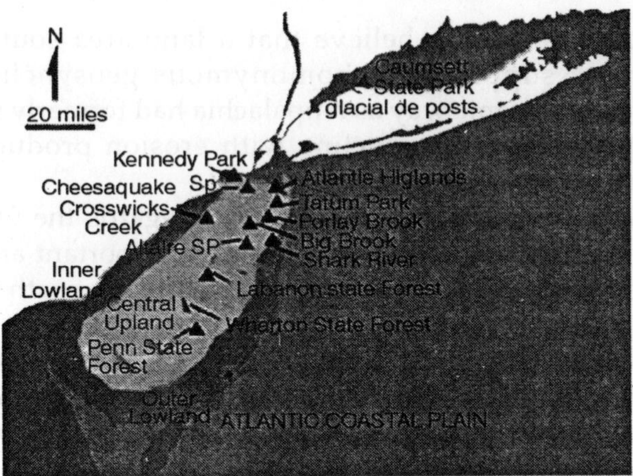

Fig. Atlantic Coastal Plain

Holtedahl pointed out that the Devonian and Tertiary sediments of Spitsbergen originated in an area further west, known as Scandia. The latter must have vanished into the Atlantic at a time when the Tertiary strata of Spitsbergen were being folded.

The Mesozoic strata provide another. indication of the presence of an area of denudation in the west, according to Frebold, and both authors agree that this land cannot be identified as East Greenland, nor that it could have been connected with the same).

The only remaining part of Scandia, or the North Atlantic Continent, as it is sometimes called, may perpaps be the Pre-Cambrian area of the Hebrides and North-West Scotland, against which the Scottish Caledonides were overthrust. But the existence of a PreCambrian block west of Spitsbergen and extending as far as Scotland need not imply that it occupied the whole North Atlantic).

That a former mountainous area existed along the West Coast of Africa can be deduced from the distribution of the sediments in the Congo basin, which grow steadily coarser towards the west).

The Melanesian Islands represent mere fragments of an extensive land-area. The presence of Mesozoic sediments in New Caledonia can only mean that important areas of erosion, all of which have since disappeared, were present in the immediate neighbourhood. Besides, it might be asked whether it would be possible to indicate at present some area of denudation which could formerly have supplied the Tertiary sediments of Viti Levu with erosion products.

The same applies to all those cases in which large quantities of pre-Tertiary and Tertiary detritus are found on small islands. We find it difficult to picture the formation of these rocks under the prevailing distribution of land and sea.

These facts, including other data, (such as petrographical) have lead to the wide-spread assumption that the above area should be regarded as a continent, or, in other words, as a sialic block, large parts of which have subsided to a considerable depth. Geologists often refer to it as the Melanesian Continent.

Marine relief of this block is extremely irregular at present. Many islands and submarine ridges show a linear or arcuate arrangement. Several authors believe that the preceding characteristics, including the long deep-sea troughs, prove that important crustal folding has also affected the submerged part of Melanesia.

The deepest deep-sea trough (with a depth of more than 10,000 meters east of the Philippines) is found in this area, though even in this case "with islands containing rocks of definitely continental types such as granite, syenite, serpentine and schists on both sides of it".

Faulting probably constituted another important feature of the foundering, or submersion, of large parts of the Melanesian Continent. Many geologists in distant parts of this region arrived independently at a similar conclusion). Chubb) describes the region as "an area of continental land, that was gradually folded, fractured and submerged in Mesozoic and Tertiary times".

The exact period of the "foundering" has not been ascertained in all cases but there are indications that in one area — in the south-eastern marginal zone — important faulting occurred during the Upper Tertiary, while in Viti Levu, to mention but one example, major movements were observed along the faults towards the close of the Tertiary, and these are thought to have accompanied the foundering of large parts of Melanesia).

The south-eastern boundary of Melanesia consists of a linear belt of volcanoes, and it is highly probable that the accumulation of these magmatic products is closely associated with a deep fault plane. This belt can be followed from New-Zealand to the neighborhood of Samoa, and constitutes an important line of demarcation of volcanic rocks, known as the andesiteline. The subsided continent, as compared to the remaining Pacific sector east of Melanesia, is now known as an andesite zone from a petrographical point.

It is difficult to say what influence was responsible for the subsidence of the blocks, but in a few areas the periods in which subsidence began, indicate that it can probably be associated with events in neighbouring geosynclinal zones. Appalachia began in fact to subside during the Triassic, i.e. after the mountain-building of the neighbouring Appalachian geosyncline.

Cascadia subsided after the Upper Jurassic folding of the Sierra Nevada. The submersion of Melanesia seems to be related to the Laramide and younger periods of disturbance in south-eastern Asia. Scandia probably subsided during the Upper Tertiary, and contemporaneous folding (including faulting accompanied by volcanism) is known to occur in Spitsbergen.

BASINS AND TROUGHS

Folded mountain-chains — those majestic features of the earth — have always attracted both the tourist and the geologist. There is another feature, however, which, though less stimulating and even monotonous, can be said to be of primary importance in the earth's aspect. We mean such formations as troughs and basin-shaped depressions, containing more or less folded strata or even sediments which have remained almost undisturbed.

As soon as we begin to examine these structures more closely, it transpires that they present heterogeneous characteristics, revealing different shapes, sizes, structures, depths, tectonic frames, periods of formation, and a varying distribution of facies and duration of development.

In studying the earth's structural history, special attention will have to be paid to the location and time of origin of the basins, without losing sight, however, of those other diverse characteristics which have just been mentioned.

The shape of these formations ranges from long troughs and elliptical depressions to practically circular basins. The thickness of the accumulated sediments varies between a few hundred meters and several thousands in often uniform shallow facies. Some basins are consequently not unlike geosynclines both as regards their form and the distribution of the facies of the sediments.

It is in fact difficult to draw a sharp line between basins, troughs and geosynclines, or to formulate some definition which would distinguish clearly between them as separate physiographic units.

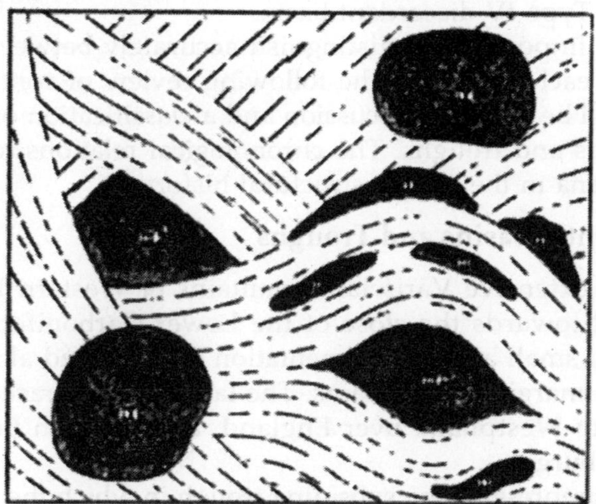

Fig. Schematic Review of Four Types of Basins.

The extremes are different enough; but the intermediate types cause difficulties. Thus we find one author calling a certain type of formation a geosyncline, while another refers to it as a basin). Basins and troughs are often bounded by faults and overthrusts, and this circumstance makes it difficult to distinguish even between a trough and a graben. To make things still more complicated, names such as "stable and unstable shelf" have since been introduced!

We do not propose to dwell upon all these different names, but will describe the features of each type of basin as it comes up for discussion and will then choose the name which seems to fit it best.

A study of continental basins and troughs has induced me to classify them into two groups, each group being subdivided as follows into two types

Group 1.

Type I marginal deep.

Type II intramontane trough.

Group 2.

Type III nuclear basin.

- With an isochronous frame.
- With an anisochronous frame.

Type IV discordant basin.

It is impossible to distinguish accurately between both types in each group. To the following review of continental types will be added a discussion and a classification of deep-sea basins and troughs. The chronological relations to other phenomena in the earth's structural history.

Continental Basins and Troughs

The extensive Variscian mountains of Western Europe emerged towards the close of the Lower Carboniferous. A relatively small area of sedimentation later formed along the northern margin of the chains. The latter can be traced from Ireland to Westphalia over England, and through Belgium and Limburg.

This represents the subsiding trough in which the paralic and coal-bearing Upper Carboniferous was deposited, accumulating to a thickness of approximately 4,000 meters in Westphalia, and to as much as 7,000 meters in Upper Silesia. A considerable amount of detritus from the Variscian mountains gathered in this trough, which constituted a marginal deep of the European Variscides.

Marine Devonian and Lower Carboniferous had already been deposited in this area previously, but the creation of the marginal deep only dates from Upper Carboniferous times. Its contents were then folded towards the close of the Upper Carboniferous, and during this same epoch (which was probably the Asturian), blocks of the older Variscian ranges in the south were overthrust towards the north as far as, and over, th sediments of the fore-deep.

We will not enter into details at this point, nor will we describe the important role which the Brabant Massif played during these events. The fore-deep should not be imagined to be a deepsea basin in this case. On the contrary. The enormous supply of detritus from the south kept pace with the

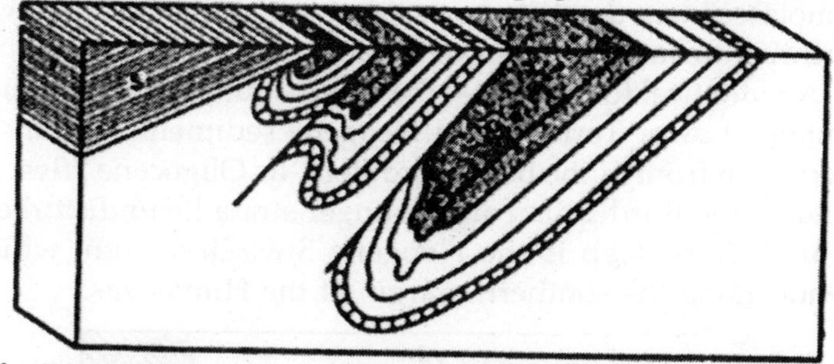

Fig. Diagrammatic Block of the Folded and Overthrust Marginal Trough of the European Variscides. S. Lower Paleozoic.

subsidence, producing a swampy zone which was intermittently flooded by a shallow sea. It was not, therefore, a deep in the morphological sense of the word, but in the structural or tectonic sense, as Stille formulated it. The mechanism of the subsidence and sedimentation of this Upper Carboniferous "mio-geosyncline" was the object of a special study by Pruvost.

While discussing epochs of compression, the principal phase in the Appalachians was shown to be the Bretonic (Acadian). It involved the eastern and most intensely folded and metamorphosed zone with its manifold intrusions of large batholiths. A marginal trough was formed along the western edge of the newly elevated mountain range following this Acadian revolution.

The marginal deep, which was filled with Carboniferous and Permian strata, was folded by the so-called Appalachian (Saalian) phase during the Permian. This is the classical Appalachian geosyncline of former authors, for which Schuchert introduced the name "monogeosyncline".

As a tectonic unit, it formed the counterpart of the marginal deep of the European Variscides mentioned above). These two examples refer to Variscian marginal deeps. We will now give a brief summary of marginal deeps of the Alpine belt.

A Miocene trough was formed along the northern front of the Alpine chains after the intensive folding of the Oligocene. A sequence of approximately 2,500 to 3,000 meters of "molasse" was deposited into this trough, and the latter was overthrust towards the end of the Miocene.

An identical formation — the Guadalquivir trough, which countains Lower Tertiary and Neogene sediments — can be observed in front of the Betic Cordillera. Its Oligocene "flysch" is folded; the Burdigalian and younger strata lie undisturbed. An analogous deep is the Neogene Siwalik trough, which extends along the southern margin of the Himalayas.

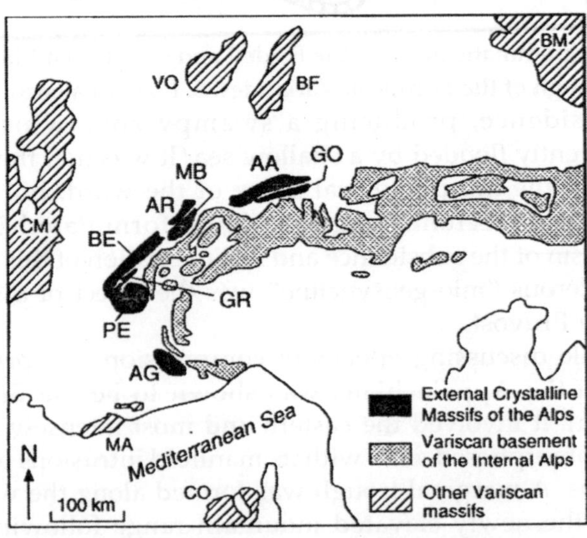

Fig. Upper Paleozoic Variscides of Europe.

On the northern side of the Alpine chains lies the depression of Karakum. The drainage basins of the Tigris and Euphrates and the Persian Gulf form the marginal deep of the Zagros and Iran chains). Recapitulating, we may say that a marginal deep can be characterized as a subsiding trough of sedimentation, which originates along the edge of a folded chain — or eu-geosyncline) — shortly after folding and mountainbuilding occurred in this zone.

This trough faces the continental Shield in the European, North American and Asiatic Variscides, and the same applies to the Guadalquivir, Molasse and Karakum troughs of the Alpine ranges. The Mesopotamian and Siwalik troughs, however, are found on the other side of the Alpine belt, and an identical situation can be observed in the Asiatic Caledonides.

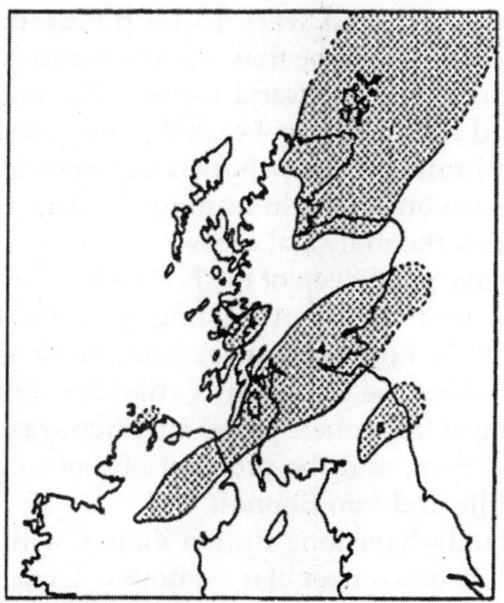

Fig. Intramontane Old Red Basins of Great Britain. 1. Lake Orcadie; 2. Lake Lorne; 3. Lake Fanad; 4. Lake Caledonia; 5. Lake Cheviot.

We propose to give the formations a neutral name, for they can be called "foredeep" in one case, and "hinterdeep" in another, while at times both terms may apply — it all depends on the way an investigator looks at it. We shall therefore speak of them as "marginal deeps".

Stille, too, refers to them as "Saumtiefe".

As stated above, the Appalachians form an example of what Schuchert called a "monogeosyncline". We want te restrict the application of the term "marginal deep", however, to the youngest and final marginal trough in the history of a given mountain-belt.

One notable feature is that this marginal deep does in fact originate along the margin of the older folded chain, whereas the migrating geosynclines of the earlier history of the belt are formed in most cases partly in the area of the preceding zone of folding and sedimentation.

Another notable feature is that the marginal deep is in many cases situated on the side facing the continental Shield, thereby opposing the general tendency of centrifugal development.

If a marginal deep were to be pictured without its contents, its shape would be that of a small and typical trough with a depth of a few thousand meters. The marginal deeps of Kusnezk and Karakum may be cited as exceptions. The latter have a more limited, basin-shaped appearance. This may merely be a secondary phenomenon, due to extensive overthrusting of the marginal areas.

Thus the marginal deep of the European Variscides is also entirely "obscured" by overthrusting from the south in the western part of the north of France). Troughs of sedimentation also formed within the European Variscides shortly after the Sudetic folding of this zone. Special attention was paid to these formations by Born, and the problem of their oigin was later studied by Stille and von Bubnoff.

These troughs have long-drawn shapes, with more or less elliptical and at times irregular contours. Their longest axes extend parallel with the trend of the Variscides, and ridge-shaped "geanticlinal" elevations separate them.

The contents of these furrows, consisting of terrigenous strata, which are in turn composed for the most part of detritus from the geanticlinal zones of the Variscides, kept up with the gradual subsidence of the bottom, and became very thick in some cases, accumulating to as much as 6,000 meters in the trough of the Saar, 2,500 meters in that of the Saale, and 5,500 meters in the Mid-Sudetic trough.

Still, precise observations have shown that the pace of this subsidence was not always the same, that the troughs were steadily enlarged in the direction of their longest axis, that they also broadened, and that there was a distinct relation between volcanic phenomena, a hiatus in the sedimentation, and warping in the strata, especially between the middle and upper part of the Lower Permian.

The beginning of subsidence varies between the lower part of the Upper Carboniferous and the upper part of the Lower Permian. Another striking feature is that many troughs (e.g. the Saar trough) originated on the boundary of a crystalline and sedimentary area.

Thus in several cases these furrows may in a certain sense be said to represent marginal deeps, and it is in fact impossible to distinguish clearly between Type I (marginal deep) and Type II (intramontane trough).

The intramontane troughs of the Caledonides of Scotland and Ireland, which were also formed in the direction of the trend of the mountains, were filled with "Old Red". These troughs are known to us as "Lakes" Orcadia, Caledonia Cheviot, Fanad and Lorne.

There are many examples of intramontane troughs in Alpine zones of folding. For instance, the Maracaibo, Orinoco and Magdalena troughs in South America, which should be regarded as such, and the Pugget trough, Ventura basin and California Valley in North America. Other examples of intramontane troughs are the "ovas" in Asia Minor.

Mention should also be made of those basins in the East Indian Archipelago and Burma to which I formerly referred to as idiogeosynclines. In Burma, Sumatra and Java these formations are bounded on one side by a pre-Tertiary area and on the other partly by a Miocene zone of folding.

A sharp division between marginal deeps and intramontane troughs cannot mbe based on the facies of the sediments, nor can the intensity of movements contribute anything towards such a division. For the fact is that the facies depends on the distance between the level of sedimentation and the corresponding level of the sea.

The idiogeosynclines of the East Indies, unlike the intramontane troughs of the European Variscides, provide examples of troughs with a chiefly marine and paralic sedimentation.

The frequent appearance of eruptive rocks in the intramontane troughs of Germany, and the complete, or almost complete absence of these rocks in a typical marginal deep would seem to constitute an important characteristic, but should in no case be generalized into a law, for volcanic products (viz. diabase, melaphyr and basalt) do in fact occur in the Kusnezk basin, a marginal deep of the Asiatic Caledonides.

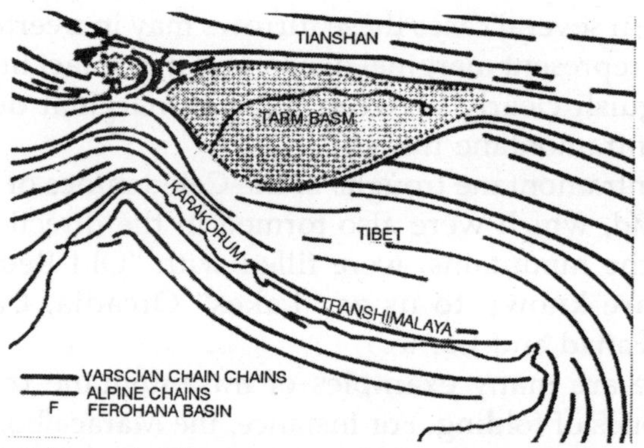

Fig. Schematic Illustration of the Ferghana and Tarim Basins and the Surrounding Mountain-Chains.

The following may be said as regards the third characteristic. The contents of the intramontane troughs of the European Variscides are hardly folded in comparison with the marginal deep of that zone but we again find examples of a moderate folding of intramontane troughs in the East Indian idiogeosynclines.

We still have to consider the location of the troughs. The name intramontane in itself would seem to distinguish this type of trough from a marginal deep, but we saw that it is in some cases difficult to distinguish clearly between the two.

One more type of basin can be found in Europe, viz. the

Hungarian, or Pannonian basin. Its basement was very problaby folded during a Variscian epoch, and the basin more than likely formed a rigid nucleus around which the Carpathian Mountains grouped themselves in the north and east, and the Dinarides in the south and west.

The Pannonian Massif itself, however, subsided, forming a basin which was filled with Lower Miocene up to Pleistocene strata, and these were subjected to folding that adapted itself to the strike of the Alps in the west, and that of the Carpathians in the east.

The Pre-Cambrian nuclei of Ferghana, Tarim and Ordos played an important part during the Variscian folding in Asia. The chains were pressed against, and arranged around the nuclei. However, the massifs later showed a tendency to subside, and became basins.

Sedimentation in the basin of Tarim began during the Permian or Upper Carboniferous and continues up to this day. The neritic, but mainly the terrestrial deposits, are probably exceptionally thick, and show several unconformities and folding of a not very pronounced character.

Therefore, the most salient characteristic of type III is that it originates on a subsiding block which, situated in a zone of folding, formerly acted as a rigid nucleus and thus produced the fore-mentioned effect on the trend of the folded chains that frame it. This is why I called them "nuclear basins".

The situation of the large salt basin of Texas is a different one in some respects, for it subsided on a Pre-Cambrian basement, as the Llano Burnett uplift in the south-east, and the Colorado Plateau in the north-west show. Its frame consits of the Variscian Marathon-Ouachita chains towards the south and east, and the older Amarillo-Wichita zone in the north.

The former western confine of the basin is unknown, but it is now occupied by the Tertiary ranges of the Rocky Mountains. This salt basin (containing Permian deposits up to as much as 4,000 meters) may in a certain sense be compared to the Pannonian basin of Europe, but the frame of the latter, unlike that of the Texas salt basin, is isochronous, as it were (the chains originated during the same period of folding).

The submarine BarentSea Massif also has an anisochronous frame The thickness of the deposited sediments remains unknown, but Frebold pointed out that numerous Mesozoic transgressions have at any rate invaded this area.

The three types discussed so far are clearly connected with the mountain chains alongside of which, or within which the basins are situated, but the same is not immediately evident in the large group of remaining basins. We can begin by stating that the most conspicous feature is that their boundaries intersect existing structures. They originate "right across everything", so to speak, and we will therefore call them discordant basins.

Table I gives a summary of the numerous basins which belong to this group. They will be found under the heading "Type IV", and their situation is shown by Plate 6. We will discuss only one example here — viz. the basin of Paris.. A glance at the geological map of France shows that the basin of Paris originated in Liassic times.

The deepest part gradually moved from the south-east to the northwest, lying near Paris during the Oligocene, and near the Channel coast during the Miocene. The subsidence represents approximately 1,200 meters in the northern sector of the basin, while the Pre-Cambrian basement — on which rests the Lias — descends to a depth of 1,532 meters in the "Pays de Bray".

Its continuation is found in Hampshire and Wight on the other side of the Channel, and here the sediments are even considerably thicker (the boring at Portsdown reached the Upper Triassic strata at a depth of 1,993 meters).

The anticlines striking west-north-west — south-south-east were formed during the Upper Cimmerian epoch. The Subhercynian phase is indicated by but faint movements, the Laramide revolution by faulting and the Pyrenean epoch by "epirogenetic" movements.

Finally, a rather faint folding occurred during the Attic phase. It would be wrong to think of one, large basin-shaped depression. Not only did the deepest part move westnorth-west, but the bowl-shaped habitus of the basin is the result of certain transgressions and of subsequent erosion, and does not seem to reflect the morphology of the basement.

Lemoine drew attention to the fact that the Liassic strata of a trough extending from Lorraine in the direction of the "Pays de Bray" are thicker than those found outside this area, and remarked that this feature also manifests itself in the foundation of the Mesozoic upper structure.

An examination of the manifold discordant basins shows that these formations originated in heterogenous basements at different periods, and that their boundaries distinctly intersect those of pre-existing structures. Yet it can be pointed out in a few cases that the original shape and boundaries of a basin are related to specific positive structural elements of their foundation.

The original shape of the Liassic trough of the basin of Paris, for instance, the migration of its deepest part, and the strike of the anticlines, clearly reveal the posthumus activity of its Variscian foundation. Similarly, the north-south boundary between the North-Sea and the WestGerman Basin, or so-called axis of Erkelenz, and the Western boundary of the North-Sea Basin — the Pennine axis, in which positive movements occurred during the Saalian epoch — probably represent very ancient trends in the structure of the basement complex.

The entire arrangement of positive and negative elements in North America is apparently also the reflection of the Pre-Cambrian structure of the Canadian Shield. Africa may be cited as another example. A north-east — south-west and a north-west — south-east strike (referred to respectively as the Somali and Erythrean trends) is inherent to the Pre-Cambrian structure of this continent, which repeatedly manifests itself in one form or another, revealing itself also in the basins, their arrangement and boundaries.

These facts show that the original location of a discordant basin if examined more closely, will probably reveal a certain "concordance" with some of the structural elements of its surroundings, and the devision of basins into nuclear and discordant basins consequently apears to be not so much a fundamental one, as one of degree.

DEEP-SE A BASINS

In their attempt to classify geosynclines, troughs and basins, some authors have excluded the deep-sea troughs and basins. To the present author such a procedure seems unreasonable.

For whether a subsiding area appears at the surface as a depression or not depends merely on the relation between subsidence of the bottom and supply of detritus derived from the adjoining areas. And this relation may even vary during the evolution of a trough.

The East Indian idiogeosynclines furnish a clear illustration. The subsidence of these intramontane troughs begins in continental regions. Thus e.g. the geosynclinal series begins with fluviatile-terrestrial sedimentation in the Barito basin of S.E. Borneo and in S. Celebes.

In other places the first deposits consist of neritic sediments of a Miocene transgressive and epicontinental sea, which extended far beyond the area of subsidence, as e.g. in Sumatra. After some time the subsidence surpassed the rate of sedimentation.

Hence the strata represent a deeper hemipelagic facies and accordingly the morphology of the area was a submarine trough. Still later, in the Pliocene, a much shallower facies developed, ending in the area becoming land. Obviously the subsidence had become slower and was surpassed by the rate of sedimentation. The submarine trough had become a swampy land surface.

These include the Persian Gulf (an example of a marginal trough), the Gulf of Martaban and Madura Straits (representing the intramontane type). The Barent-Sea and Kara-Sea stand as examples of nuclear basins, and the Black-Sea and Caspian-Sea have been cited as discordant basins.

All these cases, with the exception of the Caspian and Black-Sea, deal with shallow seas, and as most of the continental basins and troughs were filled with neritic, paralic and terrestrial sediments, a comparison between these formations and the above submarine furrows is obviously warranted.

If deep-sea basins are to be classified according to our four types, their characteristics would have to comply with the following conditions.

To begin with, they would have to correspond with the features of continental basins, both as regards their location and the time of their formation. And, in the second place, the morphology and the surroundings would have to satisfy the same conditions. Finally, the facies of the deposits in these depressions would have their counterpart in fossil basins on the continents.

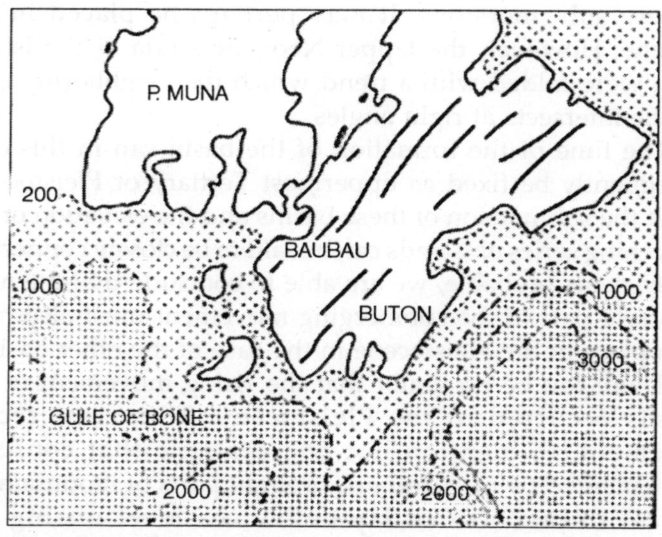

Fig. Upper Tertiary Anticlines in the Island Buton, East Indies.

The submarine relief of the East Indian Archipelago, which has become so well-known since the Snellius Expedition and Kuenen's excellent morphological analysis of this area, would seem a suitable subject for the discussion of submarine basins.

The present deep-sea relief of the East Indies must have been formed in the recent geological past. I treated this question at greater length in *"The Geological History of the East Indies"*, as well as in a later paper of the year 1938, and will therefore confine myself here to a few of its salient features. I want to mention in the first place that the trend of the folded

Miocene strata is intersected at an angle by the present boundaries of the deep basins in several places, from which it follows that the adjacent submarine relief must have originated at least after that Miocene folding.

Molengraaff explained, for example, that the Amanuban mountain-chain, in Timor, is intersected at an angle of 12° by the coast. The view that the origin of the deep-sea basins, and the elevation of the series of islands between them, probably occurred simultaneously, is now generally accepted.

It is difficult to determine the exact time of the beginning of these sub- mersions. It may perhaps be placed in the Pliocene. However, the Upper Neogene strata of the island Buton were folded with a trend which the neighboring Gulf of Bone intersects at right angles.

The time of the formation of the basin can in this case consequently be fixed as uppermost Tertiary or Pleistocene. And if the submersion of these basins and the elevation of the intervening series of islands can be said to be closely connected, which seems probable, we are able to conclude that the most important part of the submerging movement must also, have taken place in the Pleistocene in the case of the other basins.

There can be no doubt that the rising movement of the islands occurred in the Pleistocene. This most recent movement, to mention but one example that could be supplemented by many others, brought parts of mountains in Central Ceram (which lay below sea-level during the sedimentation of the marine.

Fig. Raised Reef-Terraces of the Island Kissar

Pliocene strata in the Masiwat-Bobot graben) to a height of at least 3,000 meters above the sea. The recently elevated reef limestones and terraces give a particularly good idea of the amount and intermittent character of these recent movements. The actual Pleistocene age of the elevated reef terraces has been verified in some cases.

Kuenen arrived at a similar conclusion in his geomorphological analysis of the bottom relief, and he exhaustively argues that the deep-sea basins originated recently as a result of the subsidence of "continental" (sialic) areas.

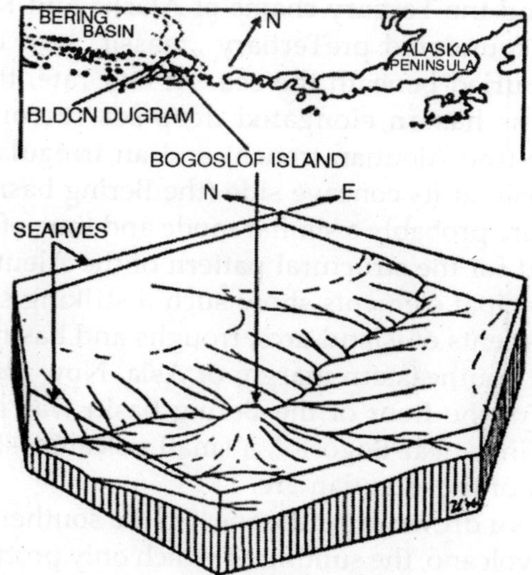

Fig. Isometric Block-Diagram of the Submarine.

In the case of the East Indies we are dealing with a double system of island-arcs concave towards the Asiatic continent. On their convex external side the islands are bounded by a nearly continuous series of long and comparativery narrow deep-sea furrows. However, much broader irregularly shaped basins with relatively steep sides and flat, horizontal bottoms; are situated at the internal side of the island-arcs. It is not the place here to go into the problem of the origin of such a remarkable pattern of islandarcs and accompanying deep-sea basins. For the moment we will only point to the well known fact that a more or less similar arrangement of morphologic and structural elements characterizes the Western and

Northern margins of the Pacific Basin proper. The same view as regards the origin of the basins and throughs along the Eastern margin of Asia has also been held by Born, Leuchs, Lawson and others. Yabe pointed out that the entire region of Taiwan, the Riu-Kiu; Japanese and Kurile islands lying within the isobath of 720 meters was a land area which was connected with the Asiatic continent until Pleistocene or still more recent times.

Leuchs and Born both assume that the arc of the Aleutians and the Tertiary chains of Alaska and Kamschatka surround a foundered preTertiary „massif" part of which is at present buried beneath the sea. At any rate, the chain of the Aleutians has an elongated deep-sea furrow along its convex side (the Aleutian trench) and an irregularly shaped deep-sea basin at its concave side (the Bering basin).

Therefore probably a similar mode and time of origin may be supposed for the structural pattern of the Aleutians which in their principal elements show such a striking similarity to the arrangements of island-arcs, troughs and basins along the eastern and southeastern margin of Asia. Now the subsiding movement of the floor of the Bering basin was revealed by echo-soundings near Bogoslof, a small volcanic islands in the central part of the Aleutian arc.

It shows a drowned landscape. On the southern "plateau" is Bogoslof volcano, the summit of which only protrudes above the present level of the sea as a little volcanic ruin. The submarine stream pattern seems to be adapted to the presence of the volcano and therefore it appears to be of subaerial origin. Smith thinks that the submarine part of the volcano was also formed under subaerial conditions.

If his interpretation is correct, it may be inferred at once from the isobaths that the region around Bogoslof was subjected to a submergence of at least 1300 fathoms, approximately. Bogoslof is situated on a sea-bottom which forms the southern slope of the deep Bering basin at the concave side of the arc.

Hence it would not be astonishing that Bogoslof and the surrounding sea-floor should have moved downward through a comparatively large distance since Pleistocene times. Neither is there anything astonishing in the revealing of a deeply drowned pattern of subaerial erosion on the sea-floor around Bogoslof.

Kuenen distinguishes between five groups of basins and troughs in the East Indies on morphological grounds. We will discuss four of them, though purposely in a different order. Kuenen's fifth group consists of grabens and other formations, to which we will not here give our attention. The dubious cases, indicated with a special notation. Kuenen's fourth group consists of the long furrows on the external side of the zone of Miocene folding, i.e. the trough west of Nias and the Mentawei-Islands, the trough south of Java (which has an occasional depth of 7,000 meters), and the Timor trough.

These troughs concur with marginal deeps of the Miocene zone of folding as regards the time of their formation, their "tectonic" location and general morphological features (Plate 8). It will be pointed out, that genetically two different types of marginal deeps have to be distinguished. Those accompanying a double arc originated in a fundamentally different way from a marginal deep in front of a single arc.

His third group includes not only considerably shallower formations such as e.g. the Mentawei trough, the trough along the south coast of Java, the Sawu-Sea and the Wetar deep, but also the Weber deep, which has an exceptional depth. They differ morphologically from the preceding group in that they present a less oblong and therefore a more basin-shaped appearance.

As regards the location of this group, the Mentawei trough and Sawu-Sea have both subsided in the Miocene zone of folding , and as the submarine ridge south of Java can be said to belong in a morphological and gravimetrical respect to the Miocene belt of folding extending from Mentawei towards Timor the same may probably be assumed of the basin situated immediately south of Java.

This conclusion also holds good for the Wetar and Weber deeps, since these formations are bounded on one side by the Miocene zone which can be followed from Timor via the

Tanimber and Kei Islands towards Ceram and on the other by the submarine continuation of the Miocene zone of the Lesser Sunda Islands. These troughs may consequently be said to belong to the intramontane type as regards their situation, morphology and formation.

It is doubtful whether the Flores Sea ought to be regarded as a marginal deep or as an intramontane trough. Kuenen included it in his fourth group. If this interpretation be correct, we would find a marginal deep on either side of the Miocene zone of folding, i.e. the Java-Timor trough on its southern side, and the Flores trough north of it. The Siwalik and Karakum troughs are placed similarly in respect to the Himalayas.

Fig. Types of Submarine Basins and Troughs in the East

Kuenen's first group is composed of basins with relatively steep sides and flat, horizontal bottoms (the Banda basins, the Celebes-Sea and the SuluSea). Dr. Fr. Weber pointed out that the southern Banda-Sea, or at least the greater part of it, had lain above sea-level during the Mesozoic.

The folded chains group themselves around the actual Banda-Sea, and the latter thus closely resembles such nuclear basins as the Pannonian, Tarim, Ordos and Kara-Sea basins as far as the time of formation and the morphological and tectonic characteristics are concerned.

The Macassar, Bone and Tomini basins make up Kuenen's

second group. These formations have a somewhat shallower flat-bottomed cross-section, and a long and irregular shape. The Upper Tertiary chains are intersected by the steep coast of the Macassar and Bone basins.

It seems most remarkable, therefore, that Kuenen's division of deep-sea furrows into four groups should correspond with our own classification of four types of continental basins. We discussed Kuenen's types in a different order because we wished to emphasize the similarity between the results. One more question remains to be examined, namely whether basins or troughs can be found on the continents, which — like the present EastIndian deep-sea basins — were filled with sediments of bathyal and even abyssal facies. This question can be answered in the affirmative, for the Indian Archipelago itself contains examples of such fossil troughs, e.g. on Timor.

One or more troughs with bathyal and abyssal sediments existed in Timor during the Mesozoic. A paleogeographic reconstruction of these troughs is not possible at the moment, for too little is known of the complicated structure of this island, but the distribution of the facies types leaves no doubt that deep-sea deposits originated in the vicinity of areas of neritic sedimentation. Thus, too, we know that troughs containing bathyal and perhaps even abyssal sediments existed in the complicated geosyncline of the Alps during Mesozoic times. The cherty to siliceous limestone and black shales of the Ouachitas and the Marathon Mountains may possibly be indicative of pelagic conditions and sedimentation in deep water. And, lastly, mention must be made of the Caspian basin a large portion of which was filled during Maastrichtian times by coccolith and globigerina ooze, such as now accumulates at a depth of from 2,000 to 3,000 meters according to the latest oceanographic data.

We have enumerated a few examples of fossil deep-sea troughs and basins, and most of the present deep-sea basins of the East Indies may consequently safely be classified according to our system. The morphological aspect of an area of subsidence merely depends on the existing relation between the speed of the subsidence and the quantity of detritus supplied. The idiogeosynclines in East-Sumatra and North-Java were filled with thick deposits because of the huge amount of detritus supplied by the extensive area of Sundaland, which lay above sea-level at that time.

The position of the Mentawei trough is a less favorable one, and this furrow is therefore not filled to a high level. The Weber deep is a very deep trough, as exceptionally few deposits accumulated within it. We mentioned already that rising movements of the surrounding areas probably accompanied the subsidence of the basins. Nor do conditions in the Indian Archipelago differ, in this respect, from those prevailing in the surroundings of the basins on the continents.

Further examples of deep-sea basins will be found in which reference is made to the basins of Eastern Asia and those of the Mediterranean, West-Indies and Southern Antilles. All these formations are situated inside continental folded chains and loops of continental islands, or immediately alongside of them. Deep-sea basins are very numerous at present. It is obvious, however, that a comparison with the past has to restrict itself to the present boundaries of the continents. Leuchs observed quite rightly that many of the marginal deep-sea troughs lie at such a distance from an area of erosi on of any importance (e.g. along the arc-shaped islands in Eastern Asia),

That it is difficult to understand how they could be filled by a thick geosynclinal sequence of strata. Another point which deserves attention is that the abyssal sediments which are at present deposited within these troughs contain little or no calcium carbonate. This lack of lime has only been observed in a few exceptional cases in fossil troughs (e.g. in Timor).

Thus, though but a few examples are known of fossil basins which are similar to our present deep-sea basins, we should not forget that these formations are only met with in exceptional cases on the continents. In addition to this the relief of our continents is at the present time particularly high, and not only did the Alpine chains undergo a "rejuvenation" in the form of a rising movement, but even many older chains partook of this process. Moreover, if we stop to consider that the subsidence of a submarine area in many regions is accompanied by a rising movement of the surrounding land, it will at once be clear that vertical movements became very intense within the crust during the latest part of the earth's history and that it created a strong relief on the continents, and many deep basins below sea-level.

A comparison between the present relief of the floor of the sea and fossil troughs would be misleading, for the continents are the only areas that supply us with data of former periods. No one can tell whether the relief of the floor of the ocean had been more accentuated during certain periods than others.

Deep-sea troughs cannot be said to have occurred very abundantly periodically; nor can the floor of the deep-sea be said to have flattened out in the intervening time. Still, it would be quite as unfounded to affirm that this certainly was not so. This is one of those typical points which are left open to conjecture.

CHRONOLOGICAL RELATIONS WITH OTHER PHENOMENA

The most striking results of an examination of basins and troughs is that the origin and history of these formations are related to certain epochs of folding and mountain-building. This is not surprising in the case of marginal deeps and intramontane troughs, if we consider their situation. The nuclear basins, too, lie inside, or in between certain folded zones, and are surrounded by mountain-chains.

It is obvious, therefore, that these basins must be subjected to the influence of processes deeper down in the earth when the latter occur underneath these particular orogenic belts. A problematical point is what happens in the substratum.

We might be inclined to associate this question with changes and displacement of subcrustal matter, but in doing so we would stray too far into the domain of speculation for the moment. One remarkable feature, however, is that even discordant basins show chronological relations with folded belts, though these basins sometimes originate at a great distance from such zones. The subsiding movement of discordant basins, too, begins after a special epoch of folding, and in addition to this, certain phenomena in the history of these basins, such as uncon formities, faulting, moderate folding and a repeated tendency to subside appear to be chronologically related to specific phases known in folded chains.

One thing and another not only points to a deep internal terrestrial process, but also shows that this process has a world-wide activity. Lastly, the greatest part of the basins (as well as of the troughs) form after certain Variscian epochs. A similar and exceptionally intense formation of basins succeeds certain Alpine phases. Those basins which are connected with Caledonian epochs as far as their origin is concerned, are undoubtedly far less numerous, and only a few can be associated with Lower and Upper Cimmerian phases.

Can it be a mere accident that an identical sequence was arrived at in the case of periods of mountain-building? The given classification of basins can no doubt be improved upon. Some readers will perhaps feel inclined to change certain notations. However, I am of the opinion that nothing can alter the remarkable result that the periods of Variscian and Alpine mountain building, which were of far greater importance than the Caledonian and Cimmerian, corresponded with periods in which a considerably larger number of basins were formed.

It certainly is not a mere accident that the periodicity and intensity of mountain-building is related in point of time to the analogous periodicity and intensity of the formation of basins. Epochs of folding, i.e. periods of increasing pressure in the earth's crust, are succeeded by periods of decreasing compression, which also represent periods of mountain-building. As soon as the intensive compression ceases, the mountains not only begin to rise, but the negative elements, too — the basins — begin to take shape.

As pointed out previously, the shape of some discordant basins is known to be determined by certain structural elements of the basement. Nevertheless so few details are known of these basins that it is in most cases impossible to say what brought about their shape and location.

On the other hand, there can be no doubt that they are the result of the whole structure of the foundation and the surrounding areas, and this is probably why subsequent phases of folding are manifested more strongly in one basin than in another. These factors are likewise responsible for the fact that the tendency of a contemporaneous basin to subside is continued for a greater length of time in one basin than in another. The same applies to geosynclines.

Generally speaking, those basins that are found in Pre-Cambrian Shields are wider than those originating within, or near, a later folded belt. I hope that the above will have succeeded in showing that the origin of basins and troughs is related to periodic processes in the earth's interior, and particularly that periods of decreasing compression in the earth's crust are characterized both by a rising movement of the folded chains, and the beginning of local tendencies to subside, — in other words the beginning of the formation of new geosynclines and basins.

Chapter 23

Crust and Subtratum

Those rocks which can probably be said to build up the earth's crust. This important question will now have to be examined somewhat more closely. It should first be noted, however, that opinions differ considerably as regards the composition of the terrestrial crust though all agree on a few fundamental points.

An imaginary cross-section of a continent would probably present more or less the following aspect. On top we would find a veneer of sedimentary strata, totalling as much as 15 km in the geosynclines, or possibly even more, but thinning out to zero at points where their foundation is exposed.

Beneath this veneer would stretch a zone of acid rocks, consisting mostly of granite and gneiss, and descending to a depth from 15 to 35 km below the cover of sediments. The thickness of the crust in the continental regions is thought to vary between 40 and 70 km. Gutenberg's latest review shows that comparatively smaller figures apply to such regions, for instance, as New-Zealand and the north-eastern part of Japan (30 km).

The largest values found are in the Sierra Nevada, California and the Alps (60-70 km). The lower side of the crust is known to form an important primary discontinuity — "the Mohorovièiæ. discontinuity". There is no one to-day who doubts that the lower part of the continental crust is composed of basic rocks with an occasional thickness of 10-25 km.

The lower boundary of the crystalline crust is within the basic material. Opinions disagree as regards the nature of these basic rocks, which are piezo-gabbro or olivine basalt according to some authors and peridotite according to others. The most important point, however, as far as we are concerned, is that this lower part of the crystalline crust is now universally assumed to consist of basic to ultrabasic rocks, the so-called sima. This chemical composition of the sima differs but slightly from that of the substratum upon which rests the crust.

The lower layer of the crust is indeed regarded as a layer of crystalline sima. Some writers, including Daly, assumed that amorphous sima — the "vitreous substratum", an overheated and very viscous basaltic liquid under high pressure — follows beneath the Mohorovièiæ discontinuity.

To avoid any misunderstanding, I want to emphasize that the terms "sima" and "sial" (which were introduced by Suess) are used in the petrographical sense, and not in Wegener's physical sense, such as rigid and elastic sial and viscous sima.

Some authors are of the opinion that one, or possibly more than one layer, with a thickness of about 25 km, extends between these two parts of the earth's crust.

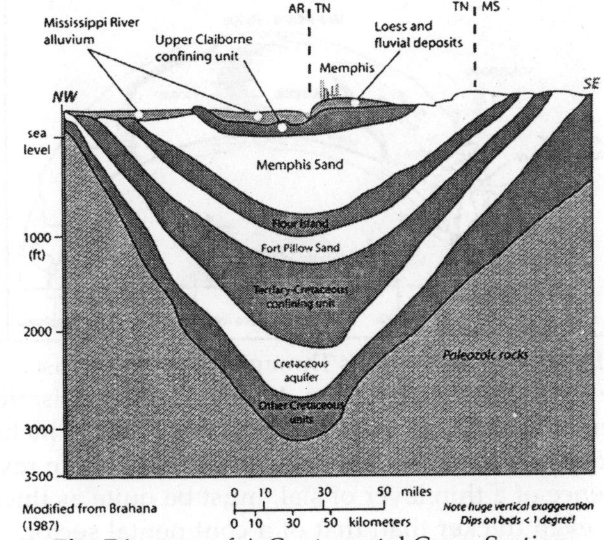

Fig. Diagram of A Contenental Cross-Section.

There are many conflicting views as regards the composition of this intermediate layer, some assuming it to consist of granite or diorite, and others of granodiorite or granulite, and even of amphibolite, olivine-free basalt or gabbro. Figure.contains a few data on petrography, density, pressure, temperature and the velocity of longitudinal waves in the terrestrial crust.

As pointed , the general opinion is that a considerably thinner layer of sial exists under the Atlantic and Indian Ocean and that no such layer is found under the Pacific east of the andesite line.

The nature of the sialic sheet of the Atlantic forms another point of uncertainty. One possibility is that the upper, granitic layer of the continents as well as one or more of the intermediate layers, are thinning out towards the oceans.

Another supposition is that one of these layers is entirely absent from the Atlantic sector. Though the floor of the Pacific differs in volcanologic, petrographic, magnetic and seismic respect from a continent, the same cannot be said of the floors of the Atlantic and Indian Ocean and the surrounding continents.

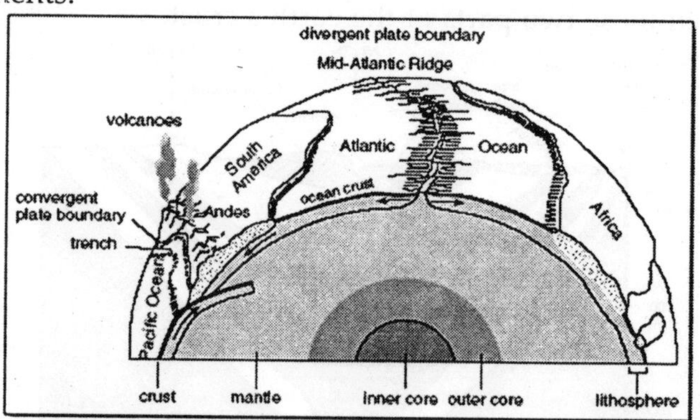

Fig. Schematic Sections Through the Earth's Crust.

The crystalline crust of the Atlantic, where seismic data (together with the evidence found in the granitic inclusions observed in the lavas of some volcanic islands have revealed the presence of a thin layer of sial, must be quite as thick and possibly even thicker than that of a continental sector.

Under the Pacific, where a sialic layer appears to be absent, the rigid crust, consisting entirely of crystalline sima, is thicker than anywhere else. Daly's view is based on the assumption that the thickness of the crust depends upon the production of heat by radioactive minerals, the sial being considered to be more radioactive than the basic rocks of the sima.

The terrestrial crust will consequently be thinner under a sialic continent, when in a state of thermal equilibrium, than under an ocean, where the sial is either thin or non-existent, and the solid crust will have to be even thinner under the additional quantity of sial of the "root" of a folded mountain-chain.

The lower side of the crystalline crust should at the same time be regarded as the upper limit of melting of the sima, since the temperature at this depth is equal to the melting point of the rocks under the prevailing temperature. The results obtained so far indicate that the temperature from radioactive minerals in the crust is insufficient to melt the rocks anywhere in the crust.

"This suggest that extensive melting of the crust is only likely to occur in geosynclinal areas where the crust is thickened, or in some other abnormal circumstances". One of these abnormal circumstances is the formation of a crustal down-buckle or mountain-root.

Rittmann's model of the structure of the earth's crust was based on arguments similar to those of Daly, and their view was lately confirmed by investigations of an entirely different kind by Vening Meinesz.

A study of gravity fields over the Hawaian Archipelago and the Madeira area has shown, amongst other things, that there cannot be any important difference in the crustal rigidity and thickness beneath both these areas , and that the positive anomalies of the isostasy strongly indicate the existence of a rigid crust, acting as an elastic sheet that is bent by the load of the volcanic islands which formed on top of it.

It was claimed by Molengraaff in 1916 that a subsiding movement of volcanic islands in the Pacific would be brought about by slow isostatic downward movements, the subsidence continuing until the islands had reached the level of the simatic ocean-bottom, which he regarded, as Wegener did, as a viscous liquid. Molengraaff thus sought to explain the ultimate formation of thick coral reefs, and the creation, in conformance with Darwin's theory, of a barrier reef, and, finally, an atoll, from a fringing reef by the gradual subsidence of the volcanic foundation.

However, no strong downward movement is apparent in many of the Pacific islands, as has since been shown by observations along their coasts. Sub-recent negative shifts of the shoreline with a height of approximately 6 meters occur in a number of these localities, and the world-wide distribution of this phenomenon proves that the sea-level has fallen over the whole world in comparatively recent times.

The discovery of an equal amount of negative shift on the simatic Pacific islands shows that the subsidence of the latter (i.e. assuming that they were really subsiding) has at any rate ceased. Vening Meinesz' geophysical investigations show that the subsidence of volcanic islands in the Pacific is confined to the down-bending of the simatic floor of the ocean, and that the latter acts as an elastic sheet, and not as a viscous liquid, as Wegener and Molengraaff supposed.

MAGMATIC CLANS

The conception of the earth's crust at which we arrived, above, explains several striking volcanological and petrographical phenomena. Only basic magma can reach the surface where the sialic layer is lacking, and this material — through differentiation — can furnish trachyte to phonolite (the so-called atlantic rocks), but no quartz-bearing acid rocks, and therefore certainly no granite.

If the basic magma of the simatic substratum suddenly invades a continental crust by a so-called abyssolithic injection into faults, the only material to extrude at the surface will again be basic material.

The question therefore arises how the acid rock-tribe comes into being. It has commonly been supposed that a great variety of igneous rocks may originate from one and the same parental magma by processes of normal, i.e. gravitational differentiation. All the volcanic and plutonic rocks have thus been considered as products derived from three types of magmatic melts, called atlantic, pacific and mediterranean respectively. The atlantic clan of consanguinous rocks evolved from basic parental material with the composition of an olivine-basalt.

In the foregoing pages we assumed it to be present in the form of a world-embracing layer below the crust. Becke, who introduced the terms atlantic and pacific provinces, chose these names because examples of the first are known from certain islands in the Atlantic Ocean, whereas more acid and calc-alkaline igneous rocks characterize most of the volcanic belts surrounding the Pacific Ocean.

Without exception, however, the volcanoes within the Pacific Basin proper are composed of atlantic lavas. The name mediterranean clan was introduced by Niggli for a group of igneous rocks in which potash is a highly characteristic chemical constituent, sanidine and leucite being the two most characteristic minerals.

In order to explain the origin of the pacific rocks Kennedy distinguished two basaltic layers in the earth's crust. According to this theory the sialic part of the crust rests on rocks of tholeiitic composition and this layer in turn rests on material with the composition of olivine-basalt.

The tholeiitic material would give rise to pacific differentiation products, whereas the atlantic lavas would derive from a melt of olivine-basaltic material from the lower layer. It would, however, be unreasonable to consider one of the layers to have originated as a differentation-product of the other.

This objection invalidates the hypothesis of Kennedy. A much more plausible theory is advocated by Backlund, van Bemmelen, Fenner, Holmes, Rittmann Walker, Wegmann and many others. According to their opinion only one kind of parental basic magma exists, having an olivine-basaltic composition. The tholeiitic tribe originates from the olivine-basalt magma as a result of its acidification by melting and assimilation of parts of the sialic crust.

Pacific rock suites might originate as primary or juvenile differentiates from a parental magma of gabbroidal composition as was supposed by Bowen. But it seems very probable that acid melts are frequently formed in a very different way.

One suggested process is that the ascending magma becomes altered into an acid melt by a process of melting and assimilation of pre-existing sialic crust-rocks. The new liquid has a composition intermediate between alkalibasalt and "pure" sial. It is called a syntectic magma.

Another process consists of the ionic or gaseous transfer of elements or mineral-components. And it seems highly probable that emanations from the substratum may convert continental crustal-rock into igneous rocks of a fundamentally different composition. Such a process is called *migmatization* and it may ultimately lead to *granitization* of pre-existing rocks of an kinds.

"Those who regard granites as having been largely formed from the preexisting rocks recognize that they passed through a stage when part of the material was mobile or fluid. The partially fluid mash has been styled *migma* by Reinhard, to distinguish it from a mash consisting of incompletely crystallized magma. Moreover, magma may be generated from migma either by the attainment of complete fluidity or, at any stage, by the squeezing out of the fluid portion".

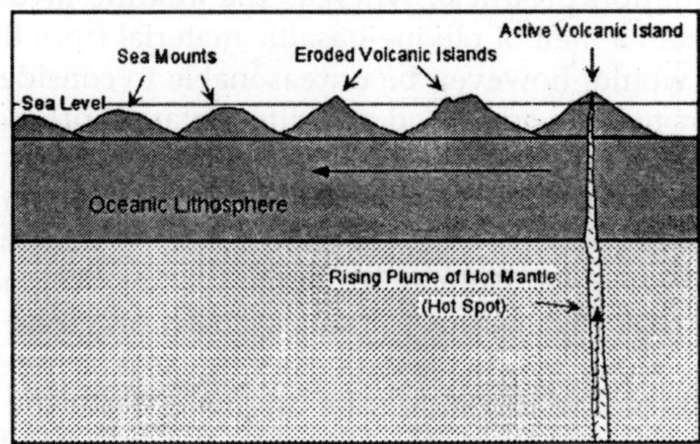

Fig. The Rising Migmatite

By the action on a large scale of one or both of these two processes preexisting plutonic rocks and sedimentary strata such as schists, graywackes, quartzites, gradually become metasomatically altered into granite-bodies of large dimensions. Probably most, if not all, of the granite batholiths are not of primary igneous origin but originated as secondary or palingenetic products of granitization. The crustal rocks are enriched by alkali-silicates and alkali-aluminates supplied from the ascending emanations.

A process of progressive metamorphism — *migmatization* — converts them into calcalkali rocks, the ultimate product being a granite body. Gradually the process migrates upward. Backlund and Wegmann speak of a rising *migmatite front*. Even higher levels of the crust are invaded by the secondary (palingenetic) granite-melt.

Immediately surrounding the granite mass is a zone of ultrametamorphism, in upward direction passing into a zone of pneumatolitic and hydrothermal metamorphism, and still further by an aureole of crustrocks that were influenced by hydrothermal processes only.

A schematic representation iwhich represents Van Bemmelen's interpretation of his observations of plutonic and volcanic phenomena in Sumatra and Java. The distribution of the East Indian magmatic provinces and their relationship to the structural history of this remarkable island-festoon.

According to Holmes the geochemical relationships involved in the granitization of the country rocks can be briefly summarized in the formula: "granite = pre-existing rock plus added material (A) introduced by and abstracted from the incoming emanations (A + x), *minus* displaced material (B) driven forward with the outgoing emanations (B + x)".

In the course of the notable investigations of the Newry igneous complex in Ireland Doris L. Reynolds has proved "that the minimum introductions (A) were sodium, calcium and silicon; while, after several intermediate exchanges that are traced in detail, the displaced materials (B) eventually carried forward consisted of aluminium, iron, magnesium, potassium, hydrogen, titanium, phosphorus and manganese.

The latter, together with some remaining sodium, calcium and silicon, became fixed in adjacent bands of hornfels which were thereby basified and transformed into rocks chemically

equivalent to certain varieties of quartz-diorite. From these results and other relevant evidence, Dr. Reynolds concludes:

- That the introduced material (A + x) cannot have been an ordinary magma, since x has left no recognizable traces in the rocks;
- That the basic material migrating from a region of granitization, besides enriching the surrounding aureole in biotite and other minerals, was probably also responsible for the igneous-looking basic and ultrabasic rocks that overlie the granitic rocks of many plutonic complexes; and
- That before a given mass of country rock was actually granitized, it passed through a preliminary stage of basification".

The origin of the igneous rocks belonging to the mediterranean or potash clan is ascribed by Daly, Rittmann and Van Bemmelen to a process of calcification of a primary basaltic magma. Magma-sclerosis as Rittmann calls it, was described by him in a monograph on the structure and history of Mount Vesuvius. During an eruption blocks from Tertiary, Cretaceous, and Triassic strata underlying the volcano, are ejected. They are found in great quantities among the tuffs of Monte Somma.

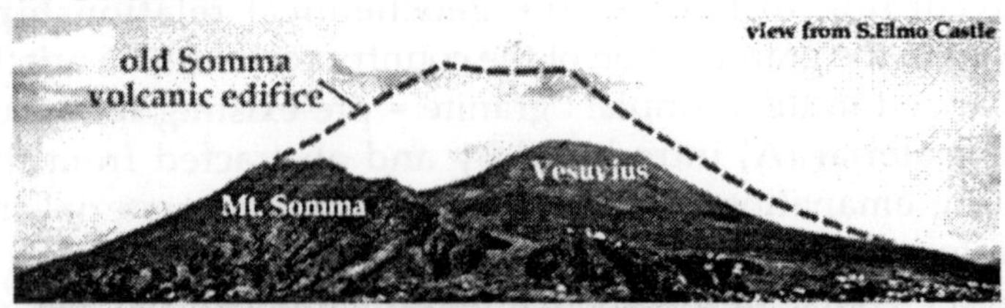

Fig. The Somma-Vesuvius Vvolcano.

The fragments derived from Tertiary strata show hardly if any magmatic influence. Blocks of Cretaceous limestone, too, are mostly unaltered. Exceptionally they are recrystallized into crystalline limestone. They have been in contact with the magma only during the short time of their ascent in the craterpipe during an eruption. In strong contrast the blocks of Triassic dolomite and limestone are throughly influenced by chemical processes.

Many pneumatolitic silicateminerals developed in them and not seldom the rock is entirely metamorphosed into silicate rock. The Triassic dolomites must have been subjected to a close and long-enduring contact with magmatic emanations.

The Mesozoic strata crop out at the surface in the peninsula of Sorrento and Amalfi. So their thickness and the angle of their northwestward dip are known.

From these data the depth of the top of the magma chamber below Vesuvius is estimated at approximately 5 kilometers below sea-level. From a closer examination of the lavas and tuffs of Somma and Vesuvius it appears that the composition of the magma was considerably modified in the course of time.

The oldest eruption products of Somma are trachytes followed by vicoites and leucite-basanites and leucite-tephrites. And the still more recent eruption products of the Vesuvius consist of leucitites abounding in plagioclases (i.e. leucite-tephrite rich in leucite).

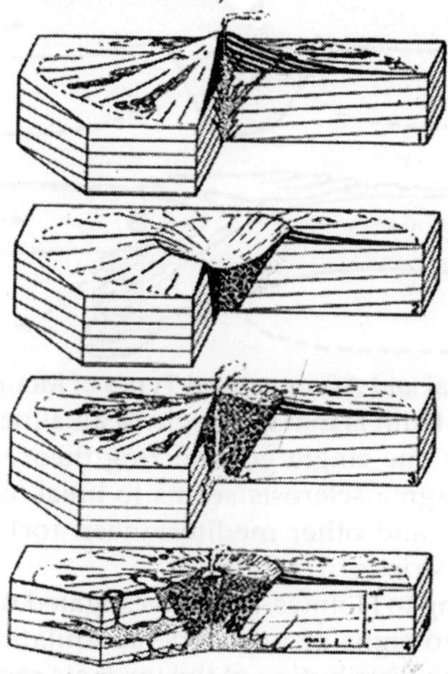

Fig. Schematic Representation of the History of the Laziale Volcano, S.E. of Rome, Italy.

The gradual transformation of the magma is ascribed to the progressive assimilation of Triassic dolomite and limestone. Assimilation of limestone is considered to have caused the hydrothermal escape of soda. Consequently the potash content of the residual melt became comparatively enriched.

The extinct Laziale volcano, SE of Rome probably may be mentioned as another example of a volcano displaying an increasing calcification of its lavas.

The frequent occurrence of alkali rocks among the products of Italian volcanoes induced Niggli to introduce the name mediterranean province.

The theory that the alkaline rocks may originate through the interaction of calc-alkali magma with calcareous rocks has been vigorously combated by several petrologists. Other theories were put forward by Bowen, Evans, Shand, and Holmes.

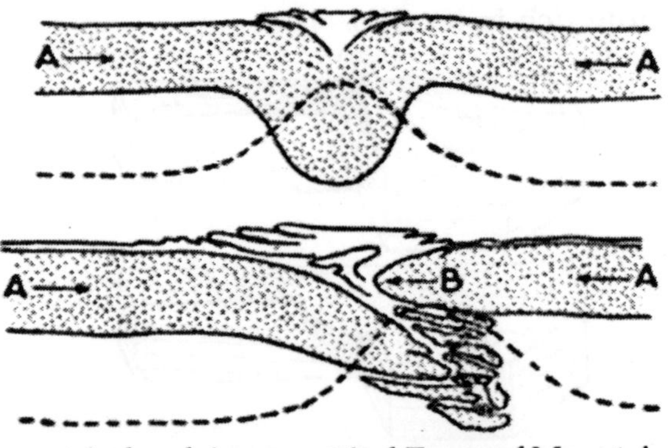

Fig. Symmetrical and Asymmetrical Types of Mountain-Roots.

Perhaps the truth is that the origin of alkaline rocks cannot be explained by one single generalizing theory. In any case the theory of magma-sclerosis seems to break down entirely for the leucitites and other mediterranean rock-types of the Bunyaruguru district of Uganda.

For according to Holmes, who investigated the rock-suites from Uganda, no signs of limestone assimilation have been detected. And the distribution of the volcanic rocks in the area is independent of that of the dolomite and limestone.

He suggests that the alkaline rocks may be due to a more intensive action of magmatic emanations.

TECTONIC AND MAGMATIC CYCLES

We now return once more to the question of a geological cycle, should the terrestrial crust be drawn into this figure, it would become clear that a thick downward bulge of acid basement rocks must have penetrated the underlying, denser basic rocks during the process of crustal folding. A root of lighter rocks must then have formed under the folded geosyncline, either symmetrically and perpendicular or asymmetrically and oblique.

In the last case the riding and overthrusting part of the crust caused the contents of the geosyncline to become squeezed unilaterally. The phenomenon is known very well from the Alps. In both cases, however, the resulting sialroot is due to the increasing compression of the earth's crust as a whole. Moreover this appears clearly to be true from the worldwide contemporaneity of the principal epochs of compression.

It ought to be possible to determine the existence of such a root by means of gravimetrical observations, as the same would constitute an anomaly of the isostatic equilibrium.

This sounds obvious, but, remarkably enough, the isostatic anomalies were in fact observed first, and from these was derived the existence of a sialic root. There can be no doubt, as Griggs remarked, that "one of the greatest contributions to the understanding of tectonics during the twentieth century has been Vening Meinesz' discovery of the great bands of gravity deficiency in the East and West-Indies". Vening Meinesz discovered abnormally large isostatic anomalies during his gravity expeditions in the East Indies in 1929 and 1930.

One of the principal features of these deviations is a narrow belt of strong negative anomalies which was found to continue for about 5,000 miles. Anomalies are expressed in milligals, i.e. the third decimal — or, roughly, millionth parts -of gravity. Anomalies of more than 50 milligal are but seldom observed, but this strip shows anomalies of more than 100, and occasionally of even more than 200 milligal. An analogous band was discovered in the WestIndies by the expeditions of Hess, Browne, Vening Meinesz, Hoskinson and Ewing.

These remarkable strips of negative anomalies are undoubtedly due to an abnormal distribution of crustal material beneath the strip. Vening Meinesz' explanation — the only one that seems to cover all the facts — is that a great isoclinal protuberance of the crust was downfolded into the substratum.

This led him to suppose "that the crust under the action of great horizontal stresses is buckling inwards, and that the folding and overthrusting of the surface layers, as found by the geologists, are an accompanying feature of this great phenomenon". It would be impossible for us, at this point, to enter into an examination of the details of either the accompanying strips of positive anomalies, or their relation to the structural history and seismic and geomorphological features of the explored areas.

We are at present particularly concerned with those phenomena in general which may be said to be associated with a subsiding and subsequently buckling crust. One characteristic phenomenon is that basic to ultrabasic rocks are found in the axis of a geosyncline, but that no acid plutonic rocks are observed at this stage of the development.

Conversely, acid intrusions enter the geosynclinal belt during and after its folding. Thus an outpouring of large flows of lava is known to have occurred in the Jurassic Franciscan beds of the Sierra Nevada geosyncline, and these flows are actually found as intercalations between the strata.

Diabase, gabbro and ultrabasic intrusions (serpentine), and, lastly, volcanic products, the latter mingling with the marine sediments, accompanied these outflows. Intensive folding ensued, during which the numerous batholithic intrusions of granodiorite, quartz-diorite and quartz-monzonite were emplaced.

Kossmat described a similar sequence in the Alpine, Variscian, Caledonian and older folded belts of Europe and in the Australian Caledonides. Stille grouped some European and American examples systematically in a table.

The Alpine, Variscian, and Caledonian Mountains of Europe had already been grouped schematically at a previous date by both Stille and Lotze. A great many examples will also be found in regional treatises.

This remarkable sequence of magmatic rocks can also be derived without the aid of strained hypothetical constructions from the previously sketched structure of the earth's crust if we assume that a zone was first subjected to geosynclinal subsidence, and that it subsequently buckled as Vening Meinesz has suggested. The following considerations. require but one assumption, i.e. that an upper layer of acid rocks rests on top of a layer of basic rocks.

The intermediate layer will be left out to make things simpler. The crystalline crust consists of a sialic layer of approximately 20 km and this is seen to lie on a basic layer of equal thickness.

It will be obvious that the ensuing reflections would hold good even if other thicknesses were concerned. In the meantime, should be regarded as rough and entirely schematic sketches of the events.

Moreover, anyone who has attempted to make such a drawing will have realized for himself that many dubious and difficult points arise in the course of such an undertaking.When an area of subsidence originates in the earth's crust, sediments may accumulate at the surface in abnormal "geosynclinal" thickness. The lower part of the crystalline crust enters an area of higher temperature and pressure and subsides to the fluidity boundary.

If that part of the crust were to melt, it would only be able to furnish basic to ultrabasic magma similar to the basic substratum already present beneath it.The strength of the crust will be exceeded when the downward movement reaches a certain level, and disruption will follow. The basic magma will then immediately (since it is subjected to high pressure) fill the fissures and faults in the crust.

As the bending of the crust is strongest in the axis of the

geosyncline, basic magma in this part will have the greatest chance of reaching the surface along faults. Moreover, that portion of the crystalline crust may be considered to have thinned as a result of subsidence, which caused the fluidity boundary to shift to a higher level.

Once the sialic crust has been bent so far that it collapses, one of the consequences will be that it will buckle inwards into the deeper layers of the earth's crust . The result will be: Folding of the contents of the geosyncline, together with the basic rocks in it. A local thickening of the sialic crust, resulting in the formation of a "root" of sialic material, penetrating into an area of progressively higher temperature and pressure.

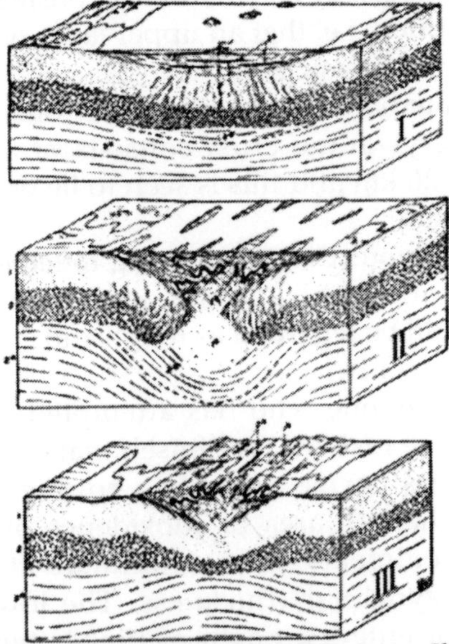

Fig. Schematic and Tentative Block Diagrams Illustrating the Relation between Tectonic and Magmatic Cycles. 1 and 2 Crystalline Crust; 1 Sialic Layer; 1a Molten Part of 1; 1b Acid Batholithic Intrusions and Metamorphic Aureole; 2 Basic Layer; 2a Basic Substratum; 2a1 Molten Part of 2; 2b Basic Abyssolithic Injections, Extrusions and Volcanism.

When the material of the root begins to melt and invade the surface strata as batholithic intrusions, plutonism and volcanism cannot but be strongly acid ("pacific") in character, since the batholiths assimilate more and more sialic material from the root, and eventually from the sediments of the already folded geosyncline (acidification).

The ten characteristics of batholiths, and the explanation which Daly gave of them , fit the diagrammatic explanation given here. As soon as the compressive forces in the sialic crust decrease, the "root", in order to resettle into isostatic equilibrium, will have a strong tendency to rise.

This effect is heightened by the expansion of the sialic root, which, through fusing, grows specifically lighter. This explains how a mountain-chain is formed from a folded geosyncline. More and more deeply situated batholiths are exposed by denudation at this stage (block III).

The upward movement will continue till the isostatic equilibrium has been re-established.

The sima will only be able to crystallize beneath he upper sialic layer to a thickness corresponding approximately with that of the sialic layer after part of the sialic root, situated at an abnormal depth, has spread out and disappeared (i.e. the crystallized sima will be thinner as the sial grows thicker).

As an accompanying phenomenon of the rising zone, sima will flow in from the neighboring strips of the earth's crust. This process may perhaps explain why a depression or "marginal deep" originates on one side, or both sides of a mountain belt.

The "zonal migration" of geosynclines can be traced distinctly in the history of some continents from the Cambrian up to our own time. Though the exact nature of events in the substratum cannot be determined, the subsidence of a geosyncline may be stated to be the result of internal terrestrial processes, whereas the sedimentation may be described as a secondary, external effect. This volcanism may perhaps be partly due to differentiation of the orogenic batholiths, but it is also partly related to the dome-shaped elevations.

A PERIDOTITE-LAYER IN THE SUBSTRATUM

Hess has pointed out that the axial part of a folded geosyncline is characterized by a zone of dunite or serpentinized peridotite (also called "ophiolite"). He considers, moreover, that in the island-arcs the peridotites were intruded only during the formation of the first great down-buckle.

Later epochs of compression implying a rejuvenated down-buckling of the sial-root, were not accompanied by a further series of peridotite intrusions. The complete absence of such ultramafic intrusions elsewhere clearly indicates the relationship between the development of the first down-buckle and the intrusion of peridotite from a primary dunitic magma-layer. Hess speaks of a peridotitic substratum and supposes it to be present at a depth of approximately 60 km.

Fig. Situation of a World-Encircling Peridotite Layer.

"During the forcing of the bottom of the downbuckle into the upper part of the peridotitic substratum, sufficient stress is present to permit squeezing off of a product of partial fusion of the peridotite substratum (this is the hydrous peridotite magma).

Under any other conditions this product of partial fusion would merely remain between the interstices of the grains of the peridotite substratum, because as a rule probably insufficient stress would be present to squeeze it off, or if squeezed off, it might not be able to penetrate the relatively plastic basaltic layer above it.

The relatively rigid and strongly deformed down-buckle allows the hydrous peridotite magma to migrate up its vertical structures, and thus the products of this magma are formed over or near the axis of the downbuckle or in the belts on either side of the downbuckles".

"That the peridotite intrusions accompany only the first deformation may be accounted for by two possible factors:

- That after the first squeezing off of a product of partial fusion insufficient low melting material is present to yield a second body of magma;
- That fusion of the bottom of the downbuckle after

the first deformation forms an impermeable cap through which the ultramafic magma cannot be intruded. Either one or both of these possibilities may be operative".

We agree with the interpretation given by Hess with the exception of one point. It seems more plausible to suppose the peridotite layer to be present at a lower level below the crust. Three reasons enduce me to prefer this supposition.

Firstly the presence of a layer of basaltic composition immediately below the crystalline crust can hardly be doubted. For it is in accordance with what is known about the world-wide distribution of basalt-outpourings.

Secondly, the absence of peridotite intrusions during later epochs of rejuvenation may be interpreted to mean that the first downbuckle probably invaded deeper into the substratum than it did in later stages.

Probably the first was more rigid and capable of giving fissures that could tap the magma Later ones would tap higher magmas.

In the third place, a deeper situation of the peridotite layer agrees better with the distribution of the fields of strong positive anomalies of isostasy in the East Indies. If the depth to which the first downbuckle invaded were known, the depth of the peridotite-layer would be known too.

ZONAL MIGRATION OF GEOSYNCLINES

Continents and Ocean-Floors

The oldest cores of the continents are known to consist of intensely folded and migmatized belts. None of the oldest areas observed so far can be regarded as a primary, i.e. an unfolded fragment of the earth's original sialic crust. That part of the which represents the seemingly simple structure of the basement of a geosyncline, or of an area of folding, can undoubtedly be stated to be very complicated for it passed through one or more geological and magmatic cycles.

This fact, if born in mind, might induce us to suppose that the continents might have originated as a result of periodical folding, culminating in a consecutive thickening of a relatively thin sialic layer originally enveloping the entire earth. We then feel inclined to enquire whether two problems cannot be explained simultaneously, viz.

- The formation of continental blocks
- The formation of sial-free parts of the earth, such as the floor of the Pacific.

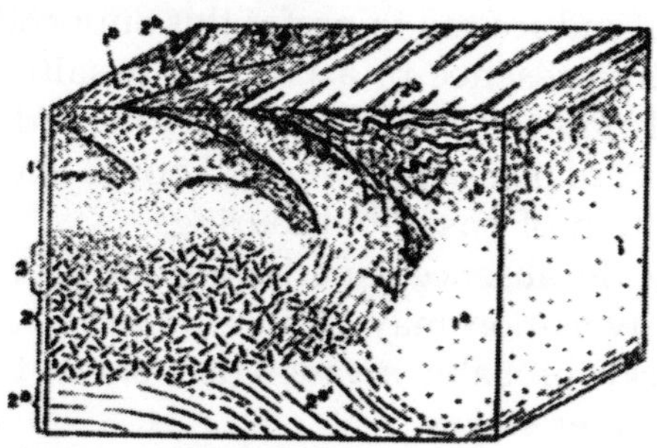

Fig. Tentative Illustration of Four Subsequent Tectonic and Magmatic Cycles in a Continental Area, Completing the Left Part of Block.

These speculative reflections differ to a considerable extent from the hypotheses of Fisher, Wegener, Schwinner, Escher and others . For, unlike these authors, the opinion may be put forward that an upper and acid sialic layer which enveloped the whole earth during a very early stage of its history was indeed stretched over a basic lower layer. In the beginning the material of both layers must have originated from a primordial melt containing the combined elements of the diverse, now contrasting, crustal and deeper shells.

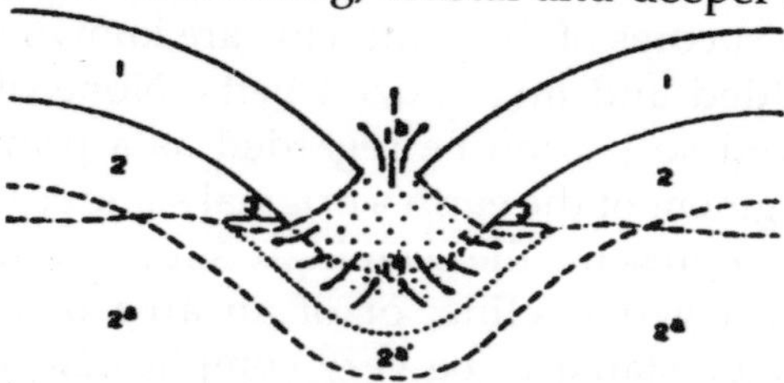

Fig. The Possible Origin of An Intermediate Continental Layer

These entirely hypothetical considerations lead directly to the important problem of the origin of continents and oceans

However, I would like to draw attention to an important deduction. If we assume the primordial crust to consist of two layers — an upper acid and a lower basic layer, would represent the acid melt formed as a result of buckling of the crust.

This liquid, however, will probably not only move upwards in the shape of batholithic intrusions but will also flow out laterally in the direction of the arrows. If the molten sial (1a) and sima (2a and 2a1) recrystallize during a later stage, i.e. during the rising of the root, a zone containing a mixture of la and 2a will have to form in the crust. This may help to explain the formation of a layer such as 3, which is intermediate to the layers 1 and 2 both as regards its position and its petrographical composition.

Volcanism in Basins

The foundation of a subsiding basin, like the down-bending crust of a geosyndine, will be affected by fracturing and become injected with magma. The magmatic products intrude as sills, and flow out at the surface as lava flows, or mingle with the sediments as clasmatic products. This can be observed in many basins, and it would certainly be worth while to investigate to what depths the bottoms of geosynclines and basins subsided before the first series of basic rocks made their appearance.

This would provide us with valuable data as regards the possible mechanical process in the earth's crust. Repeated reference is made in the descriptions of these basins to basalt, gabbro, diabase, melaphyr and other basic rocks, but no acid rocks have ever been observed so far as I know.

It need hardly be said that those basins which are closely connected with zones of folding (the basins of the intramontane type e.g.) should be considered cum grano salis, for pacific rocks in their immediate vicinity invaded the crust before and at times even during their formation.

It is certainly noteworthy that the marginal deep which originates subsequently to the principal epoch of folding of the geosyncline should, generally speaking, seem devoid of volcanism. This negative characteristic must have some connection with the deep crustal buckle which was not formed until a short while before and along which the basins and troughs are arranged.

Fragmentation and Growth of Continents

"And so the earth is growing old". This sentence appears at the end of one of Stille's numerous papers on the structural history of the continents In my opinion, however, the distinguished author to whom we owe so many valuable data and suggestions regarding periodical events in continental areas should not have closed his reflections with such an *accord en mineur*.

In his interpretation of the face of the earth Stille is obviously impressed by the breaking down in later times of large stretches of the mighty belts of Caledonian and Variscian mountains America has lost territory to the east and west in the foundered blocks of Appalachia and Cascadia. Africa, India and Australia, all of which paid heavy tribute to the ocean, are mere remnants of larger areas.

They seem to a process of continental fragmentation and decay. Along the eastern border of Asia a whole series of basinshaped depressions originated by the collapse of a corresponding series of sialic masses. Rift-valleys opened like gaping wounds in the earth's skin. They nearly cross Europe from the Vettern graben in Sweden to the border of the Mediterranean and they break across Africa from the Red Sea to the Zambesi.

There is nothing to contradict these statements. However, in spite of these and other similar facts I cannot agree with Stille in his speaking of a *marasmus sinilis* of our planet. It may be true that part of the impressive chains of Variscian times are covered by the waters of the Atlantic. But we are living in a geological epoch which has its own mighty chains of Alpine grandeur.

And we should not forget that in several districts the lofty chains of the Caledonides and Variscides ruthlessly dissected and destroyed structural units of previous periods.

The Paleozoic earth had also its wounds. And are not we witnesses of the rejuvenation of many an old pattern into a fresh-looking expression? Constantly erosive forces are wearing down the earth's surface, trying to demolish its vigour into the dullness of continental peneplaines. So long, however, as the earth disposes of the youthful power to withstand the destructive work of denudation I cannot detect any symptoms of senility in the pulse of the earth.

If we were to adopt the metaphor of a human creature we could better compare the processes of the earth's interior to processes of metabolism in a living being. For what happens is a constant change of surface expression under the influence of periodical processes in the interior.

The surface of a continent is steadily lowered by the perpetual action of denudation. Loss of matter at the surface is partly compensated by isostatic uplift rejuvenating the general relief of the continents. So, the flotation of the continents remains a feature enduring through the aeons of geologic time.

If, however, no fresh sialic matter were to be supplied to the continental bucklers, they would become ever thinner and thinner. At a final stage the areas occupied by the primordial continents would have degenerated into very thin and perfectly peneplained sial-flakes.

There would remain no possibility for them to rise or sink. Some 9/10 of their original thickness would have been removed for ever, — assuming that sial has a mean specific gravity of 2,7 and sima of 3,01).

Hence the question arises what processes may be available to counterbalance the constant lowering of continental relief by erosion and denudation. Transport of waste-products from the surrounding mountain-belts would account only for a certain retardation of the process of levelling.

Probably therefore, addition of sialic matter must come from below. And, indeed such a process is not improbable in as much as the molten part of the sialroot under a mountain-chain will spread laterally. On account of its light specific weight and high radioactive content it has an urge to migrate in an upward direction, melting its way into the crust. Convection currents may transport material from the root over long distances and, thus, far distant parts of the crust will be supplied by sialic matter.

However, one or more zones of geanticlinal updoming generally originate in the vicinity of a belt of crustal buckling. These are places par excellence for the ascènt of light and acid magmatic emanations and melts.

Obviously a sialic melt could be provided not only by the molten part of a mountain-root but also by any deeply subsided part of the crust. Subcrustal sial-melts may also originate below the bottom of a continental basin, in the subsided part of a continental margin and at the base of a deep-sea trough or of a subsided block like Appalachia.

In any case it seems clear that new sialic melts must be continually formed under certain parts of the crust. And the spreading, migration and subsequent ascent of acid melts due to periodical processes in the earth supplies the underside of the continents with the sialic material that they lost at the surface. So, far from growing old, the continents are being periodically rejuvenated.

The conception of the structure of the earth's crust, as advocated above, makes it clear that the presence of certain magmatic rocks must (assuming that a period of geosynclinal subsidence is followed by crustal buckling, as postulated by Vening Meinesz) be related to the phenomena of geosynclinal subsidence, folding and mountain-building. Thus, it is obvious, for instance, that only basic to ultrabasic rocks can occur in a geosyncline, and that intrusions of acid batholiths take place after the crust has buckled and the contents of the geosyncline been folded.

All evidence shows that periods of increasing compression of the earth's crust alternate with periods of decreasing compression. During this last stage the sialic root (which was brought about by buckling of the crust) is given an opportunity to rise until the isostatic equilibrium has been restored. A mountain-chain, in a geographical sense, then comes into existence.

This process has thickened the sialic layer in the meantime, and the continents may consequently have originated as a result of periodical folding and thickening of an originally thinner sialic layer enveloping the whole earth.

The accumulation of sediments in a geosyncline is a secondary and external effect of a subsiding crustal furrow, and this phenomenon, together with the alternating periods of increasing and decreasing compression, and the migration of geosynclines, must be attributed to some common and worldembracing, deep-seated cause, for all the phenomena are related chronologically.

The theory is summarised of a common origin of all the igneous rocks belonging to three magmatic tribes called atlantic, pacific, and mediterranean — from a parental material of olivine-basaltic composition which forms a world-encircling layer immediately below the crystalline crust. Most of the granite batholiths are thought to be of palingenetic or syntectic origin.

The genesis of rocks belonging to the mediterranean clan by a process of assimilation of dolomite and limestone is described from Somma and Vesuvius. Other theories on the origin of alkaline rocks are mentioned. At any rate the example of the Bunayaruguru district in Uganda shows that a process of magma-sclerosis cannot have taken place in that area. For no signs of limestone assimilation have been detected and the distribution of the alkaline rocks is independent of that of dolomite and limestone in that area.

A peridotite-layer is probably present in the substratum below a layer of olivine-basalt. This explains why peridotite intrudes a mountain-root only during its first deep down-buckling, whereas later epochs of compression, implying a rejuvenated down-thrusting of the sial-root, are not accompanied by another series of peridotite intrusions.

The continents are rejuvenated periodically, receiving a supply of sialic melts derived from the spreading, migration and subsequent ascent of sialic melts from the molten part of mountain-roots and the subsided parts of the crust such as basins and submerged borderlands.

Chapter 24

Oscillations of the Sea-Level

The sea has repeatedly invaded large portions of the continents and then subsequently retreated. This part of the earth's history is an extremely monotonous one, but this very monotony, this continuous recurrence of the same events, constitutes one of its most important aspects.

Why does the sea invade the continents, and why does the water later withdraw to the oceanic receptacles, only to advance again afterwards? In other words, in geological terminology, —what is the origin of the periodic trans- and regressions?

Before considering the major periodic movements, however, we should first note that there are undoubtedly examples of trans- and regressions of a more limited, regional importance.

In these cases it is but seldom possible to connect the relative shifts of the sea-level directly with one or more clearly discernible impulses.

In geosynclinal areas the phenomena may occur at a different rate as a result of their own and possibly opposed movements, and the negative or positive shifting of the sea-level may perhaps be quite counterbalanced.

It is even conceivable that transgression may occur in a geosyndinal zone simultaneously with a negative phase, and vice versa.

On the continents, too, the strand may advance or retreat over an area of greater or lesser extent as a result of different causes, but this need not necessarily indicate a world-wide movement.

REGIONAL TRANSGRESSIONS AND REGRESSIONS

The sea invaded the Baltic regions twice during the melting of the socalled Fennoscandian ice-sheet. The first to form as the ice-sheet retreated was the Baltic ice-lake, which occupied the southern part of the actual Baltic Sea.

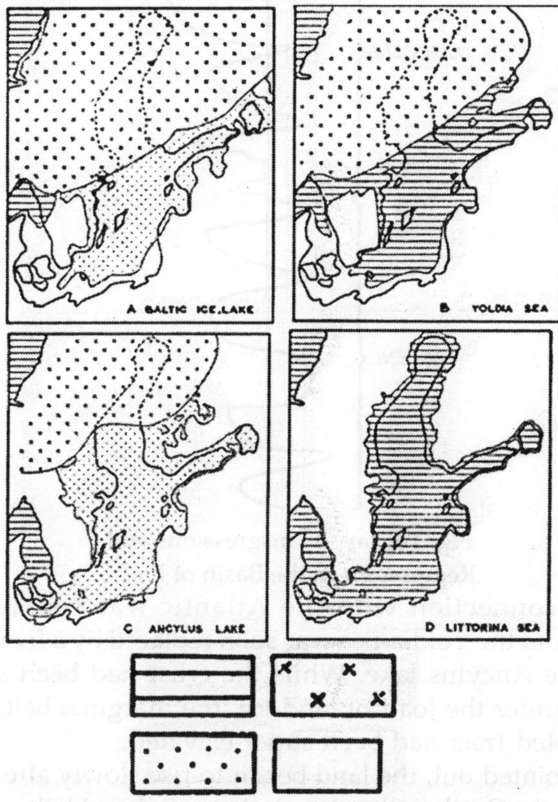

Fig. Pleistocene History of the Baltic Region.

A strip of land in the south of Sweden became free as a result of the steadily increasing recession of the ice-front. This area was so low that it was flooded by the ocean, causing the Baltic freshwater lake to be transformed into a sea — the Yoldia-Sea, which was connected with the Atlantic.

The crust of South and Central Sweden was now rid of the load of ice which had previously been pressing it down and began to rise gradually.

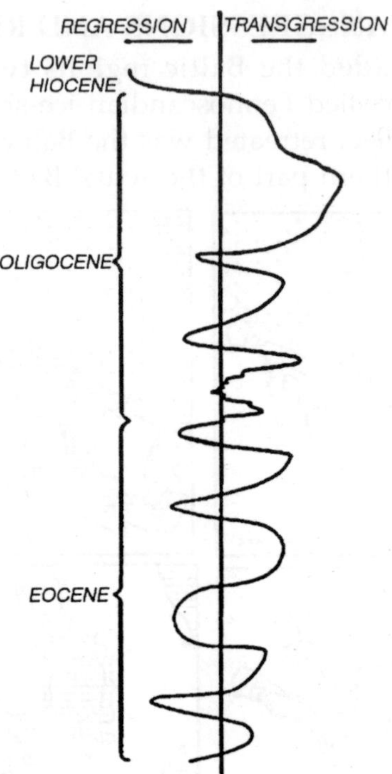

Fig. Tertiary Transgressions and Regressions in the Basin of Paris.

The connection with the Atlantic was consequently severed, and the Yoldia-Sea was soon replaced by a fresh-water body—the Ancylus lake. While the crust had been strongly basined under the load of land-ice, the marginal belt around the glaciated tract had been super-elevated.

As pointed out, the land began to rise slowly after the ice had melted. On the other hand, the peripheral belt – which includes the North of Germany and Denmark – began to subside.

This, combined with the rising of sea-level by the melting of ice, caused the connection with the waters of the Atlantic to be re-established. The Straits of the Sund, the Great Belt and Little Belt were opened, and the fresh-water of the Ancylus lake was replaced by salt water – the so-called LittorinaSea , which was gradually transformed into the present Baltic-Sea.

I mentioned this example because it clearly how the local rhythm of the alternation of fresh and salt water can be derived from one constant, universal and non-periodic factor.

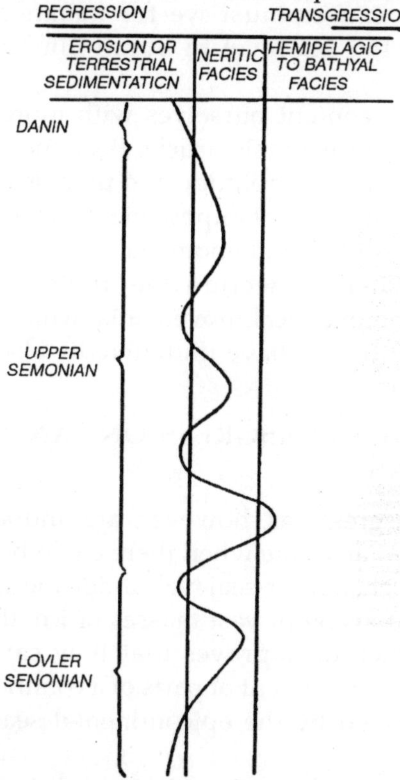

Fig. Transgressions and Regressions in the Upper Cretaceous of Limburg, in the Netherlands.

The land movements are in fact a consequence of the retreat of the Fennoscandic inland-ice, while the rise of the sea-level is due to the melting of the Fennoscandic ice-cap and other ice masses. All this is brought about by one changing factor, viz. a world-wide increase in the average annual temperature.

This example clearly shows that the repeated advance and retreat of the sea in a restricted area need not necessarily be the reflection of a universally active cause that changes periodically. If we bear in mind that the basin has a complicated history it will become clear that the latter has also found expression in the curve. In many cases the origin of trans- and regressions remains a matter for conjecture. To what cause or causes must we for instance attribute the oscillations of the sea-level in South-Limburg during the Senonian.

We have to content ourselves with more or less vague surmises. In a very readable article, Born mentions examples of great differences in amplitude and time of such rhythmical movements. It seems as yet impossible to furnish a satisfactory explanation for all these phenomena.

When considering a world-wide "rhythm", we should not forget that numerous local movements, which may have had a very different origin, have undoubtedly also played a part in several regions.

WORLD-WIDE TRANSGRESSIONS AND REGRESSIONS

Large transgressions, however, are undoubtedly known to have occurred at a time when there could be no question of a considerable change of sealevel under the influence of the melting and extension of vast masses of ice, the considerable expanse of which only proves that they cannot merely be explained by the movement of parts of a continent.An example of this is furnished by the epicontinental seas of the Upper Cretaceous.

The gradual advance of the sea into the European Continent can be closely followed. This great invasion of the sea was not only felt in Europe, but also on other continents. Conversely, a world-wide retreat of the sea prevails towards the close of the Mesozoic, and is in turn succeeded by a positive phase of the epicontinental seas in the Lower Tertiary.

Suess was the first to show that these phenomena have a worldwide importance and that the periods of transgression were far longer than the relatively short periods of regression. Suess, Haug, Stille and Grabau paid special attention to the chronological and regional distribution of the transgressions, and each in turn came to the conclusion that a great number of major transgressions took place, each being separated by periods of widespread emergence of the continents.

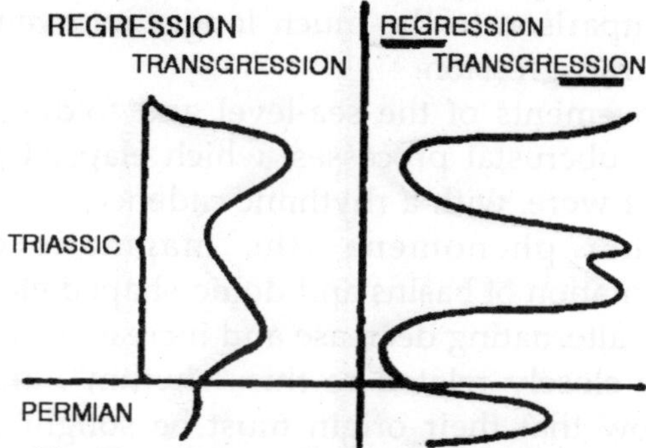

Fig. Two Curves Showing the Relative Movement of the Sea-Level in Triassic Times.

There was a rhythmic advance and retreat of the sea. We can therefore only conclude that the trans- and regressions on the continents must be ascribed solely to a world-embracing cause. Stille expressed the synchronism of the great trans- and regressions in his law of epirogenic synchronism, which Bucher formulated as follows: "In a large way the major movements of the strandline, positive and negative, have affected all continents in the same sense at the same time".

Once this rule is established statistically, the local and regional exceptions appear clearly. Thus Stille gives two curves for the Triassic, one representing the world-wide movement of the sealevel, and the second the strongly diverging and even opposed transgressions and regressions of the Triassic in Germany the problem of trans- and regressions in a paper of 1939 and will therefore deal with them very briefly at present. A study of the oscillating sea-level caused me to draw the following conclusions:

- World-wide regressions are brought about either by a periodical elevation of the continents, or by periodical subsidence of the ocean-floors, or by simultaneous but opposed movements of both continents and oceanfloors. The last possibility appears to be the most probable one, and the same applies — *mutatis mutandis* -- to the transgressions.

- Epochs of folding coincide with periods of world-wide regression, which were relatively short in comparison to the much longer intervening periods of transgression.
- Movements of the sea-level and folding are caused by subcrustal processes, which elapsed periodically, as it were, with a rhythmic cadence.
- Other phenomena (the magmatic cycles, the formation of basins and dome-shaped elevations, and the alternating decrease and increase in compression) are closely related to this "rhythm", and they, too, show that their origin must be sought deep within the earth, as pointed out in the preceding.

Explained that my conception of the structure of the earth's crust coincides with that of Daly and Vening Meinesz. This means that the earth is actually enveloped by a rigid crust. A cross-section of a continent shows that the crust is composed of sial and crystalline sima, and that a thinner sialic layer (though resting on a thicker layer of sima) is present under the Atlantic, and that a very thick layer of crystalline sima extends under the Pacific. Wegener's view that sima represents a viscous substratum in which continents would only drift as rigid blocks is unacceptable under the present day circumstances).

It may still be conjectured whether the rigid and crystalline upper layer of sima enveloping the earth at present had at some time been considerably thinner or possibly nonexistent (aside from the fact as to whether they were controlled by periodic radioactive processes, as Joly assumed, or whether another unknown internal process was responsible for them). At any rate, the periodicity of trans- and regressions shows that the differences in level between continents and ocean-floors have changed periodically!

The curve in Table II indicates the result as regards the relative movement of the level of the sea, showing the positive and negative shifting of the sea-level and its chronological relation to epochs of compression. This curve is entirely schematic. Besides, it is highly probable that certain trans- and regressions, which cannot be indicated for lack of data, had a greater amplitude than others.

A detailed study of the facies of the strata deposited on the continents by transgressive seas may produce evidence of the height of the sea-level at various periods above certain parts of the continents. But in that case local movements (e.g. of basin-shaped depressions) will also have to be taken into account, and the problem is thus far from being a simple one.

The vast epicontinental extent of Upper Cretaceous strata, which are often compared to the present cocolith-ooze found at a depth of at least 1,000 meters, would lead one to suppose that an important shift of the strandline accompanied some transgressions.

Kuenen assumed that a major transgression would result from an eustatic movement with an amplitude of some 40 meters. This amount, as he himself declared, should be regarded as a minimum value. We might add that it ought undoubtedly to be more. For Daly pointed out that if all the ice of the present ice-caps were to melt, this fact alone would cause the sea-level to rise some 40 to 50 meters).

We may safely say that the period in which the last world-wide regression reached its deepest level now lies behind us, but we are undoubtedly still far removed from the day when the prevailing transgression will have reached its maximum. During the maximum glaciation in the Pleistocene, the sea-level was approximately 100 meters lower than it is now, and the sea was already in a new transgressional stage at that moment.

Thus the maximum for a transgression must be fixed at at least 150 meters, plus an unknown amount, and the rise of the sea-level towards the maximum of the now prevailing transgressional phase will attain 50 meter above the present level, plus an unknown but undoubtedly considerable amount. Penck estimated that transgressions were caused by a relative rise of the sea-level with an amplitude of 500 meters, and Joly evaluated the amount of the eustatic change between a deep regression and an extreme transgression at 1,200 meters.

Many other factors complicate this problem. For instance,

the formation of the Mediterranean and the East and West Indian and Asiatic basins would have resulted in an eustatic lowering of the sea-level of approximately 60 to 80 meters according to Kuenen.

The most important question concerns the depth to which the sea-level was depressed in distinct periods of intensive regression, in other words the extent of the change to which the distance between the surface of the continents and the ocean-floors was subjected during the pulsating rhythm of subcrustal processes. Joly was the only one who approached this question from the geophysical side and he arrived at an order of 1,000 meters.

No one can foresee what further geophysical speculations may lead to eventually, but it is clear that this question has an extremely important bearing on various geological problems. For if the sea-level moved downwards far beyond its present position during certain periods, it would have had a strong influence on many phenomena.

In this connection may be mentioned the origin of isthmian links, and broader transoceanic land-bridges. Further, all those physiographic factors that lead to glaciation would be considerably accentuated.

Obviously, not a single marine sediment could have been deposited on the continents or within shoal-water geosynclines during such a large emergence of the continents as may be expected to have preceded the late Paleozoic and the Pleistocene glaciation. As far as I have been able to ascertain, however, the actual facts do not confirm the idea of a world-wide regression of such extent that the surface of the sea would at some time have been lowered to a few thousand meters below its present level.

THE CONTINENTAL MARGIN

Almost everywhere the real edge of the continent is situated below sea-level. Generally a shallow platform borders the continent. Its edge lies about 200 meters below sea-level and an area of 10,000,000 square miles or 13% of the continents is thus covered by shallow shelf-seas.

What is the reason of this world-wide uniformity of depth of the shelf-surface and its edge?

Echo-soundings and geological and geophysical explorations carried out during the last decade have, moreover, revealed many features hitherto unknown about the morphology and structure of the shelf and its outer slopes. From these observations it appears that the history of the shelf is a rather complicated one.

Sedimentation, abrasion and denudation all played their rôle. The area was subjected to changes of sea-level and movements of the bottom. Wind-waves and tidal currents acted upon the sediments of the shelf. The influence of each of these and still more factors in the building of the submerged part of the continental margin is still an open question.

The outer slope is notched by great valley-like trenches descending towards the abyssal depths of the ocean. The origin of these huge gullies is another problem that challenges us. The outer slope of the Atlantic shelf of the United States — often called continental slope — drops from 200 to 2,400 meters and more in approximately 50 miles. In a few sections the slope is as steep as 7 1/2° (about 13%) or 700 feet per statute miles1).

On the other hand the region inland from the coast offers a series of problems of its own. Wide-spread features of rejuvenation and a number of elevated marine terraces are characteristic of the marginal part of the continental land.

Fig. Shelf of the North Sea Showing Reconstruction of Rivers Debouching Near the Dogger Bank.

The prevalence of these phenomena in all latitudes along the continental border is impressive and requires an explanation which extends beyond local interest. How can these features be understood so as to form a harmonious supplement to those of the submerged part of the continental margin?

Generally the shallow platforms bordering the continents are classified in two categories, viz. the inner and the outer shelves. The first group comprises such regions as the shelf of the North Sea and the Sunda Sea. Bathymetric charts show them to be furrowed by river-like trenches.

And indeed, in the two examples just mentioned their sub-aerial origin as rivers extending over these regions during low stands of the sea-level in Pleistocene times was clearly demonstrated by data furnishing converging evidence. Their course can be followed towards the debouchement of present-day rivers. Moreover, the frequent finding of remains of large vertebrates as far as the Dogger Bank shows the North Sea to have been a land area in the near past.

In the case of the Sunda-shelf the congruence of such animals as fresh-water fishes clearly proves e.g. the rivers of western Borneo and eastern-Sumatra to have been connected in sub-recent times. The shelf of the Barents Sea is considered to belong to the European continent.

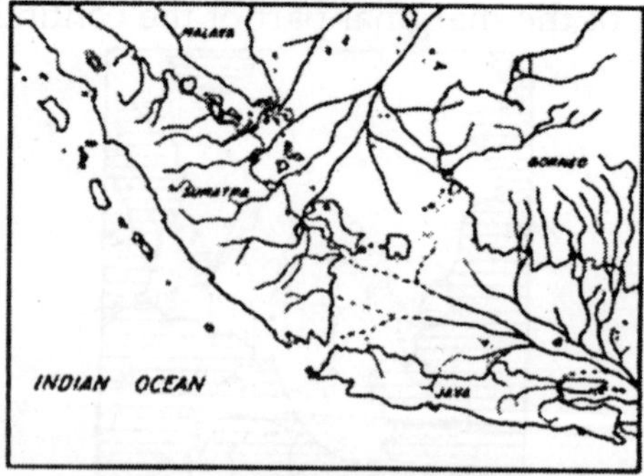

Fig. Drowned Rivers in the Java Sea and South China Sea.

In this area Spitsbergen and Bear Island portray the nature of structural units which, over large stretches, were probably truncated and covered by the sea. So, between Spitsbergen and

Scandinavia the inner shelf of the Barents Sea gradually passes towards the outer shelf of north-western Europe.

Here, too, the sea-bottom shows a drowned system of rivers. However, although highly interesting in many respects, the origin and history of the inner shelf regions will be left out of consideration here, since they do not belong to the marginal zone proper of the continents.

According to Kossinna the shelves all over the world cover 27,500,000 square kilometers, i.e. 7.6 percent of the surface of the oceans. The figure for the outer shelves is 9,900,000 square kilometers, i.e. 3.1 percent distributed as follows: Atlantic Ocean 4.6 million km2, i.e. 5.6 percent of the sea-surface

Indian Ocean 2.4 million km2,,, 3.2 ,,,,,,,,,,

Pacific Ocean 2.9 million km2,,, 1.7 ,,,,,,,,,,

The origin of the outer shelves is the first problem to be considered.

The Surface of the Shelf

Waves laden with sediment and shingle attack the shore. Their continual bombardment undermines the coast. Gradually the sea encroaches upon the land, abrades and invades it.

Undertow transports the abraded material seawards, and tidal currents assist in the submarine distribution of the sediments.

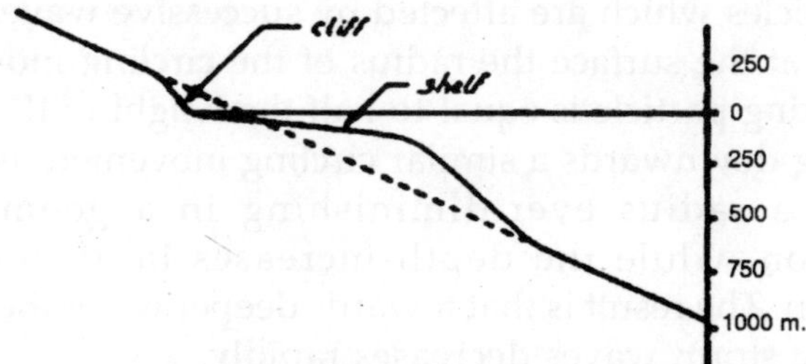

Fig. Formation of a Gradation-Plane Along the Coast of Saint Helena.

The result is clearly demonstrated in the morphologic which shows the small shelf round the island St. Helena). From the cliffed coast over some distance seaward the surface of the submarine bank consists of an abrasion-plane formed by the erosive action of the breakers. The more distant part of its surface, however, is built up by the accumulation of sediments derived largely from the land. From the edge of the shelf, situated here at a depth of 100 meters, the outer slope of the small shelf may be followed downwards to a depth of 700 meters.

At that depth the slope diminishes; the reason for this seems obvious: on that level the normal slope of the St. Helena volcano is reached, as is suggested by the broken line drawn.

The surface of the shelf therefore seems to be a plane of equilibrium2) between two antagonistic forces. One of these is the demolishing, erosive and transporting action of the sea. The other is the accumulating action of the same medium, the sediments being transported, selected and deposited by wave-action, undertow, tidal currents and other off-shore sea-currents. The surface of the shelf expresses a balance between two opposing sets of factors which in short might be indicated by the terms submarine erosion and aggradation.

One of the first doubtful points is the depth of the abrasive action of the sea. Obviously the demolishing power of waves is dependent on the intensity of wave-motion. The motion of waves following one another in trains, is expressed by the formula $L = VT$ when L is the length of the wave (measured from crest to crest), V its velocity, and T the period of oscillation of the particles which are affected by successive waves.

Now, at the surface the radius of the circling movement of a revolving particle is equal to half the height of the wave. Proceeding downwards a similar circling movement is found but with a radius ever-diminishing in a geometrical progression, while the depth increases in arithmetical progression. The result is that towards deeper water the power of even big storm waves decreases rapidly.

Jenkins1) gives the following examples: Waves 90 meter long and 3 metres high are not uncommon with strong winds in the open ocean.

"At 10 metres depth the height of the subsurface trochoid would be 1.5 metres; at 20 metres depth 0.75 metres; at 50 metres only 9 centimetres; in 100 metres not quite 3 millimetres — that is hardly perceptible". "An ocean storm-wave 600 feet long and 40 feet high from hollow to crest would have at a depth of 200 feet a subsurface trochoid with a height of 5 feet; at 400 feet (6/9 of the length) a trochoid of 7 or 8 inches only".

It is clear, therefore, that the abrasive action will hardly be able to be effective beyond the depth of average wave-base. Accordingly marine planation by abrasion at wave-base seems to be restricted to a narrow strip, parallel to the coast, in the litoral zone.

The more so since material tumbling down from the cliffed coast forms a barrier sheltering the coast from further abrasion for an appreciable lapse of time. It would seem, therefore, that the sea can invade the land by abrasive action over a large extent only when it is accompanied by a simultaneous positive shift of the coast-line.

Some authors, however, are of the opinion that the formation of the entire shelf-surface is controlled exclusively by the demolishing action of waves and currents. According to their theory the shelf is considered to be a feature which originated mainly by abrasion. Jessen is a modern advocate of the origin of the shelf by abrasion. In order to explain the frequent occurrence of the shelf-edge at a depth of 200 meters he put forward the following theory.

Firstly he supposes the abrasive forces to be effective down to a depth of 60 meters. Secondly he thinks that — as a result of sediments being transported from the land oceanwards — a worldwide rise of the sea-level of 30 meters should have taken place since the beginning of the Pleistocene. Finally, a glacial lowering of sea-level would account for the remaining amount of 100 meters.

In this hypothesis Jessen confuses the movement of the water which under normal conditions is able to displace sand-grains over th surface of the shelf, with an abrasive action at wave-base. The amount of 30 meters is chosen to suit his hypothesis, and is not based on well founded estimates.

Moreover, granted even for a moment that these figure were correct, a glacial lowering of sea-level to an amount of

100 meters is inadequate to explain the features of the shelf-surface. Novák, too, is convinced that the shelves were carved out of the continental blocks.

He considers they were formed mainly by submarine peneplanation during a low stand of sea-level, but marine abrasion and subaerial planation are admitted by him as co-operating factors. Novák, however, believes that the duration of the Pleistocene stages of 'low sea-level did not last long enough for the shelf-surface to be formed entirely by these factors.

He therefore thinks it inevitable to look back in geological time to a period when sea-level stood 100 m or so lower than at present for a time sufficiently long for the cycle of the present shelf to be completed. For that reason he places the formation of the present shelf in Maeotian-Pontian times! Moreover his hypothesis implies a subsequent rising of the sea during the Pliocene to some 200 meters above its present level, in order to explain the higher marine terraces which occur along the shore of the Mediterranean (Calabrian terrace).

Finally, a world-wide lowering of the sea-level would have taken place during the Pleistocene, but was interfered with by glacial oscillations. For several reasons his hypothesis is untenable. Firstly, paleogeographic conditions at the end of the Pliocene are opposed to the view that sea-level was 200 meters above its present level. On the contrary, the upper Pliocene was a time of wide spread regression of the epicontinental seas, which were then confined to a few basinshaped depressions.

Moreover the whole idea of the shelves being carved into the continental blocks — which was also the gist of Buchanan's hypothesis — is rendered out of date since the geophysical explorations carried out by Ewing, Crary and Rutherford. For they made it highly probable that the Atlantic shelf of North America is composed of an accumulation of sediments ranging from Triassic to recent times resting on the submerged part of the former continent and attaining a thickness of 4000 meter at a distance of 60 miles off the shore-line.

Similar results were obtained by Bullard for the shelf to

the west of the English Channel. He also found the surface of the continental basement — consisting of crystalline rocks — to slope steadily downwards to abyssal depths on receding from the land. At the 100 fathom line the surface of the basement is over 8,000 feet below sea-level and covered by a prism of off-shore deposits.

So it appears at any rate that the continental shelf must not be regarded as a huge step incised in the continental block. Wegener's interpretation of the shelf, too, must be discarded for the same reasons. And this applies also to analogous views of Du Toit.

However, apart from the problem as to the role played by several factors in the formation of the shelf-surface, there is a question of primary importance, viz. at what depth would a shelf-surface and its edge be situated when formed entirely under present conditions? The process of grading of the shelf implies that there exists a certain hypsometric relation between sea-level and shelf-surface.

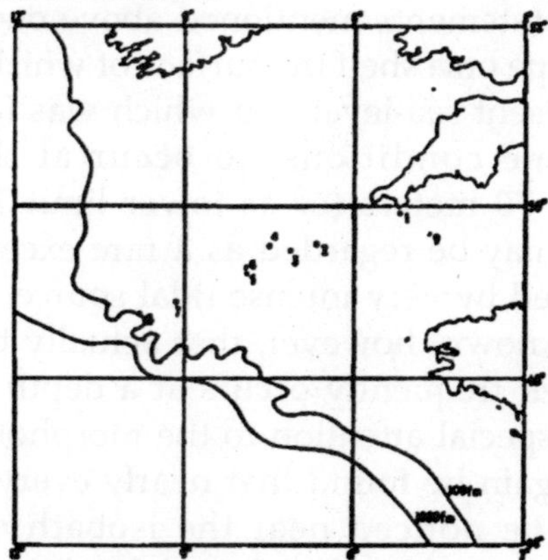

Fig. Seismic Investigation on the Continental Shelf

The example of the island Saint Helena makes it probable that the edge of the plane of equilibrium — in other words the outer edge of the shelf — is formed at a depth of about 100 meters below the level of the sea. At least for the southern part of the Atlantic this does not seem an exceptional figure. The island of Ascension for example, shows exactly the same feature.

"As usual", Daly wrote), "the horizontal distance between the shore-line and the 50-fathom isobath is considerably greater than that between the 50-fathom and 100-fathom isobaths. The break of slope at the outer edge of the coastal shelf is located close to the 50-fathom line".

In his paper on the morphogenesis of Ben Lomond Rode) mentions that along the Californian coast the submarine action of waves and currents extends down to a depth of at least 50 meters, but not surpassing 100 meters.

In his general description of the American shelves Shepard summarises: "steepening beyond the shoal platforms is initiated at an average depth of about 65 fathoms". In some areas, however, the plane of balance between aggradation and erosion, decidedly occurs at a slighter depth.

According to Stetson for example, "60 to 70 meters can probably be regarded as the limiting depth to which the storm waves of the Atlantic, when aided only by normal tidal currents, can agitate the bottom deposits."

From the statements mentioned above one might expect the break of slope on a shelf the surface of which is completely adjusted to present sea-level and which was formed entirely under the same conditions, to occur at the isobath of approximately 70 meters (or as lower limit 100 meters). A deeper figure may be regarded as a rare exception. It might locally be caused by very intense tidal source.

It is well known, however, that actually the edge of the continental shelf frequently occurs at a depth of 200 meters. Bourcart paid special attention to the morphological features of the shelf. Again he found that nearly everywhere a break of slope may be noticed near the isobath of 200 meters. Respecting the eastern border of the Atlantic he mentions only two exceptions.

Along the coast of Angola and off the Ivory Coast the isobath of 1500 to 2000 meters approaches the coast so nearly that no shelf at all is present. The Atlantic shelf off North America shows lower figures, ranging from 60 fathoms east of Chesapeake bay to 80-fathoms north of the.

Hudson and on Georges Bank. Hence, the problem has to

be solved as to how the shelf-surface below the isobath of 70 meters (40 fathoms) came into being).

The Influence of Changing Sea-Level

In order to explain the break of slope so frequently occuring at a depth of 200 meters the possible influence of eustatic changes of sea-level and movements of the bottom need to be examined. Undoubtely changes of sea-level rank among the principal agencies which have affected the surface of the shelf in the near geological past.

One factor is the changing sea-level associated with transgressions and regressions caused by periodical processes in the earth's interior. The last regressional stage of the sea not controlled by changes in the volume of ice-caps happened to occur towards the end of the Pliocene, simultaneously with the intensive folding and overthrusting which characterises so many mountain-chains belonging to the Alpine belt of folding. Since then a transgressional stage has begun.

But at yet data are lacking which could show us the amount of the transgression since the regressional stage of the Pliocene. It may well be that it is rather of the order of ten or tens of meters than of a hundred or hundreds of meters. And of course the extent of the transgression at a certain stage of the Pleistocene must be less. Owing to lack of data this amount has to be left out of account.

Apart from the positive shift of the shore-line considered just now, the sea-level swung downwards and upwards four times during the immediate geological past under the influence of the growing and melting of the icecaps. The amount of the lowering of sea-level at the time the ice-sheets reached maximum size, may be estimated at 75 meters, possibly as a lower limit 90 to 100 meters1).

A certain amount has to be added in as much as these figures are based on estimates of the volumes of ice in excess of the present volume, which existed when the Pleistocene ice-sheets reached their maximum extension.

From these estimates one might perhaps feel justified in considering the shelf-surface between the isobaths of 70 (100) and 200 meters a plane of equilibrium adjusted to one or more stages of low sea-level in the Pleistocene. In doing so, however, the implicit assumption has to be made of a stable bottom.

Possible this holds good in a few special cases, viz. in such oceanic islands as Ascension and Saint Helena. For in all probability these volcanic islands date from at least pre-Pleistocene times. According to Daly who summarized the different data bearing on this question, Saint Helena dates from pre-Glacial, and probably from pre-Pliocene time. "The stupendous cliffs of St. Helena seem to represent a duration of wave-attack which may be reasonably guessed as more than 2,000,000 years".

The sea-level related to the strongest glaciation and the corresponding abrasion and shelf-surface has been drawn tentatively. It will be seen at once that the glacial shelf-body is entirely buried by the more recent shelf-formation and that its present morphology is the result of both bodies being fused together.

In reality the situation must necessarily be still more complex because of the repeated alternations of glacial and preglacial sea-levels. In other regions present circumstances might be such as to cause the recent profile of equilibrium not to cover the Pleistocene shelf-surface entirely. In that case one or more older surfaces might be exposed in the outer area of the shelf between the isobaths of 70 and 200 meters. However, the outer parts of the continental shelves present further problems worth of special attention.

Subsidence of the Shelf-Area

It has been mentioned already that the preceding considerations are based on the implicit assumption of stable conditions of the shelf-area in Pleistocene times. There are reasons, however, for saying that for the outer shelves of the continents such a surmise is not justified. The problem is much more complicated. One remarkable feature of the outer portion of the shelf is the frequent occurrence of coarse sediments, sometimes even called "rock bottom" It was noticed on both sides of the Atlantic.

On more than one occasion Shepard reported it from the shelf along the eastern coast of the United States. And Bourcart made a thorough study of this phenomenon in shelves of North Africa and southern Europe. Shepard reported it also from off the Californian coast.

This feature is of great importance as it appears that the sediments on the continental shelf form an exception to the well-established rule of decreasing grain-size with increasing distance from the shore. The conclusion of Shepard and Cohee reads as follows):

"The study of a large number of samples from the shelf off the MidAtlantic States has shown that there is the same absence of outward decreasing gradation of grain-size which had been suggested previously, on less substantial evidence, for the shelves of the world in general. The sediment is, for the most part quite out of harmony with present-day conditions and was evidently deposited during the Pleistocene states of lowered sea-level".

In a later paper Shepard) advocated: "the concept of an older generation of sediments formed on the shelf when it was exposed to subaerial conditions during the last glacial epoch and left uncovered since the rise in sea-level, because the currents may have removed some of the finer portions of the sediments".

This quotation from Shepard leads us to the following remarks. A theory based on changing level of the sea only, might account for a Pleistocene plane of equilibrium situated at a deeper level than the 70-meter isobath1) and a shelf edge at about 200 meters. This would correspond to the possibility of subaerial erosion at a depth of (at maximum) 100 meters below present sea-level.

Hence, if the level of the coarse sediments be situated below the isobath of about 100 meters it never could have been exposed to subaerial conditions unless it be assumed that — apart from a rise of sealevel — a simultaneous or subsequent downward movement of the bottom has also taken place! Conversely, if convincing arguments could be presented to show the formation of the coarse sediments to have taken place either in the litoral zone or under subaerial conditions, we would — for the areas under consideration — have to suppose a subsequent downward movement of the bottom superimposed on the changing sea-level.

Now, according to Bourcart the outer shelves, at least those of southern Europe and North Africa, which he studied in some detail, can have no other interpretation. Indeed, they provide good reasons for believing in the widespread influence of a subsidence of the bottom as an additional factor of importance.

For he found the continental platform strewn with a variety of boulders of rather large dimensions. Often their diameter is 50 or 60 centimeters. The possibility of their being brought from the land towards their actual place by sea-currents is out of the question.

Moreover, to many of them Bryozoa and other delicately branched organisms have attached themselves and the very fragility of these organisms prove the boulders not to have been displaced or even moved by submarine currents. A further asset of importance is their occurrence in regions — e.g. off the low coasts of southern France — which were unaffected by glaciers. In general the dimensions of the boulders surpass those of boulders carried to day by rivers in the coastal area.

Apparently they were transported by more powerful rivers in the Pleistocene. For the reasons just enumerated the boulders have to be considered as orginally deposited in the litoral zone. And remarkably enough, in many places especially off the coast from Gibraltar towards Senegal the boulders were found to occur in elongated tracts of increasing depth, parallel to the coast.

One between 0 and 50 meters, a second between the isobaths of 50 and 100 meters, a third between 100 and 200 meters, and off Azila an elongated accumulation was even met with at a depth of 400 meters below sea-level.

So, the deepest must be considered much older than those which frequently were found down to 200 meters and which are supposed to be of Mousterian age *sensu lato* by

Bourcart. Apart from the 0 to 200 surface Bourcart distinguishes a second surface along the oceanic slopes of the shelf at a depth of 200 to 500 meters and a third between the isobaths of 500-1000 meters.

THE MARGINAL FLEXURE OF THE CONTINENTS

In clear contrast to the positive shift of the shelf area the land-side of the continental margin shows wide-spread evidence of a movement in the opposite direction. Zones clearly demonstrating a rejuvenation of the relief are known along the greater part of the oceanic coasts under consideration. It is interesting to note that in their well-known textbook of Geology, Chamberlin and Salisbury wrote as early as 1906):

"Nearly every coast is bordered by inlets which are almost invariably submerged valleys; but, followed inland, these inlets usually graduate into deep sluggish rivers, and these, farther inland, are very often replaced by rapids or falls, or at least by steepened gradients. When the continental borders are examined throughout their full extent in all latitudes, the prevalence of this phenomenon becomes impressive.

The chief exceptions are the great rivers which drain interior basins through broad gaps in the elevated tracts that so generally border the continents, as the Mississippi, which issues through the great gap between the Appalachians and the mountains of Arkansas and Indian Territory, and the Amazon, that issues between the Parima and the Brazilian mountains.

A critical study of the gradients of the normal coast-border river-channels, embracing at once the submerged portions, the inlet portions, and the high-gradient portions, indicated a warping rather than a simple uplifting and depression, such as is implied in the epeirogenic conception. This is an important factor in the alternative interpretation".

Similar ideas were published a few years later by Jessen who devoted a voluminous memoir to this subject. His treatise gives much additional evidence of the widespread occurrence of upward movements of the present continental edges — resulting in a rejuvenation of the relief along the continental border — and a downward movement of a zone parallel to it, now submerged below the level of the sea.

Rejuvenation of the relief is a characteristic feature of the recent geological past of the whole globe including continental and oceanic sectors. The facts assembled by Jessen offer some very welcome points of view regarding rejuvenated zones surrounding continental and submarine basins. But we will restrict ourselves at present to the borders of the continents. According to Jessen the thickenings of the continental margin may be classified according to three types, viz.

- Warping accompanied by faults,
- Updoming accompanied by faults and slight folding,
- Folding and overthrusting, accompanied by underthrust of the oceanfloor, giving origin to a deep-sea trough1).

It is clear that a subsequent landward migration of the hinge of the marginal flexure causes a shifting of the shoreline2). This phenomenon may perhaps account for the local absence of a marginal zone of elevation as for example along the Atlantic coast of southern France.

The notion of the continental flexure is also found in a publication by Veatch and Smith of the year 1937. Veatch found that the Tertiary and Cretaceous beds of the coastal plain near the mouth of the Congo have a thickness of over 3000 meters and rest on a Cretaceous peneplained surface sloping seaward.

The uplifted part of the peneplain has been eroded to a new peneplain in about Mid-Miocene times. This Mid-Miocene peneplain was uplifted and warped by Mid- to Late Miocene disturbances.

Probably, however, successive post-Miocene uplifts and warpings have tended to form several additional partially developed peneplains. And these involve "probably not only upwarping landward but accompanying downwarping seaward". The result is: "a steepening of the continental slope initiated by the warping of the.

Miocene peneplain, both by subsequent marginal deposition and the upwarping responsible for the younger partially developed peneplains". However, the accurate charting of the Atlantic shelf of North America revealed some complicated features of great interest.

One feature, as revealed by the excellent charts published by Veatch and Smith shows a gradual deepening of part of the shelf-surface from the south towards the north including the edge of the shelf.

The shelf broadens from 95 kilometers east of Cape Henry to 150 km near the Hudson channel and it remains approximately 125 to 135 km wide from off Long Island up to Georges Bank. The related characteristics of the surface and edge of the shelf are compiled.

Obviously the outer part of the shelf shows a downward tilt to the northnortheast. This phenomenon is revealed.

- By the lowering of the shelf edge from 60 fathoms down to 80 fathoms.
- By the deepening of a zone of narrowed isobaths called "Franklin Shore"— from 40 fathoms to 60 fathoms over the same distance, i.e. between the southern limit of the charted area east of Cape Henry to south of Nantucket.

The downward northerly tilt was clearly recognized by Smith who considered it as "probably the residual tilt due to incomplete recuperation from ice-loads"1). The Franklin Shore is regarded by Smith "as the result of a stillstand in the sea level during one of the intermediate glaciations".

I cannot agree with this interpretation of the Franklin Shore. Both the shelf edge and the Franklin Shore show the same amount of notherly tilt, viz. 20 fathoms, and both diverge from the shore in a northern direction. Therefore both were probably in existence when the bottom movements happened to occur.

However, remarkably enough the distance between the present shore-line and the 40 fathom isobath remains approximately constant (80-90 km). For obvious reasons the area occupied by the Hudson delta forms an exception.

As a tentative explanation of these features I suggest that the region east of the Franklin Shore is an older shelf surface which was brought downward by a movement along the continental flexure. After this movement a new shelf-surface formed adjusted to the then prevailing conditions of submarine. Erosion and aggradation. This surface is still to be seen in the region between the isobath of 40 fathoms and the Franklin Shore which may be considered as the seaward end of the new surface.

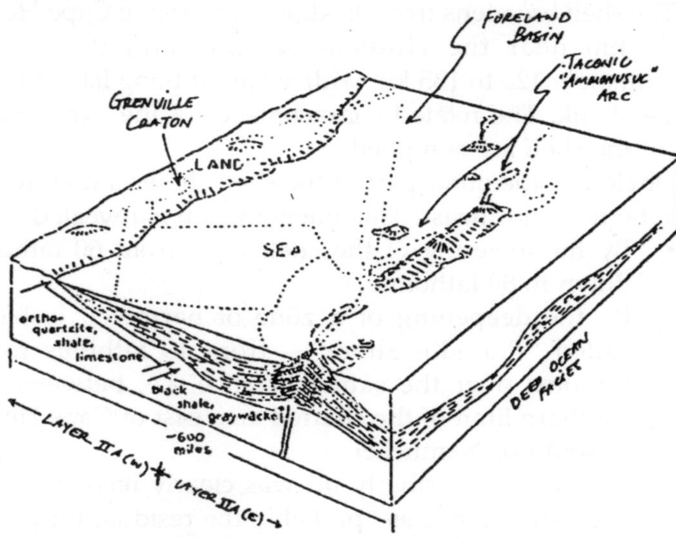

Fig. Schematic Block-Diagram of the Atlantic Shelf

Then a northerly tilt of the shelf-area occurred, probably due to ice-load and the subsequent beginning of a partical isostatic restoration of the region due to vanishing of the ice-masses, as suggested by Smith. Finally the shelf-surface between the present shore line and the isobath of 40 fathoms formed as a surface adjusted to conditions prevailing at present.

Summarizing, our suggestion will make clear that the morphology of the Atlantic shelf of North America shows the influence of two different sets of movements, one being a spasmodic seaward tilt due to a movement along the continental flexure, the other being an additional notherly tilt due to local processes of loading and unloading of a Pleistocene ice-mass.

The age of the respective movements cannot be fixed exactly for lack of data. Nor is a correlation of the surfaces to those postulated by.

Bourcart attempted. We may say only that if the northerly tilt be due to the load of ice then the most recent ice-mass was the Wisconsin ice-cap which is known to have advanced as far as Long Island.

Now the Wisconsin Ice Age was at its maximum approximately 50,000 years ago and the recession of the ice-cap started approximately 30,000 years ago. Hence, if the northerly tilt was caused by the Wisconsin ice-mass, then the shelf-surface between the 40 fathom-isobath and the Franklin Shore would date at least from pre-Wisconsin time.

And this would hold good *a fortiori* for the lower surface of the shelf between the Franklin-Shore and the shelf-edge. On the other hand the shelf surface between the shore and the isobath of 40 fathoms would have been modelled during post-Wisconsin time, i.e. during the last 30,000 years.

Marine Coast-Terraces

The most recent part of the movement of the coastal tract is demonstrated by the elevation of marine terraces along the continental border. In *"The Changing World of the Ice-Age"* Daly has given a clear review of the intricate problems inherent to the age and correlation of the marine coast-terraces.

Although locally marine terraces were found at exceptional heights) the majority of the coastal tracts show four terraces, the uppermost one — called Sicilian in the Mediterranean — occuring at an altitude of 100 meters above sea-level.

In such far distant regions as the eastern border of the Pacific, New Zealand, South Africa, Chile, Alaska and Japan the present altitude of marine terraces has been tentatively correlated.

By various authors with the well known scheme of four marine terraces occuring in the Mediterranean.

It would be a too naive idea to surmise eustatic emergence

causing a constant height of the terraces all over the world. For as a result of the shallowing of the ocean during glacial stages the coast-terraces would not now be of exactly uniform height.

Moreover, even the melting of all the land-ice now existing would rise the sea-level only about 40 meters. Hence the present altitude of the Sicilian terrace cannot be explained by eustatic movements alone. The same holds good for the Milazzian terrace which stands at 55 to 60 meters above present sea-level.

The possibility of matching the marine terraces with the glacial stages was put forward by Daly who pointed to the fact that the fossil molluscs of the 30-meter Tyrrhenian terrace lived in water considerably warmer than the present Mediterranean. He therefore suggested the correlation of the Tyrrhenian with the Yarmouth (Mindel-Riss) interglacial stage, which is known to have been the longest and warmest Pleistocene stage

Possibly the climate of the whole world was then so mild that little land-ice was left and the sea-level was raised some 30 meters above its present level i.e. to the height of the Tyrrhenian terrace. If this assumption be accepted as correct it follows that an elevation of the coast must have taken place in post-Sicilian, and even in post-Milazzian times.

Now the Sicilian terrace is usually dated as pre-glacial which means that the sea-level of that time stood 40 meters higher than at present. According to this line of argument the total elevation of the Sicilian terrace can be estimated at 90 to 100 minus 40, i.e. at 50 to 60 meters.

If we suppose the Milazzian terrace to have been formed under conditions approximately equal to the present so called "post-glacial" time, an elevation of 55 to 60 meters — the present altitude of the Milazzian terrace — must have taken place in post-Milazzian though pre-Tyrrhenian times. And this amount corresponds exactly with the amount of elevation of the Sicilian terrace!

However, everyone will feel that the problem is beset with many doubtful questions and numerous pitfalls). For, in the first place the inter-regional correlation of terraces is a difficult and delicate procedure in itself. Secondly the fossil molluscs of the Sicilian terrace give evidence of a sea cooler than the present Mediterranean.

On the other hand the fauna of the Milazzian terrace seems more in accordance with the idea of warmer conditions than are found in the Mediterranean waters of our days. Unless one would try to explain these features by supposing paleogeographic conditions to have been responsible for these facts, they present difficulties which do not fit in the above given reconstruction of events. Hence, future research may lead to quite different results!

A further difficulty is presented by the marine terraces of Hawaii Stearns found them to occur at the same height which is characteristic of those found by Cooke in S. Carolina and they were correlated to the well known scheme of Dépéret for southern Europe and that of Krige for South Africa.

The Hawaiian islands present a difficulty in so far as they form a comparatively small area which, moreover, has no sialic substratum—according to now prevailing ideas. Hence one ought to explain why in these islands the same coastal features are met with as in the marginal zone of the continents — notwithstanding the apparent differences just mentioned.

If, however, a pre-Tyrrhenian elevation of the coastal tract should prove to be the general rule, no synchronism would seem to be established between the movement of the shelf-area and the adjacent coastal strip. Obviously the determination of the ages of the Pleistocene movements is still far from being a settled question. We have to wait for many more data which only can be assembled by detailed geological surveys of the continental margin.

THE GEOPHYSICAL SIDE OF THE PROBLEM

In the preceding sections an endeavor has been made to picture the events that seem characteristic of the continental margin in its superficial parts. More difficult is the problem of the processes underlying these phenomena in deeper layers of the earth.

In a schematic section through the continental margin of

eastern Australia Jensen expressed his opinion that a displacement of deep-seated material ought to take place from the down-faulted blocks of the shelf and the deep-sea towards the continent.

Probably currents in the simatic substratum have to be considered as the primary agencies. Their action appears to involve the movements along the continental flexure — in this case the origin of the faults bordering the shelf of eastern Australia and the subsequent movements along these faults. The attempt to unravel the mutual connections of these phenomena means entering the domain of mainly theoretical speculations.

The theory of continental flexure implies the periodical occurrence along the continental margin of movements of a rather small though uniform amount causing a subsidence of the shelf-areas and an elevation of the adjacent coastal tracts. The question arises whether such processes can be accounted for from a geophysical point of view.

Fig. Geological Section Across Eastern Australia and the Shelf of the Great Barrier Reef..

A further question is by what mechanism isostatic equilibrium is re-established in the deeper parts of the down-bent or down-faulted portions of the earth's crust. *Mutatis mutandis* the same problem arises for the periodically elevated part of the continental margin.

The features of the continental border Chamberlin) supposed the ocean basins to have sunk and to have crowded somewhat upon the land sectors. Hence at their junction the sea-bottom would have tended to sink and at the same time to have been pushed under the land, "while the latter tended to rise relatively, and perhaps even to spread above toward the ocean-basin".

According to his opinion the submerged portions of the shelf-area would be carried obliquely out of the water during epochs of a strong world-wide compression of the outer shell of the earth.

After such an epoch of compression accompanied by the bowing up of the continental border a slow reverse movement would set in and this would again submerge the now strongly eroded coastal region.

Similarly an alternating sinking and elevation of the ocean-bottoms as a whole, accompanied by simultaneous epochs of upheaval and downward movement of the continental border is ascribed by Jessen) to hypothetical processes in a presumed sheet of "salsima" which is supposed by him to underly the sial of the ocean-bottom and to terminate at the sialic thickenings of the continents.

The facts known at present do not seem in favour of these hypotheses. In general the shelf-area is intermittently subjected to a downward movement, whereas the continental side is only elevated periodically. The rejuvenated relief of the continental border is worn down by subaerial erosion in the course of time and the denudation products are transported seawards where most of it accumulates in the shelf-area.

Gravity observations at sea carried out at right angles to the coast, generally show an increase in the isostatic anomalies of 30 to 100 milligals, when passing from the shelf to deep water. This rather sudden jump in the anomalies mostly causes positive values at the ocean side of the profiles.

In all 26 complete profiles have been obtained by Vening Meinesz on board submarines of the Royal Dutch Navy; three at the end of the Channel, one near Lisbon, four along the coast of W. Africa, two near the mouth of Chesapeake bay four between Panama and San Francisco (one incomplete), six off the east-coast of S. America, two off the west-coast of Australia, one off the south coast of Ceylon and one incomplete near Socotra.

From a study of several methods of isostatic reduction of the gravity stations Vening Meinesz concludes that the most probable explanation of the data is to assume one or more of the sialic layers of the earth's crust to thin out rather suddenly at the continental margin. His solution is in agreement with the generally accepted idea — based on seismological and petrographic data — of the presence of a thick granitic layer in the continents and a rather sudden thinning out at the continental margin.

A much thinner sialic layer is thought to be present under the Atlantic and Indian Oceans. And probably this layer is totally absent in some other oceanic sectors including the North Polar Basin and a large part of the Pacific Ocean inside the so-called andesite line. The question why these upper layers of the earth's crust are distributed in such a remarkable manner involves the whole problem of the origin of continents and ocean floors. It forms one of the most fundamental and baffling problems of geology.

However, possibly the peculiar distribution of sialic and simatic masses in the border zone of continental and oceanic areas causes the intermittent action of convection currents and these, in turn, cause the periodical movements along the continental flexure.

In order to explain why the submerged parts do not emerge during a period of rest of the convection system one has to admit that sialic material of the deeply submerged lower part of the continental margin is transported continentward. It thus supplies new sialic material to the continent from below, whereas erosion on the surface tends to the opposite effect.

Finally, if the intermittent submergence of the shelf-area has to be considered as a world-wide phenomenon continuing from primeval times, we would have to conclude that a complete series of sediments lies buried in the deposits of the shelf-body. Chamberlin and Salisbury already laid emphasis on this aspect of the situation. However, from the few facts at hand such a sequence seems only partly to be realised. For example, the shelfsediments accumulated on the sunken borderland Appalachia seem to cover a sequence ranging from Mesozoic to recent times.

It appears that Appalachia began to subside during the Triassic, i.e. after the mountain building of the neighboring Appalachian gesosyncline. Similar time-relations between the beginning of the submergence of a borderland and certain tectonic phenomena in adjacent geosynclinal belts were mentioned in a previous section.

In addition it may be mentioned that the subsidence of the present shelf-area of southwestern Europe cannot possibly have begun earlier than after the Variscian chains of Europe had been formed. Hence, the shelves that existed in these regions previous to those the origin of which can be dated approximately, must necessarely have been situated farther off the present continental margin.

What was the origin, situation and history of these presumed shelves of a very remote past is a problem that cannot be attacked at the moment for lack of reliable data.

Submarine Valleys

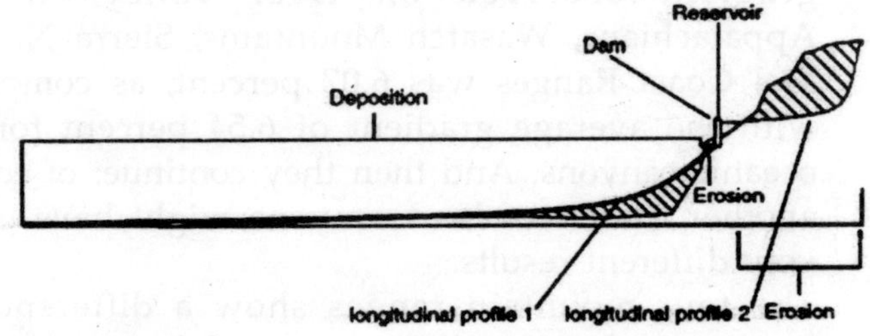

Fig. Longitudinal Profilesof Canyons

The continental margin shows a geomorphological feature of high interest in the wide spread occurrence of submarine canyon-like valleys extending from the shelf downwards to two or three thousand meters. It is not yet possible to furnish a well-founded classification of these furrows on the upper surface and outward slopes of the shelf since our knowledge of the submarine relief is still far from complete. However, the following three types may be fairly well distinguished and tentatively classified.

- As a first group submarine gorges originating near the edge of the shelf and running downwards to great depth have to be mentioned. In many places they were found crowded together in great numbers. Their course is in general uniformly parallel, and perpendicular to the edge of the shelf: They are cut back only a short distance into the platform of the shelf, their headward extensions being seldom more than 5 to 10 miles.

  As an additional feature these gorges only exceptionally show branching of appreciable extent and have comparatively few tributary gorges. Respecting steepness and depth they are comparable to large subaerial canyons. Cretaceous and Tertiary fossils dredged by Stetson in 1936 from the walls of some of these deeply submerged marginal notches of the shelf bordering eastern North America seem to prove that at least the upper part of the canyons were cut in Late Tertiary or Pleistocene time.

  According to Shepard and Beard) the average gradient observed on river valleys in the Appalachians, Wasatch Mountains, Sierra Nevada and Coast Ranges was 6.07 percent, as compared with the average gradient of 6.54 percent for the oceanic canyons. And then they continue: of course another Choice of land canyons might have given very different results.

  The four mountain ranges show a difference in gradients) Land valleys are well know to show decreasing gradients along their length. The submarine canyons have the same tendency by the averages of 11.62; 6.63 and 4.76 for the head, center and end portions. Furthermore it is true of both land valleys and submarine canyons that most of the large rivers have much lower gradients than the small." The same author was of the opinion that the high average slope, the steepness of the gradients of the streams of the American continental slope and the probable uniformity of the Cretaceous and Tertiary beds of the region are responsible for the characteristic tendency of the valleys to be so uniformly parallel).

- A few gorges of the type mentioned just now extend across the continental platform, their much shallower headward extensions reaching the vicinity of the shore near or at the debouchement of a large river. According to Bourcart the gradient of the longitudinal profile of these gorges rather surpasses the gradient of a normal river profile in its lower or even median course.

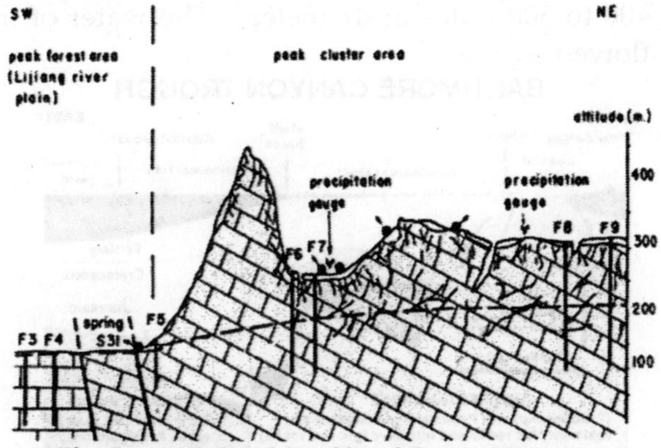

Fig. Longitudinal Profile of Congo Canyon.

These figures are however, widely diverging from those given by Shepard and Beard, viz.: the submarine slopes of the Congo the Indus, and the Ganges having, according to them gradients of 1.1, 1.0 and 0.8 percent respectively. The canyon is incised at the deepest as much as 1480 meters (820 fathoms) below the general slope of the shelf.

The Hudson canyon is cut 1120 meters below the rim near the edge of the shelf. These are no extraordinary figures if compared to the maximal depth of 1830 m incision below the rim of the Colorado canyon and 2370 meters of the Hell's canyon of Snake River in Oregon and Idaho).

Two cross-sections through the Hudson canyon and should be compared to those of the submarine canyons off the Californian coast. Veatch regards the Congo valley as of postglacial origin. According to this interpretation the main erosion of what is now the elevated interior Congo basin took place when the Congo river had a much more northerly course and reached the sea several hundred miles north of the present mouth.

He asserts that "faulting along the western part of this original course of the river caused the water to pond in the interior basin forming a great lake about 400 to 500 miles in diameter" The water of the lake flowed over a

BALTIMORE CANYON TROUGH

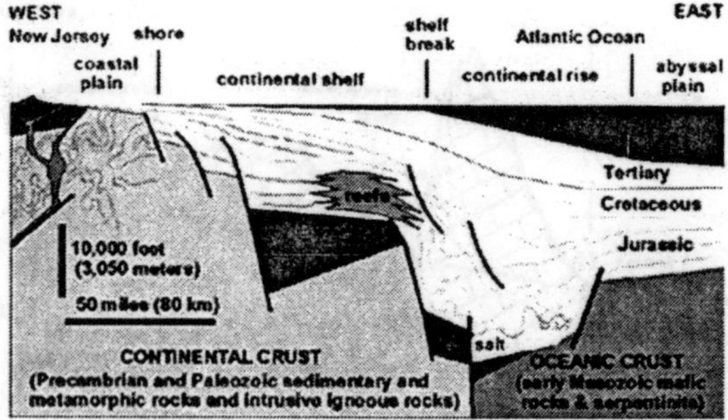

Fig. Cross-Section of Hudson Canyon.

low point of the surrounding rim and started the Congo on its present course to the sea. Now the time of this lake is fixed by the abundant human artifacts found beneath the sills of the lake — and regarded as Mousterian, some of them even as Aurignacian and Solutrean. Hence, the present course of the Congo seems of post-Mousterian age. Most authors regarded the origin of the submarine canyons of groups

- as contemporaneous. And since its appears that the gorges are cut in Cretaceous and Tertiary beds, it seems certain that they are at least younger than Pliocene. This conclusion applies to the Congo canyon as well as to those off the eastern coast of the United States).
- Another type of submarine canyon Was revealed by sonic soundings off the coast of California. They are characterized by a dendritic river-like pattern, deeply incised in the surface of the continental shelf and thence continuing towards the great depth of the Pacific Ocean.

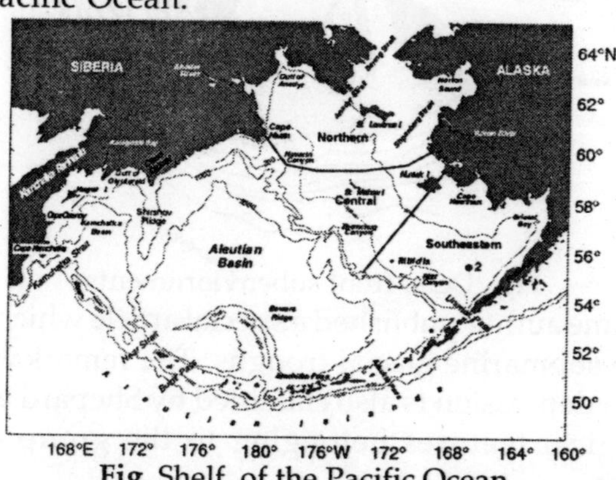

Fig. Shelf of the Pacific Ocean.

It is to these submarine notches incised in the surface of the continental shelves and running down along their outer slopes towards the abyssal regions of the oceans that the following considerations will be devoted. For it would be erroneous to classify all elongated submarine depressions under the same head and to assume their history to be genetically the same.

For example, a structure which will be left out of consideration here is the submarine depression occuring southwest of the delta of the Missippi. It looks like a scar formed by the sliding down of a part of the shelf, possibly along faults. Shepard is of the opinion that its origin was connected with the many submarine saltdomes found in the vicinity of the edge of the shelf.

The submarine trough-like bottom configuration in the Bahamas described by Hess, will also be left out of consideration here. Whatever its origin may be the morphological features seem clearly contrasted with those of the gorges.

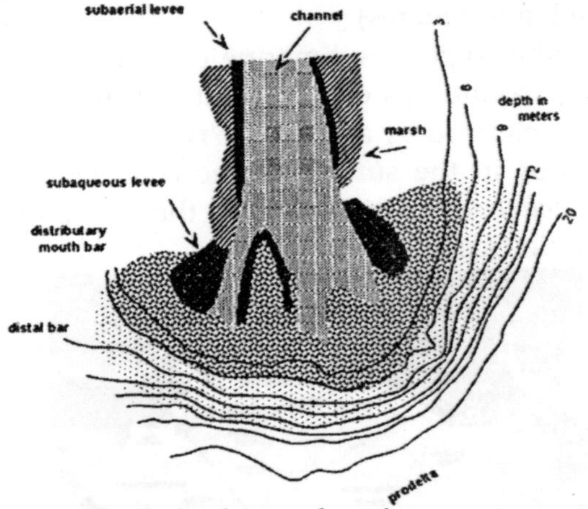

Fig. Delta front subenviornments

The same author published a special article which is mainly devoted to submarine glacial troughs. The remarkable South-Norwegian depression is also classified by Shepard among the morphological features belonging to the group of glacial phenomena.

The characteristic difference of cross-sections as compared to a submarine canyon of the "river-type"

A further case which has to be discarded from the problem of the submarine canyons is the submarine relief which was revealed round the island Bogoslof Smith, who published a contour-chart of the surroundings of this little volcanic island in the

Aleutians, was probably quite right in arguing that the relief of the bottom originated under subaerial conditions and that it was subsequently drowned to great depths. In the case of the relief surrounding Bogoslof, however, there is again no shallow outer shelf platform.

All the problems connected with the presence of the outer shelf are simply lacking and Bogoslof therefore will be left out of consideration. To include the case of Bogoslof in a discussion on the general problem of submarine gorges along the surface and edge of the shelf — as has been done by some authors — would be to compare unequal. situations.

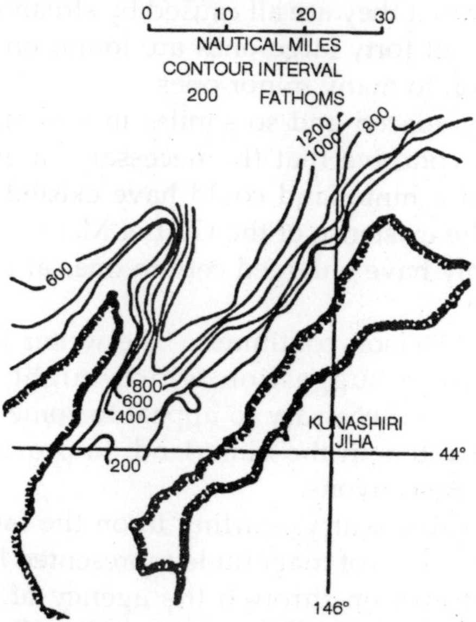

Fig. Type of Submarine Valley off the Coast of Japan.

The Notched Shelf Edge

Submarine gorges of the type mentioned under:

- were charted in great abundance along the outer slope of the shelf of Georges Bank and from Cape Hatteras to the Grand Banks.

  These gorges have been traced to a depth of more than 8000 feet. They display, however, the remarkable feature that their headward extension is seldom cut back into the surface of the shelf more than 5 to 12 miles.

  The exceptions will be mentioned later on. In the shallower parts of the shelf (less than 50 fathoms) all trace of them is lost. Their head and side-walls are steep; v-shaped in crosssection, and from 2 up to 6 miles in width.

THEORIES OF SUBAERIAL ORIGIN

In his article on the geology of Georges Bank Stetson wrote: "the abundance of the canyons themselves precludes the possibility that they are all caused by streams crossing the shelf. Upward of forty large ones are found on the southern face in addition to many minor ones.

So closely spaced and so similar in size are they that it is difficult to conceive that the necessary number of rivers flowing out of a hinterland could have existed side by side, even before the existence of the Gulf of Maine. Stream piracy would certainly have enlarged certain ones at the expense of the others".

And then Stetson continues: "The writer is indebted to Kirk Bryan for a suggestion which might explain this situation. It is not necessary to appeal to some ancient river, with its headwaters in the hinterland, to provide the means for cutting these canyons.

In a plateau country, starting upon the face of a scarp, canyons of the order of magnitude represented here can grow by headward erosion through the agency of groundwater sapping. Many formed by this means are well shown on the United States Geological Survey's topographic sheet of the Bright Angel quadrangle, Coconino Country, Arizona" etc.

Indeed the similarity is very striking and many authors have tried to explain the submarine gullies by suggesting a subaerial origin and a subsequent submergence of the area. Consequently the continuation of the above quotation of Stetson's reads as follows:

"Elevation of the continental margin would provide ideal conditions for erosion of this sort. If Georges Bank now stood about 7000 feet above sealevel, its southern face would look not unlike the scarp near Cameron".

Here, however, lurks the unsurmountable difficulty viz., the improbability of an oscillation of thousands of meters either of the continental margin or of the sea-level since

Pleistocene times. According to the usual estimates, Pleistocene changes of sea-level would account only for an oscillation of about 100 meters.

And even Shepard who once claimed a glacial lowering of sea-level of some 3000 feet by suggesting enormous icecaps about four miles thick over the Arctic and Antarctic regions — was unable to account for the lower 3000 to 5000 feet of the submarine notches.

Hess and MacClintock even went so far as to put forward a hypothesis which holds that under the influence of some stellar body the earth's rotation might have been decreased causing a depression of the sea-level in low latitudes and raising it in high latitudes.

A sudden change in the speed of rotation is, however, highly improbable from a mechanical and astronomical point of view. Such a happening would moreover bring about formidable changes in other processes on the surface of the earth. Nothing of the kind is known to geologists.

On the other hand a theory surmising that the continental margin could have been uplifted to a height of 3000 or more meters above its present level and that subsequently it could have been submerged by a similar amount, has to be rejected on very sound arguments.

Elevation of the shelf-area accompanied by comparative stability of the land would undoubtedly have resulted in a very characteristic drainage system of the dammed-up rivers at the land side. Again no effect of the kind is to be noticed in the present coastal areas of the continents.

Moreover, if rivers be assumed to have cut the many trenches, then — as Daly remarked — "the master rivers of the continent must have simultaneously induced physiographic changes (either cutting broad valleys with floors now far below sea-level or else making local, voluminous deposits of sediment) on a much larger scale and visible in the existing morphology and structure of the continental land. No drastic effects of the kind have been found".

Remarkably enough difficulties of this short have been overlooked even by a few very serious workers. Bourcart and

Francis-Boeuf for example are of the opinion that during the low levels of the Pleistocene when the shelfsurface was largely situated at or above sea-level, the upper part of the canyons were incised.

Applying his theory of an intermittent down-bending of the shelf-area one would have to ascribe the formation of the deeper parts of the canyons to a more remote past. This theory would be adequate if the depth of the gullies at the edge of the shelf were a negligible factor. However, near the shelf-edge (in the region of the 200 meter isobath of the shelf-surface) most of the gullies show a depth of more than 1000 meters. In order to account for an erosion-base at that level one would have to admit the hypothesis of an oscillation of at least over 1200 meters!

A great marginal uparching followed by a coastal downwarping of still greater magnitude is indeed postulated by Du Toit.

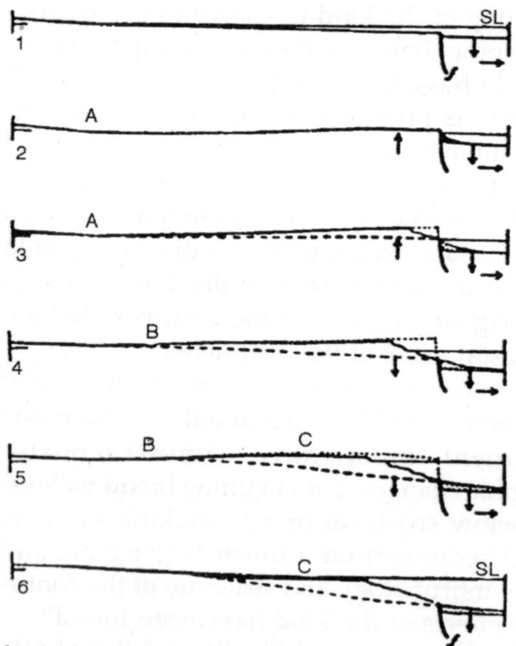

Fig. Hypothetical Stages in the Evolution of a Submarine Canyon; The Broken line Indicates the Floor of the Canyon.

His scheme involves both faulting and warping with trends more or less parallel to the continental margin. In his opinion a process of periodic tension, due to continental drift and operating from the Late Mesozoic onwards, was responsible for the origin of one or more coastal faults accompanying the downwarping of the continental rim along the Atlantic and Indian oceans.

The same process of thinning of the oceanic sial-layer would cause the continental side of the fault-block to bend upwards "in accordance with isostatic and paramorphic principles" while the rivers maintained and deepened their courses forming subaerial canyons. In order to explain the present submarine gorges a coastal up-arching of some 4000 feet is postulated. Then:,,continued or renewed tension causes further sagging of the ocean floor with downdragging of the rim".

The result would be a great marginal down warping of the furrowed coast, the drowning of the gorges, and finally, the formation of a marine abrasion plane. This plane would constitute the continental shelf which in most cases would have been cut right across the drowned gorges that mostly became severed from their subaerial parts.

All that has been set forth in the previous sections on the origin of the shelf is in strong contradiction to such a theory. For, the characteristics of the shelfsurface only allow for a subsidence of the shelf combined with changes of sea-level to a total amount of some 300 meters since Pleistocene time, certainly not for an amount more than four times as large. Nor do the marine coast-terraces fit in such a theory.

We are, therefore, compelled to look for submarine agencies which might have caused the remarkable phenomena. However, before entering into this aspect of the problem, it may be remarked that some other difficulties which were inherent in older theories are explained very nicely by Bourcart's theory of the continental flexure.

In the first place his view accounts for the occurrence of a break of slope at 500 and 1000 meters and perhaps at deeper levels below the sea. In the mean time it explains the steepening of the outer margin of the shelf and finally it might perhaps explain why the average gradient of the submarine canyon-profile is steeper than those generally found in rivers on land.

Theories Involving Submarine Agencies

The problem, however, of the origin of these submarine trenches remains untouched by Bourcart's considerations. Therefore, in order to explain these submarine furrows attention must be given to submarine agencies as a possible cause of notching of the edge of the shelf. Several factors have been advocated by various authors.

The late D. Johnson brought them under the following headings: submarine landslides, submarine currents, subterranean river outlets, foundering of subterranean caverns, solution along faults either by uprising subterranean waters or by downfiltering marine waters, non-deposition along faults due to uprising subterranean waters and finally — sapping by submerged artesian springs (the latter hypothesis being put forward by Johnson himself).

It is clear from such an *embarras de choix* that the problem of the notched shelf-edges is far from being solved. Indeed: "Fifty years of discussion, and much new knowledge of the characteristics of the canyons, have not, however, brought agreement among geologists respecting the genesis of these forms.

Many hypotheses have been advanced, but none has yet attained the standing of a generally accepted theory ". We may refer anyone interested in the variegated diversity of these theories to Johnson's review and the more recent literature cited at the end

An exception, however, will be made in respect of Daly's theory in which appeal has been made to the widespread influence of turbidity currents. For even his opponents will avow that it represents a fine example of a well considered working hypothesis.

And, moreover, it has so far received more attention from other investigators than has any rival hypothesis involving a submarine origin of the trenches. According to Daly waves beating upon the loose sediments of the continental shelf during epochs of a lowered sealevel of the Pleistocene, charged the sea-water with much sediment held in suspension.

These turbid waters tended to flow down over the shelf-edge as density-currents which carved the trenches. During Kuenen's laboratory-experiments concerning Daly's hypothesis I more than once witnessed the fascinating view of a suspension running down into the deeper water of the tank along the artificial "canyon" previously carved by him in the outer slope of the shelfmodel.

It is, however, a crucial point, whether currents of muddy seawater of glacial times were able to erode shelf-notches to a depth of 3000 and more feet. Daly and Kuenen are hopeful in so far as they think that further experiments and field data will bring a full confirmation of the theory.

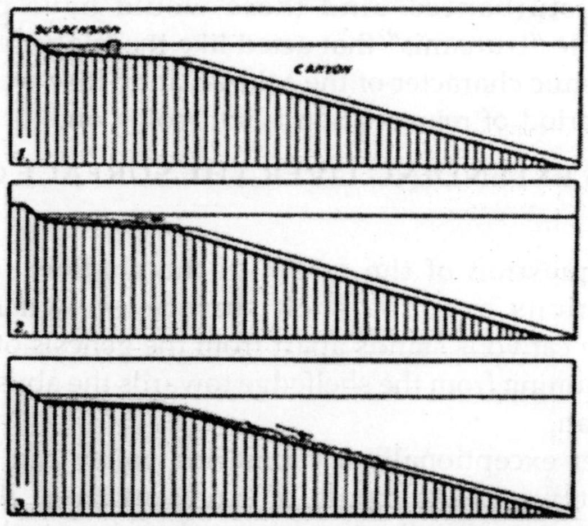

Fig. Kuenen's Experiments Showing Density Currents Running Down the Slope of a Shelf-Model

However, if the inner walls of the canyons are composed of solid rocks, instead of loose sediments, it seems hard to believe in Pleistocene density or turbity currents as the agency which was competent to notch the continental shelves. Now, according to Stetson the walls of the submarine canyons of Georges Bank actually consist partly of strongly lithified rocks and the same is held true by Bourcart regarding some trenches off the coast of Portugal and southern France.

Both groups of data are worthy of further examination and we have to wait for many more facts which can only be gained by future investigations.

A different theory was offered by Bucher. He agrees that the cause of the submarine erosion must lie in movements of the ocean water. He suggests, however, that the action of seismic sea waves (tsunamis) has to be regarded as the principal factor capable of etching valleys along the continental slopes.

Bucher pointed out that the Upper Tertiary period of rejuvenation affected also such regions as the Mid-Atlantic Rise, as is shown by elevated deposits of Upper Tertiary age on most of the oceanic islands of the Atlantic.

This rejuvenating movement was accompanied by strong seismic disturbances. And these would have produced innumerable "tsunamis" that acted like the trigger effect. The highly seismic character of the Mid-Atlantic Rise would show that the period of rejuvenation is still going on.

GORGES EXTENDING OVER THE SURFACE OF THE SHELF.

The question of the origin of the shallow headward prolongations over the shelf-surface, shown by a few submarine canyons stands apart from the genesis of the deep notches running from the shelfedge towards the abyssal region of the ocean.

Rather exceptionally a submarine valley can be traced upwards towards the mouth of a present river. The Congo canyon provides a well-known example, whereas the head of the Indus canyon is situated 20 kilometers off the coast formed by the delta of the river after which it is named. It crosses the broad shelf at the edge of which its depth is about 1000 meters.

Out of the mouth of the river leads a shallow and rather broad submarine valley connecting New York Harbor with a submarine canyon which extends down the slope beyond. The outer gorge has been cut into the shelf platform twice as far as the neighboring canyons and those of Georges Bank, viz. 18 miles.

And instead of being straight the course of the gorge is s-shaped. The shallower valley on the platform dies out near the margin of the shelf and the canyon heads about six miles northeast of the end of the shallow valley.

Traces of an ancient Hudson delta are found near the continental edge. According to Shepard : "tracing the canyon down the slope it is shown to make a double bend and come into line With the inner valley for most of its length.

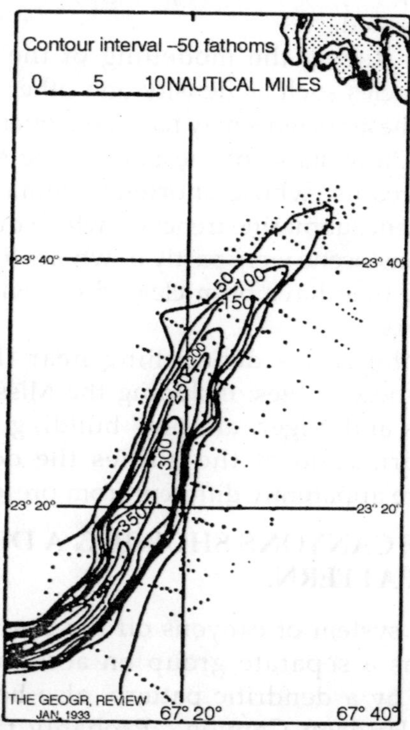

Fig. The Submarine Valley off the In-Dus Delta.
Each Dot Represent a Sounding.

This suggests the possibility that the two were formerly connected, but that they were separated as a result of a complex series of sea-level oscillations which allowed the filling in of the head of the canyon and the production of a new head as a result of a diversion of the inner valley".

A few students of the problem have assumed that many

more submarine gorges, which now appear to be incised only a short distance into the shelf, must once have had a greater headward extension.

If these supposed extensions did exist, they must have been obliterated by marine planation and filled during certain epochs of the Pleistocene. And only if they were dependent on a large stream like the, Congo river the obliterated channels were partly cleared in "post-glacial" times or a new channel was formed.

Obviously during the modelling of the shelf-surface submarine agencies such as density and other currents have been at work. These currents may have held open gullies which originated at a low stand of the Pleistocene sea-level from consequent streams, whose shortened remains still exist. Moreover, ancient submarine trenches, which during previous stages of the Pleistocene were partly or wholly filled with loose sediments may then have been cleared out wholly or partly by such agencies.

Some of the rivers debouching near the headward extensions of these gullies, including the Missisippi, Fraser, Indus, Ganges and Niger, are now building deltas. Hence during the formation of the gullies the oceanographic conditions were apparently different from present conditions.

SUBMARINE CANYONS SHOWING A DENDRITIC RIVERLIKE PATTERN.

The great system of canyons off the coast of California was classed as a separate group on account of its being characterized by a dendritic pattern. also be compared to those of the Hudson Canyon. Probably the submarine topography of the Californian canyons may indeed be considered as a subsided marginal block of the continent and a drowned system of canyons originally formed under subaerial conditions. As a matter of fact Maxson argued that the crust of northern California may probably be regarded as a fault-coast.

Daly thought it "quite possible that a few trenches, including perhaps more than one off the coast of California, were actually river-cut far back in geological time, then drowned, afterwards more or less completely filled with sediment, and finally, recently, re-excavated by the same' process which caused the majority of the open furrows now being discovered by soundings".

Investigations by Emery and Shepard give strong support to the theory of the subaerial origin of these canyons. Rock was obtained from 132 localities. Great quantities of fragments were obtained in a single haul. The fragments were often of large size, angular and with fresh fractures.

These features suggest that at any rate a great deal of the samples come from the bedrock outcropping at the sampling locality. The submarine topography much resembles the structural features of California. Several faults could be traced in the surface of the shelf. The rocks on the sea-floor include sedimentary, igneous, and metamorphic types.

Triassic, Jurassic, Cretaceous, Eocene, Miocene and Pliocene are all represented and their distribution on the sea-floor "conforms to the paleogeographic maps given by Reed and Reed and Hollister, though changes are necessary for the Middle Miocene and the Pliocene maps...." On the wall of one canyon consolidated Pliocene sediments were found, indicating that at least this canyon was formed in postPliocene time. At any rate this conclusion holds good for the upper part of the canyon.

- From the available data it appears that the gradation-plane of the surface of the shelf, so far as it is adjusted to present conditions, is formed as a surface gradually deepening from the coast to a depth of about 70 meters (40 fathoms) below sea-level. This figure is based on observations on the Atlantic and Pacific shelves of North America. For two islands in the southern Atlantic, Saint Helena and Ascension, a slightly higher figure -nearly 100 meters — seems to be the most probable estimate.

  Generally, however, the break of slope which is mostly called the edge of the shelf is situated lower, often near the isobath of 200 meters. The question therefore, arises how the shelf-surface below the isobath of 70 meters originated. In order to explain this situation the possible influence of eustatic changes of sea-level and movements of the bottom have to be examined.

- At first sight it would seem that the problem may be solved in a simple way by surmising the deeper part of the shelf-surface to have been graded by submarine agencies during Pleistocene epochs of a lowered sealevel and to be adjusted to the existing hypsometric relation between seasurface and shelf-surface.

Although eustatic changes undoubtedly must be regarded as a factor of importance, the problem is more complicated. For, on both sides of the Atlantic the occurrence of coarse sediments and boulders was observed on the outer parts of the shelf and these features bear undoubted evidence of their being originally deposited in the littoral zone.

Therefore, apart from eustatic changes (which during the Pleistocene may have caused the deposition of littoral deposits at a depth of maximally about 100 meters below present sea-level) an additional widespread subsidence of the bottom in the order of 100 meters should be assumed.

The same author distinguishes a second surface occurring along the oceanic slopes of the shelf at a depth of 200 to 500 meters, and a third between the isobaths of 500 and 1000 meters, as noticed in many regions off the Atlantic coasts studied by him. Apparently, in the course of time the original continental margin has sunken deeply below the level of the sea. A prism of sediments (for the greater part erosion products from the continents) accumulated on it. Such a situation was revealed by the seismic refraction measurements of Ewing,

Crary and Rutherford in the case of the sunken borderland Appalachia, which is now covered by a pile of shelf deposits ranging from Triassic to recent times and attaining a thickness of 12000 feet at a distance of 60 miles off the present coastline.

- In clear contrast to the positive shift of the shelf-area, the land-side of the continental margin shows widespread evidence of a bottom-movement in the opposite direction, a zone parallel to the oceanic coast being characterized by rejuvenation of the relief. Chamberlin and Salisbury and in recent times especially Jessen and Bourcart paid attention to this phenomenon. Locally a more recent post-Mousterian tilting has been noticed.
- The periodical action of convection currents which are supposed to display themselves below the continental margin as a result of the peculiar distribution of sialic and simatic masses in the border zone of continental and oceanic areas, possibly accounts for the periodical movements along the continental flexure. Thus, the phenomena of the continental margin are possibly correlated to other periodic events occuring in the earth's crust and its substratum.
- The submarine canyon-like trenches which are incised in the outer slopes and — occasionally — in the surface of the shelf may tentatively be classified in three groups, viz.
  - Slightly ramified notches in the shelf-edge running down towards great abyssal depth,
  - Gorges of the same type which however, have headward extensions over the surface of the shelf,
  - Submarine canyons showing a dendritic river-like pattern.

Near the shelf-edge the trenches of group:

- Show a depth of 1000 meters and more. In order to account for an erosion base at that level one would have to admit an oscillation of at least 1000 meters either of the bottom or of the sea-level. However, such a hypothesis is incompatible with the data presently known about the geological history of the continental margin. Therefore theories involving a subaerial origin of the notched shelf-edge have to be abandoned. And hence their formation must be due to submarine agencies. The gullies of group

- Probably originated as oceanward extensions of large consequent rivers existing in Pleistocene and more recent times and debouching in notches of the shelf-edge. During the Pleistocene and more recent modelling of the upper shelf-surface the gullies probably were held open by submarine processes. The canyons of group
- Probably formed in pre-glacial times under subaerial conditions. Subsequently a subsidence of the bottom caused them to become drowned. Afterwards they were more or less completely filled with sediments and finally they were reexcavated by the same processes suggested for the digging of the furrows of group (a).

Chapter 25

Island-Arcs

Island-arcs are among the most challenging features of the earth's surface. Island-arcs border the eastern margin of Asia and appear in the West Indies and the so-called Southern Antilles which form a discontinuous festoon between South America and Antarctica.

At the same time these arcuate structures represent one of the most intricate problems of structural and historical geology. When considering a geographical globe, one wonders how these remarkable island-festoons have developed into their actual shapes. A further question which arises is, why similar features are conspicuously absent from other areas such as e.g. the eastern border of the Pacific, the entire African continent, Europe, Greenland and Australia with the exception of its northern part.

On the other hand a geological examination of the earth's surface shows that arcuate structures are by no means confined to certain border-regions of continental and oceanic areas. A single glance at Plates that they form one of the most conspicuous features in the structural pattern of such continents as Asia and Europe.

To some extent they also played an important part in the geological history of North and South America and eastern Australia. Again we are struck by the fact that they are lacking in some other continental areas, at least since Cambrian times, i.e. since the moment from which the surface history of the continents can be deciphered to some appreciable degree, thus far..

It will be remembered that among the characteristics

mentioned with some emphasis were the following two. Firstly, the fact that some of the arcuate belts are arranged consecutively around a central nucleus of respectable antiquity, spreading in ever widening arcs.

Secondly, it may be stated that the younger the age of the belt the greater its distance from the Pre-Cambrian nucleus. Again, however, it had to be added at once that other continental areas do not reveal such a regular centrifugal configuration of tectonic belts.

On the contrary, in some districts the sequence is irregular and non-concentric and may even display the reverse of the time-relations just mentioned. Elsewhere certain structural belts boldly intersect older zones.

Once more we learn that be earth's crust is not built up by a single dynamic activity that can the expressed in a simple formula or "law".

Islands-arcs are associated with several other remarkable features. They are the sites of conspicuous girdles of still active volcanism. Zones of strong seismic activity can be traced in their immediate vicinity.

Deep-sea furrows are situated along their external side; deep basins appear at their concave side. Belts of large disturbances of isostatic equilibrium accompany them. The phenomenon of deep-focus earthquakes seems to be related to these zones. And, finally, terrestrial magnetism probably shows noticeable deviations from its normal values in these arcuate island-zones.

However, once more it should be added at once that by no means all the characteristics enumerated so far are confined to these structures alone. It is true that the different types of basins and troughs accompanying island-arcs have their counterpart on the continents proper.

And, to mention only one more striking feature, the structural zones of the East Indian festoon can be followed northwestward into similar structural zones in the Asiatic continent, viz. in Burma, whereas the Aleutian arc continues into the North American continent and the Kurile arc into Kamschatka.

Accordingly, problems of the island-arcs are for evident reasons linked to those of their counterparts on the continents

THE PROBLEM

At this stage it is worth while to formulate in a few words the principal lines that will be followed in attacking the problematic origin of islandarcs.

At the outset the question why island-arcs occur in one district while they are conspicuously absent from others will be left out of consideration.

Let us not begin by asking why the western border of the Pacific shows a series of island-festoons from the Aleutians to the East Indies, whereas no such arcs occur along its eastern border.

Doubtless the presence of the Antillean arcs amidst the arc-less coasts of the Atlantic is a striking phenomenon. But to ask for an explanation of their absence from Europe, Africa and South America raises a problem which may well be treated apart from the question as to how the arcs originated in the districts where they actually exist. And it is to the last named question that the following sections are principally devoted. At the end it will be useful to return once more to the problem of their absence from other areas.

However, in order to avoid misunderstanding, the subject needs even further precision. It should be added that the problem treated in the following pages is restricted

- To those single island-arcs which are associated with volcanism and show a marginal deep-sea trough in front of their convex side;
- To double island-arcs, the inner one being a volcanic arc, the outer one a non-volcanic arc with a marginal deep in front of it. Hence, per definitionem a single row of islands with a more or less arc-like shape, like those in front of the coast of British Columbia, fall outside the subject under discussion as the festoon is not associated with a deep-sea trough.

It is evident that the origin of single and double arcs is one of the principal problems to be attacked. We want to know how the row of volcanoes came into existence and why they are absent from the outer are of a double festoon. Among the other points that come up for discussion are the origin of the marginal deep and other deep-sea troughs and basins which are associated with the island-arcs.

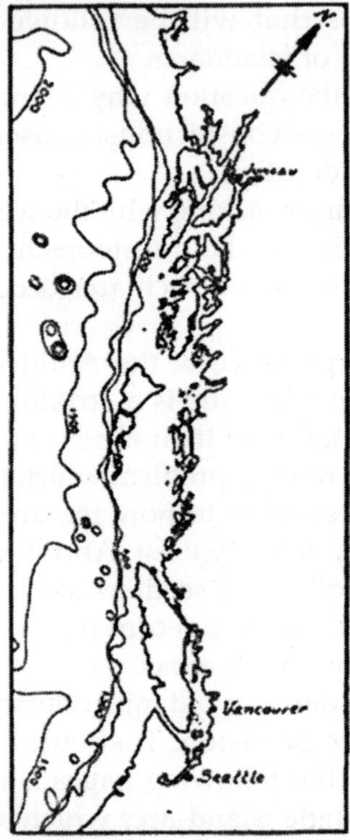

Fig. Island-Festoon in Front of the Coast of British Columbia.

And the same holds good regarding the accompanying belts of strong seismic activity, zones and fields of isostatic anomalies, and certain features of structural geology, plutonism and volcanism.

If possible the relations between these various groups of phenomena have to be revealed and an attempt at synthesis has to be made. A further question which deserves special attention is whether there is evidence for the theory that island-arcs differing in physiographic aspect represent at the same time different evolutionary stages in the history of a tectonic belt or not. Finally, the unsolved problems or questions for which no satisfactory explanation can be given, have to be outlined as sharply as possible.

The various groups of data will be considered in systematic order. It will be seen that, in doing so, gradually — step by step —, our insight into the genesis of the arcs changes from a chaotic assemblage of data into an attempt at a harmonious synthesis.

PHYSIOGRAPHIC FEATURES

Older Theories

Island-arcs might be classified according to different principles. Some arcs are single, others double. Some have a slight curvature, others are more strongly arcuate. In the Aleutians the outer non-volcanic arc is only slightly developed along its eastern side, near the coast of Alaska.

On the other hand the outer festoon is much more strongly developed in the East Indies. It seems an obvious thing to suggest a genetic connection between the various types and to assume an evolutionary sequence from a simple stage towards a more complicated one.

According to Hobbs the evolution of an arc would "pass through a progressive series of changes marked by ever increasing curvature". It is moreover maintained that the succession of changes through which an arc passes is the result of a pressure caused by the progressive settlement of the ocean floor, the rock floor of the sea acting as a girder. In a youthful stage the arc would show a curvature of large radius and a low elevation above the surface of the sea.

From this early stage the arc would pass through a series of movements in both horizontal and vertical directions. It is well known that Argand considered an island-arc like the East Indies to be a mountain-chain *in statu nascendi*. In a future epoch a mountain belt comparable to the Alps or the Himalayas would grow out of these embryonic structures. Then, once more the continent would have been enlarged at the cost of the oceanic areas.

The idea of a youthful stage of the island festoons apparently finds support in the occurrence of a strongly accentuated relief of the surrounding sea-floor, the seismic activity in the vicinity of the arc, the presence of a belt of active volcanoes, etc.

Fig. Reconstruction of the Alpine Geosyncline in the Mesozoic

is attractive by its combining so many different problems into a single synthesis. But, remarkably enough the island-festoons seem all to be situated inside the andesite line. A sialic crust even seems to reach as far as the festoon of the central volcanoes of the Aleutians. In the case of the Himalayas one might point to the large block of India pushing the chains towards the Asiatic continent.

But what is the pushing power in such festoons as the Aleutian and the Mariana-Pelew arcs bordered as they are by the deep lying floor of the Pacific? It may well be that the comparison of the Alps in a Mesozoic stage with the island-festoons, especially with the East Indies, is misleading.

The evidence offered by our knowledge of the structure and history of the islands-arcs has to be examined without prejudice and then the question will again be taken up whether or not the different physiographic types of island-arcs can be arranged so as to represent a series of corresponding stages in an evolutionary sequence.

It is not my intention to mention all the theories bearing on the origin of island-arcs, which in the course of time have been presented by various authors. For we should then be compelled to treat and weigh separately more than thirty opinions of widely divergent character and importance. Many of these theories are out of date.

There would, for instance, be no sense in dwelling any longer on hypotheses involving large horizontal movements of island-festoons in the way suggested by Wegener, Du Toit, Wing Easton, Smit Sibinga a.o.

Horizontal movements and crustal shortening of

geosynclinal belts are phenomena associated with epochs of compression. The movements are, however, not of the sort suggested by drift-hypotheses which envisage sialic blocks floating in a simatic syrup.

The earth is surrounded by a world-embracing rigid crust and the tangential pressures in the crust have to be described and figured in a quite different manner. These aspects of the problem are dealt.

Accordingly these older theories may be left out of consideration for the present. Moreover, the reader who feels inclined towards a more detailed study of older views will find surveys of these theories in two papers, viz. one published in 1935 by Kuenen, and one shorter review by the present author in 1934). It is true that more than two dozen of the hypotheses considered in these publications bear especially on the East

Indies, but it goes without saying that all theories presenting view-points of general importance may be found incorporated among them. Therefore I will restrict myself here in focussing the general attention only on those speculations that seem to me of special importance in the light of our present state of knowledge.

Deep-Reaching Thrust Planes and Shear-Zones

In this respect special relevance should be given here to the theories of Sollas, Lake and Lawson. The gist of Lake's theory can be summarized in a few lines. In his own words these read as follows): "In 1908 Professor Sollas showed that many mountains and island-arcs are truly circular, so far as a large-sized globe will show, but he had no explanation to offer.

We cannot infer from the shape that there must be a thrust plane at the base of each, but we can say with confidence that a thrust plane at the base would account for the shape and that no other explanation is so simple and complete."

"Moreover if there is a thrust plane at the base, it is easy from the form of the arc to determine the dip of that thrust plane at its outcrop, for the angle which the thrustplane makes with the surface is equal to the angle subtended at the centre of the Earth by the radius of the arc. The thickened portion of the line BT represents a thrustplane cutting through the outer crust of the earth. P is the pole of the arc made by the outcrop of the thrustplane. The angle *ATB* made by the thrustplane with the surface, is equal to the angle *AOT* subtended at the centre by the radius of the arc. All that is necessary is to determine the pole of the arc and measure the radius in degrees and minutes of a great circle."

"If the mountain arc is not circular a negative inference is possible. There is not now a plane thrust-surface at its base. It is possible, however, that the range was formed upon a true thrustplane and that there has been subsequent deformation. Or the surface along which movement took place may not have been a plane or the movement may have been of a different character altogether for there are other ways in which a mountain-range may be formed."

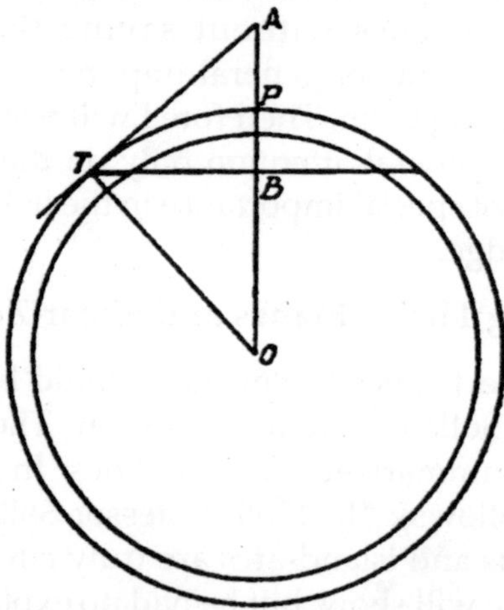

Fig. Outcrop, Angle and Dip of a Major Thrustplane Through the Earth's Crust.

Lawson is one of the modern exponents of similar ideas about the association of islandarcs, large thrustplanes and seismic activity, with subsiding sea-floor along the concave side of the arc and an elongated deep-sea furrow along its convex side. We shall turn to his theoretical considerations in the section on deep-sea troughs.

A difficulty which arises in the practical procedure of

drawing the form of an arc is to choose the right place for the circular line on the globe or geographical map. Another point that is open to personal interpretation is the fixing of the terminal points of the arc.

But let us proceed and consider some of Lake's further remarkable constructions — and conclusions -which are strikingly lucid and simple. Sollas had pointed out that the poles of his circular arcs lie upon the same great circle. Lake, surmising these arcs to coincide approximately to the outcrops of large thrust-planes, concludes "that the thrustplanes are all at right angles to the plane of the great circle that passes through their poles"!

In order to examine this question more closely he proceeded as follows. After he had chosen the lines representing the arcs the longitudes and latitudes of three points upon each line were determined, one at each end and one in the middle. "From these data the position of the pole of a circle passing through the three points was computed.

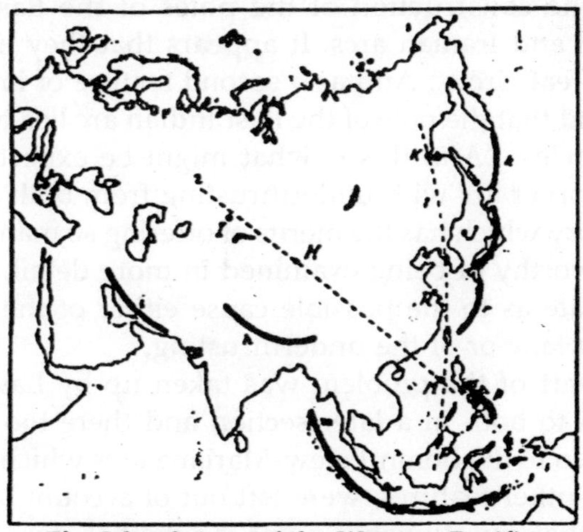

Fig. Island-Arcs of the Western Pacific.

The positions of the poles are shown on the map, and in each case a circular arc — circular upon the sphere, not circular upon the map — has been drawn through the three selected points". The Japanese arc is not a good circular arc either geologically or topographically. Therefore the position of its pole can notbe defined clearly. Hence, "the Japanese arc" should be left out of consideration for the moment.

The other four poles, however, lie so nearly on a great circle that it can hardly be a question of mere fortuity. This result is the more striking in as much as the arcuate lines were determined not on actual outcrops of the thrustplane but on various geological and topographic features which do not necessarily coincide exactly with the true outcrop.

Accordingly Lake suggested that the Asiatic continent as a whole is being pushed over the floor of the Pacific upon a series of large thrustplanes and that the direction of the movement is at right angles to the great circle indicated by the line KP on the map . More accurately the phenomenon has to be described as a continentward underthrusting of the Pacific area.

It will be seen that Lawson laid special stress on this point and speaks of a landward creep of the Pacific sima. But there are still two other remarkable features in Lake's map. One concerns the construction of the poles of the East Indian, Himalayan and Iranian arcs. It appears that they lie upon a common great circle . And as a second feature of importance it was found that the pole of the East Indian arc lies practically on both circles. "And this is what might be expected in the case of a corner arc with underthrusting from both sides."

A theory which has the merits of offering so many striking results is worthy of being examined in more detail. Lake did not speculate as to the possible cause either of the origin of the thrust plane or of the underthrusting.

This part of the problem was taken up by Lawson. We will return to both in a later section and there too attention will be given to the Bonin-Pelew-Mariana arcs which, together with some other features, were left out of account by Lake.

A further questionable point is whether deep-reaching thrustplanes are known to us from older mountain-belts on the continent. Lake and Lawson are of opinion that the great boundary fault which is known to occur along the southern border of the Himalayas might be considered as an example of such a deep reaching surface.

It generally lies between the Siwalik foot-hills and the pre-Siwalik rocks that constitute the Himalayan belt proper.

And the main ranges were pushed up from north to south along this and several other thrust-planes. According to Lake the agreement of the dip of the main boundary fault with the amount of the angle of about 14° as inferred from of the arc can hardly be an accidental coincidence. Reference was also made by Lake to the structure of the Carpathians.

In most places the foreland formations are not only independent of those of the Carpathians, but they even run almost at right angles to the border of the mountain-belt where they are abruptly cut off. Similarly the Carboniferous formations of the marginal trough along the northern front of the Variscides in northern France and Belgium has been over-ridden by older formations resting on a major thrustplane.

The great thrustplanes upon which the Caledonides of northwestern Scotland were pushed over the rocks of their foreland. Schematically the same phenomenon by the geological sections through the Sierra Nevada geosyncline.

Personally I am not convinced that the marginal thrustplanes of the regions just mentioned should be considered as the outcrops of deep-reaching shear-planes similar to those that are supposed to border the western Pacific. Probably most of the continental examples — not all — are no more than superficial features resulting from the squeezing-out of the contents of a geosyncline over the rocks of its "foreland".

In such a structure the dip of the thrustplane is largely due to the dip of the earth's crust originating from its downward buckling beneath the folded ranges. Kuenen is of a similar opinion. In his discussion of Lake's paper he wrote1): Lake's thrusts and tectonic thrusts are as different from each other as stratification is from the layers of discontinuity in the earth's crust. They may of course develop into a normal thrust-plane at the surface, as in the case of the Himalayas, in the same way as for instance a river may change to a tidal estuary near its mouth". As a matter of fact, however, these examples formed the starting point of Lake's theory.

I quite agree with Kuenen's criticism. But apart from our doubts about the validity of some of the examples on the continents, Lake's theory remains a working hypothesis of high interest. Remarkably enough, deepfocus earth-quakes were recognised in the same areas and their actual distribution induced some seismologists to construct large continent ward dipping shear-zones which reach a depth of more than 600 kilometers!

For the moment the existence of deep-reaching shear-zones connected with the location of island-arcs of the Pacific seems hardly open to doubt.

Now, obivously, a shear-zone in the substratum cannot possibly be a thrustplane of the type known from surface geology. For such a plane can not possibly persist in the plastic layers below the crust. However, if shearphenomena happen to occur they appear to be bound to that special continentward dipping seismic zone. For the sake of clearness it may be called a *potential zone of shear*.

Moreover, it will be seen that the angle between the shear-zone and the earth's surface greatly surpasses the angle that might be expected from Lake's physiographic theory. So, *if* the curvature of the arc is due to a thrustplane this plane must have been restricted to the earth's crust. It must have had a gradient corresponding with the curvature of the arc and, finally, it must have been disturbed and obliterated by crustal processes of later epochs.

The only reason for adhering to such a *hypothesis ad hoc* is the simple explanation offered by Lake's hypothesis which, however, will have to be abandoned as soon as a more plausible explanation of the shape of the island-festoons has been given. Up to now this is not the case.

CLASSIFICATION OF EARTHQUAKE ZONES

Jeffreys, after a thorough discussion of the available data, concluded in the year 1928 that the focal depth of the great majority of all tectonic earth-shocks does not surpass 35 kilometers. In the meantime Turner had argued as early as 1922, that in addition to earthquakes of the normal type, i.e.

with foci in the earth's outer mantle, shocks might also originate at much greater depth. An analogous suggestion had been made by previous workers, though based neither on adequate data nor on scientifically sound determinations.

However, in 1928, the year of publication of Jeffreys' conclusion, Wadati clearly showed that apart from the normal shocks, other earthquakes with focal depth of several hundreds of kilometers undoubtedly occur in the surroundings of Japan. Since then earthquakes with foci down to 700 kilometers below the earth's surface have been established in several other areas. Gutenberg and Richter distinguish three classes, viz.

- Normal shocks at depths not exceeding about 60 kilometers.
- Intermediate shocks at depths from 60 to 250 kilometers.
- Deep shocks from 250-700 kilometers. The latter are named plutonic earthquakes by Macelwane. The classification info three groups should be regarded only as a preliminary attempt which is not always appropriate.

  Taking the evidence from all parts of the world together, it may be said that earthquakes are known to originate at practically all levels ranging from the surface down to depths of approximately 700 kilometers. No earthquake has been located with satisfactory reliability from a depth much in excess of 700 kilometers. This well-established phenomenon has been correlated by Bullen with three other changes of properties at a depth of approximately 700 km below the earth's surface, viz;
- A rapid increase in density.
- A rapid increase in velocity of seismic waves.
- An increase in electrical conductivity. When the individual districts of seismic activity are considered separately, it appears, however, that there is no uniformity in the limits between the earthquake classes. The limits may vary regionally and nearly every district shows its own characteristic depth

zones of seismic activity. Thus in the Japanese area deep foci are known almost continuously from 300 to 650 km, whereas in South America, no shocks are known to occur at depths between 300 and 600 km. However, we should bear in mind that the considerations on shocks of the deeper classes are based on comparatively scanty data covering no more than 30 years at the most. It may be that the absence of shocks from certain depth zones is a feature of real significance. If so, we must leave an explanation of the phenomenon to the combined efforts of seismologists and other geophysicists.

The Mechanism

A consideration of the technical foundations of the seismic results does not come within the scope of the present subject. One exception, however, has to be made, although in this case the subject will be treated concisely and suited to the general understanding. Seismologists agree that the cause of the deep-focus earthquakes is in no essential respect different from tectonic earthquakes with foci situated comparatively near the surface. This conclusion is of paramount importance to geological science.

The cause cannot, for example, be of an explosive character, due to sudden and local recrystallization of unstable mineral forms initiating explosive chemical reactions such as violent atomic readjustements, overheating, or overcooling accompanied by a change of volume. The large shear-waves observed in those earthquakes leave no doubt on that point.

For if an earthquake were of explosive origin, the initial movement in the surrounding rocks would be everywhere outward, and this would be so recorded in all the seismographs, regardless of their geographic site and distance from the epicentre. On the other hand, if the movement at the focus represents a sudden slipping displacement along a shear-plane the initial movements recorded will differ in a peculiar manner at different stations, some seismographs recording an intial movement toward the focus others away from it.

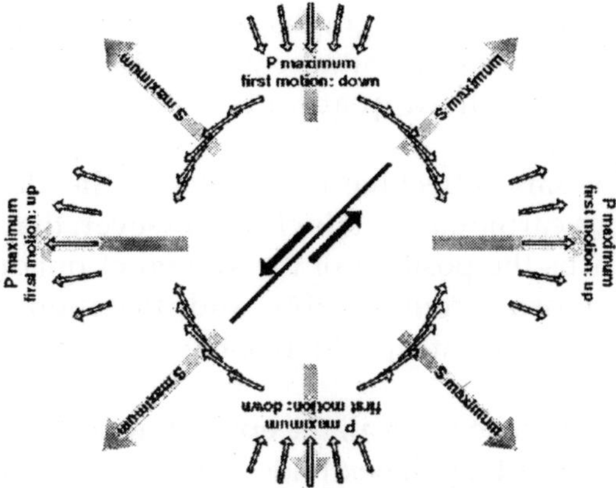

Fig. p and s wave radiation from faulting

Conclusive proof is given by a comparison of two patterns of initial dilatations and compressions, one of an earthquake that originated in the surface layers, the other one of a deep-focus earthquake. Earthquakes of tectonic origin are known to cause a very characteristic distribution of the directions of the first shocks arriving at the surface.

The direction of the first shock as recorded by the seismograph depends on the character of the event in the focal area. The first movement recorded by the apparatus will be upward, i.e. in the direction of wave propagation, provided the initial movement in the focus was of the same character.

It is called compressive. The reverse type of shock in which the initial movement is directed towards the focus, is called a dilatation. The arrow direction of the initial shocks caused by a tectonic earthquake in western Europe, each arrow representing a seismic station.

The tectonic earthquake originated along a fault of approximately *NW-SE* direction. The movement of the block along the north side of the fault was directed towards the south-east. The southern block moved in the opposite direction, i.e. northwestward. It appears that if a movement along a fault-plane takes place the surroundings of the fault have to be divided into four quadrants, in each of which the initial shock has its own direction.

As a result of the opposite movements of the blocks along

the fault-plane the seismographs of the eastern and western quadrants recorded an initial compression; in those of the northern and southern quadrants the initial shock was a dilatation.

Now the same effect has been noted in the identifications of deep-focus earthquakes. It is clear, however, that in the case of a deep focus the position of the system of quadrants (with their characterising compressions and dilatations) may vary endlessly. In a horziontal position of the fault-plane the surface features agree.

If its position be vertical, the epicentral area will not show four quadrants but it will be divided into two areas of opposite character, according to the line *AB*.. To the left of *AB* initial compression will be recorded. On the other hand the initial shock will be a dilatation in the area to the right of *AB*.

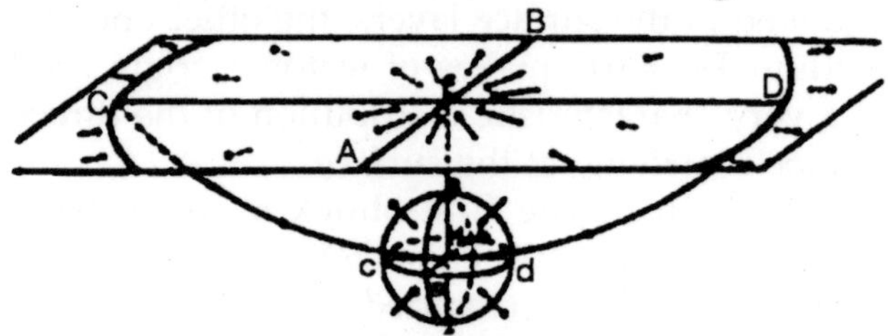

Fig. Distribution on the Surface of Initial Dilatation

A seismic shock starting in the horizontal direction *Hd* gradually curves upward along the route *HD*. Shocks reaching the surface in an area situated outside of a circle with a radius *ED* come from below the surface *abcd*. The left of the nodal line *AB* they will cause initial dilatation or motion away from the focus, whereas to the right the initial motion will be of the compression type, i.e. directed towards the focus. Hence outside of the circle — the radius of which depends upon the depth of the hypocentre *H* — the seismic effect is the reverse of what is observed inside the circular region round the epicentre *E*.

Patterns of initial shocks agreeing with these two types of theoretically deduced systems have been recorded by Honda and Ishimoto in numerous examples of Japanese earthquakes. The two examples that are offered, may serve as convincing illustrations of the important conclusion to which seismologists have arrived, viz. that both deep-focus and normal tectonic eartquakes originate by the same mechanism, viz. by shearing resulting from the sudden release of stresses accompanying a faulting movement.

Geographic Distribution of Deep-Focus Earthquakes.

Notwithstanding the apparent incompleteness of our present knowledge the examination of the geographic distribution of intermediate and deep foci has brought to light some very remarkable features.

It appears that earthquakes of the intermediate class have been identified in all regions in which shocks of the normal class are of frequent occurrence. One district is of special interest viz. a comparatively small area in Hindukusḥ. Repeatedly shocks from about 220 km have been recorded over a period of some 30 years.

Concerning the class of deep-focus earthquakes it may be said that they are known only from areas bounding the Pacific basin. A second feature of importance is the arrangement of foci in zones of ever increasing depth continentward. So they give the impression of originating along deep-reaching shear-zones, sloping inland from the Pacific border.

This idea was presented by Wadati for the first time in a map of the Japanese area, showing the contour lines for foci of equal depth. For the present we will only stop to notice the parallelism of the volcanic belts to both the deep-sea troughs and the zone of normal shocks. The zone of intermediate shocks lies more or less under the islands with their superimposed rows of volcanoes. The possible relations of these various phenomena deserve special consideration in a later section.

Any one inspecting these maps will notice the remarkable distribution of epicentres. It appears that the principal seismic activity is associated with the deep-sea troughs lying off the Japanese, Kurile- and Bonin islands. Foci of intermediate shocks lie further continentward, under Japan, and the class of deep foci still further away from the Pacific basin. It appears, moreover, that a zone of deep foci extends along the Bonin islands towards the Marianas. And another zone may be seen to run parallel to the Kurile arc and the Riu-Kiu Islands.

More or less analogous results have been obtained from other districts. the distribution of deep foci along the Aleutian arc and their relation to the deep-sea trough in front of it).

Another district known for the frequent occurrence of deep-focus earthquakes extends from New Guinea and the Solomon islands towards the region of the New Hebrides and the Loyalty islands. Once more we see that it is bordered by deep-sea troughs such as the Tonga and Kermadec deeps. Again the same associations hold good in the Philippines, the East Indies and South America.

The last named district deserves special attention because of the absence of island-arcs, which in the other districts appear to be associated with the occurrence of intermediate and deep foci along their concave side and with foci of the normal class in a zone extending continentward from the deep-sea trough on their convex side. In South America the epicentres again possibly lie on a plane sloping from the Pacific eastwards until it reaches a depth of more than 600 km east of the Andes

The East Indies present a complicated geographic distribution of deep foci apparently due to the situation of the district along the southeastern margin of the Asiatic continent and in the vicinity of Australia. The region will be considered more closely in a later section. For the present, however, it may be said that the less complicated area between Borneo and the deep Java trough again shows the same arrangement of foci along a continentward sloping zone.

A further important feature is the well-established absence of intermediate and deep shocks along the north-eastern border of the Pacific. Neither are deep foci known from the Antarctic and south-eastern Pacific. Remarkably enough deep-sea troughs of the type called marginal deeps are also conspicuously lacking along the coasts of these regions.Outside the Pacific border earthquàkes originating at depths exceeding the thickness of the earth's crust are of the intermediate class.

No true deep-focus shocks are known to be associated with large thrustplanes on the continents. This holds good even for the notable district of Hindukush.

Deep-Reaching Shear-Zones and Convection Currents

It is clear that these seismic results imply several aspects of geological importance. One question is whether the notion of shearing zones down to a depth of several hundred kilometers may be reconciled with the geophysical theories of isostasy and of convection currents in the earth's interior. It appears from modern researches that the question may be answered in the affirmative.

According to Griggs, experiments at high pressures showed in a convincing manner that „when a rock enters the region of plastic flow, it will not deform indefinitely, but will rupture if the deformation is carried far enough". Similar results were obtained by Bridgman and by Haskell.

The principal features mentioned in our short review of such seismological results as may be regarded of special importance to geology may be summarized as follows. Seismic data imply the existence of deep-reaching shear-zones along the Pacific border. And it seems hardly open to doubt that most of these shear-zones are intimately connected with the site of island-arcs.

It appears that in general the dip of the shear-zone is much greater than the amount of the dip that was calculated from the physiographic theory. Now, the theoretical construction of the curvature of the arc was rather arbitrary. In our present state of knowledge it would seem preferable to measure the curvature of the arc along the axis of the zone of negative anomalies of isostasy.

One will notice, however, that the curvature may vary considerably in one and the same island-festoon but even the preliminary data show that the dip of the shear-zone varies also within comparatively short distances. More seismic data are needed in order to get a better idea of the shape and variable dip of the shear-zone and its relation to the degree of curvature of an island-arc or parts of it.

There remains however one problematic point viz. the discrepancy between the curvature of the are and the angle the deep-reaching shear-zone makes with the earth's surface.

Possibly two sets of shear-zones formed simultaneously at the time of the origin ol the arc, viz. one in the crust and another in the substratum.

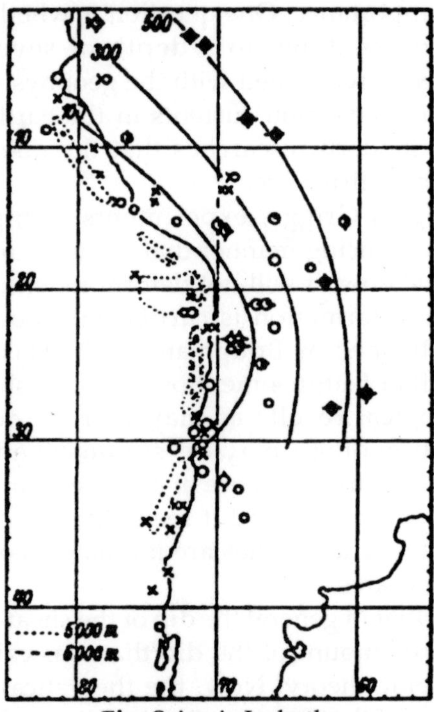

Fig. Seismic Isobaths

The shear-zone in the crust may have had a gradient of some 15 degrees corresponding with the curvature of the arc. This shear-zone, however, was disturbed and partly obliterated by crustal processes of later epochs.

On the other hand the potential deep-reaching shear-zone in the subcrustal matter was rejuvenated time and again. This tentative suggestion tries to make clear why it will be in vain to look for a correlation between the degree of curvature of the arc and the dip of the deepreaching shear-zone, whereas the latter shows a striking relation to the site of the arc.

Probably the potential shear-zone in the substratum is caused by convection currents which are in some way related to the site of the island-arcs. Island-arcs occur on the boundary of crustal areas of different composition. Hence it seems clear that the location of the island-arcs as well as the convection currents and the shear-zones all depend on the distribution of crustal matter.

Originally the present author) thought the shear-phenomena would occur on the boundary of two systems of convection currents, a stronger system on the continentward side meeting a weaker one at the oceanic side, as suggested by Holmes). In turn the shear -zone would seem the place *par excellence* for the origin of a crustal down-buckle. However, according to this theory one would expect a shear -zone in a nearly vertical position, which is not the case.

In a very recent theory Vening Meinesz made an attempt at an explanation of the obliquity of the shear-zone. He argues that owing to the higher temperature of the belt of down-buckling, a convection current will originate. It will show a rising column under the belt and a sinking column at the continentward side, viz., in the area of the deep-sea basin behind the arc. The largest values of shear are found in planes under 45° with the vertical at the four turning points of the system.

But he thinks that, probably, shear-phenomena will only take place under the influence of a trigger effect. And this effect is supposed to be caused by a rising movement of the zone of buckling.

Therefore shear-phenomena would occur especially in a zone dipping continentward at about 45° from the sialic mountain-root, which at the same time is revealed as a belt of strongly negative anomalies of gravity.

According to Vening Meinesz the curvature of the arc is the cause of the absence of another convection system of equal strength along the convex side of an island-arc. This is a weak point in his interesting hypothesis. For, the Kermadec and Tonga zone shows not an arcuate but a linear arrangement of volcanoes and deep-sea troughs.

Therefore a symmetrical set of shear-zones on either side

of the straight line might be expected according to the theory. However, they were found only on the west side.

And in South America deep-focus earthquakes were found at the continentward side of structural elements of deep-sea and coastal districts which are decidedly concave towards the Pacific.

Moreover Vening Meinesz suggests that the downward movement of such basins as the Banda Sea basin would be caused by the downward moving column of a convection system below the basin.

However, the absence of a deep-sea basin in the area of the Java sea, where deep-focus earthquakes were found as well as in the Banda Sea is not in accordance with his hypothesis. Another attempt at an explanation for the origin of the deep-sea basins will be put forward later on.

One thing will be clear from this short survey of recent theories viz., that many questions regarding the remarkable deep-reaching shear-zone still remain problematic. However, the relation between this zone and the zone of down-buckling seems unmistakable.

TERRESTRIAL MAGNETISM

According to Visser anomalies of terrestrial magnetism might be regarded of interest to the present problem in two respects. In the first place the anomaly-fields reveal a rough relation to the regions in which deep-focus earthquakes occur. In the second place they appear to be in agreement with the theory of convection currents in the earth.

However, both the sources of evidence are vague and the conclusions have the character of provisional hypotheses which stand in obvious need of confirmation by more detailed investigations.

The "zero-meridian", being the great circle through the magnetic poles, and the magnetic equator. The two circles intersect in Central Africa and in the central part of the Pacific. The other curves represent positive and negative anomalies of the vertical component of terrestrial magnetism.

The principal areas of deep-focus earthquakes are

indicated on the map by shading. It appears that they are confined to two great areas of positive deviations of the vertical component, viz. the large field which extends from the Asiatic continent into the western and southwestern part of the. Pacific, and the part of another large field that extends as far as South America.

However, no deep-focus earthquakes are known in another part of the same field that extends over the southern Atlantic and Indian Oceans.

A field of less intense positive anomalies may be seen to occur on the North American continent. A few indications of deep shocks have been recorded from the area of the south-eastern extension of this field in the Caribbean region. If these also were taken into account the arrangement of the principal deep-focus areas might appear to be roughly symmetrical with respect to the zero-meridian of the Pacific, as is the case with the fields of positive anomalies.

It hardly needs saying that the relations suggested by Visser are not yet very convincing. He concluded: "If indeed the connection between anomalies of terrestrial magnetism and deep-focus earthquakes can be proved to be real, it is by no means a simple one, as appears when we compare the fairly capricious behaviour of the deep-focus areas of the southwest Pacific with the great magnetic anomaly of the same region."

Regarding the second question Visser presumes that convection-currents transport material with special magnetic properties from the deeper realms towards the more external parts of the earth's interior and, accordingly, cause positive anomalies of terrestrial magnetism at the surface.

The small area of negative anomalies in the Pacific might be mentioned as a weak point of Visser's theory. And we may stop to notice that the field of negative anomalies over the African continent is the reverse of what one would expect from the theory.

At any rate Visser has drawn the attention to a source of data which is worthy of detailed examination when more data on terrestrial magnetism become available. Finally it should be mentioned here that in another paper, Visser reported that he had found some indications suggesting a relation between the magnetic anomalies in the East Indies and the gravity anomalies of the same region.

VOLCANISM AND PLUTONISM

One of the two fundamental postulates, which were formulated in the introduction, pertained to the invariable association of the island-festoons with volcanism. As a general statement this point needs further precision. An arc of the single type is crowned by a majestic row of volcanoes, like those of the Pelew- and Mariana islands.

In double arcs the inner one bears a crown of volcanoes, while in the outer arc volcanism is conspicuously lacking. Geological exploration, however, may reveal the presence of volcanic and plutonic rocks in the outer arc as well, but dating from more remote times. Acid batholiths may occur in both of them. And according to Hess serpentines and other ultra-basic rocks are characteristic of the outer arc.

It will be the aim of a separate section to deal with their mutual relations to the structural history of the whole region. Such a procedure seems reasonable since a discussion of these features has to be preceded by a discussion of a few other sources of data. For the same reason, however, a treatment of deep-sea troughs and isostatic anomalies necessarily has to be preceded by a short enumeration of the volcanic and petrographic characteristics of the arcs.

The Andesite Line

In the first place we may recall that the true boundary of the Pacific Basin proper is generally represented by at hypothetical line, known as the andesite line. On the continental side of the line the petrographic composition of volcanic rocks implies the probable existence of sialic material in the crust.

On the other hand volcanic rocks from the Pacific Basin situated inside the boundary of the andesite line, invariably point to a simatic ocean floor, devoid of a cover of sial. Some doubt exists concerning the Albatross plateau off the coast of

South America. Possibly a thin film of sialic matter may be present in that area. It is clear, however, that the andesite line is generally drawn as an immediate boundary of the volcanic arcs and the coastal districts that are known to belong to the andesite zone.

But nobody knows whether the sialic crust ends exactly along that imaginary line. On the contrary it would seem much more probable that in many places the true extension of the sialic part is somewhat beyond the line, and that the true limit between sialic and basaltic sea-floor is presumably much more irregular than the nicely curved andesite line on Plates 4 and 7 would suggest. It will be seen in the following section that it is well to bear this point in mind when Lawson's speculation on the origin of the arcs is examined.

Petrographic Provinces

In strong contrast to the uniformity of the basalts of the Pacific the island-arcs revealed a diversity of rock types belonging to two magmatic clans. Mostly the petrographic and petrochemical character of the rocks is of the calc-alcali or so-called pacific suite. Igneous rocks of the potash or so-called mediterranean type are also frequently met with. Their distribution in the East Indies distribution of active volcanoes in the same region.

In all 1220 rock-analyses have been reviewed by Willems in a paper especially devoted to the subject. Van Bemmelen has discussed the origin and history of both the pacific and mediterranean magma-provinces in two papers. His conclusion is that the igneous rocks of the calc-alkali type, characterizing the geanticlinal belts of the volcanic inner arc, are connected with certain orogenetic cycles.

Fig. Distribution of Rock Clans in the East Indies.

But they did not originate by differentiation of a primary or juvenile tholeiitic magma. Probably magmatization of pre-existing crustal rocks by ascending emanations from greater depths was a fundamental process.

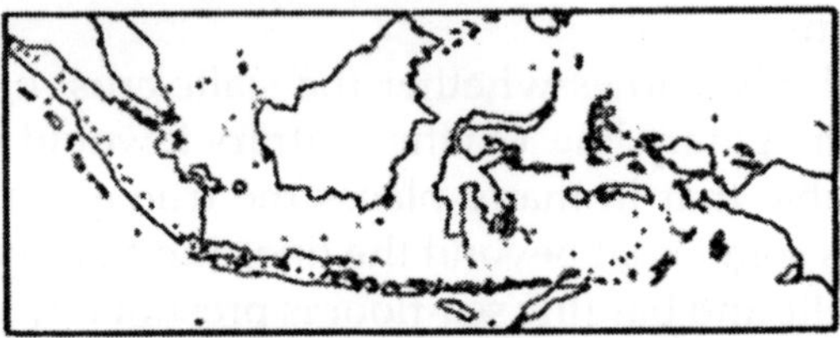

Fig. Distribution of Active Volcanoes in the East Indies.

Besides, hybridisation or contamination of the ascending magma by assimilation and migmatization of sialic crustal rocks played a role of importance. If the plutonic rocks were entirely the products of juvenile origin and normal gravitational differentation they would be more basic in the deeper parts.

Instead of such a sequence the rocks appear to be more acid towards the depth of the batholith. And this supports the view that the granitization was effected by migrating materials emanating from a rising migmatite front, thereby causing a progressive metamorphism of the existing crustal layers.

The alkaline rocks of the potash province are considered by Van Bemmelen as "pathological" products of which the origin is doubtful.

Their formation may be the result of assimilation of comparatively large quantities of limestone and subsequent loss of soda in the hydrothermal phase. A process of this kind in connection with the lavas of the Vesuvius but according to other views, the alkaline rocks may be due to a more intensive action of magmatic emanations.

Igneous Rocks in Geanticlinal Belts

It will be seen presently that the upward migration of magmatic emanations from the depth was especially active during epochs of increasing compression. During these epochs the geanticlinal areas tended to rise and apparently this process stimulated and activated the underlying earthmaterial.

Batholithic intrusions, mainly granodiorites and granites of Upper Neogene age, are known to occur on certain islands of the inner arc, viz. Sumatra, Java, Flores, Lirang and Wetar, and obviously the present volcanic action might be genetically related to such bodies.

We owe to Van Bemmelen a clear insight into their mutual relations. It is worth while to quote a passage which summarizes the historical succession of igneous activity in the inner arc. Before doing so it should be mentioned that previous to the Upper Miocene a series of calc-alkaline volcanics called the Lower- and Mid-Miocene "old andesite formation" had accumulated in Sumatra.

"During the younger Miocene"— Van Bemmelen—"an orogenic phase occurred. The geanticlinal zone was lifted above sea-level and at the same time it has been block-faulted and intruded by acid magma. These intrusions nowadays are exposed as batholiths, stocks and bosses of coarsemedium grained granite and granodiorite with contemporaneous offshoots and dikes of dacitic and liparitic appearance.

The whole old-andesite formation has been more or less altered by hydrothermal processes (such as propylitisation) and locally gold-silver ores originated.."

"Thereafter, the geanticline sank down again and in some places this system of old-andesite formation with acid intrusions has been covered unconformably with younger Neogene (Pliocene) marine sediments."

"After this orogenic phase during the younger Miocene, with its concomitant acid intrusions and explosive paroxysms, in some places, the basicintermediary volcanism temporarily came to rest, at other places however it seems to have continued uninterrupted.

But soon it gathered again enough force to cause the general, formidable volcanic activity of the geanticlinal zone during the Plio-Pleistocene." "These numerous, younger strato-volcanoes of basic-intermediary, calc-alkaline composition and Plio-Pleistocene age form an unconformable capping of the geanticlinal zone. This so-called younger andesitic series has not been altered (or only locally) by hydrothermal processes." "In southern Sumatra this period of younger andesitic eruptions was followed by a second uplift of the geanticlinal zone, which was accompanied by a second suite of acid (dacitic — liparitic) eruptions. These eruptions occurred chiefly along the remarkable rift-graben on the top of the geanticline ,the so-called Semangko-graben."

Why these details? Continually we are concerned in an attempt at understanding the challenging features of the remarkable island-festoons. It is clear, however, that an adequate theory of their formation should also account for the varied succession of phenomena presented by plutonic and volcanic rocks and the tectonic movements to which they seem to be related.

It appears from this short review that since the Upper Miocene the geanticlinal ridge of western Sumatra has been subjected more than once to a rising movement. During these epochs the arching geanticline became faulted and fractured. In its present state Sumatra shows longitudinal faults and a central rift-valley. Volcanism was very active along these faults and in many places its activity has continued along the tectonic fractures up to the present day.

The acid pacific magma is of the strongly explosive type. Ascending along the faults and graben of the updoming geanticline it gave origin to volcanotectonic phenomena which are very impressive. A whole series of calderas and large volcano-tectonic depressions came into being on the crest of the geanticline.

IGNEOUS ROCKS IN GEOSYNCLINAL BASINS

It is of importance to consider a different area of volcanic activity. represents the historical succession of events in the Karangkobar region of Central Java. Apart from the rising migmatite front accompanied by an increasing magmatization of the crust, the figure clearly demonstrates the principal epochs of tectonic activity.

The region is situated at the northern side of the geanticlinal belt of southern Java, where the latter is bounded by a geosynclinal trough. Volcanism started in Upper Miocene times with the formation of the submarine (geosynclinal) volcano Penjatan. Its vents, cutting through the Miocene Merawa-shales are now exposed by erosion.

They are composed of gabbro-dioritic rocks. Pliocene volcanism produced the volcanic products of the Ligung series which is of andesitic composition. The formation of the Pleistocene and more recent volcanoes. In the same region, but further to the north in the geosynclinal basin there occur few intrusions which belong to the alkali or mediterranean series of igneous rocks.

Finally it should be noted that several epochs of tectonic After subsidence of the bottom during the Miocene, a warping movement took place in the Upper Miocene. Then after continued subsidence of the trough the region was subjected to a rising movement in the Pleistocene. A theoretical synthesis of the formation of the East Indian island-arcs ought to account for these various expressions of plutonism, volcanism and tectonics.

Problems to be Solved

The principal problems involved in the manifold manifestations of plutonism and volcanism during the origin and history of the island-arcs may now be shortly enumerated. A synthesis which aims at explaining the structural history and present condition of the island-arcs has to elucidate at least the following eight points.

- It has to be made clear why in a double festoon a volcanic arc is always the inner one, i.e. why it is always on the concave or inner side and never on the convex or outer side of a non-volcanic arc.
- We want to know why there were two phases of elevation and rejuvenated magmatic phenomena in the region of the East Indies inner arc, since Miocene times.
- The situation of the mediterranean rocks on the concave side of the inner arc is a further problematic point.
- It has to be explained, moreover, why active

volcanoes are lacking on the islands of the outer arc.

- Then, why is active volcanism absent on the inner row north of the island Timor.
- Another feature which calls for explanation is the occurrence of a double row of volcanoes in the northern Moluccas, one on the northern arm of Celebes and thence continuing into the Sangi islands, the other on Halmaheira and some neighboring islands.
- A further problematic point is the presence of volcanic rocks of Upper Tertiary age on Buru, Ceram, Timor and Sumba.
- Finally, the occurrence of serpentines and other ultrabasic rocks on many islands has to be explained.

DEEP-SEA BASINS AND TROUGHS

A feature intimately related to island-arcs as defined in the introduction, is the presence of a deep and elongated furrow in front of the festoon. It belongs to the type called marginal deep. Among them are the deepest furrows known in the earth's crust and it seems obvious that their formation is due to some deep-reaching process of crustal deformation.

A further characteristic is the presence of deep troughs along the concave side of the outer arc, and finally a series of deep-sea basins which generally occur on the continental side of the inner arc.

The different types of troughs and basins associated with a double arc like the East Indian festoon were classified and compared with the continental basins. The comparatively recent time of their formation was demonstrated for the basins bordering the Asiatic continent in general, and the East Indies and the Bering basin more in particular.

It will be seen, in the section under the heading "Synthesis", that converging evidence from various groups of data reveals some essential features of their origin and history. For the moment, however, we will stop to examine a hypothesis which was proposed by.

Lawson in the year 1923. For although Lawson's theory

proves to be unsatisfactory the mechanism proposed by him has many interesting aspects. Compensation for relief of load by denudation of the continental areas is held to be effected by a continentward inflow of heavier subcrustal matter from surrounding regions.

At the boundary of continent and sea-floor the zone of flow would be lowered and the coast-region would become a shear-zone, the continent riding on the simatic floor which would thus creep landward. "On this rupture zone the crust of the sea floor would be thrust under the continental margin and elevate it as a ridge").

The arcuate shape of the island-festoons would arise from the fact that a thrust plane slicing the earth's surface is necessarily arc-shaped. In addition the landward under-flow of the oceanic sima would result in the formation of open fissures which would be intruded by dunite from deeper parts of the sima.

And because of the higher specific gravity of dunite the fissures would remain as hollows that are known to us as the marginal deeps in front of the island arcs. In order to explain the deep basins between the arcs and the present coast of the continents Lawson once more assumes a landward creep of the sima from beneath that region.

He supposes the region to have been a broad coastal belt that had been approximately peneplaned, whereas back of this low continental margin a region with high relief would have existed, rising steadily in response to removal of load by erosion.

Again the compensation for the erosionalloss would have been effected "by an indraft of heavy rock in depth from the coastal region, causing' the low lands and shallow epicontinental sea-floors to sink." Again the replacement of basalt by the heavier dunite would account for the submerged part of the continent lying deeper here.

As regards the four arcs of the Bonin-, Mariana-, Yap- and Pelew islands, which are arranged *en échelon*, each with a marginal deep in front, Lawson remarks: "it is not easy to regard this great compound arc as having been at one time the continental margin. To do so would be to claim that the sial extends out to east longitude 146 degrees in the latitude of Manila; and this would in some measure invalidate the hypothesis which has been advanced for the origin of the arcs nearer the mainland."

Hence Lawson tries to explain the origin of these arcs in two fundamentally different manners. Moreover, the sialic crust probably does extend so far out as these outer arcs. We need only recall the situation of the andesite line, while arguments based on the gravity-field will be mentioned in the next section.

Apart from these difficulties it is assumed by Lawson that the edge of the former continent should have had the shape of the thrust planes. No explanation is given for the complicated double arc of the.

East Indies, bordered at its south-eastern side by Australia instead of by a simatic ocean-floor, while the western part of the same arc is bounded by the sialic floor of the Indian Ocean. Another striking feature of the arcs is the presence of volcanoes on the crest of the inner arc. Again Lawson's theory offers no explanation for this conspicuous phenomenon.

Finally one of the most serious objections against the theory is of a geophysical nature. Lawson assumed the marginal deep-sea furrows to be in isostatic equilibrium. On the other hand a defect of mass in their crosssections would be rather what one might expect, having regard to their shapes as abnormal deep depressions of the crust. In order to reconcile these two opinions.

Lawson assumes intrusions of the heavier dunite from a deeper layer of the substratum into the sectors below the deep-sea furrows.

The replacement of basalt which has a specific gravity of about 3.05 by dunite with a specific gravity of about 3.3 would account for the re-establishment of isostatic equilibrium of the furrow. However, since the announcement of the hypothesis, measurements of gravity have been carried out over some of the deeps.

The results obtained revealed a situation which is not in

accordance with Lawson's theory and compel us to accept a quite different view regarding the origin of the basins and troughs, as will be seen presently.

ISOSTATIC ANOMALIES

The interesting results of the gravity expeditions at sea carried out by Vening Meinesz from 1923 to 1932 in submarines belonging to the Royal Netherlands Navy are of much importance for more than one of the problematic points under discussion. One aspect concerns the boundary between continental and simatic areas.

In addition, the results of the gravity expeditions at sea have largely contributed to a better understanding of several other phenomena. Among these are the deeper structure of the island-arcs including the origin of deep-sea troughs and volcanic belts accompanying them. These subjects will be considered subsequently.

The Boundary of the Pacific Basin

Speculations on the course of thermal convection currents under the earth's crust were advanced by Ampferer, Schwinner, Holmes, Pekeris, Escber, Griggs and Vening Meinesz. Entering into the various suggestions of their interesting hypotheses would take us too far out of the scope of the present subject. For the moment one asset of Vening Meinesz's notable results is of, special interest.

The question arises whether the boundary between areas of basaltic and sialic character coincides with the deeply dipping shear-zones that were deduced from physiographic and seismic data. Now indeed Vening Meinesz found that boundary "on the right place".

On a crossing of the Pacific from San Francisco to the Philippines, via the Hawaian-islands, the gravity field shows positive anomalies until the islands Guam is reached. Here rather strong negative anomalies reveal the presence of a crust of sialic character.

It is here too that the Pacific basin proper is bordered by the Nero deep, and that the outcrop of a deep-reaching shearzone was assumed on physiographic and seismologic grounds. It is the same boundary which from petrographic evidence is marked as the andesiteline.

It is worth while considering in some detail the gravity profiles at the border of the Pacific Basin proper.

Single Arcs

Single island-arcs at the Pacific border were crossed in two places. Both the gravimetric profiles show a similar phenomenon, viz. rather strong negative anomalies over the marginal deep-sea trough. The geological interpretation of the profile across Guam and the Nero-deep suggests the existence of a root of light crustal matter which must have buckled downward so as to replace the pre-existing heavier matter beneath.

A geological interpretation is represented by the schematic block. The anomaly curve is asymmetrical and hence seems to be best explained by assuming that the eastern sector was thrust under the western part. As a result the Nero deep originated. The riding part of the crust bowed up and gave rise to the formation of faults dissecting the crust.

Possibly a down-faulted part of the vault caused the second, though slighter downward bending of the anomaly-curve. Moreover the upward arching of this crustal part caused a relief of pressure. And accordingly emanations from the substratum ascended in these parts and caused volcanism at the surface. It is clear that the lavas will mainly belong to the pacific suite, since a sialic crust is present, and so, in fact, they are.

The place of the underthrusting was probably predestined by the boundary between two crustal areas of different composition. The location of the shear-zone, which is the site of numerous deep-focus earthquakes, is possibly due to the action of convection currents. And finally, these are probably also controlled by the different character of the ocean-floor on either side of the andesite line.

The second gravimetric profile across the Pacific border is also represented. It runs through the island Yap and the adjoining deep. Again the negative anomaly is over the deep-sea trough. Unlike the profile across Guam the anomaly-curve is symmetrical and no secondary root of sialic matter is indicated under the island Yap.

Accordingly a geological section must show a symmetrical root of the sialcrust. The site of the crustal buckle again appears to be accompanied by a deep-reaching shear-zone.

We mentioned already the symmetry of the gravimetric profile across the Yap-deep as contrasted with the asymmetrical character of the curve over the Nero-deep. This feature is explained by Vening Meinesz by the assumption that this part of the arc "encloses a greater angle with the direction of the compression".

Double Arcs

Much stronger are the deviations of isostatic equilibrium in the complicated areas of the West- and East-Indies. In the latter region Vening Meinesz made observations at 281 submarine stations. One of the most conspicuous results of his explorations is the establishing of a belt of strong negative anomalies of isostasy.

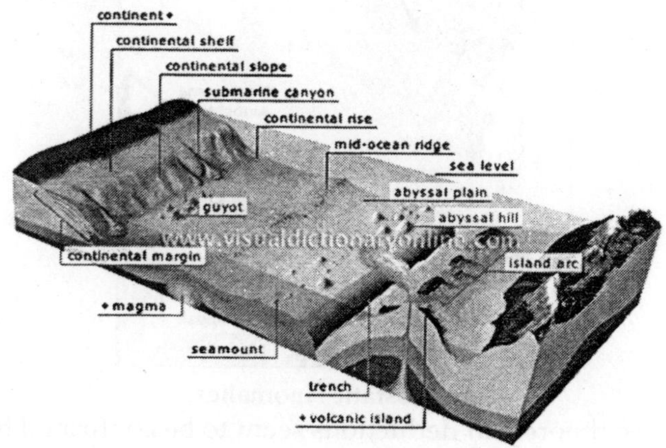

Fig. Submarine Topography

The relation between submarine topography and isostatic anomalies in the East Indies is shown by the profiles . It is clear that the situation of the negative belt is different from the Pacific border in one fundamental respect, viz. the presence of a sialic crust of much greater thickness. If the sialic part of the crust be comparatively thin and hence situated below sea-level — and moreover far from continental land — a deep-sea trough will come into being above the sialic root and a negative anomaly will be found over the deep.

The result will, however, be radically different if the sialic crust protrudes above sea-level or at any rate if it is mainly of continental thickness. For large quantities of waste-products from the land will accumulate in the downward moving crustal wave.

And when the crust buckles to form a downward root of sial, the contents of the furrow will necessarily become crumpled and squeezed out so as to form a ridge-shaped elevation above the zone of buckling. It goes without saying that in the first case considered a comparatively thin veneer of bottom-sediments was squeezed out, as well. But it did not form a factor of importance in the surface-topography. In the other case, however, a double arc is born.

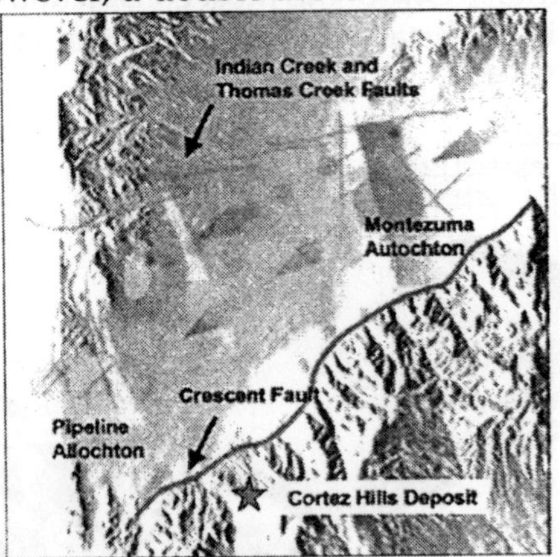

Fig. Isostatic Anomalies

These theoretical deductions seem to be confirmed by the situation of island-arcs and troughs in the East Indies. A comparison between topography and gravity-field clearly shows the belt of strong negative anomalies to follow the islands west of Sumatra and to continue over submarine ridges towards Ceram and Buru, via

Timor and the Tanimber islands. And this forms a striking contrast to the location of the anomalies along the Pacific border in the vicinity of Guam and Yap where they coincide with the marginal deep.

A more or less similar pattern of anomalies was revealed in the West Indies by the explorations of Ewing, Hess and Vening Meinesz. An interpolation of the results may Remarkably enough it appears that the negative zone partly coincides with a deep-sea furrow, viz. the Brownson trough north of Puerto Rico.

But to the east and southeast the axis of the negative zone lies over a ridge, which even partly projects above sea-level in the islands of Barbados, Tobago and Trinidad.

A proper understanding of these opposed relations will become possible only after a thorough study has been made of all available data bearing on the stratigraphy and structural geology, plutonism and volcanism, submarine topography, palaeogeography and geophysics of the West-Indies.

It is worth mentioning, however, that Hess has given the following preliminary explanation of the remarkable features of the negative belt. It will be seen that his interpretation agrees with the above given theory for the island-arcs of the Pacific and the East Indies.

In 1939 Hess wrote: "During Mesozoic time, the continent of South America to the south may have afforded an abundant source of sediments. These sediments might be expected to thin and die out northward. Thus at the time of buckling presumably at the end of Middle Eocene, there may have been a thick layer of incompetent material on the crust in the area of the southern part of the present arc, but very little at the north end.

When buckling occurred, the incompetent material might be expected to be squeezed up out of the core of the down-buckle, as in Kuenen's experiments, forming the ridge toward the south, but toward the north, where there was little such incompetent material, the axis of the down-buckle is actually indicated by a deep narrow trough."

The following section will be mainly devoted to the East

Indian arcs and the isostatic anomalies will be amply taken into consideration together with other available sources of data.

SYNTHESIS

Various sources of data have been discussed in the previous sections. Finally the structural history of the island-arcs requires to be examined. All such information will be considered in the present section and at the same time all other appropriate sources of data will come up for discussion, for the different phenomena discussed in previous sections are intimately correlated. Up to the present the highly interesting region of the East Indies has been more thoroughly examined in the fields of geology and geophysics than any other double island arc.

At the outset I shall recall the following geophysical theory of Vening Meinesz. During epochs of increasing compression the earth's crust reacts by the development of large waves of two to four hundred kilometers in length. Increasing compression will cause an increasing amplitude of the crustal waves till in one of them the strength of the crust was surpassed.

The crust then broke at the weakest place of a downward wave and there buckled inwards so as to form a sialic root penetrating the simatic rocks beneath.

The relation between seismic and gravimetric phenomena will have to be considered in the first place. Then in consecutive order the corelatedness will be demonstrated of other characteristic (and problematic) features of the outer arc, the accompanying deep-sea furrows, the volcanic inner arc and the basins on its concave side. Displacements in the substratum as revealed by fields of positive anomalies of isostasy form the next subject.

Site of the Outer Arc

The relation of the belt of negative anomalies of isostasy to the zone of normal earthquake-foci appears to be rather unmistakable in the western part of the East Indies, which is less complicated than the eastern part.

It would seem that most of the strong earthquakes

belonging to the normal class originate in the zone of negative anomalies and the accompanying deep-sea troughs. Other strong foci are known to be associated with some well established faultzones on the islands of the inner arc, for example in Java and Sumatra.

The most recent review of deep-focus earthquakes has been compiled by Gutenberg and Richter. Apparently the distribution of the foci under Java and the Java Sea is not unfavourable to the assumption of their occurrence along a deep reaching potential zone of shear.

This zone would seem to intersect the earth's surface in the area of the zone of negative anomalies and its adjoining troughs, which at the same time is a belt of seismic unrest at normal depths. At an earlier stage of our knowledge Berlage constructed a map showing seismic isobaths. On his construction the seismic surface dipping under Java and Borneo makes an angle of some 50 to 55° at the surface.

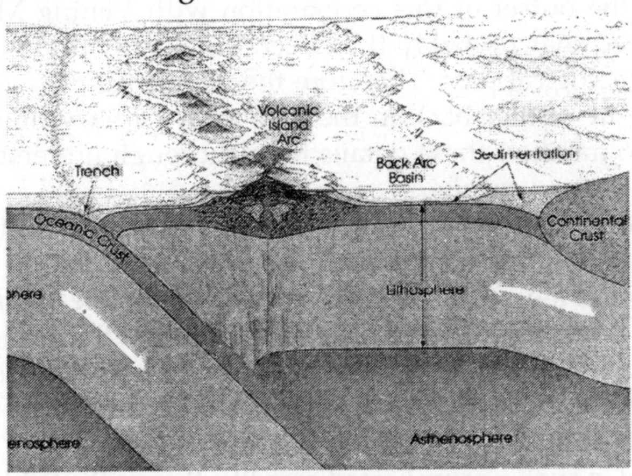

Fig. Deep-Focus Earthquakes in the East Indies .

A map differing in some details from Berlage's was published by Visser in 1943, but the angle of the seismic surface to the South of Java is also nearly 50°. Koning found about 55° for the area of the Banda Sea.

Apparently the crustal root — and hence the site of the islands arcs — was predestined to originate in a downward wave of the earth's crust which coincides with the outcrop of the deep-reaching potential zone of shear. Some theoretical suggestions on the origin of the shear-zone At any rate this zone and the zone of buckling appear to be correlated phenomena.

Time of Origin and Epochs of Rejuvenation: The downward buckling of the sialic root had some far reaching geological consequences. For the islands situated in the belt of negative anomalies may be expected to reveal intensively folded and crumpled strata. Neighboring areas might show folding pertaining to the same time. A geological determination of the age of the epoch of compression would reveal the time of formation of the sialic root.

These considerations induced the present author to take up a critical examination of the available data in order to fix the stratigraphic sequences, including the epochs of compression, for the whole region of the East Indian Archipelago.

At the outset of this coöperation with Vening Meinesz, we agreed that probably the effect had come into being in a recent geological past. Therefore the geological evidence was not sought further back in the past than Tertiary times. It is useful to mention these details for a thorough understanding of the obtained results.

And for the same reason a few more historical notes may be of some interest. congruence was not found between eastern Celebes and the axis of negative anomalies, for the latter leaves Celebes untouched and turns right on towards the Philippines, coming from Ceram and Buru. In the mean time the isostatic reduction of the gravimetric stations were made at the Bureau of the United States Coast and Geodetic Survey.

And in such a complicated area like the southern Moluccas some surprising results — deviating from the provisional estimates — might be expected. Indeed, when the final results became available the congruence between geology and gravity-field appeared to be established in a convincing way, not the least so for eastern Celebes.

The gist of the final report, published in 1934 may be seen in Plate which should be consulted constantly in the course of the ensuing pages. It shows the results of the gravimetric survey combined with a synopsis of the structural history of the region.

Moreover summarizes the stratigraphic results and the different epochs of compression for the different zones.. Repeated reference will be made to them, for the following pages may be regarded as an elucidation of Plate 8. Let us stop to consider first a few remarkable coincidences.

It appeared that the most recent epoch of compression in the islands belonging to the zone of negative anomalies occurred in a certain stage of the Upper Miocene called Tertiary f_2 in the stratigraphic taxonomy of the East Indies: notation I on Plate 8. In addition intense folding and overthrusting had been demonstrated from many of these islands.

On the other hand the same epoch of folding was demonstrated from the islands of the inner arc., e.g. Sumatra and Java, or more correctly: mainly from their western and southern parts respectively. Sumba too bears the same notation II on the map. But the folding was moderate in these regions. No crumpling and overthrusting took place comparable to that of zone I.

Among some other remarkable features the idio-geosynclinal basins should be mentioned. The moderate folding of their sedimentary contents took place at the end of the Tertiary. Again there is a striking congruence with the expectations of the geophysical theory.

For their situation on Sumatra and Java clearly shows that they happen to occur in the downward wave that originated parallel to the wave which buckled downward. Some of the idio-geosynclines began to subside at the time of the Miocene epoch of compression, others already originated in the Early Eocene but these too reveal the influence of the Miocene epoch of compression.

The belief in a comparatively early re-establishment of the isostatic equilibrium seemed to oppose any assumption of a still greater antiquity for the phenomenon. To be sure an Upper Miocene origin seemed already a greater age than was expected at the outset. For it was generally admitted that, geologically speaking, the crust reacts swiftly to loading or unloading. And it is well known that the isostatic equilibrium is being comparatively quickly re-established in such areas as Scandinavia. In response to the melting of its ice-load that area regained half of its isostatic and elastic recoil during the last 20,000 years.

On closer inspection, however, it appears that the experience gained from studying of the process of ice-loading is of no value for conclusions concerning isostatic anomalies that were produced by the formation of a mountain-root. For a quite different situation arises when a crustal down-buckle is formed.

Not only does a sialic root protrude beneath the lower surface of the normal crust, but also a bulge of the lighter crust-material is forced down into the heavier layers of the crust. And even if the root disappears by melting and spreading the sialic bulge in the crust still remains and can be indirectly observed by the negative anomalies, as Bucher and Kuenen pointed out).

Indeed, a year after the publication of the final Report I had taken up a similar critical study of the pre-Tertiary history of the East Indies). And now the results, unexpectedly, opened a new aspect regarding the time of formation of the zone of buckling.

Attention should be given here only to one interesting feature, viz. the zone indicated by letter A, the Banda geosyncline. It will be noticed at once that it coincides with part of the present belt of negative anomalies of isostasy. It now transpires that the structural history of zone A was fundamentally different from its immediate surroundings during the Mesozoic. It constituted a complicated geosyncline with an intricate structural history.

During the Triassic wasteproducts were supplied to it from the present area of the Sulu islands, New Guinea and Australia. A second large area of denudation may be noticed in the western part of the archipelago. We will, however, not go into details here inasmuch as the only point of interest regarding the problem under consideration is the discovery of the Mesozoic Banda geosyncline, the site of which corresponds to the belt of negative anomalies of gravity of our own days!

Hence our opinion with regards to the time of origin of the zone of buckling has to be revised. It appears that, at several epochs the crust was buckled downwards along the same zone of weakness. The first downbuckling of the crust in this region may have happened in a very remote past.

Hess pointed out that serpentinized peridotites are always present in the most intensely deformed part of the strongly negative strip.

They represent an ultramafic magma which was intruded during the first great crustal down buckling. Probably a peridotitic substratum is present below the basaltic layer. Assuming this, a very deeply penetrating root would reach the peridotitic magma, while at the same time, the latter would get an opportunity to invade the sialic root.

Hence a determination of the age of the serpentine-belt would fix the age of the first and greatest down buckling movement. In the West Indies this happened at the end of the Mid-Eocene).

In the East Indies, however, the serpentines and other ultra-basic rocks which were found on many islands of the outer arc, probably date from Triassic times in some places, whereas they appear to be more recent in other islands. Hence the first down-buckle may have occurred in a very remote past.).

Following upon the first buckling movement a new epoch of increasing compression caused a rejuvenation i.e. a further down-buckling of the sialic root). And this process may have occurred repeatedly, at least three or four times since the beginning of the Tertiary.

As a further inevitable consequence we are led to the conclusion that the deep-reaching potential shear-zones also originated in a very remote past) and were rejuvenated time and again during the later history of the belt.

Origin of the Double Arc

A further inspection of the structural history of the area

will elucidate many other interesting features of the intricate pattern of the East Indian Archipelago. At the same time it will furnish some striking confirmations of the geophysical theory mentioned at the outset.

To the left two parallel waves, are drawn, one upward and one downward. Let us suppose that they respectively represent zones II and III of Plate 8 and, to begin with, let us confine our comparison between theory and facts to the western part of the archipelago. It may, moreover, be assumed that profile A represents a Tertiary epoch of compression and rejuvenation.

Whether the intervening belt of folded strata (zone I) appears as a submarine ridge or as an island-festoon, depends on the quantity of strata that was squeezed out. In this case a double island-festoon has come into being, an inner volcanic arc and, parallel to it, an outer arc which at this stage is nonvolcanic.

In all probability the downward folded root did not remain intact. It is reasonable to assume that the sialic matter of the root began to melt and spread.. An actual depression of this kind is known as the central "graben" or "geosyncline" on Timor, containing a sequence of a few hundred meters of Pliocene sediments. Similar features are known from the Kei islands Tanimbar islands , and Ceram.

The ensuing epoch of compression at the end of the Pliocene, caused a moderate folding of the contents of zone III. The shallow depression on zone I was only slightly influenced at this stage, certain tectonic features at the margin of the shallow "geosyncline" being the only phenomena that have been noticed. The plutonic and volcanic processes of zone II became more active.

The last stage, represented by profile *E* the most recent period of decreasing compression. Again zone I rises isostatically. The contents of the central depression are now raised above sea-level, in places several hundreds of meters, and locally even more than a thousand). Zone III also rises isostatically but to a slighter degree, and the plutonism and volcanism of the geanticlinal belt II decreases to its present though still active phase. Again the root melts and spreads laterally in the substratum. A comparison of the gravimetric and bathymetric maps shows that the sialic root has spread, e.g., below the site of the Weber deep in the vicinity of the Kei islands.

According to the considerations just mentioned the topographic and bathymetric features of the East Indies should be of comparatively recent origin. They should all have come into being since at least the last epoch of strong compression in the Miocene. And this is what had been concluded from geological and morphological data).

It would appear from theory, moreover, that the deep-sea furrows Ia and Ib are of a still more recent origin than the depressions of zone III. In the same way the most recent elevation of zone I began later than the last rising movement of of zone II.

And the upward movement of zone is of a fundamentally different nature from that of the geanticlinal updoming of zone the height of raised coral limestones and fluviatile terraces in the southern Moluccas. No data are available to fix their age exactly, nor would it be possible at the moment to discern such comparatively slight differences of age as are postulated by the theory. Mostly they are all designated as "Pleistocene" or "Plio-Pleistocene".

In order to check these theoretical deductions one must examine the pattern of a pseudo-single arc. The western and central parts of the Aleutians consist of a single row of islands but the eastern part of the festoon — near Alaska — is double. Now the bathymetric chart of the North Pacific Ocean clearly shows that the distance between the volcanic inner arc and the marginal deep increases). This part of the arc shows the transition between the two types of marginal deeps. A similar phenomenon can be noticed with the Kurile and Japan trenches.

One of the consequences of the foregoing theoretical deductions concerns what might be called the problem of the island Sumba. Obviously the island does not belong to the volcanic inner arc. Neither in stratigraphy nor in structural history is it intimately related to the islands of the outer arc, such as Timor and Roti. Now compare the submarine relief with the gravimetric results.

A submarine ridge partly emerging above sea-level in the shape of islands like Mentawei and Nias can be followed from off the coast of Sumatra to the neighborhood of Sumba. The axis of the belt of negative anomalies coincides with the ridge, which on either side is accompanied by a deep-sea trough. Along the external side is the marginal deep, while another series of troughs lies along the inner side, among which is the Java deep.

The belt of negative anomalies terminates southwest of Sumba, only to reappear once more southeast of Sumba, continuing northeastward over the island Timor. And there a ridge is again accompanied on either side by deep troughs, viz. the Timor trough and the Sawu trough. So, the site of the island Sumba as well as the submarine topography in its vicinity seems to be intimately related to the interruption of the zone of negative anomalies). Of course the question might be put in two different ways.

One would be to ask whether the presence of the island Sumba caused the zone of buckling to become interrupted to the south of the island, and a further point would be to examine the cause-and-effect relation. The other attitude regarding the situation of Sumba might be formulated as follows. If the zone of buckling were continuous so as to unite the Timor ridge to the submarine ridge which is known to terminate S.W. of Sumba, the island Sumba would not exist, its site being occupied by a deep-sea trough!

According to our opinion Sumba stands as the exceptional evidence of a sort of terrain that elsewhere subsided so as to form the bottom of one of the series of deep-sea furrows between the outer and inner arcs I

The subcrustal migration of sialic material on either side of the root ought to cause a rising movement of the bottom of both the troughs Ia and Ib. The depth of the fluidity boundary of the sima increases towards the regions of Ib, due to the thinning out of the sialic crust-layer towards the Indian Ocean.

Therefore the rising movement of the bottom of trough Ia ought to surpass that of Ib. Now a glance at the bathymetric map shows the Ia trough between Sumatra and the Mentawei Islands to be much shallower than the marginal deep off the Mentawei Islands.

The same holds good for the Ia and Ib troughs south of Java. At first sight these data seem in remarkable agreement with the theoretical deductions.

However, between Sumba and New Guinea the bathymetric map shows exactly the reverse, the marginal deeps being shallower than the Ia troughs, although a symmetrical spreading might be expected due to the presence of the Australian continent.

In this part of the East Indies the marginal deep is bounded by the large shelf between Australia and New Guinea, which was an extensive area of denudation during most of Tertiary and Pleistocene times. In the western part of the East Indies an extensive area of denudation existed along the opposite side of the arc, bordering the Ia troughs!

It appears therefore that the bathymetric effect due to the sideways spreading of sub crustal sial is a factor of minor importance if compared with the effect due to sedimentation. Indeed, the amount of the movement may expected to be much smaller than e.g. the rising movement of zone I if the spreading of the sialic material of the root takes place in the form of thin films and filaments as suggested by profile C.

THE VOLCANIC INNER ARC

We started our disquisitions by assuming the formation of crustal waves some one to four hundred kilometers from crest to crest. The upward wave accompanying the zone of buckling on its continental side is more strongly developed than the corresponding wave on its convex side. It is moreover always a volcanic arc. The following explanation may be proposed for this state of affairs.

Subcrustal sima was pushed aside by the downward penetrating root. The curved shape of the arc involved a centripetal crowding of the sima on the concave side of the arc, whereas it could expand more freely on the convex side of the arc. This caused the arching of the geanticlinal belt to be sustained and augmented by the accumulating sima.

Hence it became an ever more pronounced belt of active plutonism and volcanism, according to the mechanism described on pp. 72, 167 and 174. No such process happened to occur on the convex side of the buckled belt. And this may explain why a volcanic arc always developes at the concave side of the non-volcanic arc).

According to our theoretical deductions plutonism and volcanism in the geanticlinal zone II would start during epochs of increasing compression. For arching of the zone induced relief of pressure at the underside of the crust and subsequent rising of a migmatite front as described on p. 72 and 174. Now the Miocene epoch of compression was followed by another one at the of the Pliocene and accordingly two periods of increased volcanic activity might be expected in zone II, since the beginning of the Miocene. And in fact these have been found by field observations.

Quite different phenomena may be expected from the zone of buckling. For during an epoch of compression no geanticlinal arching or relief of pressure occurred. On the contrary a root of mobile sialic matter penetrated downwards and the belt was strongly compressed. No manifestation of volcanism is to be expected under such circumstances. During the ensuing period of decreasing crustal compression, the melting and expanding sialic root might give rise to ascending batholiths.

Obviously, however, the mechanism of their formation was fundamentally different from that operating in the geanticlinal belt II. Seldom do the batholiths penetrate to a high level, so as to become revealed at the surface now exposed by the erosion of the arc. (The Adamello massif in the Alps may be cited as an example of a „high-level" batholith).

So, active volcanism is conspicuously absent in the outer arc of an island festoon. But volcanic rocks of Upper Tertiary age have been found on Timor, an island of the outer arc. It may be that these were derived from plutonism belonging to the buckled belt. Possibly, however, these Upper Tertiary volcanics originated in a different way.

Originally the buckled zone was bordered immediately by rising crustal waves. The geanticlinal belt was appreciably narrowed by the formation of the deepsea trough Ib. Hence the volcanism of the remaining geanticlinal zone was also restricted. But we might expect to find volcanic rocks on the bottom of the trough Ib and even on the other side, even as far as the present belt I. And we might add

- That these igneous rocks should date from Upper Tertiary times.
- that their occurrence should be restricted to the concave part of zone I.
- That they might possibly be of submarine origin, and
- That they may be expected to occur on the bottom of trough Ib as well. Now, volcanic rocks and tufts have indeed been found on the island Timor, and exactly where they might be expected to occur, viz. north of the central basin.

  Volcanic rocks of Upper Tertiary age have been found also on Sumba. But here they are widely distributed over the whole island. Once more this is in accordance with the view that Sumba has to be considered as a crustal part which, unlike the areas east and west of it, has not subsided so as to form the bottom of a deep-sea basin.

  The theory is supported by the fact that volcanic rocks on Buru and Ceram are restricted to the southern coastal district whereas they are widely distributed on Nusa Laut, Saparua Haruku, Amboina, and Amblau which are all situated to the South of Ceram and Buru.

Another remarkable feature is the double row of volcanoes in the northern Moluccas. The belt of negative anomalies is very strong between Celebes and Halmaheira. Anomalies of more than -200 milligals were found by Vening Meinesz. In addition the root shows a contour which is, concave towards the west as well as to the east. Remarkably enough one row of volcanoes runs from northern Celebes to the Sangi islands; on the opposite side are the volcanoes of Halmaheira and the neighboring islands.

The presence of a deep-sea furrow on either side of the zone of buckling would be in accordancewith the theory. And they would both simultaneously represent the marginal deep type and the intramontane trough type. Matching the inference, deepsea furrows are actually found to be present in the expected places. Kuenen marked them as marginal deeps on his map.

Finally, a feature of volcanological interest concerns the distribution of active and extinct volcanoes in the southern Moluccas. On many occasions Brouwer pointed out that volcanic action is extinct in those islands of the inner arc that are situated nearest to the island Timor, which belongs to the outer arc. No active volcanoes are found on Alor and Wetar. Proceeding from these islands towards the more western or eastern islands of the inner arc, the distance from the outer arc increases, and the number of still active volcanic vents increases as well.

Probably this phenomenon is caused by the strong compression which the crust is undergoing here, as indicated by the formation of the marginal Timor trough and more especially by the subsiding Sawu trough. For in this area the subsiding troughs had to become adapted to the limited space between the inner arc and the Australian Continent, as contrasted to their free development more to the west. This phenomenon is clearly revealed by the narrowing deep-sea relief on the bathymetric chart.

BASINS AND TROUGHS BEHIND THE INNER ARC

We may now proceed to consider the downward wave on the continental side of the volcanic geanticline or inner arc. Two possibilities should be considered separately in respect of a supply of waste-products from the surroundings.

One possibility is that the quantity of waste-products equals or even surpasses the rate of subsidence of the bottom. The trough will then gradually become filled up and appear as a geosynclinal basin. During an ensuing stage of increasing compression the contents of the furrow will become folded.. They are called idio-geosynclines and constitute oil-basins of high economic value. The strata underwent a moderate folding at the end of the Tertiary.

A comparsion of clearly reveals that most of the igneous rocks of the mediterranean suite are found in the idiogeosynclinal basins. Possibly the ascending magmatic emanations reacted with much lime-material from the marls and limestones that occur in comparatively great abundance in the geosynclinal basins and 168).

On this view the location of the alkali-rock suites, mainly along the concave side of the inner volcanic arc, seems to be controlled by the occurrence of the limestone in the geosynclinal basins, the sites of which are in turn predestined by the formation of crustal waves accompanying the down-buckled belt.

The other possibility is that the rate of subsidence of the bottom surpasses the supply of sediments. In that case a deep-sea basin will originate and persist. The Flores trough might be cited as possible example. In the eastern part of the East Indies the situation is more complicated on account of the double curvature of the arcs.

However, the deep basins may be explained along the same lines of argument, their peculiar shapes being the result of a process of interference set up by the vortex-shaped arrangements of the buckled zone.

POSITIVE ANOMALIES OF ISOSTASY

A field of positive anomalies — with an average of + 20 milligals -covers the whole area of the East Indies outside the belts of negative anomalies. According to Vening Meinesz it is caused by lateral compression of the crust.

Strips of stronger positive anomalies run parallel to the negative zone, south of Java. Vening Meinesz pointed out that the readjustment of equilibrium of the negative zone must have involved a tendency to rise. The adjoining zone would have a share in the same movement and would therefore also rise. The tendency to rise would explain the positive anomalies. A different suggestion was offered by Vening Meinesz concerning the fields of positive anomalies in the Moluccas. These fields were ascribed to downward convection currents in the substratum.

So the fields of strong positive anomalies were explained along two different lines of speculation. One would feel more content if a single theory covered the whole of the problem, the more so since the distinguishing of the two separate types of positive fields seems rather artificial.

In a more recent paper Vening Meinesz recognized the difficulties inherent to the dualistic character of his first attempt at explanation. He, therefore, tried to interpret the regular succession of positive and negative strips by a mechanical hypothesis. The alternating belts would have been brought about by the formation of crustal waves.

The waves would give origin to alternating deviations from the isostatic equilibrium of the crust, the upward wave causing positive anomalies and the downward wave negative anomalies. Continuing his considerations Vening Meinesz thinks that his theory finds support in the topographic features of the East-Indies.

For he writes: "..... the profiles of the ocean-bottom in many of the gravity profiles west of Sumatra and south of Java show a regular wave-like topography; the amplitude of the wave, i.e. the height of the topography, corresponds to the amplitude of the wave in the gravity anomalies when assuming no isostatic compensation, and this is of course as it ought to be according to our supposition because the wave is entirely a deviation from the isostatic equilibrium.

We do not find this same relation to the topography as clearly elsewhere in the archipelago, but this may be explained by the complicated deformations of the surface layers and by the other surface effects of erosion sedimentation and volcanic activity."

As a matter of fact no such agreement between topography and positive anomalies is found in the eastern part of the archipelago. However, even the most complicated part of the Moluccas shows unmistakable evidence of a parallelism between belts of positive and negative anomalies. Here too their striking interrelation is a fundamental feature which calls for an explanation combining both negative and positive anomalies in a single synthesis.

The most important features may be summed up as follows. Comparing the gravimetric and bathymetric contours, one may notice some coincidences. A deep basin like the Celebes Sea shows positive anomalies of more than 50 milligals.

The same holds good for the deep Makassar strait, the Gulf of Bone (between the southern and south-eastern arms of Celebes) and its continuation to the southern Banda Sea. From a further inspection, however, the following striking features may be noticed.

- The relation to the submarine relief is not more than a rough approximation.
- Neither the marginal deeps nor the troughs of the intramontane type show marked positive anomalies.
- In the western less complicated part of the archipelago zones of positive anomalies may be seen to run parallel to the strip of negative anomalies; nevertheless, the positive strips are not over the deep-sea furrow but roughly coincide with elevations of the bottom.

  It seems obvious that the local coincidence of a ridge and a belt of positive anomalies is purely fortuitous.

Hence the cause of the positive anomalies has to be sought not in crustal but in subcrustal phenomena! We shall therefore try to give another explanation. The following speculation may perhaps elucidate the problematic features just enumerated. The sial-root penetrated downward into the heavier substratum.

Consequently the heavy masses of the substratum were pushed aside. If it be assumed, however, that the deep-seated layer of heavy dunite was influenced in much the same way as the subcrustal layer of basaltic material, the result of the displacement of the heavy dunite would be revealed as a belt of positive anomalies on either side and at a comparatively short distance from the negative belt. The great depth of the process would explain why in one place the belt of positive anomalies corresponds to a ridge-shaped elevation of the crust, and elsewhere to a deep depression by the tentative block.

The general principle of this speculation may account also for the distribution of gravity in the complicated area of the Moluccas. Here a kind of interference between the interwoven negative zones might be expected.

And indeed, the highest positive values were found in the vicinity of the Sulu-Islands, i.e. in the immediate vicinity of the greatest negative values. The sialic root is exceptionally broad and deep here. The adjoining positive field is also broader and the positive values higher than anywhere else. And this is what might be expected according to the theory.

It will be remembered that the occurrence of the volcanic belt on the inner side of the non-volcanic arc was explained as an effect due to the curved shape of the negative zone. Probably a similar process of centripetal crowding influenced the subcrustal ultrabasic sima.

And therefore the positive anomalies are strong and appear in crowding and partly interfering belts along the concave inner side of the negative zone. But they appear only in much fainter degree at its convex outer side, e.g. west of Sumatra and south of Java.

The Inverted Arcs of Celebes

Doubtless the whirl-shaped pattern of the Moluccas is due to the Presence of the continental block of Australia, including the Arafura sea and New Guinea. However, a mechanical interpretation of the puzzling islandarcs of the Moluccas is not yet possible. Our geological and geophysical knowledge of the adjacent regions is too scanty.

This concerns especially New Guinea and Halmaheira on one side, and Borneo on the other side. From a geological point of view the island Celebes belongs to the Moluccas. Its remarkable four-armed morphology is due to a double arc which -unlike all the other arcs along the border of Asia —

has its convex side turned towards the Asiatic continent. Obviously, the inverted position of the Celebes-arcs is caused by the sum of the forces interacting in the region between Asia and Australia. Accordingly, for the time being a mechanical interpretation of the inverted position of Celebes.

However, a few remarks may be made on some features which are intimately related to the inverted position of the arc and its position in the complicated pattern of the Moluccas.

The theoretical deduction formulated was substantiated by the structural history of the western part of the East Indies and a large part of the Moluccas. It will be seen that the main features of several other island-arcs are in accordance with the theory. At first sight, however, the theoretical scheme seems to break down if applied to Celebes.

A belt of negative anomalies of isostasy coincides with the eastern and southeastern arms of the island. These arms form an arc with its concavity towards the east. Hence — according to the theory — a volcanic "inner" arc ought to be present parallel to it somewhere in the Banda Sea.

But instead of a volcanic arc a deep-sea basin has developed east of Celebes and a volcanic zone came into being on the opposite side. For the southern and northern arm of Celebes, as well as the uniting central part are characterized by numerous manifestations of volcanism dating from Upper Tertiary to sub-recent times.

The same zone is moreover characterized by the abundance of granodioritic rocks and gneisses. Similar rocks are conspicuously absent from the eastern arc, where serpentines, lherzolites and other basic or ultrabasic rocks have a wide distribution.

A central zone of the island, running north-south, formed a depression with marine sedimentation, during the Upper Tertiary. It is now characterized by a graben (the Posso graben) and separated from the western part of Celebes by a zone of mylonites.

Finally three remarkable deep-sea basins have to be mentioned. On morphological grounds Kuenen united them in a special class. Two of them, the Tomini and Bone basins, separate the eastern and western zones of Celebes; the other one — the basin of Makassar straits — lies in front of the "volcanic arc".

In an attempt to explain the unusual arrangement of these morphologic and tectonic elements, attention should be focussed on the exceptional features of the Moluccas as a whole. We pointed out that the formation of the basin-shaped depression of the North Banda Sea was probably caused by interacting processes in the crust, due to the whirl-shaped arrangement of the zones of buckling.

The accumulation of displaced dunite masses was thought to be responsible for the large fields of positive anomalies in the same region. Probably these processes hampered the development of a volcanic "inner" arc east of the negative zone of Celebes. For the deeply subsiding crust under the Banda Sea basin acted as an antagonistic and dominant factor. And probably this effect was strengthened by the exceptionally strong accumulation of dunite in the same region.

So, as a tentative explanation I suggest that these cooperating factors suppressed the formation of an "inner" arc in this area.

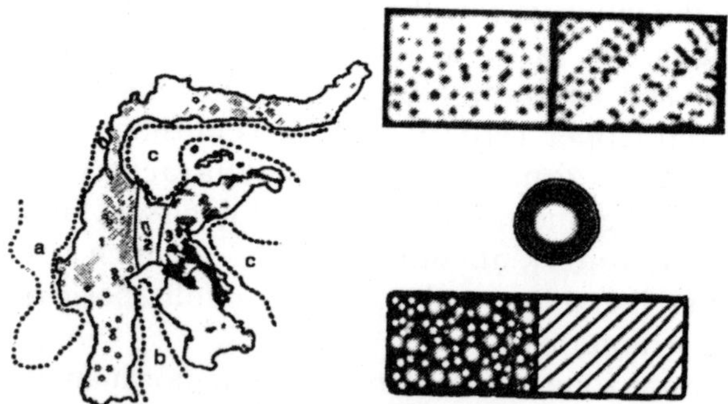

Fig. Structural Zones of Celebes.

Hence, the basaltic material of the substratum that was pushed aside by the downward movement of the root under the eastern part of Celebes had to flow towards the opposite side, i.e. westward. Thus exceptionally the total volume of sima that was displaced by the root had to flow towards the convex side of the arc, where it set going the genetic processes responsible for the numerous granodioritic intrusions and volcanic manifestations found in the western part of the island.

The site of the dunite masses in the substratum is revealed by the fields ands belts of strongly positive anomalies. In order to explain their distribution along the convex side of the negative zone the influence of all the major tectonic elements in the surroundings would have to be taken into account. Unmistakably the strongly negative zone between Celebes and Halmaheira had a paramount influence on the site and shape of the positive field in the Celebes Sea and its connection with the positive belt in the straits of Makassar.

However, in order to understand the total aspect of the positive fields as a whole we ought to take into consideration the gravimetric data on land, on Borneo, as well as on Celebes. Accordingly a complete analysis of Celebes and its surroundings has to be postponed until more data become available. This applies also to the three deep-sea basins , the presence of the idiogeosynclinal basins in East-Borneo and S. Celebes, and finally to the different aspects of the zone of graben in central Celebes as compared with the deep-sea basins in its northern and southern continuation.

Evolution

Now that the origin and development of the East Indian arc has been sketched in its principal lines, the problem of some different types of islandarcs of the Pacific will be considered more closely. The question was raised whether different types of island-arcs represent different stages of evolution of a mountain-chain in statu nascendi. It will now be possible to give a well-founded answer to this question.

It was pointed out that the marginal deep of a double arc differs genetically from the marginal deep of a single arc. It was argued, moreover, that the quantity of waste-products from the land has to be considered as the principal factor causing the formation of either a single or a double arc which is composed of elements like those of the East Indian festoon, more especially the western part of the latter in which the complicated pattern of the Moluccas is lacking.

According to the theory the Riu-Kiu islands and the East Indies should have passed through a paleogeographic stage of extensive land areas competent to produce great quantities of waste-products.

As to the East Indies the theory is supported by what is known of its paleogeographic evolution in Tertiary times. The region of the archipelago was probably an extensive land area at the end of the Mesozoic and the beginning of the Eocene.

A western land-area, including large parts of Borneo, Malaya, the present Java Sea and South China Sea, and Sumatra is still to be recognised in the Eocene. A second large land-area existed in the south-east including northern Australia and the present Arafura Sea. Is it mere fortuity that the zone of buckling appears as islands exactly where it runs along the areas that for a long time resisted the invasion of marine transgressions?

Similar data are not available from the Riu-Kiu arc and its "hinterland". However, we might point out the following features. A deep basin with a maximum depth of 1248 fathoms is comparable to the idiogeosynclinal troughs on Sumatra and Java, or to the Flores deep. More continent-ward the sea is shallow, its depth seldom surpassing 50 fathoms. Hence a comparatively slight rising of the bottom or lowering of sea-level — or both combined — would provide extensive land areas behind the arc.

The Riu-Kiu chain continues southward towards Formosa and northward into Kyushu which is one of the "continental" islands of Japan. Accordingly, the whole area between the arc and the present coast of Asia may be regarded as a "continental" region, which means that the sialic part of the crust is of continental thickness.

The outer arc is largely composed of strata ranging from pre-Carboniferous to Cenozoic. Parts of these sediments are composed of detritus, which could not possibly have been deposited under conditions of land- and sea-distribution such as prevail at present.

Accordingly, the former existence of areas of denudation which have since sunk below sea-level has to be assumed. As a matter of fact Yabe pointed out that the area of the Riu-Kiu, Japanese and Kurile islands including the region from the 720 meter isobath to the present coast of the Asiatic continent, was dry land until sub-Recent times.

A similar conclusion holds good for the region behind the Kurile arc. The southern end of the arc is linked to Hokkaido, the northernmost island of Japan, whereas the arc continues northward into Kaṃchatka. Again there is a deep basin behind the arc with a maximal depth of 1836 fathoms, but the depth of the remaining part of the Okkotsh Sea does not surpass 50-100 fathoms.

The Aleutians represent another example of an arc with a "continental" sea-floor behind the arc. For, apart from the deep Bering basin, which again has to be regarded as a hollow comparable to those of zone III of the East Indies, the remaining northern and eastern part of the Bering Sea is mainly less than 100 fathoms deep.

Remarkably enough the Kurile arc is a single arc, and the Aleutian arc is single over its largest extent. But its eastern end, where it continues into the American continent, is a double arc. The outer arc has developed over a comparatively short distance from Kodiak island towards the Kenai peninsula of Alaska, and it is separated from the volcanic inner arc by Shelikof straits and Cooks Inlet. The volcanic arc, too, continues into the continent by way of the Alaska Peninsula.

Both the Kurile and Aleutian arcs might be called pseudo-single arcs, for the Kurile-festoon, too, is double over a short distance, viz. near Japan. For the Kurile and Aleutian arcs no paleogeographic data are available for making a comparison between the type of arc and the previous conditions of sedimentation, as was possible for the East-Indies. But, the arcs so far considered have some striking features in common.

For in a certain sense they might all be called "continental" arcs, which means that the sea-floor between the arc and the present coast-line of the continent lies comparatively high and corresponds to a sial-layer of continental thickness. The average depth of the sea-floor behind the arc is of the order 2000-2500 fathoms or still more. This means that the sialic crust is thin, when compared with the areas considered so far. For, the deeply situated sea-bottom extends over an enormous distance, as much as 1500 miles from the central Marianas to the Philippines and the Riu-Kiu Islands.

Hence sedimentation in this area is free of waste-products from the continent. Geographic conditions are very different from the regions considered so far because the sial-layer is much thinner. And it is not imaginable that the situation was fundamentally different even in earlier times. Under certain paleogeographic conditions a double arc of the Riu-Kiu type might have developed in the place of the Kurile arc or the Aleutians.

But a double island-festoon of the RiuKiu type could not possibly come into being in the area of the Marianas, nor will it ever be able to develop in the future! The only possibility is the development of two or more volcanic arcs more or less parallel to each other owing to several shear-planes dissecting the crust near each other. This is what probably happened in the area of the Vulcan and Bonin islands.

Possibly the site of the shear-zones, as well as the location of the arcs, correspond with the boundaries of sial-layers of different thickness. So the Marianas and Bonin arcs developed near the boundary between the Pacific basin proper — which has no sial-layer — and the thin sial-layer extending from the Marianas as far as the Philippines and the Riu-Kiu islands.

The latter developed on the transition of the thin sial-layer to one of continental thickness, whereas the Kurile and Aleutian arcs came into being on the boundary between a crust of the continental type and the basaltic floor of the Pacific.

For some distance the East Indian zones may be traced into the Asiatic continent, viz. in Burma. For lack of sufficient detailed knowledge, it is not yet possible to follow them to the Himalayas, western Asia and Europe. The few facts that are known, however, justify the conjecture that this will probably be possible in the future.

From the continuation of the East Indian zones into the Asiatic continent it appears, however, that the different physiographic aspects presented by an island-festoon and a continental mountain-chain do not represent different evolutionary stages.

Whether an island-festoon or an arcuate structure on the continent, whether a deep-sea furrow or a "fore-deep" like the Siwalik trough, etc, merely depends on the geographic conditions that prevailed during the structural history of the region, more especially during the period preceding the last epochs of crustal compression.

The thickness of the sialic part of the crust and the available quantity of waste-products from the land are the principal controlling factors. It was shown from the known history of the outer arc of the East Indies that the latter started as a geosynclinal belt in a remote past, at least as early as the Triassic.

Moreover, Mesozoic sediments of bathyal and abyssal facies types are known from the island of Timor, demonstrating a very strong crustal relief in the vicinity of the present islands in a remote past. Hence, the region of this special belt went through a long and very complicated history which in itself shows some points of resemblance to the history of such mountainchains as the Himalayas and the Alps.

But the same facts make it clear that the East Indian festoons considered as a whole should not be interpreted as the embryonic stage of a mountain-belt which when fully grown, would represented by the Alps and the Himalayas. For these regions are different in the same way as the Marianas differ from the East Indies: not because of their representing different evolutionary stages of one continuous process in the formation of mountain-chains.

Theoretically, i.e. in the ideal case of strictly unchanging conditions of crustal thickness and quantity of sediments, an island-arc of the Mariana type might come into view and vanish periodically, according to the alternating rhythm of epochs of increasing and decreasing compression in the earth's crust.We have progressed far enough into the realm of speculation and refrain from picturing the future stages of a much more complicated area like the East-Indies.

The scanty data known from its past history — e.g. the

occurrence of deep-sea deposits in central Borneo, the presence of intense Mesozoic folding and overthrust strata in

Sumatra and Java — clearly show that the past history was not a simple pulsating rhythm in loco and therefore we are not justified in expecting a simple repetition in the future of the Upper Tertiary rhythm that was pictured in the foregoing pages. But are we not equally unable to predict the future development of any other region, say southern Asia, Alpine Europe or western North America?

We may once more point to the structural analysis of the continents. According to our present state of knowledge some regions seem to reveal a regular sequence of events, but in other areas the succession of tectonic belts seems capricious.

Geologists have to admit frankly their inability to understand the cause- and- effect- relations of the observed successions.

Perhaps when more is known of Pre-Cambrian events the apparent capriciousness will be gradually replaced by a better understanding of past stages which will make possible a well-founded prediction of the future development of continental mountain-belts and arcuate island-festoons.

Chapter 26

The Floor of the Oceans

The antipodal distribution of continental blocks and oceanic receptacles ranges among the most peculiar features of the earth. The deep-sea area of the North Pole is situated directly opposite the Antarctic Continent. The six non-polar continents are grouped in pairs, with a roughly meridional orientation, and form an elongated triangle with its apex pointing southwards.

Conversely, the oceanic sectors extend in an opposite direction. The broad bases of the continents, arranged in a nearly continuous ring around the North Polar Basin, have their antitype in the broad, circum-Antarctic ring of water.

The counterpart of the Indian Ocean is found in North America and that of Australia in the Atlantic north of the equator, etc.Of course, there are exceptions to this antipodal distribution of land and water and there is no question of a rigorous crystallographic symmetry, but the fact remains that only 1/20 of the whole landsurface is antipodal to land). *The distribution of the continents is not hap-hazard but displays a certain regularity*.

This exceptionally prominent feature of the globe, in which land masses are opposed to oceanic areas, *reminds* one of a tetrahedron, i.e. a crystallographic body in which a rib is always situated opposite a plane. It has given rise to a great deal of speculation, but it need hardly be said that this characteristic is such a remarkable one that any valid theory as regards the origin of continents and oceanic receptacles ought to be able to furnish an explanation of their tetrahedral arrangement).

It is clear, however, that this question involves the whole intricate problem of the origin of continents and ocean-floors, and there are many conflicting views on this subject. Some are of the opinion that the continents and oceans represent permanent features.

Another hypothesis, diametrically opposed to the preceding one, is that the oceanic receptacles originated as a result of the submergence of land-masses.

Others associate their formation with a drifting-apart of continents, while a fourth group claims that at least two of the ocean-floors were formed by stretching of continental blocks. All these possibilities but it appears advisable to begin with a brief review of all that is known at present of the bottom relief of the oceans, while a comparison with the results of geological and geophysical research is obviously warranted at this stage.

THE MAJOR CHARACTERISTICS OF THE BOTTOM-RELIEF

As more and more data become available, it becomes increasingly evident) that the relief of the ocean-floors presents some very curious characteristics. A comparison between the various oceanic sectors will make it immediately clear that the relief of the. Atlantic and Indian Oceans — while differing to a considerable degree from that of the Pacific, especially in the case of the North Pacific basin proper, i.e. the area limited by the andesite line and a southern demarcation running approximately from the Fiji to the Galapagos Islands — concur in many respects.

The morphology of these areas is matched by their geological structure, for seismic and petrographic data both clearly show that the bottoms of the Atlantic and Indian Oceans are very different from that of the Pacific. The generally accepted view at present is that no sialic layer exists beneath this Pacific sector, though such a layer is in fact assumed to extend under the Atlantic and the Indian Ocean.

These broad outlines make it possible to divide the morphological characteristics into four major groups:

- The Atlantic Ocean, and the western part of the Indian Ocean, with their numerous and comparatively flat-bottomed basins, separated by relatively narrow and steep ridges. The large North Pacific basin, with its frequent linear and at times intersecting ridges and troughs.

  No mention will be made of the manifold basins in eastern and southeastern Asia, since these formations are situated inside the andesite line. Nor will we discuss.

  Melanesia, which comprises the East Indies from a geological and morphological point of view. Most authors assert that the andesite line represents the true boundary of the Pacific basin proper, and the real problem may thus be stated to begin in this area. Thus too, when dealing with the problem of the Atlantic, no mention will be made of the Mediterranean and West Indian basins.
- The eastern part of the Indian Ocean, the south-western part of the Pacific and the three Antarctic basins. These basins are in so far similar that a plain bottom-relief with hardly any subdivision is found in these areas. The three elongated deep-sea basins around the Antarctic continent may be referred to briefly as the Atlantic, the Indian and the Pacific Antarctic basins).

  The above conception of the morphology of the basins may possibly prove to be more complicated than we suppose it to be, as none of them has been examined in detail.

  The eastern part of the Indian Ocean and the south-western part of the Pacific form vast receptacles with a symmetrical arrangement in respect to Australia and Melanesia, the concave sides being turned towards one another.
- The North-Polar basin. Little is known of the details of this receptacle (cf. the basins of group (3), which it would appear to resemble in some respects), and

no further reference will consequently be made to this basin).

THE WESTERN PART OF THE INDIAN OCEAN

We will begin with a summary of a series of features which the significant congruence of these oceanic sectors and the surrounding continents, showing that in many respects the topographic limits between continental and oceanic areas do not correspond with structural limits. The regions appearing under group.

- Are characterized by a great number of deep-sea basins with comparatively flat bottoms surrounded by relatively steep ridges. The whole structure is in many respects very similar to that of the intervening African continent. This similarity is further accentuated by the fact that Africa, besides containing many more basins than South America, is also flanked by far more submarine basins than the American sector west of the Mid-Atlantic Rise.

  The morphology of the southern Atlantic blends, as it were, with that of the Atlantic Antarctic basin. The Mid-Atlantic Rise bends eastwards in the same way as the loop of the Southern Antilles and ultimately forms the boundary of the Antarctic basin, joining the Kerguelen Rise. In the Indian Ocean an analogous ridge bends eastwards around the Indian Antarctic basin.

A series of ridges with a roughly NS. trend, separating the continental and submarine basins in NS. alignements), can be observed on the South American an d African continents, parallel to the Mid-Atlantic and the Mid-Indian Rise. The thresholds between the continental basins consist mostly of Pre-Cambrian rocks.

Two main strikes dominate in Africa since the early PreCambrian: a NE.-SW., or so-called Somali strike, and a NW.-SE., or socalled Erythraean strike. Both played an important part in the whole later history of the continent. The same strike is again apparent in the narrow submarine ridges separating the deep-sea basins. The same holds good for the northern part of the Atlantic. Bucher), for instance, observes: "The map of the North Atlantic shows that the rises which separate the individual basins consist of lines of swells comparable to a certain extent to that which in eastern North America runs from southwestern Ontario through the Cincinnati and Nashville domes, separating the Appalachian from the eastcentral sedimentation basin.

The whole pattern of the oceanfloor seems, in fact, comparable to that of the basins and swells of the continental areas outside the great orogenic belts, although the scale is larger both horizontally and vertically on the oceanic surfaces".

Mention should be made here of the opinion of Cloos), who asserts that a domeshaped elevation with a fault-system, accompanied by volcanism and similar to that observed on the European and African continents, exists in the Azores, and this leads him to conclude that there can be no fundamental difference between this part of the Atlantic and a continent. Another striking feature, as pointed out by Stille is the symmetry of the northern and southern Atlantic sectors.

These may be described as forming one another's counterpart. This applies to the submarine characteristics (e.g. the convex are of the Mid-Atlantic Rise bends eastwards twice), as well as to the features above sea-level (e.g. the arc of the northern Antilles with the shelf of the Bahamas and the arc of the southern Antilles with the Falkland Islands shelf, etc.).

Symmetrical features are also found in the Indian Ocean, and deserve special attention, as half of the symmetrical structure appears on the African continent. The new chart of the John. Murray expedition shows the existence of a series of ,widely-curving ridges in the north-western part of the Indian Ocean.

The best-known elements are the Seychelles Bank, the double Carlsberg Ridge and the Murray Ridge. Wiseman and Sewell report that the ridges form the almost exact mirror-reflection of the famous tectonic rift-valleys in East Africa. This analogy is emphasized by the fact that the Carlsberg ridge is also a belt of tectonic earthquakes and volcanism. This consequently constitutes another point of similarity between sub-oceanic and continental sectors. Wiseman and Sewell assume that an UpperTertiary age may probably be attributed to the Carlsberg Ridge.

The Murray Ridge probably fuses with the Carlsberg System, and this probably connects in the south with the ridges upon which the Lacadive-, Maldive- and Chagos-Islands rest. The latter appear to unite as the MidIndian Rise, which can be traced to Antarctica via New Amsterdam, the Kerguelen, and the Heard Islands.

It will be dear that, in case the basins of the northwestern area correspond with old structural elements, their frames should at any rate have been subjected to a "rejuvenation" (a phenomenon also observed on the continents) in comparatively recent times. The same applies *mutatis mutandis* to the relief of the Atlantic.

Young fault-systems, graben, Upper Tertiary strata and Pleistocene vulcanism are found in Iceland, the Azores and St. Paul. Unfolded Upper Tertiary sediments occur in the first two areas. Other young volcanic regions are Ascension, Tristan da Cunha, St. Helena, Gough Island and Bouvet Island. Moreover, the seismic unrest of the Mid-Atlantic Rise proves that the movements are still being continued).Petrographical data obtained in the oceanic areas stress the results of geological and seismic observations and point to the sialic composition of the floor).

So far attention has chiefly been paid to the pattern of the submarine relief. Yet it is obvious that valuable information can also be obtained from, the geology of the surrounding continents.

We will consequently give a brief summary) of a few of the most important points that call for special attention when dealing with the problem of the ocean-floors.

- Scandia. This area lay above sea-level till the LowerTertiary according to Holtedahl and Frebold. Its former extent is purely hypothetical.
- Appalachia extended at least 200 miles beyond the eastern coast of North America, and now lies

submerged to a depth of at least 3,000 to 4,000 meters. Geophysical data seem to imply that this area began to founder in the Triassic and that this process continued until at any rate the Tertiary

- The seismic method of investigation was also applied by Bullard and Gaskell to the submarine geology of the eastern part of the Atlantic i.e. along a line extending 170 miles WSW. from the Lizard. These authors revealed the presence, in this area, of a submerged block probably consisting of igneous rocks covered by Triassic strata.
- In his monograph on the Congo basin Veatch writes as follows: "A consideration of the source of the sediments forming the Lubilash of the western Congo, likewise, points to a more elevated, though not necessarily mountainous, land mass occupying the present depressed coastal plain region and extending into what is now the Atlantic Ocean".
- Faulting of presumably Pliocene or post-Pliocene age occured along the coasts of Africa, Arabia and India around the Arabian Sea, and movements of a similar age caused part of the Oman- Indus arc to break down into the Arabian Sea; according to Wiseman and Sewell.
- The southern part of the South African Variscides and the probable connection with the Cedar Mountains disappeared into the sea. Nothing is known of the original extent of this land mass.

In none of these cases need one think of blocks of enormous continental extent, though many investigators are indeed known to have jumped at this conclusion, while others arrived at a similar result on the plea that the folded chains (the Pre-Cambrian, Caledonian, Variscian and Alpine belts) terminate abruptly along the coasts of the Atlantic. Still others base their conclusions on the analogy of the submarine morphology and that of the continents).

Many coasts are indeed formed by faulting). Although these areas have been discussed previously the possible consequences of the idea of tremendous subsided blocks will be dealt with subsequently. Biogeographical data will play no part in this discussion, as these should never be allowed to influence the solution of any geological problem *a priori*, even though such data may prove to be valuable if compared with geological results.

Recapitulating, a comparison between the submarine relief of the oceans and the geology of the surrounding continents may therefore be said to lead to the following conclusions:

- The bottom of the Atlantic and the floor of the western part of the Indian Ocean are built up of similar (sialic) material as the continents.
- The same events (i.e. the formation of basins, ridges, and rift-valleys) occured in the oceanic sectors as well as the bordering continents.
- The structural lines determining the location of the continental and submarine basins date from very early Pre-Cambrian times.
- The submarine relief, however, indicates that it underwent the same process of "rejuvenation" in comparatively recent times as the continents.
- The submarine basins need not have originated simultaneously, any more than the continental basins, which are known to have formed in very different periods.
- The pattern resulting from the arrangement of the three oblong basins around Antarctica, the blending of the submarine relief of the adjoining areas and the symmetry of the bordering oceanic sectors must surely have some special significance. Cloos attributed this constellation to the influence of the rotation of the earth. It would in any case appear to reflect the activity of subcrustal forces, which would then have produced this pattern in the early Pre-Cambrian.
- The arrangement of the submarine basins and their increasing depth and width towards the south until

confronting the Atlantic and Indian Antarctic basins on a broad line, reveal two large wedge-shaped oceanic sectors with their apexes pointing northwards, while the South American, African and Australian continents point in the opposite direction. There is no doubt in my mind that this situation can be said to constitute a fundamental feature, resulting from primordial and deep-rooted forces.

The same applies to the tetrahedral arrangement of the continents and their antipodal relations to the oceanic depressions, as mentioned in the introduction.

- Several blocks of unknown size have foundered very deeply. Many coasts are bounded by faults. Nevertheless, the characteristics mentioned under 1-7 should not induce us to suppose that the whole floor of either the Atlantic or the Indian Oceans constitute foundered continental blocks. However, this possibility will be examined below.
- The oceanic and continental sectors differ chiefly as regards the thickness of the sialic layer. This layer is thinner under the former sectors, and situated at a great depth. It can be stated to constitute one of the major problems of our globe.

THE PACIFIC OCEAN

The large North Pacific basin, situated north of an imaginary line running from the Fijis to the Galapagos Islands is characterized by many linear ridges and troughs which occasionally intersect one another, thus giving rise to a pattern which is met with in no other oceanic area.

This linear arrangement, which is such a prominent feature of many Polynesian Islands , is believed to have been caused by the rising of magma along faults and fissures in the floor of the ocean, which in this case appears to be composed of basic rocks (crystalline sima).

The lineaments have relatively straight courses and are extremely long. In the opinion of Betz, Hess and others the fissures have to be considered as transcurrent faults. Many volcanic peaks in the central part are crowned with atolls, and nothing is consequently known of the nature of the eruptive rocks in their foundation.

A quantity of data could nevertheless be obtained outside this area, upon which many investigators — especially Lacroix) — based conclusions of a general kind. There seems no doubt that the bottom of the Pacific Basin proper, i.e. the area inside the petrographic boundary known as the andesite line consists of olivine basalt.

In spite of the above, there would seem to be no doubt that sialic blocks have foundered in some parts of in the Pacific. A few of these masses.If we exclude such areas as Choco and the muchdebated region of Burckhardtland, there would appear to be ample justification for the assumption that Cascadia represents a "border-land" of unknown extent, which probably acted as an area of denudation to the geosyncline of the Sierra Nevada up to Jurassic time, supplying it the while with detritus.

It has probably sunk to a depth of 3,000 to 4,000 m below sea-level). In addition it should be noted that a land-mass of unknown dimensions has undoubtedly sunk into the Pacific off the coast of South America). The assumption that the eastern margin of Melanesia is a faultzone, on top of which formed the series of islands extending from New Zealand to Samoa, would logically imply that a block must have foundered east of this area.

In fine, blocks of unknown extent may undoubtedly be said to have sunk in the marginal zones of the Pacific, but there can be no question that these constituted large sialic blocks, as enormous stretches of the floor of the Pacific appear to be built up of rocks of basaltic composition.

The Distribution of Isobathic Areas

The distribution of isohypses on the continents and isobathic contours in the oceans has led to interesting speculations. A graphic representation for the whole world, i.e. a so-called hypsographic curve of the earth's surface, shows two frequency maxima, corresponding with two favoured levels, one occuring at 100 m above sea-level, a second at a depth of 4,700 m below it. In other words the continental surfaces are situated at n average altitude of 5,000 meters above the ocean-floors.

Wegener argued that the lower level should be regarded as the surface of the sima, whereas the upper level was that of the sial. These two levels of equilibrium would give rise to two frequency maxima, which would simply be controlled by Gauss' law of errors). I wish to draw attention, however, to a few salient features of the submarine relief as known to us at present. The following table shows the distribution of isobathic areas in the Atlantic and the North Pacific Basin, according to Kossinna), The areas are given in millions of square kilometers, the depth in km.

Depth	*0.0-0.2*	*0.2-1.0*	*1-2*	*2-3*	*3-4*	*4-5*	*5-6*	*>6*	*Total area in m*	*Average depth*
N. Atlantic Ocean	2.6	1.8	1.8	3.0	6.7	11.5	8.8	0.6	36.8	3,788
S. Atlantic Ocean	2.0	1.5	1.2	3.2	9.2	16.2	13.1	0.2	45.6	4,036
N. Pacific Ocean	1.1	0.8	0.8	1.9	8.8	21.2	33.2	3.0	70.8	4,753

The most frequent depth of the Atlantic appears to occur at the 4-5 km level, whereas that of the North Pacific is generally one km deeper!).

It seems to me that the explanation is obvious, viz. the deep level of the Pacific is controlled by the simatic nature of its bottom, the 1,000 m higher level of the Atlantic floor is due to the presence of a sial-sheet and the still higher level of the continents is the result of the fact that the latter consists of sial-slabs of a considerably greater thickness than the sial-sheet under the Atlantic.

We pointed out that the morphological features of the Atlantic and the Pacific are fundamentally different. If we glance at the new bathymetric chart of the Atlantic Ocean by Stocks and Wüst, it appears at once that the system of the Mid-Atlantic Rise and its adjoining ridges is situated at an average depth of 4 km or less (from 2-4.5 km), but the average level of the basins is 5 km (from 4.5-6 km)! Which one of these two is the original level of the Atlantic floor, the higher or the deeper one? If the basins are to be considered as secondary subsided formations, the primary surface of the Atlantic floor would be the higher level!

Moreover, it seems very probable that the major positive relief-features of the Pacific Basin proper, as compared to those of the Atlantic, are largely due to later accumulations of lava on the primary surface, which seems to lie at an average depth of 5.5 km below sea-level.

These considerations stress the fundamental differences between the average depths of the Atlantic and Pacific oceanfloors. On the other hand, the rejuvenation of the relief is a factor which should be taken into account. Thus, the result which we arrive at here is that at least three distinct levels of primary importance may be observed. These are represented by the surface of the continents, the floor of the Atlantic and that of the Pacific.

The Problem

Now that the major characteristics of the sub-oceanic relief have been considered and before entering into a discussion of the various hypotheses which try to explain these features, the main points of the problem can be summarized as follows.

- The "tetrahedral arrangement" of the continents, their antipodal relations to the oceanic receptacles and their peculiar shapes
- The occurrence of at least three favoured levels on the earth's surface and their respective vertical distances.
- The occurrence of a thin sial-sheet in the Atlantic sector and part of the Indian ocean.
- The presence within these areas of a remarkable structural pattern dating from the early Pre-Cambrian and the occurrence of certain tectonic features which have their counterpart on the adjacent continents.
- The congruence of orogenic belts on either side of

the Atlantic ocean, terminating abruptly on the coast.

- The absence of a sial-sheet over the floor of the Pacific Basin proper as well, probably, as in a few other oceanic areas mentioned above under groups 3 and 4.

Continental Drift

Alfred Wegener's much-debated hypothesis of continental drift seemed to give an elegant and simple solution of many problems of the earth's crust, but is it clear that this original conception cannot be made to agree with our present knowledge of the floors of the Atlantic and Indian Oceans, for these are now generally believed to be sialic layers.

Moreover, the structural pattern of these ocean-floors and their striking similarity to that of the surrounding 'continents cannot be explained by the drifthypothesis. However, it might on the other hand be assumed that our continents occupied their present site as a result of sliding, which occurred in such a way that a sialic block of continental thickness was stretched. The stretched portion would then simultaneously have sunk and thus formed the bottom of the Atlantic and Indian Oceans.

This deduction characterizes du Toit's hypothesis of drift. In 1937 this same author wrote the following in a publication on the submarine relief of the Atlantic: "A pattern like this over so gigantic a region asks for a single controlling cause, and finds its proper answer only in continental sliding". To this he added a year later:

"The relief of the Atlantic bottom shows all the characters of a stretch basin – particularly in the symmetrically-set, though crooked, Mid-Atlantic Rise with its lateral branches that reach out to either shore following north-easterly and north-westerly trends."

Yet I agree with Cloos that the striking similarity of the structures of the basins and surrounding ridges as compared to those of the adjacent continents is incompatible with the idea that these originated as a result of drift.

Nor do I find it possible to reconcile any of the above conclusions as regards the Atlantic and Indian Oceans with the hypothesis of sliding continents, since the features of the bottom relief at all events make it clear that if the floors of said oceans originated as a result of stretching *this Process must* have occurred during the Pre-Cambrian.

The same holds good regarding several other types of drift-hypotheses, including those of Gutenberg, Grabau, and Chelikowsky).

The above would be quite sufficient to reject the hypothesis of du Toit and others, but I would like to examine another argument, which has recently appeared in several publications. Biological and paleontological data may be able to furnish more or less convincing evidence in favour of trans-oceanic land-bridges, but not — generally speaking — of the nature of such land-connections.

We must make an exception for those data which seem to indicate that the distance between the continents had originally been a shorter one.

Gerth, for example, argues that the similarity of the marine Devonian strata of South America and South Africa pleads for a shorter distance between Africa and America during the Devonian. 50% of the Lower Devonian species of the Falkland Islands and 30% of the Lower Devonian species of Brasil and Bolivia correspond with those of the South African Bokkeveld beds.

Gerth claims that this can only be explained if we assume that South Africa lay nearer South America in former times). Still, these facts cannot be said to contribute any arguments in favour of the idea of continental drift.

The same applies to the assertions of Gerth, du Toit and others as regards a stratigraphic and structural similarity of the Gondwanides and Cape Mountains. It should be noted that the existence of analogous structures and stratigraphical sequences were reported by Holtedahl in such widely-separated regions as Spitsbergen and Scotland.

Yet no one would think of continental drift in this last case! Moreover, I wonder whether this hypothesis would be used to explain the following: the Maastrichtian fauna of Western Australia is closely allied to that of the Arigalus beds of the Trichinopoli district) of India; and Spath) recently described the Ammonites of Geraldton, Western Australia, and revealed that they closely resembled those of the Bajocian in England!

The distribution of the Gondwana-flora, including plant remains like Glossopteris and Gangamopteris has often been considered as in favour of the theory of continental drift. However, Longwell has given an example of similar problems presented by the distribution of modern plants belonging to the primitive dogwoods of the genus *Cornus*.

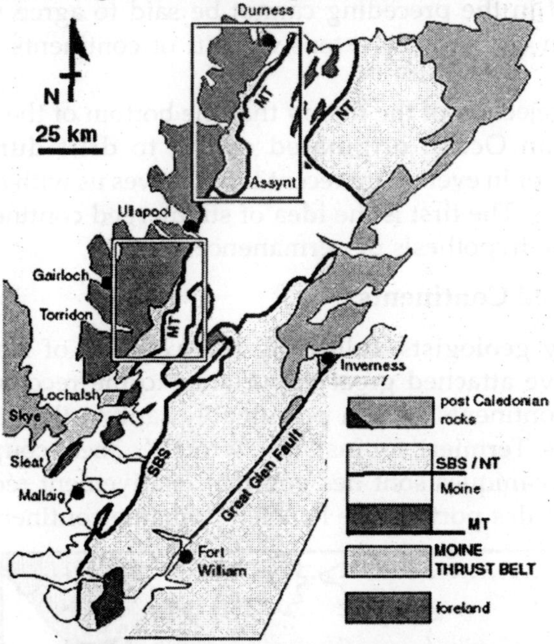

Fig. Geological Structures of Scotland.

The distribution of the genus, containing about 40 species and its near allies, "poses greater problems than that of the *Glossopteries* flora, since it involves greater oceanic distances and more formidable climatic -barriers.

Moreover, the hypothesis of continental drift does not help explain the dispersal of dogwoods — rather it increases the difficulties. Obscure ways in which plants and also some animals must have been distributed during long geologic time intervals are suggested by other examples from the vast Pacific region.

When these numerous problems are viewed together, emphasis that has been placed on the *Glossopteris* flora as "compelling evidence" of oncecontinuous lands seems dangerously near the unscientific procedure of selecting evidence to support a favored theory."

No other examples will be mentioned, for it would take us too far to discuss all the difficulties inherent to the hypothesis of drifting continents). Besides, the conclusions arrived at in the preceding cannot be said to agree with the idea of an enormously extensive drift of continents *since the Paleozoic*.

The rejection of the theory that the bottom of the Atlantic and Indian Ocean originated owing to drift during the Paleozoic, or in even more recent times leaves us with two rival hypotheses. The first is the idea of submerged continents, the second the hypothesis of permanence.

Submerged Continents

Many geologists, following the example of Haug and Suess, have attached great importance to the reconstruction of large continents which presumably sank to the bottom of the oceans. Termier), for instance, wrote: ".... beaucoup de nos abimes océaniques sont des gouffres relativement récents où se cachent des portions de l'ancien domaine continental.

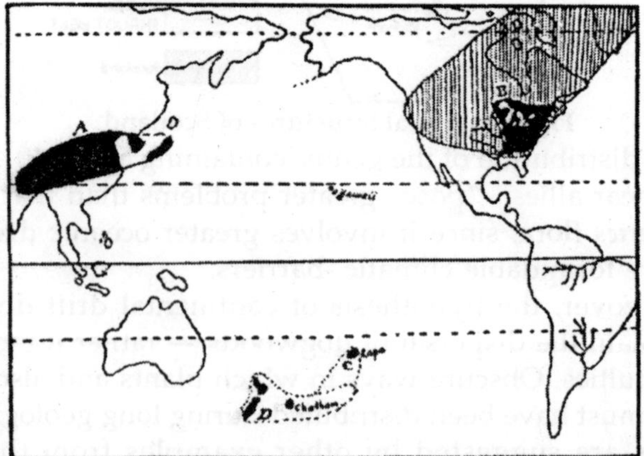

Fig. Hypothetical Reconstruction of Ttrans-oceanic Continents

Yet many geologists have refused to accept the idea of enormous submerged continents, as, apart from objections of a geophysical nature which will be discussed later, such a hypothesis would complicate the watereconomy of the oceans in a hopeless manner.

The continents would (assuming that the volume of water of the oceans has remained constant) need to have been permanently flooded prior to the submersion of these huge blocks, but this conflicts with all available data.

The alternative would be that the volume of water would have increased steadily and to an enormous extent since the foundering of continental blocks. Yet this seems highly improbable. This point was clearly by Kuenen in the accompanying graph , which includes Walther's and Twenhofel's opinion.

Kuenen writes in this connection): " Walther held that there was probably no deep-sea until the end of the Paleozoic because no Paleozoic faunal elements have been found in the present deepsea faunas. But this assumed absence of old deep-seas is only one of several possible causes that might explain what is after all a rather vague conception.

It necessarily implies that the earth had an unimportant hydrosphere for some 1500 million years and then acquired its present volume of sea water at the rate of a few km^3 per year or a few times the speed at which volcanic rocks are produced. Add to this that salts had to be furnished on the same gigantic scale and the impossibility of Walther's conclusion becomes manifest.

" Twenhofel's assumption, though less extreme is still unacceptable. He says:

'In this inquiry it is assumed that Pre-Cambrian and Paleozoic deep-seas existed, and that if the deep-seas of all time are considered they would be equivalent to deep-sea, with area like that of the present, extending as far back as the beginning of the Devonian'.

"In our graph, the volume of the oceans is plotted against geologic time, showing the views of Walther, Twenhofel and the writer. In Twenhofel's view the area abc must be equal to.

"There are many arguments for and against permanency of the oceans. In the opinion of the writer the most forcible are: the absence of fossil deep-sea sediments from the continents and the impossibility of sudden production of water and salts since the juvenile stage of the earth.

For post-Algonkian times doubt as to the existence of permanent ocean basins is quite unreasonable; although their position and shapes may have varied through continental drift and although isthmian links may have arisen and disappeared".

This question is closely related to that of the total amount of sedimentation in the deep-sea and to the whole complex of geochemical processes of the continents and oceans. So far, however, there seems to be no evidence of an annual increase of a few km^3 in the amount of the oceanic waters, and it is quite as improbable that the salinity increased at the same rapid rate.

Walther's curve would therefore appear to be too extreme, and — unlike du Toit, whose point of view is expressed in a publication of 1940 on the subject of the Pacific Ocean — the opinion that Kuenen's graphic representation is probably more in accordance with the facts than Twenhofel's curve.

Two conflicting deductions are found side by side if we adhere to the hypothesis of submersion. The first asserts that vast continental blocks foundered in the course of time. The second claims not only that the total amount of the oceanic waters has not increased proportionately since the Cambrian, but that it has even remained almost constant.

Should one nevertheless cling to the theory of submerged continents, the only alternative would be to assume that while vast blocks were being submerged in one area, parts of the ocean-floor of an almost identical size were being elevated in others, i.e. — to cite one example — that the bottom of the Pacific which was originally situated at a far greater depth, rose while blocks sank into the

Atlantic and Indian Oceans, and that the total amount of water and its surface level consequently remained almost wholly unaltered. The line A-C, on the other hand, the idea of a relatively stable floor of the Pacific. In addition to this, both graphs show the periodic cadence of the floor of the Pacific, as deduced from the rhythm of transgressions and regressions.

It is not quite clear, however, why such opposed movements should have occurred in areas of almost equal extent. Nor is it clear why these movements should have occured in such a way that the sea-level remained comparatively stable.

Besides, the above assumption gives rise to further complications, for if the phenomena can be attributed to this process, it might be asked what happened in the substratum. To begin with, it remains entirely problematical (in spite of Barrell's speculative though interesting reflections on this subject) why a given area such as, for instance, the present floor of the Atlantic should have been submerged, while the neighbouring blocks remained standing as continents.

Moreover, this process involves more than just submersion, and cannot merely be compared to a block of wood which is pressed down more deeply into the water. Seismic data show that the blocks must have grown much thinner during the act of submersion.

This might be attributed either to stretching, or else to the melting of sialic material from the lower part of the crust, which subsequently vanished. The first alternative was discussed in the preceding paragraph.

This stretching-process might possibly have occurred during an early stage of the earth's history, but it contributes nothing towards the solution of the problems which have arisen as regards

Cambrian and later times, as the features of the sea-bottom (which were dealt with extensively above) appear to contradict the idea of a more recent process of stretching. The second contingency would only force us to accept some additional *hypothesis ad hoc,* with no sound basis whatever.

The above clearly shows that the hypothesis of submerged continents leaves many points unsolved, and we consequently proceed to the third hypothesis with the unpleasant sensation of having been twice disappointed.

THE HYPOTHESIS OF PERMANENCE

The aspect of the continents and oceans has not escaped frequent alterations. We need only think in this connection of the East Indian basins, Melanesia and Appalachia. Some parts have been submerged, even though such regions as Cascadia, Appalachia and Scandia were no larger than "border-lands", as Schuchert called them.

The idea of permanence is consequently untenable in its extreme sense. But just as Schuchert and Willis proclaimed themselves advocates of the theory of permanence, in spite of the border-lands and isthmian links which they themselves reconstructed, so, too, we can speak *cum grano salis* of permanence as regards the oceans as long as we refrain from reconstructing huge submerged continents.

This hypothesis will have to be based on three premisses:

- That comparatively thin sialic layers existed in the floors of the Atlantic and the Indian Ocean since at least Cambrian time;
- That in spite of this, relief-features such as basins and ridges resembling those of the intervening African continent originated within these blocks, and that these were brought about by the same subcrustal processes;
- That the amount of oceanic waters has not changed to any important degree since the Cambrian. Among the features which are liable to change in one or perhaps several respects are the details of the submarine relief.

For example, here too we need only think of such foundered blocks as Appalachia, the possibility of widely diverging periods of origin of the basins of the Atlantic and Indian Oceans with their possibly later movements, and the "rejuvenation" of the relief, which is so strikingly by the Mid-Atlantic Rise and the Carlsberg System.

These movements remind one at once of the "isthmian

links" of Schuchert and Willis. Reflections of a highly interesting though speculative nature were made by Willis and Nölke on the emersion and submersion of narrow trans-oceanic isthmi.

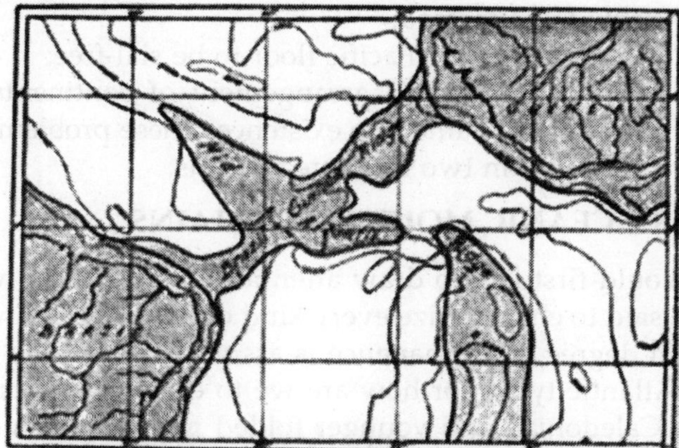

Fig. A Transatlantic Isthmian Link .

Their origin and submersion will probably remain a mystery for some time to come, but the inference that such features had in fact been formed and then vanished would not seem to be an improbable one. Cloos), too, when dealing with the analogy of the characteristics of the continental and oceanic sectors (Atlantic and Indian) drew attention to the fact that the narrow zones surrounding the continental basins can look back upon a particularly agitated history, with important vertical upward and downward oscillations.

I consequently incline towards the view that no essential objection can prevent us from accepting the idea of isthmian links as postulated by Schuchert and Willis. Of course, it cannot be denied that the existence of land-connections during certain given periods cannot be proved geologically. Each so-called reconstruction of a trans-oceanic land-bridge must necessarily retain a hypothetical character.

Yet such land-bridges need not be discarded *a priori* as mere products of the imagination. Another changeable factor is the periodical variation of the distance between the floor of the sea and the surface of the continents. A few fundamental questions still remain to be solved, however, viz. how to explain.

- The congruence of the abruptly ending folded belts on both sides of the Atlantic;
- The way in which the relatively thin and deeply situated floors of the Atlantic and Indian Oceans came to be formed;
- What caused the Pacific floor to be sial-free;
- How the antipodal arrangement of continents and ocean floors came into existence. These problems will be treated in two separate sections.

TRANS-OCEANIC MOUNTAIN-CHAINS

I would first like to draw attention to a difficulty which may be said to characterize every kind of hypothesis in which a certain degree of permanence is assumed in oceanic floors of the Atlantic type. For how are we to explain the abruptly ending Caledonian and younger folded mountains on either side of the ocean? Stratigraphy and tectonic data show that a strip of unknown extent is lacking. Yet the morphology of the floor of the ocean contains no indication of submerged Caledonian or other folded chains.

Moreover, Haug remarked quite rightly that the theory of trans-oceanic mountain-chains requires, amongst other things, that a transoceanic geosyncline, including an area of erosion on at least one of its sides, should have existed prior to the said mountain-chains.

It was this that induced Haug to reconstruct those continents, the existence of which we rejected on the grounds mentioned above. Yet even so, another fact remains, viz. that the abruptly ending belts are situated approximately opposite one another on either side of the ocean.

The following suggestions may help to overcome these difficulties. Let us, for example, suppose that, owing to the influence of subcrustal processes, the earth's crust had reached a stage where it became possible for a geosyncline to form within a certain zone.

Let us also assume that this zone continued under an area such as that of the Atlantic. In that case a geosyncline and folded chain would potentially be able to originate within a

northern belt of triangular continents alternating with a southern belt of oceanic units distributed "like a pair of cog-wheels with interlocking teeth"?). When, on page 230, the problem of the ocean-floors was outlined in six different points, this question was mentioned under number one.

Thus far, however this question remained unsolved, even untouched. As a matter of fact thinking of the terrestrial processes that must have operated in the earliest days of the infancy of our globe, means to enter inevitably a domain which is characterized by very scanty available data and is therefore wide open to unbridled speculation. But why should we not enter it if every one who wants to join us on our geopoetic expedition into the unknown realm of the earth's early infancy is warned at the beginning that probably not a single step can be placed on solid grdund?)

It is believed by some geologists, especially by petrologists like Lawson and Rittmann, that the Earth's surface in its initial stage was composed of a basaltic melt. It will be interesting to see how they deduced sialic continents from a primary homogeneous surface-layer of basic composition.

A second group of theories lays stress on a happening of cosmic dimensions: the birth of the moon out of the body of the earth. In this group will have to be considered the theories of Osmond Fisher, Pickering, Escher a.o.

They start with a globe surrounded by an outermost layer of sialic rocks of continental thickness, and they try to deduce the origin of continental bucklers and oceanic depressions from the cosmic catastrophe which gave rise to the genesis of the moon in a way suggested by G. H. Darwin.

Then, in the third place, we will start anew with the idea of a worldencircling layer of sial and try to deduce the desired distribution of continents and oceans without the aid, however, of the hypothetical events of the lunar catastrophe. Some efforts in this direction have been presented by the present author and by Vening Meinesz.

In 1932 Lawson formulated a theory to account for the formation of continents from a crust which originally consisted of sima only. In his opinion an upper layer of basaltic rocks covered a deeper substratum of dunite. Basaltic continents were produced by orogenic stress.

Erosion products from these basaltic continents were carried to the primeval oceans, the basic parts went in a state of solution, while the sialic elements became concentrated, and when these waste-products sank sufficiently they were fused together so as to form acid, granitic rocks. By this process the continents became continually more sialic. Then, the sialic material differentiated into a double layer, of granite above and diorite beneath.

Seven years later the Swiss petrologist Rittmann introduced a theory which in its main elements is closely akin to that of Lawson. It differs from it, however, in one respect. According to Rittmann the sialic wasteproducts which assembled in the primeval oceans were migmatized by gases emanating from the substratum.

By this process they grew specifically lighter. To restore isostatic equilibrium the sialic floors of the primary oceanic receptacles emerged to form sialic continents. On the other hand the original bucklers of alkali-basalt subsided to form the bottom of the present oceans.

A weak point in both these theories is the formation of the primary basaltic continents. For, in the first place, it is not clear how these protuberances could have formed under the preservation of isostatic equilibrium. Nor is it clear how these units could have risen isostatically during the long continued rise above sea-level necessary to provide the primeval oceanic depressions with thousands of meters of erosion products).

Moreover, the primary disturbance causing the genesis of continental domes out of the basaltic surface layer comes like a *deus ex machina*. Without accepting this initial stage, the further deductions of the theory would become impossible. Finally, the theories leave the question of the tetrahedral or antipodal plan of terrestrial elements untouched. Nor does it explain the existence of three hypsometric levels, nor the remarkable structural pattern of the bottom of the Atlantic, etc. etc.

Disappointed we return, and at once we depart again to share another group of excursionists in the realm of the earth's infancy. It was G. H. Darwin who announced the hypothesis of the moon's origin not as a little sister but as a daughter of the earth. Several geologists have tried to imagine the consequences of such a hypothetical birth upon the structural history of the earth's surface-layers.

If we accept the resonance theory of the moon's origin it seems highly probable that a great part of the earth's outer shells entered the moon's formation. But what happened to the remaining sialic matter of the earth? There are two possibilities, viz. either the terrestrial remnants of sial were able to flow out and unite again as one continuous shell, enveloping the entire world, or they were unable to do so, being too rigid.

The last possibility is basic to the theories of many authors. In their opinion the earth was originally enveloped by a sialic shell of continental thickness. After the moon's disruption the remaining sial-fragments of the earth would have formed our continents, floating on the heavier simatic substratum that forms the bottom of the oceanic receptacles.

Long ago the question was raised whether the Pacific might not be considered as a scar brought about by the separation of the moon. Our continents should be compared to the acid slags floating on the surface of the heavier melt in a blast furnace.

Practically the same idea can be found in the theories of Osmond Fisher, Taylor, Wegener, Schwinner, Mohorovèiæ and Escher. The two last named authors even tried to calculate the thickness of the lunar sial by computing the mass of terrestrial sial that originally filled the gaps of the oceanic sectors. (Both assumed an original crust of continental thickness and no account was taken of the presence of the sial-flakes of the Atlantic and Indian oceans).

It is at all events clear, however, that neither the continents nor even their so-called nuclei (representing the innermost structures of the continents) should be regarded as undisturbed remnants of that distant and turbulent period in our planet's infancy.

The preceding that these "acid slags" had from the early

beginning a very complicated history and that the continents (parts of which have been breaking down until quite recently) formerly occupied a more extensive area than they do at present. We know, moreover, that the history of the submarine relief of the oceans is a far from simple one. Another point is that the bottom of the Atlantic and Indian oceans differ considerably from the floor of the Pacific

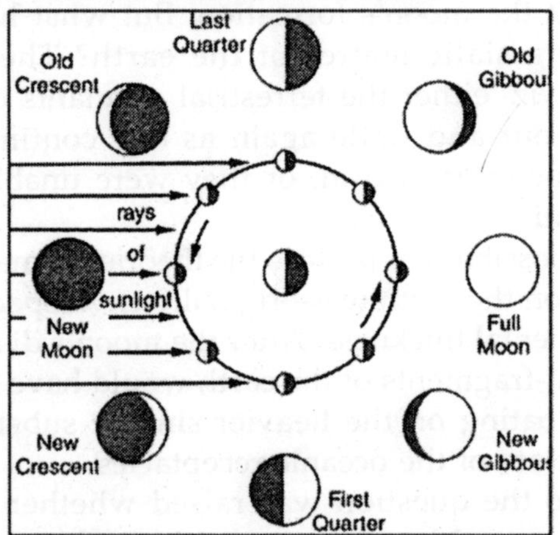

Fig. Internal Constitution of the Moon and the Earth.

This last important point is, however, taken into account by Pickering's hypothesis. For, firstly, his supposition that the scar left on our globe's surface when the moon was torn from the earth's body must be represented by the Pacific, explains why that ocean-bottom is free of sial

Secondly, he supposes the scar originally to have been much larger than the present Pacific. ".... three-quarters of this crust was carried away" — he wrote -"and it is suggested that the remainder was torn in two to form the eastern and western continents." So we could very well imagine that *at this stage* some continental masses drifted apart and that between them the floor of the Atlantic and Indian oceans formed as a result of stretching. Here Pickering's theory is much to be preferred above Wegener's and du Toit's theories of continental drift.

For in a previous section much stress was laid on the convincing arguments which show that provided the floor of the said oceans originated as a result of stretching this process must have occurred during the early Pre-Cambrian and decidedly not since Cambrian or still less in post-Carboniferous times, as

Wegener*cum suis* wished us to believe. But again the question arises, why these sial-flakes drifting apart like large ice-floes, came to be settled down in their peculiar tetrahedral arrangement.

Another question of primary importance, however, is: why should the sialic material have flowed out to form a continuous shell on the moon and not on earth? The following points should not be forgotten in this connection. It is questionable, although quite possible that prior to the postulated disruption, the sialic material had differentiated into the outer parts of our globe, but was not yet in a solid state.

Every one will admit that the outer shells of the earth were of such a fluid composition that they readily responded to the tidal forces which deformed the earth's geoid before as well as after the moon's separation.

It will also be admitted that part of these outer shells shifted towards the moon and that the scar of the separation was closed smoothly. Moreover, there does not seem to be the slightest doubt that when the earth, recovering from its amputation, was moulded into its new shape, the simatic layers — even the uppermost -formed a fluid mass uniting into a continuous shell.

It seems, therefore, quite unreasonable to suppose that the thin upper layer of sial would have been the only one incapable of reacting in the same way). And further if the Pacific has to be considered as the scar that originated by the birth of the moon what can be said about the North Polar Basin which probably also has no cover of sial? Finally, the preceding considerations are based on the resonance theory of the moon's origin and are valueless if we accept the opposite view, viz. that the moon and the earth formed simultaneously as two separate bodies.

For the third time we start on a trip in the realm of speculation. Our considerations will in this case have to be based on the supposition that the earth was initially enveloped by a continuous sialic layer gradually solidifying and floating on a denser basic substratum. It may be that this layer formed after the moon's disruption, but one might similarly assume that the moon was not severed from the earth.

According to the present author, the only plausible way in which continental blocks might have originated from a world-encircling sialic layer, growing gradually cooler and solidifying, is that a process of thickening occurred, resulting from folding and drifting owing to the action of convection currents).

This process contracted the sialic layer, and sial-free parts were therefore formed originated through periodical buckling and drifting of an originally thin sialic layer enveloping the whole earth, owing to the action of convection currents In several of the preceding paragraphs it was pointed out that the Atlantic floor might have formed as a result of stretching in the sense.

Bucher and du Toit attached to this process, but special emphasis was laid on the fact that in that case it would have occured at any rate in the Pre-Cambrian, as shown above by morphological and geological arguments.

It may be that during the process of drifting of these primordial continents a sial-sheet was torn into two or more parts, each moving in a different direction and producing one or more thinly stretched sial-flakes between them. In this way (and only during these primordial times) a thin sial-sheet like that which probably exists under the.

Atlantic may possibly have originated as a result of stretching of either a primordial continent or an intervening primary sial-layer.

If the sial of our present continents were imagined to spread out as thickly as the sial-layer of the Atlantic, it would probably cover a larger area than that of the present sial-free ocean bottoms. A floor such as that of the Atlantic cannot therefore be regarded as a remnant of the primary world-encircling sialic shell.

On the contrary, it would seem to have grown thinner as a result of stretching during the early Pre-Cambrian. It would be impossible to say just now whether a fragment of the primary sialic layer of original thickness exists in some part of the world to-day.

It is obvious that this process must have attained its most important effect in the early Pre-Cambrain, as the whole later process of buckling and folding of the continents is known to have evolved only in these existing Pre-Cambrian bucklers). originated in a basement which had already been folded previously).

Hence we cannot escape the conclusion that some event of fundamental importance must have occurred during the early part of the earth's history causing the surface history of primeval times to differ from that of subsequent periods. By this we mean that from a certain critical moment onward continental growth in a „horizontal" direction ceased.

The most plausible explanation seems to be that the world-encircling solid crust, had become so thick that subsequent growth was no longer possible.

The earth had attained a physical state more or less resembling that which it has at present. From that time onward the terrestrial forces were imprisoned and became operative as *subcrustal* processes.

At that time the crust consisted of basic rocks in the sial-free parts of the surface, just as it does now, and it may have contained both sialic and simatic rocks in those regions of the world where sial-flakes were present. It must have been of varying thickness, though it need not of course have equalled its prsent-day thickness. We will call this state the "consolidated surface of the earth".

Prior to the formation of a world-encircling crust, the terrestrial dynamics probably evolved with the same periodicity as later on. Yet the effect produced thereby on a non-consolidated surface was apparently a very different one than that produced on a solid crust by the "deeply imprisoned titans". Before the formation of a consolidated surface, the sialic upper layer was folded into continental thickenings.

This was the period of continental growth. With the

solidification of the earth's surface this process came to an end. Complete homogenity must, if ever it existed, necessarily have been disturbed by several factors. One needs to think only of the earth's flattening at the poles, the excentricity of its orbit, the inclination of the earth's axis, and the action of the tides.

A few years ago Bowie formulated his opinion regarding a primeval process of crustal buckling as follows: "It is impossible on a molten earth gradually cooling to have the light material which must have been present around the whole earth, pushed together into great masses to form continents. There is no force available inside or outside the earth which could have caused such a segragation of surface matter".

As mentioned before, however, the process of folding and drifting of the sialic surface layer is ascribed by us to the action of convection currents. Their periodical occurrence must at one time have found the sial-sheet cooled and solidified to such a degree that the result was the formation of continental thickenings.

If the above hypothesis may be said to contain a germ of truth, the problem of the continents and ocean-floors would consequently be associated with periodic phenomena of crustal folding which had, however ,already produced its main effect in the early Pre-Cambrian. A few points will now be examined more closely.

Let us consider the moment that these primeval surface-features became petrified, as it were, by the formation of a world-encircling solid crust of sufficient thickness. Among the first topics to be mentioned was the noteworthy structural pattern of the floor of the Atlantic and the surrounding continents.

It was shown that this pattern had probably originated in the early Pre-Cambrian and attention was drawn to the part which these old structural lines repeatedly played in later times. We arrived at the same conclusion when dealing with the continental basins.

It might therefore be asked whether it is not obvious that this ancient pattern dates from the period of consolidation of a solid crust. In the preceding paragraphs we came to the conclusion that three favoured levels occur on earth, represented.

- The continental surface., the floor of the Atlantic
- The bottom of the Pacific Ocean. It need hardly be said that these results fit in well with our working-hypothesis. The difference in the level of the primeval floors of the Pacific and Atlantic is seemingly due to the difference of the materials composing each floor.On the other hand, buckling of the primordial sialic crust was responsible for the continental thickenings.

  Their remarkable thickness and high level, as compared to that of the bottom of the Atlantic, is controlled by five factors, viz.

  - The thickness of the original world-wide sialic layer,
  - The compressive forces during the periodic phases of early Pre-Cambrian buckling,
  - The leveling effect of subaeral erosion,
  - The Pre-Cambrian stretching of sial-sheets of the Atlantic type.
  - The local thickening of the continental basement-complex as a result of the formation of later geosynclines and mountain-belts.

The above considerations indicate the main outlines of the new workinghypotheiss. I do not wish to suggest that this hypothesis solves the intricate problem of the ocean-floors. I merely followed a line of thought which had hitherto been neglected, and as all the previous hypotheses are difficient in many respects, the new "working hypothesis" may stimulate others in their efforts to unravel the mystery of the ocean-floors. I fully realise how many difficulties are inherent in this new aspect of the old problem.

It will be seen that some of the most attractive aspects of the theory were explained as well by Pickering's trend of thought. And the weak points in both the theories are again the same.

For no adequate explanation was given for the antipodal distribution of the major units of the earth's surface, nor for the triangular shape of these elements). Quite unexpectedly, however, a mathematical and geophysical treatise by Vening Meinesz seems to throw some light on this side of the problem.

In the year 1944 he published a preliminary paper on the distribution of continents and oceans. The main line of his argumentation runs as follows. It is supposed that a system of convection-currents must have come into existence during the primordial phase of cooling of the earth.

One possibility is that the rising currents carried sialic components to the surface where these acid constituents solidified *in loco,* no further displacements taking place on the surface. Another possibility is based on our above-mentioned theory of an originally continuous layer of sial envelopping the whole world.

The sialic layer then was thought to have been thrust together by the action of subcrustal convection-currents. Now, Vening Meinesz tried to prove that the system of convection-currents had to become arranged in a very peculiar manner, viz. in such a way that the breadth of a current is double its height.

Admitting that the terrestrial system, at least in a primordial stage of cooling, reached down to a depth of 2900 kilometers, Vening Meinesz found as their most probable distribution: eight currents, four rising ones alternating with four descending currents.

Hence, each current ought to occupy an octant of the outer parts of the globe and it seems plausible to suppose that the axis of two of these octants coincided with the rotation-axis of the earth. The resulting in Mercator-projection. This remarkable result suddenly seems to make clear how the antipodal distribution and the triangular shape of the major terrestrial elements possibly came into being.

For the figure clearly shows how these features are graphically represented in the scheme as a necessary result of the mathematical formulae.

Facts, many more facts are wanted. Future investigations

undoubtedly will bring more unexpected data — and theoretical points of view.

For the moment we may say that undoubtedly some remarkable results have been obtained, but we may safely say also that the nuclear point, i.e. the question of the origin of continents and oceanic depressions is and remains a baffling problem.

Science progresses and every one will be convinced that one day scientists will be able to dissipate more and ever more the fog that at present makes us stumble when we try to penetrate — along various paths -into the sacrosanct of the unknown.

Chapter 27

The Pulse of the Earth

of phenomena are genetically related to one another, and that they have a common, deeper cause in a figurative as well as a literal sense, to which we referred repeatedly by means of such neutral designations as "processes in the substratum", "changing conditions in the interior of the earth", "deep-seated forces", "pulsating rhythm of subcrustal processes", etc.

These vague descriptions designate the paramount source of all subcrustal energy, which manifests itself with a periodicity observed in a whole series of phenomena in the earth's crust and on its surface, viz. the alternating decrease and increase of compression of the earth's crust and the closely related epochs of folding, the process of mountain-building and submersion of borderlands, the periodic formation of basins and dome-shaped elevations, the magmatic cycles and rhythmic cadence of world-wide transgressions and regressions, the pulsation of the climate, and — lastly — the pulse of Life.

CHRONOLOGICAL RELATION OF PERIODIC EVENTS

The object recapitulate the above phenomena, and to give a condensed synopsis of their correlation. The latter finds its explanation which contains a schematic survey of the chronological relation of the periodical processes of one cycle. An arbitrary lapse of time has here been subdivided into five stages, the division being based on a certain chronological order represented by A, B, C, D and E, and this is followed by a new cycle.

A. This first stage is marked by a world-wide regression of the epicontinental seas, resulting either from the elevation of the continents, or the subsidence of the ocean-floors, or even — and this seems the most probable course of events — from the simultaneous but opposed movements of both the continents and the ocean-floors.

A period of intensive erosion ensues on the continents, and the climatological zones begin to show a greater differentiation than they did previously. This period is also characterized by an increased compression of the crust, which consequently buckles in geosynclinal areas. One of the consequences of this event is that the contents of the geosynclines become folded, and this is in turn accompanied — and later succeeded — by the intrusion of acid batholiths.

A simultaneous phase of movement generally occurs in those basins which were already in existence at that time. We are at present concerned with the surface of the earth in its consolidated state. The hypothetical processes responsible for the origin and arrangement of continental bucklers and oceanic depressions during the early Pre-Cambrian will not be considered here.

B. The first stage is followed by a period of decreasing crustal compression. By this time the processes of folding and regression have already attained their full effect. The folded belts now emerge as mountain-chains. Basins and new geosynclines begin to form. Dome-shaped elevations originate, and the sea-level begins to move in an opposite direction.

The interrupted line represents a period of only slight regression, and the dot-dash line and full-drawn graph in the same column depict periods of extensive and exceptionally extensive regression respectively.

C. The mountain-belts have grown to their full height, and the relief of the continents is very accentuated during this stage. The formation of dome-shaped elevations, rift-valleys and contemporaneous volcanism are now completed. The newly formed basins and geosynclines have already subsided to a considerable depth.

The three curves representing the more or less pronounced

deviations of the climate from normal conditions correspond with the greater or lesser intensity of the processes indicated by the three series of curves on the left. An extreme regressional stage and period of very marked and widespread mountain-building will thus correspond with a considerable differentiation of climatic zones and the occurrence of glaciation. An additional cooling power of the ice-sheets causes them to extend very rapidly.

Cosmic factors are now responsible for the division of the glaciation into a series of ice-ages. Naturally, this curve only represents the glacial and interglacial stages very schematically. These very abnormal climates have been observed twice since the Cambrian, i.e. during the Upper Paleozoic and Pleistocene, and it was during these two periods that the differentiation of the flora was so much more marked than during the intervening time.

The same major periodicity of approximately 250 million years is reflected in the tremendous Variscian and Alpine mountain-building and the formation of basins and troughs. The Upper Paleozoic phase of mountain-building lasted for about 50 million years, and the Alpine belts began to rise 200 million years later.

During the intervening time periods of minor activity occurred, with an average cadence of 50 million years, and were accompanied by lesser manifestations of an abnormal climate. Our galaxy rotates in about 200 million years.

Other figures can be found in astronomical literature, viz. 300 and 250 million years. This cosmic cycle may thus be said to correspond on the whole with the major periods of terrestrial activity. It is difficult to decide for the moment whether this coincidence is a mere question of chance, but it is obvious that this point will have to be taken into consideration in the further development of science.

D. The above period of abnormal climatological conditions ends even more rapidly than it began. The mountains have been eroded to a considerable degree during the intervening time, and the sea-level has risen slowly but surely.

C. At this stage we find a low relief on the continents, together with vast epicontinental seas and a mild and equable climate extending to high latitudes. A maximum transgression means a minimum difference in the height between the continents and the ocean-floors. Moreover, this stage is preeminently a period of tectonic quiet.

In the meantime, however, certain belts and basins continue very gradually to subside. Abyssolithic injections of basic melts break through the crust and adjust themselves into the steadily accumulating strata of the geosynclines and basins as eruptive and extrusive products.

The chronological correlation of A, B, C, D and E as represented is wholly diagrammatic. A, B and C combined cover a much shorter lapse of time, however, than the longer stages under D and E.

Our own cycle has already advanced to some extent into stage C.

A Fundamental Problem

Behind the results arrived at so far there still lurks one essential question, viz. to what deeper impulses must we attribute these phenomena? In what manner might these problematic subcrustal processes produce a periodic recurrence of phenomena such as those represented schematically in Table II?

We might compare our quest to that of the detective in criminal fiction, who attempts to discover the essential qualities of the mysterious culprit whose traces all lead in one direction. We are no longer content with the classical image of Hades, the old and crippled god of the subterranean realm. We know the culprit's haunts as well as some of his more salient features.

Very little has been ascertained as regards his restless infancy, but it is possible to show the main outlines of his cardiogram for the last 500 million years as derived from the movements of the terrestrial crust and the sea-level. His fingerprints are supplied by the magmatic phenomena. What creature is this that breathes so heavily once in every 250 million years, and why does its pulse beat approximately four times during the intervening period?

Why are the wrinkles in its face arranged according to the intricate pattern in Plates 1 - 5? Its skin is not only lined by mountain-belts, but is also pock-marked by numerous basins. Yet nothing is known of the events that brought about these marks during certain specified periods. And there are several hypotheses to explain what was responsible for one of its most striking features — the oceanic depressions.

To return from this geopoetical paraphrase to geological prose: can the subcrustal periodicity be said to be of a physico-chemical nature? Might we attribute it to a periodic system of convection currents, or might it be the result of a combination of these diverse activities? Other processes, too, of which nothing is known at present, may be active in the interior of the earth.

Everyone will have to admit that all views on this subject must necessarily retain a speculative character.

Most readers will be acquanited with the interesting suggestion which Joly and Holmes put forwards some years ago, when they attempted to derive periodic phenomena from radioactive processes, and will probably also have read the opinions of such authors as Schwinner, Escher, Holmes, Pekeris and Vening Meinesz on convection currents.

Some may have come across Grigg's publication in which this author tries to show that convection currents may in fact occur intermittently, and will have admired the ingenious model by which he sought to demonstrate the activity of cyclic convection currents. In short, geologists and geophysicists are already striving for a solution and the hypotheses that have appeared from time to time may perhaps already contain a germ of truth.

Nevertheless, the mysterious interior of the earth involves more than just a single periodic process. It is far more complicated than the tentative theories have supposed so far. The accompanying graphs will have made this clear. Geology will have to proceed hand in hand with Geophysics in a combined effort to unravel this most absorbing problem.

A solution will only be found when it will be possible to indicate that a logical and necessary correlation of cause and effect exists between certain events in the earth's interior and those diverse phenomena to which I referred briefly in the title as *the Pulse of the Earth*.

ICE-AGES

The earth's aspect is a very recent one from a geological point of view. Everyone knows that Greenland and Antarctica are at present covered by extensive ice-caps. Yet only one tick back on the geological clock large parts of America and Europe were covered by analogous ice-sheets of enormous dimensions. In Europe the ice-front extended from the British Isles to the Netherlands, and from there to Germany and North Russia, covering the whole of Scandinavia.

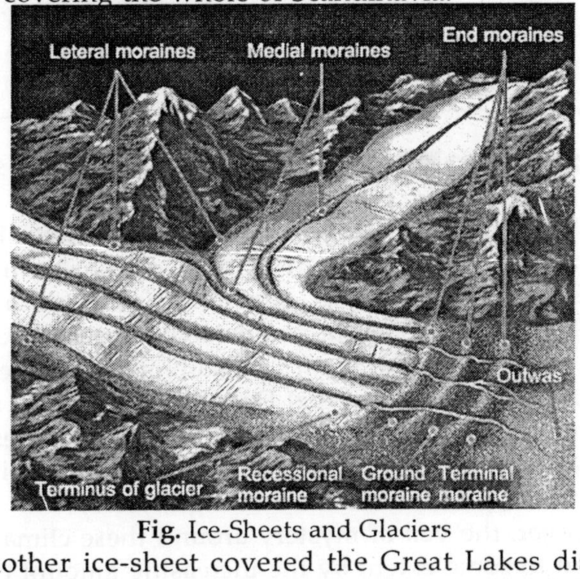

Fig. Ice-Sheets and Glaciers

Another ice-sheet covered the Great Lakes district of North America and practically the whole of Canada. Indisputable traces of large glaciations in South Africa, South America, Australia and a few other regions date from a still more distant past — the Upper Paleozoic — or even remoter times, the Pre-Cambrian, when plant-life had not yet imposed itself upon the land, and the seas were only populated by

Invertebrates. Rocks whose origin cannot be explained unless we assume that they were formed under the same conditions of heat and drought as are now found in the Sahara are met with in many places in Europe and America. On the other hand, the fossil remains of a luxuriant vegetation are still observed in such bitterly cold regions as Spitsbergen and Greenland.

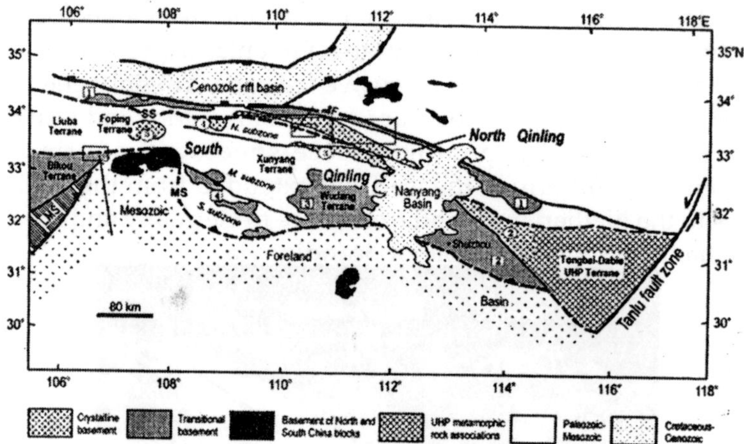

Fig. Types of Precambrian

The most absorbing topic of changing climates has not only attracted geologists, but also investigators in other branches of science such as astronomers, meteorologists, paleontologists and biologists, and the hypotheses that attempt to find a solution for the earth's climatological problems are so numerous and so varied that Daly was induced to remark in this connection: "Geologists do not know the causes", and somewhat further: "At present the cause of excessive ice-making on the lands remains a baffling mystery".

However, the veil of mystery around these climates has been lifted to some extent by the increasing amount of data, and the minute studies and lucid suggestions of a large number of investigators. An attempt will now be made to unravel, the main outlines of the chaotic mass of, data and opinions and special attention will be paid to those which I consider to be the most important.

I hope to be able to show the approximate state of present

knowledge on the subject of the climates, and will also emphasize those aspect's of the problem which are still wrapped in much uncertainty.

The Periodicity of Climates

If noted in a proper order on the geological time-scale, periods of glaciation and periods that were characterized by a mild climate up to high latitudes will be seen to give rise to a curious phenomenon, viz. a large climatic periodicity.

Some 250 million years elapsed between the beginning of the Upper Paleozoic glaciation and the Pleistocene ice-ages. Evidence of another extensive glaciation i.e. towards the close of the Pre-Cambrian, is found if we reach back an additional 250 million years.

Holmes showed that this periodicity can probably be retraced several times in still remoter times, though there are as yet too few data to enable us to speak of a phenomenon recurring with perfect regularity). A periodicity of 250 million years would appear to be a rough, but fairly accurate estimate.

The duration of a glaciation can only be indicated in one case, viz. in that of the Upper Paleozoic, which continued for about 25 to 30 million years. Should the series of ice-ages, beginning with that of the Pleistocene be attributable to a phenomenon with a cause similar to that of the Upper Paleozoic glaciation, the present period can consequently be said to constitute an intervening stage of partial deglaciation.

Many glacial and interglacial stages may follow the present situation before a new period is reached with conditions comparable to those which existed from the Upper Permian until the Pleistocene.

Speculation on what might happen if these ice-caps were to grow again excessively is one that cannot but fill one with dread. Prosperous agricultural and economic centers, thriving cities and important ore deposits would all disappear under the irresistible march of the ice, and an alarming decrease in the areas fit for population would ensue. Yet let us suppose that the reverse were to happen, i.e. that the ice were to melt entirely. The resulting state of affairs would not be a much happier one. It can easily be calculated that the level of the sea would be raised some 40 to 50) meters if the ice were to melt over the whole world, and 12% of the land would consequently be flooded.

Proud cities such as Amsterdam and Batavia would sink down in a muddy sea-bottom, and the lower parts of the sky-scrapers of New-York would be turned into shelters for lobsters. However, we are not concerned with what might happen in the future, but with what happened in the past, and it is these last problems which we now intend to examine.

An intervening period between two glaciations (e.g. from the Cambrian to the Upper Carboniferous, and from the Upper Permian to the Pleistocene) is marked by the absence of strongly differentiated climatic zones. The shape of the terrestrial globe has, of course, always been responsible for the fact that the polar regions constitute the colder areas of the earth, and the equatorial zones the warmer ones.

The inclined rotation-axis of the earth has similarly always been responsible for more or less pronounced seasons. On the other hand, the seasonal changes during some periods were so small that several writers prefer to speak of a uniform climate for the whole world. The climate would appear to have been much milder up to a short distance from the poles in former times. This is shown

- The luxuriant Tertiary vegetation in such high northern latitudes as Greenland, where the only existing plant life at present is the tundra.
- The reef corals that flourished up to much more northern and southern latitudes. Though the latter are only found in warm seas, they are known to have existed as extensive coral reefs in the Tertiary seas of North America, and were distributed as far as Maastricht (and even Denmark) in the Senonian.

Special phenomena that occurred during certain periods show that the climate was less differentiated than during the major periods of glaciation. This is how some authors interpret the tillites, which may be observed in the Eocene beds of North America and Antarctica, the Jurassic of California , the Upper

Silurian and Devonian of South Africa and Alaska, the Ordovician of Europe and North America , and possibly a few other localities and periods. No one regards these tillites as the products of former ice-sheets.

They are thought to constitute marks left by more or less local mountain glaciers and piedmont glaciers. It may still be questioned whether the formation of the glaciers in different regions did not in each case result from a combination of local circumstances.

There is reason to believe, however, that these old moraines of Eocene, Jurassic, Devonian, Upper Silurian and Ordovician time justify the assumption that minor periods of cold prevailed during the long interval of time that separated two major periods of glaciation. This enables us to draw a *schematic* curve of the climate such as that in Table II.

A mild and equable climate, affecting almost the whole world and extending even as far as polar latitudes, prevailed during such aeons of our earth's history that this condition may be considered to have represented the earth's normal climate. The major glaciations, on the other hand, and the minor periods of cold which we have just mentioned may be assumed to form "abnormal" and relatively short deviations.

The Earth's Normal Climate

What was the earth's aspect during normal climatic conditions? The remains of a Mesozoic vegetation are found in Greenland, and as far as Grahamland, also between (and including) New Zealand and North America. Darrah) elucidates this question as follows: "The Jurassic floras of such widely separated regions as the Antarctic continent, England, India and North America have so very many plants in common that the essential uniformity of the vegetation all over the earth at that period is well known.

The succeeding flora of the Cretaceous displays almost as great a uniformity in the composition. Not only were these floras homogeneous, but they appear to have flourished under more equable conditions than those now prevailing over most of the earth". "The rapid radiation of the diversifying flowering plants (of the Upper Cretaceous) certainly is indicative of an equable climate and an absence of barriers of migration".

Recent discoveries have made it clear that the climate need only change a little to bring about a recurrence of a luxuriant growth up to such high northern latitudes as Greenland. Since the retreat of the ice towards the beginning of this century, the remains of an autochthonous vegetation and of settlements dating from the eleventh and twelfth centuries have been brought to light in this region, and indicate that at that time the transitory melting of the ice-caps took place on a much larger scale than during the last forty years.

Under the mild climatic conditions of the Mesozoic, the land-fauna, including organisms such as the giant reptiles, were able to migrate from Central Asia to America and Africa without meeting any unusual barriers of temperature.

Such an evenly distributed climate also influences the atmospheric and oceanic circulation, producing less intensive horizontal and vertical movements. One of its consequences was that the deep-sea received a smaller supply of oxygen. It may be assumed that marine animals that had adapted themselves to colder water could only have thrived in the northern and southern polar seas.

These have indeed been described from Jurassic and Lower Cretaceous deposits, and are known as the boreal fauna. A specific geographical position may cause this boreal fauna to extend locally up to a shorter distance from the equator than elsewhere. The remaining marine fauna is almost entirely uniform.

Everyone will agree that though Neumayer distinguishes between four types of Mesozoic marine faunas: the MediterraneCaucasian, Himalayan, Japanese and South-Andine "provinces", his classification merely divides the biocoenoses, of the warm seas into separate geographical provinces.

The Cambrian Trilobites, too, for instance, could be subdivided into an Atlantic and Pacific province, but this does not mean that they lived under different climatic conditions.

The reef-building Rudists of the Upper Cretaceous have often been cited as examples of warm-water organisms. It is noteworthy that dwarfforms have been found in addition to the larger normal types in their most northern occurrences. Some attribute this to the influence of the boreal zone.

A like phenomenon is observed in their southern extension, in East Africa, where they occur from lat. 5° to 20° S (lat. 50° to 55° N. on the other side of the equator). It is clear that the existence of a southern cold seacurrent coming from Antarctic regions must be taken into account if their distribution is to be explained.

It would not surprise us if those zones of the earth which may be said to be predestined to become belts of arid degradation had had a greater extent during a "warm period" than they have at present. There can hardly be anything unusual in the distribution of the deserts during a period of normal geological climates if we remember that the present distribution of land and sea still exerts a strong influence on the disposition of and areas.

Brooks pointed out at great length that the earth's normal climate need not be the result of an important change in the position of the continents and poles. Berry) this statement as follows as regards the distribution of the flora: "Fossil plants have been found at six general horizons in the Arctic, namely: Late Devonian, Mississippian, Late Triassic, Late Jurassic and early Lower Cretaceous, Upper Cretaceous, and Eo-Oligocene.

They are frequently associated with coal seams, and this, together with roots in place in the underlying clays, the presence of fresh-water diatoms, mollusks and aquatic beetles, proves that the plants grew near where they are found fossil and were not drifted from lower latitudes.

The distribution of the localities around the poles conclusively discredits the hypothesis, now fashionable in certain quarters, that there has been a change in position of the pole.

A botanical analysis of the floras shows clearly that the term tropical, so frequently applied to them, is highly improper, and that they are, for the most part, distinctly cold temperate. The associated petrified woods, with the single exception of the *Dadoxylon* from the Mississippian of Spitsbergen, show well marked seasonal rings.

"Detailed comparisons of these Arctic floras with contemporaneous floras from lower latitudes, especially in the case of the younger, post-Paleozoic floras, show unmistakable evidence for the existence of climatic zones.

It was further shown that these northward extensions of temperate floras in the past were all contemporaneous with or immediately followed times of unusually widespread continental submergence and sea extension, and that they lived under the domain of oceanic as opposed to continental climatic conditions.

"It was concluded that lowlands and expanded seas were amply sufficient to account for the necessary ten to fifteen degrees northward swing of isotherms demanded by the botanical character of the floras. This, with the concomitant alterations in wind and current circulation, combined with Brooks' meteorological results, are considered to account for all of the observed facts".

Nor need we revert to the idea of continental drift to explain the occurrence of periods of abnormal climate or glaciations, to which we will confine ourselves in the following paragraphs.

The Abnormal Character of Present Conditions

It should be remembered that we are living in an abnormal climatological age, albeit in an intervening stage of milder glaciation or partial deglaciation. The actual distribution of fauna and flora in distinct biogeographical units determined by the climatic zones is a phenomenon of very recent origin, and merely dates from the close of the Tertiary.

The extreme specialization and differentiation of the flora and fauna form another unusual aspect of our time. For example, flowering plants (Angiospermae), which are first observed in the Upper Mesozoic, dominate the plant-world, while the adaptive radiation of the mammalian fauna only began to develop in the Tertiary.

What do we know of the ecological and biological conditions necessary for the maintenance of a biocoenosis during the Mesozoic and Paleozoic? It has frequently been shown that an attempt to base conclusions as regards the climatological significance of fossil finds on present conditions is bound to meet with failure.

The rhinoceros, and a type of elephant — the mammoth — populated north-western Europe during the bitterly cold Pleistocene. Yet the near relatives of these animals are now found exclusively in the tropics!

This in itself is quite enough to show that the greatest care should be taken by investigators before basing conclusions on the present situation. It was shown above that a very different atmospheric and oceanic circulation must have prevailed formerly, during periods of a normal climate. It will also be obvious that far more intensive denudation must have influenced the land during a period like the Lower Paleozoic, for no vegetation protected the earth at that time.

These examples may help one to obtain some idea of the many and occasionally very difficult problems which confront the paleontologist and climatologist in their attempts to interpret the geological past.

This point needs emphasizing when an attempt is made to find a solution for the climatological changes in the earth's history. For many authors put the problem in such a way that two wholly different questions have to be solved simultaneously. The first deals with the cold ages, the second with the warm climates.

These scientists begin by using the present climatic conditions as a basis upon which to build further conclusions, and then try to explain both the glaciations and the cause of the climatic change which transformed the climate into an almost mild and equable one all over the world.

The problem should be put quite differently, however, for in following the above method we assume the present situation to be the normal one and to characterize the whole geological past, and thus take it for granted that any other situation must be an abnormal one. Investigators such as Köppen and koppenl.

Wegener undoubtedly lost sight of the abnormal character of actual conditions when determining the climatic zones of former geological periods, for they sought to reconstruct climatic zones equivalent to our own for all these formations, and any conflicting evidence in the past was simply attributed to hypothetical changes in the position of the poles and continents. They thus construed a continuous, nonperiodical process, instead of the periodicity of abnormal climates.

THE RELATION BETWEEN PERIODICITY AND PHYSICO-GEOGRAPHICAL FACTORS

Many authors have tried to connect climatic changes with other events within the earth itself, or with processes upon its surface. The Pleistocene ice-ages succeeded the folding of such belts as those extending between the Alps and the Himalayas. The Upper Paleozoic glaciations appeared as the Variscian Mountains began to emerge, and the glacial phenomena of the Pre-Cambrian coincide with the so-called Huronian period of mountainbuilding.

The level of the continents was a comparatively high one compared to that of the sea. Glaciations can thus be said to be observed during periods in which we find much land and few epicontinental seas.

The intervening ages, however, are characterized by a relative tectonic quiet and low continents, large parts of which were flooded by the seas. These periods were marked by a mild climate up to high latitudes.

The relation of epochs of mountain-building to glaciations was revealed by Leconte in 1899, and in the same way Wright, Upham, and especially Ramsay and Ruedemann attempted to explain the phenomenon of glaciation.

These authors assert that the ice-caps originated as a result of the extension and particularly the altitude of land masses. Before going any further, however, a few other phenomena, which are likewise related to periods of mountain-building and oscillations of the sea-level, will have to be mentioned, since these enable us not only to form a particularly clear picture of the peculiar intricacies of the problem, but also of the various lines along which a solution has so far been sought for the occurrence of glaciations. Several writers stress the possible influence of a changing amount of carbon-dioxide in the atmosphere.

This gas, which is transparent to light, though opaque to heat, would act in the atmosphere like a glass pane in a hot-house and retain the heat; and the temperature would then increase proportionally to the quantity of CO_2 present in the air (Arrhenius).

Chamberlin claims that a Considerable amount of CO_2 was locked up in the formation of limestones and dolomites during periods of continental emersion (regression), and that the climate was consequently transformed into a cold one. On the other hand, periods of submersion, or peneplanation and transgression would have set this CO_2 free, and this would then have produced a milder climate.

The course of the sea-currents has undoubtedly changed (and the salinity of the oceans oscillated to some extent) in the course of time. Changes have doubtlessly also affected the atmospheric circulation and the cloudiness of the sky. All these variations were due to the repeated alterations in the distribution of land and sea.

Other investigators (especially Humphreys) argue that a great quantity of fine dust is flung into the atmosphere — and even the stratosphere -during violent volcanic eruptions, and that the dust continues to linger in these higher regions during a considerable lapse of time, scattering and reflecting the solar radiation.

Since these eruptions are known to have occurred with exceptional frequency during certain periods, it follows that the atmosphere may have been cooled during these stages.

Many factors had an undeniable influence on the climates during the changes which were constantly affecting the face of the earth. It is almost impossible, however, to describe each one's influence separately, or to estimate their combined total effect quantitatively, one of the chief difficulties being that our paleogeographical maps are incomplete in several respects, and also very schematic.

Yet it cannot be denied that mountainbuilding and the major periods of glaciation are in fact correlated in one way or another.

Still, intensive folding and subsequent mountain-building are known to have occurred during several other periods – e.g. the Caledonian, the Early and Late Cimmerian, and the Laramide. As stated above, a few of the less important glacial deposits may possibly be related to these periods.

The tillites of Ordovician time and the Silurian and Devonian have been assumed to be related as such to Caledonian mountainbuilding, and those of the Jurassic and Eocene to the Late Cimmerian and the Laramide revolution respectively.

The Variscian and Alpine Mountains were far more extensive than the Caledonian, Cimmerian and Laramide chains. Though it is extremely difficult – and at times almost impossible – to reconstruct a satisfactory and comprehensive map of the above periods, the following conclusions may safely be drawn:

- That the periodicity of 250 million years as indicated by the major periods of glaciation is also manifested in the epochs of mountain-building
- That a rhythm of minor amplitude, as observed in the less accentuated deviations from normal climatological conditions in the Ordovician, Silurian, Devonian, (Triassic?), Jurassic and Eocene can probably be said to coincide with epochs of mountain-building of far lesser geographical importance than those which accompanied the major rhythm of 250 million years referred to above.

The Deeper Causes of Periodicity

A comparison between the various processes of folding and movements of the sea-level shows that these heterogeneous phenomena are distinctly correlated. There also appears to be some relation to magmatic cycles; and the alternating increase and decrease of compression of the earth's crust seems to be directly connected with the above events, as explained in the preceding . In short, everything shows that a whole series of different phenomena reflect the influence of a deeper and widespread impulse.

The term "deeper" should be taken in its figurative and literal sense, for the phenomena are attributed to deep-seated forces. It would take us too far to examine this difficult and intricate problem at this point, but it should be noted that several diverging hypotheses have been put forward on this subject by Joly, Holmes, Bucher and — recently — Griggs, which though differing in many respects, agree on one point, for each attributes the origin of the phenomena to subcrustal forces.

This vague designation will have to suffice for the moment. However, this much can be said now: the periodical processes in the earth's interior may be thermal cycles, i.e. they may be accompanied by periods of an increased accumulation of heat alternating with periods of decreasing temperature. Folding would then have occurred in the last phase. This hypothesis implies that it cannot be a mere accident that these phenomena preceded periods of mountainbuilding and the abnormal "cold" climates.

It should not be assumed, however, that a fall in the temperature of the atmosphere is due exclusively to these subcrustal conditions. For there still remain the changes in the earth's physiographical aspect (mountain-building, regression, a decrease in the amount of CO_2 in the atmosphere, a less intensive circulation, etc. all of which have a considerable influence on atmospheric conditions).

To these should be added a further factor, which is known to intensify the fall of the annual mean temperature once the latter has begun to decrease. This factor will now be examined.

The Rapid Growth and Subsequent Retreat of An Ice-Cap

Brooks argues that the ice-cap has a cooling effect on the atmosphere above and along its edge and that this will become an important factor once the ice-cap has begun to grow. The additional cooling influence will cause the ice-cap to extend beyond the boundaries which would have existed without such an effect, i.e. it will cause it to expand beyond the polar regions where an initial fall in the temperature would result in the formation of an ice-cap.

On the other hand, the temperature is known to rise steadily in the areas between the poles and the equator, and a state of equilibrium will consequently be reached at a given moment between both counteracting factors. An exhaustive review of this subject will be found in Brooks, but I would like to draw attention to a few results of this author.

The ice-cap extends at an increasing rate once it acquires a given size, for its area increases by the square of its radius. The opposite effect, however, — i.e. the rising temperature from the poles to the equator -is a gradual one and proportional to the distance from the pole (normal horizontal gradient).

Observations in polar regions seem to bear out this statement, revealing that the cooling effect of an extending ice-cap will in fact increase almost proportionately to its surface till the growing circumference of the ice-cap has attained a radius of approximately ten degrees. The more distant regions cease to exert their full effect in case of a larger area. For though the cooling effect will continue to increase, the average effect — expressed in km^2 — will begin to diminish

Radius of ice-cap, in °	*Total cooling on edge of ice-cap, °*	*F Rise of temperature due to normal horizontal gradient, °F*
0	0.6	0.0
1	1.05	0.9
2	2.44	1.8
3	4.6	2.7
4	7.8	3.6
5	11.8	4.5
6	14.5	5.4
8	17.2	7.2
10	18.2	9.0
15	20.4	13.5
20	21.3	18.0
25	22.2	22.5

Brooks concludes that if the initial winter cooling of an open polar ocean drops to a mere 0.6°F. (0.35°C) below the freezing point of the water, the result will be the formation of an ice-cap, which at first extends rapidly to a latitude of approximately 78° and then more slowly to about 65° (i.e. assuming that no land influences or warm sea currents arrest its growth), and the ultimate lowering of the winter temperature, owing to the initial fall of 0.6°F., will then amount to 45°F. (25°C.). This applies to ice-sheets on sea as well as to those on land. Brooks attached such importance to the cooling power of an ice-cap that he devoted.

Fig. An ice cap

This much seems clear, however: the annual mean temperature will, once it has begun to fall as a result of some cause, be lowered still further the moment an ice-sheet is created.

The same so-called cooling power of an ice-cap may probably also explain the extremely rapid growth and subsequent equally swift decay of the Pleistocene ice-sheets. The retreat of the last Scandinavian ice-cap has been depicted. The effect in this case is seen to consist of a slow and subsequently faster retreat of the border of the ice-cap. It should be noted, however, that another factor helped to accelerate the retreating movement.

For the ice-front began to yield at an increasing rate as the ice-cap melted and grew thinner. The extreme right of the curve finds its answer in one more factor. The land began to rise as it was freed of the load of the ice-sheet, and the highly elevated remnants of the ice-cap consequently melted at a much slower rate.

The following constitutes another important point, in my opinion. An extensive retreat of the sea is one of many factors that initiate glaciation. Once the ice-caps begin to extend, however, they will cause the sea-level to drop. The maximum lowering of the sea-level by glacial control must have attained approximately 100 meters during the Pleistocene.

We know now that a great many shallow seas were reclaimed by the continents during this period, and that these same areas were drowned once more in post-glacial times as the ice-caps began to melt. An example of this is found in the drowned rivers of Pleistocene Sunda-land. This extension of landareas as a result of glacial lowering of the sealevel favours glaciation and therefore this has to be considered as another autocatalytic factor.

Interglacial Periods

The third aspect of the problem —viz. the interglacial periods that divide one long period of glaciation into a series of ice-ages —will now have to be examined. It is possible to distinguish between four separate ice-ages, — including three interglacial stages and our own "postglacial" period (this may appear to be an euphemism!) — from the Pleistocene up to the present day. The investigations by Milankovitch and Spitaler are of particular importance for this part of the climatic problem. Milankovitch calculated the radiation of the sun during the last 600,000 years.

The heat which different zones of the globe obtain from the sun depends on three astronomical variables — viz. the obliquity of the ecliptic, the eccentricity of the orbit of the earth, and the cyclic variation of the perihelion. The obliquity of the ecliptic is responsible for the seasons. The greater the obliquity, the more contrast we find in the seasons. The limits of variation vary between 22° and 24 1/2° in 40,400 years according to Milankovitch.

The orbit described by the earth around the sun is an ellipse, with the sun at one of the foci. The distance between the focus and the centre of the ellips, is called the earth's eccentricity and, expressed in terms of the major axis, varies between zero and approximately 0.07 in about 100,000 years. The third variable is the cyclic variation of the perihelion with a period of 20,700 years, i.e. the so-called precession ot the equinoxes. Should the maximum effects of these three variables coincide the result would be a minimum radiation.

It was on this basis that Milankovitch calculated and constructed his radiation-curve for the sun. This curve contains a number of small peaks and several higher ones. Some authors -particularly Köppen, Zeuner and Soergel— claim that a remarkable similarity exists between this curve and the conclusions regarding glacial and interglacial stages arrived at previously by geologists.

Nevertheless, it should not be concluded that the problem can be considered to be definitely solved, for several circumstances will dampen our enthusiasm in this respect. To begin with, the curve is far from simple, and can be interpreted in several ways. It makes no reference to the way in which the combined atmospheric factors (such as circulation, cloudiness, evaporation and absorption) react to a greater or lesser amount of radiation.

Köppen, for instance, supposes that periods with a less than normal summer temperature and short, mild winters favour the development of ice-sheets, but Croll advocated the exact opposite — i.e. that periods with exceptionally cold winters and short, warm summers provide the most favourable conditions for the evolution of such sheets. Besides, Spitaler's curve deviates to such an extent from that of Milankovitch's, that both may be said to agree only superficially).

These authors thus arrive at very different results as regards the chronology of the ice-ages. The following table shows the astronomical time-scales of the four glacial stages as reconstructed by Köppen (Zeuner) and Spitaler, and includes a list based on geological evidence, after Brooks. The figures are in thousands of years B.C.

	Spitaler	*Köppen*	*Brooks*
Fourth Ice-Age	250-150	181-66	50-28
Third Ice-Age	600-450	236-183	140-110

Second Ice-Age	1,000-800	478-429	430-380
First Ice-Age	1,350-1,150	592-543	—

The problem of radiation was likewise examined by Brooks. The relation of solar radiation to the appearance of sunspots is undoubtedly a complicated one. Moreover, it is difficult to predict the effect which a change in the radiation of the sun will produce on a climate. Notwithstanding these difficulties, Brooks may safely be said to have shown a distinct connection between solar variation and general atmospheric changes.

A great sunspot maximum marked the year 1870, and since then (apart from certain fluctuations, such as an eleven-year cycle) there has been a gradual decrease of sunspots. During this time the average annual temperature and pressure rose in North Asia, Europe, Iceland and part of North America. Moreover, the glaciers and ice-caps have, on the whole, been retreating — especially since the beginning of the twentieth century.

In fine, all that can be said at present is that glacial and interglacial stages may possibly, and even probably, be attributed to changes in the amount of radiation dependent upon several cosmic variables, and that a correct interpretation of the climatic effect of these variable factors, and an accurate reconstruction and interpretation of the radiation — curve, must remain reserved for the future.

Of course, these astronomical factors were already active before the beginning of the Pleistocene. Yet, the fact that they only produced ice-ages during certain distinct periods proves that glaciation must have been initiated by one or more additional causes. The latter were discussed above, and it was pointed out that the major periods of abnormal climates are controlled by geographical changes and deep-seated terrestrial processes.

THE GEOGRAPHICAL LOCATION OF AN ICE-CAP

As mentioned above, the last part of the problem deals with the site of an ice-cap once glaciation has become possible.

A few examples were given of normal climates, and it was

seen that and regions, for instance, will change with corresponding alterations in the distribution of land and sea.

If the earth's surface were homogeneous, with the exception of an ice-cap stationed in the present areas of Greenland and the South Pole, the atmospheric circulation would be approximately equivalent to that shown by Hobbs in the accompanying diagram. The actual situation deviates from this theoretical conception because of the special distribution of land and sea.

In this schematic diagram two zones of high pressure will be seen to run parallel to the equator, and these are predestined to become belts where the climate is subjected to an and degradation. The present geographical conditions, however, are responsible for the fact that this is only partly true.

To mention two examples: they caused the and zone to shift northwards in Central Asia, and the same conditions were responsible for the absence of an ice-cap in Siberia, though this last area is known to constitute the coldest region on earth. The distribution not only of the temperature, but also of pressure, water vapour and the whole complex of oceanic and atmospheric circulation has a most important bearing on these deviations from the schematic diagram.

This will also help to make it clear why the large Pleistocene ice-sheets are found in the northern hemisphere, and Brooks even tried to show why they originated in those localities in which they are actually known to have occurred. To understand his arguments, it will be necessary to return once more to the influence of solar radiation.

As mentioned above, Brooks revealed that a decrease in the frequency of sunspots since 1870 has clearly been accompanied by a corresponding increase in the mean annual temperature and pressure over specific areas in North America, Iceland, Europe and northern Asia.

With the aid of observations obtained from these regions and the logical premiss that a growing number of sunspots would have a contrary effect, Brooks constructed two maps , showing the distribution of the temperature and pressure in accordance with a hypothetical increase in spottedness up to an average of 200. This "relative sunspot number" means that approximately one five-hundredth part of the sun's visible disc is covered by spots.

The figure for the sunspot maximum in 1870 was 139, and 32 for the eleven-year period from 1904 to 1914. The pressure over the continents, and an area of increased pressure over the oceans. These reconstructions present conditions which may very well have prevailed towards the beginning of the Pleistocene ice ages.

It should not be forgotten that these results are based on the present distribution of land, sea and mountain belts. There is indeed good reason to suppose that the position of the continents during the Pleistocene did not differ from their present situation. However, let us reach back somewhat further into the past. The plant-bearing deposits of Eocene time causes Chaney to conclude that the position of the European continent in respect to America has remained unaltered since as far back as the Eocene. The accompanying maps make it easy to follow his argumentation.

Those Eocene plant units that bear evidence of the formerly existing climates have been grouped, and "similar fossil floras, which are considered to represent essentially similar climatic conditions" have been connected by lines which Chaney referred to as isoflors. These are also felt to represent Eocene isotherms. which includes presentday isotherms for the month of January and the modern flora-types.

The similarity between our isotherms and those of Eocene time is apparent, and Chaney consequently wrote: "Present departures of isotherms from coincidence with parallels of latitude are interpreted from the tempering effects of ocean currents on the western sides of continents in middle and high latitudes, and from the increasing continentality toward the east. It seems entirely reasonable to assume that a similar distribution of vegetation and temperature during the Eocene resulted from the same controlling factors, and to conclude that land and sea relations, as well as planetary circulation, were essentially like those of to-day.

Brooks shows a map of part of Europe, after F. von Kerner, with a similar trend of Tertiary isotherms, based on wholly different evidence".

Like Berry, whose view was mentioned above, Chaney concludes that all paleobotanic data plead against the idea of a shifting of continents and poles since the Eocene, and his final verdict is summed up as follows: "The pattern of land vegetation around the Northern hemisphere indicates that the factors controlling air and water circulation have been essentially the same since the Eocene — that North America and Eurasia have stood in their present positions since the dawn of the Cenozoic.

Evidence of land plants of earlier times, less completely known and interpreted, seems also to refute the hypothesis of continental drift. Forrests under compulsion of climatic change, rather than the continents on which they live, appear to have been the wanderers during the history of life upon the earth."

This brings us to the problem of the position of the Upper Paleozoic ice-sheets. Chaney's results, together with Brook's noteworthy investigations, make it clear that the unusual distribution of Upper Paleozoic and even more ancient glaciations could in all probability be explained without continental drift if only accurate paleogeographic maps could be reconstructed for those distant ages.

However, some authors regard the distribution of the tillites in the Upper Paleozoic'as one of the most important arguments in favour of the hypothesis that vast changes affected the position of the continents and poles. It was explained above that the reconstruction of climatological zones, based on the present situation, as put forward by Köppen and Wegener, was a fundamentally wrong one for several reasons.

For, to begin with, these investigators assume that the present conditions are the normal ones; but — as Coleman remarked — "that is far from being the case." In the second place, they do not explain — nay, even obscure — one of the most prominent climatic features, viz. the periodicity of abnormal climates. There can be no doubt about the occurrence of this last phenomenon, it takes place in conjunction with other periodical processes in the earth's crust.

The earth's rotation-axis has undergone any significant change, nor does the astronomer find it possible to suggest a cause for such a hypothetical occurrence. Moreover, the comparatively rapid genesis of glaciation would require equally rapid hypothetical changes, but since the earth is a gyroscope and thus has a powerful tendency to keep the axis of its rotation constant, any sudden alteration in its direction would only bring about a world-wide catastrophe. Historical geological data prove that nothing of the kind. has ever happened.

On the other hand, it might be assumed that the position of the earth's crust has undergone some changes in respect to the core. This would result in a different distribution of the continents in respect to the axis of rotation, and the position of the latter would then undergo a slight alteration owing to this process.

It was this type of "shifting of poles" that was used by Köppen and Wegener when constructing their paleogeographic maps. Milankovitch published a mathematical treatise on this subject, showing that the North Pole lay in the vicinity of the Hawaiian Islands in the Paleozoic, and that since then it had migrated to its present position *via* Alaska.

Mathematical calculations should always be welcomed when dealing with geological problems, for they frequently sift faulty arguments and incorrect suppositions. Yet an entirely accurate and unimpeachable mathemathical calculation may sometimes lead to wrong conclusions. As Huxley observed :

"Mathematics may be compared to a mill of exquisite workmanship, which grinds your stuff to any degree of fineness; but nevertheless what you get out depends upon what you put in; and as the finest mill in the world will not extract wheat-flour from peascods, so pages of formulae will not get a definite result out of loose data".

The premisses of Milankovitch's calculations are such that

its results can be said to be the reverse of what might be termed definite proof. For one of the assumptions is that no change has affected the sialic shell, and another that an activity, referred to by Milankovitch as a "Polfluchtkraft", was in existence, etc. Yet these conjectures are so questionable that his conclusions may be said to lose all value, at great length by Schwinner.

Deposits in the southern hemisphere are incompatible with Wegener's view of moving poles. Another point is that tillites occurred in the northern hemisphere, particularly in North America, north-western Siberia, (where the ice probably radiated from the existing Kara Sea Massif) and Central Asia (where it descended from the region of Tarim). It should be remembered that the existence of an enormous ice-cap in Asia during the Pleistocene only became known quite recently.

The Paleozoic of Asia may consequently hold more surprises in store for us in this respect! The presence of tillites has been reported in numerous other areas. Some may have been brought thither by local mountain glaciers or floating ice-masses.

Others, such as those existing in Texas, are probably tectonic breccias, but the so-called Squantum tillite near Boston is so thick and widespread that geologists have not hesitated to assume that these fossil moraines very probably constitute remnants of a vast ice-sheet. In Wegener's reconstruction of Pangaea, the latter is situated in his equatorial rain-zones of Carboniferous time and in the centre of his Permian desert belt.

These tillites consequently conflict with the theory of drift, and Wegener therefore attempts to "explain them away tacitly". Brooks expressed the opinion that the Squantumtillites might be explained without a shifting of poles and continents, and assumes that they were formed under the influence of a cold sea, extending from the Arctic region as far as the eastern part of America Paleozoic time and also showed that a possible westward drift of both the American continents in respect to Europe could not affect his solution in any way, though the change in the position of other continents would only augment the difficulties of the problem.

As a matter of fact, Brooks not only pointed to the defects of the hypothesis of drift, but also attempted to furnish a constructive and extensively argumented explanation of the question of geological climates, in which "epeirophorese" played no part whatever.

We are living in an age of ice-caps. The distribution of land, sea and mountain chains, etc. are responsible for the position of the ice-caps as seen above. It was likewise shown that the site of the Pleistocene ice-caps can be understood without a hypothetical shifting of continents. The emplacement of the Upper Paleozoic ice-caps, too, could probably be explained if we had all available data as regards the distribution of physiographic factors in the distant past at our disposal.

My own conclusion may be said to concur with that of Schuchert, which the latter expressed 1923. When asked what had brought about the Upper Paleozoic glaciations, this author replied : "Certainly never the impossible Wegenerian hitching together in close embrace of the present masses of South America, Africa, India, and Australia.

Let us return to a tangible geology. We all now know that the earth underwent one of its greatest crustal disturbances during the Carboniferous, beginning late in the Lower Carboniferous (toward the close of Viséan and Chesterian time), attaining a first climax in middle Upper Carboniferous time, and a second maximum early in the Permian.

The supercrust was undergoing one of its periodic revolutions, and the world was then as scenically grand as it is to-day.

It is in the youthful topography, the enlarged continents and the peculiar connections of the continents that seemingly are to be sought the reasons for the Permian ice-age." To this he added, some nine pages further on:

"If it can be shown that Australia was connected in Permian times with Antarctica, then this holding in of so much of the cold waters of the Antarctic Ocean, combined with moist climates in the southern hemisphere and the general highland condition of so much of the world in early Permian time, will be the explanation for the peculiar position of the continental ice-masses of the southern hemisphere".

The supposed distribution of land and sea, as well as that of cold and warm ocean currents, and tried to explain the Upper Paleozoic glaciation without the occurrence of alterations in the present position of continental blocks.

These authors, like Brooks, are of the opinion that a solution must take into account the existence of former land connections between Africa and Australia, and between Africa and South America. Many investigators have based similar conclusions on other grounds (i.e. on various geological arguments, and the distribution of land plants, animals and marine faunas).

The preceding explained that though each reconstruction of transoceanic land-bridges retains a purely hypothetical character, the idea that land-bridges existed in former times should not be wholly discarded.

At some time in the future, when science will have progressed still further, it will perhaps be possible for geologists to draw a complete and accurate paleogeographic and paleoclimatologic map of the Upper Paleozoic. A more profound knowledge of the geological history of the oceans will perhaps have been acquired by that time.

The Restricted Influence of Cosmic Factors

The influence of cosmic phenomena was met with twice in the foregoing paragraphs. Huntington and Visher, in their "solar cyclonic hypothesis", went so far as to regard a great increase in sunspots as the only primary factor capable of bringing about ice-ages. Yet there are two objections to this theory:

- That a periodicity of the fluctation of solar radiation, such as is sought when attempting to explain the earth's abnormal climates, is, of course quite unknown, and that its existence has never been made acceptable;
- That abnormal climates, as previously stated, are

connected with internal terrestrial processes (mountain-building, magmatic cycles, etc.).

The influence of sunspots on subcrustal processes and its effect on epochs of folding, mountain-building, regressions, etc. would therefore have to be shown first. Huntington and Visher furnish tables purporting to show the apparent relation of an increase in spottedness to an accompanying increase in tectonic earthquakes, and retrace this relation *via* changes in the atmospheric. pressure.

They derive the hypothetical and major periodicity of sunspots from another theoretical process, in which the sun is supposedly affected by changes in the position of other heavenly bodies. Their combined views thus attribute a direct influence on the climate, and an indirect one on subcrustal events to cosmic phenomena.

However, this figure also includes the effect of geographical factors, since a hypothesis which attempts to explain everything solely by means of cosmic phenomena seems to me to be hopelessly incompatible with present geological data.

Nevertheless, it may be that the influence of cosmic phenomena, as represented in this scheme, is not sufficiently restricted. Joly had, indeed, already observed that it was not improbable that the periodicity of subcrustal processes was stimulated by a cosmic effect and Schwinner subsequently argued that after the supposed disruption of the moon from the earth a resonance effect.

Would have occured as many as eleven times (i.e. the period of the tides would have coincided eleven times with the natural free period of vibration of the earth), and he assumes that this would have caused the periodic phenomena of mountain building.

Yet, it makes a considerable difference, from a fundamental point of view, whether we restrict the cosmic influence in this manner, or attribute the dominating role to astronomic factors, as in the "solar cyclonic hypothesis".

Besides, even when restricted thus, the above astronomical influence can be stated to constitute no more than a hypothetical factor. Prof. Oort recently assured me that the only exceptionally long periodicity established so far in astronomy is the rotation of our galactic system, which revolves in approximately 200 million years.

This figure agrees fairly well with that of the major geological rhythm of 250 million years. This congruence is indeed remarkable, but for the moment it is hard to see how the astronomical cycle could be connected with atmospheric and subcrustal processes.

Index

A

B

C

G

H

I

K

L

T

U

V

W

X

Y

Z